A Course in

COMPLEX
ANALYSIS

A Course in
COMPLEX ANALYSIS

SAEED ZAKERI

Princeton University Press
Princeton and Oxford

Permission is gratefully acknowledged from the National Science Teaching Association for use of a quotation by William Thurston on page v of this volume. The quotation originally appeared in Quantum magazine's pilot issue 1, January 1990, page 7 ©National Science Teaching Association

Published by Princeton University Press
41 William Street, Princeton, New Jersey 08540
6 Oxford Street, Woodstock, Oxfordshire OX20 1TR

press.princeton.edu

Library of Congress Cataloging-in-Publication Data

Names: Zakeri, Saeed, author.
Title: A course in complex analysis / Saeed Zakeri.
Identifiers: LCCN 2021009726 (print) | LCCN 2021009727 (ebook) |
    ISBN 9780691207582 (hardback) | ISBN 9780691218502 (ebook)
Subjects: LCSH: Functions of complex variables. | Analytic functions. |
    BISAC: MATHEMATICS / Complex Analysis | MATHEMATICS / Applied
Classification: LCC QA331.7 .Z35 2021 (print) | LCC QA331.7 (ebook) | DDC 515/.9—dc23
LC record available at https://lccn.loc.gov/2021009726
LC ebook record available at https://lccn.loc.gov/2021009727

British Library Cataloging-in-Publication Data is available

Editorial: Susannah Shoemaker and Kristen Hop
Production Editorial: Nathan Carr
Text and Jacket/Cover Design: Wanda España
Production: Danielle Amatucci
Publicity: Matthew Taylor and Kate Farquhar-Thomson
Copyeditor: Bhisham Bherwani

This book has been composed in Minion Pro and Optima

Printed on acid-free paper. ∞

Printed in the United States of America

10 9 8 7 6 5 4 3 2 1

" *When I was a child I took pride in how many pages I read in an hour. In college I learned how foolish that was. When reading mathematics ten pages a day can be an extremely fast pace. Even one page a day can be quite fast. On the other hand, if you already understand something, you may get more by skimming than by reading every word. You need to be alert and suspicious; you need to question and think about what you're reading in your own way. . . . Don't be afraid to stop in mid-paragraph or midsentence when something surprises or puzzles you. Speed isn't the issue. Don't assume something is obvious just because an author treats it that way. What you work out on the side, even though it takes much more time, will have immensely more value than what you read straight through.* "

—William P. Thurston, *Quantum Magazine*, January 1990

Contents

Preface

This book is based on the lectures that I have given since the early 2000s at the University of Pennsylvania, Stony Brook University, and the City University of New York. It offers enough material for a year-long graduate-level course and serves as a preparation for further studies in complex analysis and beyond, especially Riemann surfaces, conformal geometry, and holomorphic dynamics.

The presentation is guided by a desire to highlight the topological underpinnings of complex analysis and to accentuate the geometric viewpoint whenever possible. This is evident from the following sample of special topics that are treated in the book: The dynamics of Möbius transformations, boundary behavior of Riemann maps à la Carathéodory, Hausdorff dimension and holomorphic removability, conformal metrics and Ahlfors's generalization of the Schwarz lemma, holomorphic (branched) covering maps, and the uniformization theorem for spherical domains. To remain loyal to the scope and spirit of the project, I have resisted the temptation to discuss Riemann surfaces.

The primary audience of the book is beginning graduate students with a solid background in undergraduate analysis and topology. A basic knowledge of complex variables is helpful even though it is not formally assumed. Elementary measure theory (Lebesgue measure and integral, sets of measure zero, the dominated convergence theorem, etc.) shows up on a few occasions, but it is not a key prerequisite. Numerous worked-out examples, illustrations, short historical notes, and more than 360 problems have been incorporated throughout to make the text accessible for independent study by a strong and motivated student. Above all, I have strived to make the treatment of every topic appealing even to the more experienced readers with prior exposure to complex analysis.

The bulk of the book can be covered over two semesters, with enough remaining for reading projects if desired. A possible plan that I generally adhere to is to cover most of chapters 1–7 in the first semester followed by the "essential" material from chapters 8–13 in the second semester. Of course what is considered essential depends on one's taste and point of view, and the organization of topics allows quite a bit of flexibility in this respect. Occasionally I postpone a few items from chapters 1–7 until the second semester when students have developed more knowledge and skill. Examples are §6.3 on the boundary behavior of Riemann maps, or §4.4 and §4.5 on conformal metrics and the invariant form of the Schwarz lemma, which can be presented ahead of Ahlfors's generalization in §11.4.

It would be hard to overstate how much this work has benefited from the existing literature on complex analysis. My sources have been gathered in the bibliography, but I must especially acknowledge the influence of the books by Walter Rudin [**Ru2**] (even though it famously lacks geometric flavor) and Reinhold Remmert [**Re**]; the latter also contains a wealth of historical remarks which have been helpful in several of my marginal notes. In writing a volume of this size I may have inadvertently misplaced an attribution or omitted a reference, in which case I would be grateful to be notified of the error.

This project was a long time in the making. It owes a great deal to my teachers and colleagues who have shared their insights into the subtleties of complex analysis and in doing so shaped my own view of the subject. I'd like to thank John Milnor for his guidance and inspiration, mathematical and otherwise, over the years. I'm also grateful to Christopher Bishop, Araceli Bonifant, Adam Epstein, Fredrick Gardiner, Linda Keen, Mikhail Lyubich, Bernard Maskit, Yair Minsky, and Dennis Sullivan. I'm indebted to my students, too many to name here, whose incisive comments and clever questions during my lectures have improved the presentation of this book. My final thanks are due to math editors Vickie Kearn and Susannah Shoemaker, and the entire production team at Princeton University Press, for their encouragement, helpful suggestions, and expert advice.

Saeed Zakeri

A Course in

COMPLEX
ANALYSIS

1

Rudiments of complex analysis

We begin by recalling some standard definitions and notations. Throughout this book the complex plane will be denoted by $\mathbb{C}$. Every $z \in \mathbb{C}$ has a unique representation of the form $z = x + iy$ in which $x, y \in \mathbb{R}$ and i is the imaginary unit, so $i^2 = -1$. We call x and y the **real** and **imaginary parts** of z and we write

$$x = \operatorname{Re}(z) \qquad \text{and} \qquad y = \operatorname{Im}(z).$$

The **complex conjugate** of z is the complex number defined by

$$\bar{z} = x - iy.$$

The relations

$$\operatorname{Re}(z) = \frac{1}{2}(z + \bar{z}) \qquad \text{and} \qquad \operatorname{Im}(z) = \frac{1}{2i}(z - \bar{z})$$

are easily verified. Note that $z \in \mathbb{R}$ if and only if $z = \bar{z}$.

The **absolute value** (or **modulus**) of $z = x + iy$ is the non-negative number

$$|z| = \sqrt{x^2 + y^2}.$$

Evidently $|z| > 0$ if and only if $z \neq 0$, and the relation

$$|z|^2 = z\bar{z}$$

holds.

If $z \neq 0$, the complex number $z/|z|$ has absolute value 1 and can be represented by the complex exponential $e^{i\theta} = \cos\theta + i\sin\theta$ for some $\theta \in \mathbb{R}$, called an **argument** of z, which is unique up to addition of an integer multiple of 2π. The expression

$$z = |z|\, e^{i\theta}$$

is called the **polar representation** of z.

We will reserve the following notations for the open disk of radius r centered at p and the unit disk centered at the origin:

$$\mathbb{D}(p, r) = \{z \in \mathbb{C} : |z - p| < r\} \qquad \mathbb{D} = \mathbb{D}(0, 1).$$

Unless otherwise stated, when we write $z = x + iy$ we mean x, y are real, so $x = \mathrm{Re}(z)$ and $y = \mathrm{Im}(z)$. Similarly, for a complex-valued function f, when we write $f = u + iv$ we mean u, v are real-valued, so $u = \mathrm{Re}(f)$ and $v = \mathrm{Im}(f)$.

1.1 What is a holomorphic function?

Our point of departure is the notion of complex differentiability which is fundamental to everything that follows.

> **DEFINITION 1.1.** Suppose f is a complex-valued function defined in an open neighborhood of some $p \in \mathbb{C}$. We say f is **(complex) differentiable** at p if the limit
> $$f'(p) = \lim_{z \to p} \frac{f(z) - f(p)}{z - p}$$
> exists. The number $f'(p)$ is called the **(complex) derivative** of f at p.

As usual, differentiability implies continuity. In other words, if $f'(p)$ exists, then f is continuous at p:

$$\lim_{z \to p} f(z) - f(p) = \lim_{z \to p} \frac{f(z) - f(p)}{z - p} \, (z - p) = f'(p) \cdot 0 = 0.$$

The basic rules of differentiation that we learn in calculus hold for complex derivatives.

THEOREM 1.2.

(i) *Suppose f and g are differentiable at p. Then the sum $f + g$ and the product $f g$ are differentiable at p and*

$$(f + g)'(p) = f'(p) + g'(p)$$
$$(f g)'(p) = f'(p)g(p) + f(p)g'(p).$$

Moreover, if $g(p) \neq 0$, the quotient f/g is differentiable at p and

$$\left(\frac{f}{g} \right)'(p) = \frac{f'(p)g(p) - f(p)g'(p)}{(g(p))^2}.$$

(ii) *Suppose g is differentiable at p and f is differentiable at $g(p)$. Then the composition $f \circ g$ is differentiable at p and the "chain rule" holds:*

$$(f \circ g)'(p) = f'(g(p)) \, g'(p).$$

The assumption $g(p) \neq 0$ in (i) combined with continuity of g at p implies that g is non-zero in an open neighborhood of p, so the quotient f/g is well defined in that neighborhood.

PROOF. (i) The results for the sum and product are easy to prove. For the quotient rule, first consider the special case where $f = 1$ everywhere. Writing

$$\frac{\dfrac{1}{g(z)} - \dfrac{1}{g(p)}}{z-p} = -\frac{g(z) - g(p)}{z-p} \cdot \frac{1}{g(z)g(p)},$$

letting $z \to p$, and using continuity of g at p, we obtain $(1/g)'(p) = -g'(p)/(g(p))^2$. The quotient rule now follows from this and the product rule applied to $f/g = f \cdot 1/g$.

(ii) Define

$$\varepsilon(w) = \begin{cases} \dfrac{f(w) - f(g(p))}{w - g(p)} - f'(g(p)) & w \neq g(p) \\ 0 & w = g(p). \end{cases}$$

Then ε is continuous at $g(p)$ and the relation

$$f(w) - f(g(p)) = (f'(g(p)) + \varepsilon(w))\,(w - g(p))$$

holds throughout an open neighborhood of $g(p)$. Setting $w = g(z)$, it follows that

$$\frac{(f \circ g)(z) - (f \circ g)(p)}{z-p} = \big(f'(g(p)) + \varepsilon(g(z))\big) \frac{g(z) - g(p)}{z-p}.$$

As $z \to p$, $g(z) \to g(p)$ by continuity, so $\varepsilon(g(z)) \to 0$. Hence the right side tends to $f'(g(p))\,g'(p)$. □

EXAMPLE 1.3 (Polynomials). It is immediate from the definition that the identity function $f(z) = z$ is differentiable everywhere and $f'(z) = 1$ for all z. By repeated application of Theorem 1.2(i), it follows that every polynomial $f(z) = \sum_{n=0}^{d} a_n z^n$ is differentiable everywhere and $f'(z) = \sum_{n=1}^{d} n a_n z^{n-1}$ for all z.

EXAMPLE 1.4. The smooth function $f(z) = z\bar{z} = |z|^2$ is complex differentiable only at the origin. In fact, for $z \neq 0$,

$$\frac{f(p+z) - f(p)}{z} = \frac{(p+z)(\bar{p}+\bar{z}) - p\bar{p}}{z} = p\frac{\bar{z}}{z} + \bar{p} + \bar{z}.$$

But $\bar{z}/z$ does not have a limit as $z \to 0$, since $\bar{z}/z = 1$ if z is real while $\bar{z}/z = -1$ if z is purely imaginary. It follows that the right side of the above equation has a limit as $z \to 0$ if and only if $p = 0$, and that $f'(0) = 0$.

Under the canonical isomorphism $\mathbb{C} \to \mathbb{R}^2$ given by $z = x + iy \mapsto (x, y)$, every complex-valued function $f = u + iv$ can be identified with the map into the plane

$\mathbb{R}^2$ given by $f(x,y) = (u(x,y), v(x,y))$. To understand the relation between the complex derivative of f as defined above and the real derivative of f as a map into the plane, let us first introduce a few useful notations. The partial differentiation operators $\partial/\partial x$ and $\partial/\partial y$ acting on smooth real-valued functions can be naturally extended to complex-valued functions. Explicitly, if $f = u + iv$, we set

$$(1.1) \qquad \frac{\partial f}{\partial x} = \frac{\partial u}{\partial x} + i \frac{\partial v}{\partial x} \qquad \frac{\partial f}{\partial y} = \frac{\partial u}{\partial y} + i \frac{\partial v}{\partial y}.$$

For complex-variable computations, it will be convenient to introduce two new differential operators defined by

$$(1.2) \qquad \frac{\partial f}{\partial z} = \frac{1}{2}\left(\frac{\partial f}{\partial x} - i\frac{\partial f}{\partial y}\right) \qquad \frac{\partial f}{\partial \bar{z}} = \frac{1}{2}\left(\frac{\partial f}{\partial x} + i\frac{\partial f}{\partial y}\right),$$

so

$$(1.3) \qquad \frac{\partial f}{\partial x} = \frac{\partial f}{\partial z} + \frac{\partial f}{\partial \bar{z}} \qquad \frac{\partial f}{\partial y} = i\left(\frac{\partial f}{\partial z} - \frac{\partial f}{\partial \bar{z}}\right).$$

It is important to keep in mind that the operators $\partial/\partial z$ and $\partial/\partial \bar{z}$ are not defined as partial differentiation with respect to z and $\bar{z}$. After all, z and $\bar{z}$ are not independent variables!

EXAMPLE 1.5. By the definition (1.2),

$$\frac{\partial}{\partial z}(z) = \frac{1}{2}\left(\frac{\partial}{\partial x} - i\frac{\partial}{\partial y}\right)(x+iy) = 1 \qquad \frac{\partial}{\partial z}(\bar{z}) = \frac{1}{2}\left(\frac{\partial}{\partial x} - i\frac{\partial}{\partial y}\right)(x-iy) = 0$$

$$\frac{\partial}{\partial \bar{z}}(z) = \frac{1}{2}\left(\frac{\partial}{\partial x} + i\frac{\partial}{\partial y}\right)(x+iy) = 0 \qquad \frac{\partial}{\partial \bar{z}}(\bar{z}) = \frac{1}{2}\left(\frac{\partial}{\partial x} + i\frac{\partial}{\partial y}\right)(x-iy) = 1.$$

Since it is easy to verify the product rule for $\partial/\partial z$ and $\partial/\partial \bar{z}$ (see problem 3), it follows that these linear operators act on polynomials in z and $\bar{z}$ in the following way:

$$\frac{\partial}{\partial z}\left(\sum_{j,k} a_{jk}\, z^j \bar{z}^k\right) = \sum_{j,k} j a_{jk}\, z^{j-1} \bar{z}^k \qquad \frac{\partial}{\partial \bar{z}}\left(\sum_{j,k} a_{jk}\, z^j \bar{z}^k\right) = \sum_{j,k} k a_{jk}\, z^j \bar{z}^{k-1}.$$

Observe that these are the answers we would have obtained if we had taken "partial derivatives" with respect to z and $\bar{z}$.

EXAMPLE 1.6. The operators $\partial/\partial z$ and $\partial/\partial \bar{z}$ act on $\log |z|$ as follows:

$$\frac{\partial}{\partial z}\log|z| = \frac{1}{4}\left(\frac{\partial}{\partial x} - i\frac{\partial}{\partial y}\right)\log(x^2+y^2) = \frac{1}{2}\frac{x-iy}{x^2+y^2} = \frac{1}{2z},$$

$$\frac{\partial}{\partial \bar{z}}\log|z| = \frac{1}{4}\left(\frac{\partial}{\partial x} + i\frac{\partial}{\partial y}\right)\log(x^2+y^2) = \frac{1}{2}\frac{x+iy}{x^2+y^2} = \frac{1}{2\bar{z}}.$$

If we write $\log|z|$ as $\frac{1}{2}\log(z\bar{z})$ and take "partial derivatives" with respect to z and $\bar{z}$, we obtain

$$\frac{\partial f}{\partial z} = \frac{1}{2}\frac{\bar{z}}{z\bar{z}} = \frac{1}{2z} \qquad \frac{\partial f}{\partial \bar{z}} = \frac{1}{2}\frac{z}{z\bar{z}} = \frac{1}{2\bar{z}},$$

which agree with the previous computations.

In both of the above examples, we could formally consider z and $\bar{z}$ as independent variables and compute $\partial f/\partial z$ and $\partial f/\partial \bar{z}$ by "partial differentiation" with respect to the corresponding variable, pretending the other is fixed. Such formal computations are not totally meaningless and in fact there are practical situations where their legitimacy can be justified. We will provide such a justification at the end of this section.

We continue identifying $f = u + iv$ with the map into the plane given by $f(x,y) = (u(x,y), v(x,y))$. By definition, this map is real differentiable at p if there is a necessarily unique linear map $Df(p): \mathbb{R}^2 \to \mathbb{R}^2$, called the **real derivative** of f at p, such that

$$\lim_{(x,y)\to(0,0)} \frac{\|f(p+(x,y)) - f(p) - Df(p)(x,y)\|}{\|(x,y)\|} = 0.$$

Here $\|(x,y)\| = \sqrt{x^2 + y^2}$ is the Euclidean norm which agrees with the absolute value of $x + iy$ as a complex number. Equivalently, we can express this condition as the first-order Taylor approximation formula: For all $(x,y) \in \mathbb{R}^2$ sufficiently close to the origin $(0,0)$,

$$(1.4) \qquad f(p+(x,y)) = f(p) + Df(p)(x,y) + \varepsilon(x,y),$$

where the "error term" $\varepsilon(x,y)$ satisfies $\|\varepsilon(x,y)\|/\|(x,y)\| \to 0$ as $(x,y) \to (0,0)$. It is easy to see that in the standard basis of $\mathbb{R}^2$, the linear map $Df(p)$ is represented by the 2×2 matrix of partial derivatives

$$(1.5) \qquad Df(p) = \begin{bmatrix} \dfrac{\partial u}{\partial x}(p) & \dfrac{\partial u}{\partial y}(p) \\[2ex] \dfrac{\partial v}{\partial x}(p) & \dfrac{\partial v}{\partial y}(p) \end{bmatrix}.$$

For convenience, let us use the subscript notation for our differential operators. Thus,

$$f_x = \frac{\partial f}{\partial x}, \qquad f_y = \frac{\partial f}{\partial y}, \qquad f_z = \frac{\partial f}{\partial z}, \qquad f_{\bar{z}} = \frac{\partial f}{\partial \bar{z}}.$$

Suppose f has a real derivative at p so (1.4) holds. Using the matrix (1.5) for $Df(p)$, we see that

$$Df(p)(x,y) = \begin{bmatrix} u_x(p) & u_y(p) \\ v_x(p) & v_y(p) \end{bmatrix} \begin{bmatrix} x \\ y \end{bmatrix} = \begin{bmatrix} xu_x(p) + yu_y(p) \\ xv_x(p) + yv_y(p) \end{bmatrix},$$

which by (1.1) and (1.3) can be identified with the complex number

$$(xu_x(p) + yu_y(p)) + i(xv_x(p) + yv_y(p)) = xf_x(p) + yf_y(p)$$

$$= \frac{1}{2}(z + \bar{z})(f_z(p) + f_{\bar{z}}(p)) + \frac{1}{2}(z - \bar{z})(f_z(p) - f_{\bar{z}}(p))$$

$$= zf_z(p) + \bar{z}f_{\bar{z}}(p).$$

Hence, in our complex-variable notation the Taylor formula (1.4) reads

$$(1.6) \qquad f(p + z) = f(p) + zf_z(p) + \bar{z}f_{\bar{z}}(p) + \varepsilon(z),$$

where $\varepsilon(z)/z \to 0$ as $z \to 0$. If $f_{\bar{z}}(p) = 0$, we obtain

$$\frac{f(p + z) - f(p)}{z} = f_z(p) + \frac{\varepsilon(z)}{z}.$$

Letting $z \to 0$, it follows that the complex derivative $f'(p)$ exists and is equal to $f_z(p)$.

Conversely, suppose $f'(p)$ exists, so

$$f(p + z) = f(p) + f'(p)z + \varepsilon(z),$$

where $\varepsilon(z)/z \to 0$ as $z \to 0$. Then the complex multiplication $z \mapsto f'(p)z$, viewed as a linear map $\mathbb{R}^2 \to \mathbb{R}^2$, satisfies the condition (1.4). Hence the real derivative $Df(p)$ exists and $Df(p)(x, y)$ can be identified with $f'(p) \cdot (x + iy)$. If $f'(p) = \alpha + i\beta$, we have

$$f'(p) \cdot (x + iy) = (\alpha x - \beta y) + i(\beta x + \alpha y),$$

which shows $Df(p)$ is represented by the matrix

$$(1.7) \qquad Df(p) = \begin{bmatrix} \alpha & -\beta \\ \beta & \alpha \end{bmatrix}.$$

Comparing with (1.5), we see that $\alpha = u_x(p) = v_y(p)$ and $\beta = -u_y(p) = v_x(p)$. In particular,

$$f_{\bar{z}}(p) = \frac{1}{2}(f_x(p) + if_y(p)) = \frac{1}{2}((\alpha + i\beta) + i(-\beta + i\alpha)) = 0.$$

Let us summarize our findings in the following

THEOREM 1.7. *For a given complex-valued function $f = u + iv$ defined in an open neighborhood of $p \in \mathbb{C}$, the following conditions are equivalent:*

(i) The complex derivative $f'(p)$ exists.

(ii) The real derivative $Df(p)$ exists and

$$f_{\bar{z}}(p) = 0.$$

(iii) The real derivative $Df(p)$ exists and

$$u_x(p) = v_y(p) \qquad u_y(p) = -v_x(p).$$

Under any of these conditions, we have

$$f'(p) = f_z(p) = f_x(p) = -i f_y(p).$$

EXAMPLE 1.8. The polynomial $f(z) = z^2 = (x^2 - y^2) + i(2xy)$ has the complex derivative $f'(z) = 2z$ for all z. Furthermore,

$$f_x = 2x + i2y = 2z \qquad\qquad f_y = -2y + i2x = i2z$$

$$f_z = 2z \qquad\qquad f_{\bar{z}} = 0,$$

which are consistent with Theorem 1.7.

EXAMPLE 1.9. Consider the continuous function $f : \mathbb{C} \to \mathbb{C}$ defined by $f(z) = z^5/|z|^4$ for $z \neq 0$ and $f(0) = 0$. Write $f = u + iv$, where for $(x, y) \neq (0, 0)$

$$u(x, y) = \frac{x^5 - 10x^3y^2 + 5xy^4}{(x^2 + y^2)^2} \qquad \text{and} \qquad v(x, y) = \frac{y^5 - 10x^2y^3 + 5x^4y}{(x^2 + y^2)^2}$$

and $u(0, 0) = v(0, 0) = 0$. Thus

$$u_x(0, 0) = \lim_{x \to 0} \frac{u(x, 0)}{x} = 1 \qquad \text{and} \qquad v_y(0, 0) = \lim_{y \to 0} \frac{v(0, y)}{y} = 1$$

and similarly

$$u_y(0, 0) = \lim_{y \to 0} \frac{u(0, y)}{y} = 0 \qquad \text{and} \qquad v_x(0, 0) = \lim_{x \to 0} \frac{v(x, 0)}{x} = 0,$$

so the pair of conditions in Theorem 1.7(iii) holds. However, the complex derivative $f'(0)$ does not exist since

$$\frac{f(z)}{z} = \left(\frac{z}{|z|} \right)^4$$

does not have a limit as $z \to 0$. For example, this quotient tends to 1 when z tends to 0 along the real line, while it tends to -1 when z tends to 0 along the line $\text{Re}(z) = \text{Im}(z)$. Note that this example does not contradict Theorem 1.7 since $Df(0)$ does not exist.

Here is another important observation: Suppose $f'(p) = \alpha + i\beta$ so $Df(p)$ has the matrix form (1.7). If $f'(p) \neq 0$, then $\det(Df(p)) = \alpha^2 + \beta^2 > 0$, which means $Df(p)$ is orientation-preserving. Moreover, the matrix (1.7) can be decomposed as

$$Df(p) = \begin{bmatrix} \sqrt{\alpha^2 + \beta^2} & 0 \\ 0 & \sqrt{\alpha^2 + \beta^2} \end{bmatrix} \begin{bmatrix} \dfrac{\alpha}{\sqrt{\alpha^2 + \beta^2}} & \dfrac{-\beta}{\sqrt{\alpha^2 + \beta^2}} \\ \dfrac{\beta}{\sqrt{\alpha^2 + \beta^2}} & \dfrac{\alpha}{\sqrt{\alpha^2 + \beta^2}} \end{bmatrix}.$$

Geometrically, this is a rotation by the angle $\arccos(\alpha/\sqrt{\alpha^2 + \beta^2})$ about the origin, followed by a dilation by the factor $\sqrt{\alpha^2 + \beta^2}$. Alternatively, the action of $Df(p)$ can be identified with the complex multiplication by $f'(p)$, which amounts to a rotation by the argument of $f'(p)$ followed by a dilation by the factor $|f'(p)|$. This geometric description shows that $Df(p)$ is an angle-preserving linear transformation in the sense that the angle between any two non-zero vectors v_1, v_2 is the same as the angle between their images $Df(p)v_1, Df(p)v_2$. Such linear maps are often called *conformal* because they preserve shapes (but not necessarily scales).

COROLLARY 1.10. *Suppose f has a non-zero complex derivative at $p \in \mathbb{C}$. Then the real derivative $Df(p): \mathbb{R}^2 \to \mathbb{R}^2$ is an orientation-preserving conformal linear transformation.*

For several alternative characterizations of conformal linear transformations in dimension 2, see problem 9. We will return to the issue of angle preservation in chapters 4 and 6.

> **DEFINITION 1.11.** Let $U \subset \mathbb{C}$ be non-empty and open. A function $f: U \to \mathbb{C}$ is called *holomorphic* in U if $f'(p)$ exists for every $p \in U$. The set of all holomorphic functions in U is denoted by $\mathscr{O}(U)$. Elements of $\mathscr{O}(\mathbb{C})$ are called *entire functions*.

Coined by Cauchy's students C. A. Briot and J. C. Bouquet, "holomorphic" comes from the Greek ὅλος (whole) and μορφή (form). According to R. Remmert, the widespread adoption of the notation $\mathscr{O}$ for holomorphic appears to have been purely accidental.

It follows from Theorem 1.2 that sums, products, and compositions of holomorphic functions are holomorphic. In particular, pointwise addition and multiplication of functions make $\mathscr{O}(U)$ into a commutative ring with identity.

EXAMPLE 1.12 (Ratios). By Theorem 1.2(i), if $f, g \in \mathscr{O}(U)$ and $g \neq 0$ in U, then $f/g \in \mathscr{O}(U)$. An important example is provided by *rational functions*: If f and g are polynomials in z, g not identically zero, and if $p_1, \dots, p_n$ are all the roots of the polynomial equation $g(z) = 0$, then the rational function f/g is holomorphic in $\mathbb{C} \setminus \{p_1, \dots, p_n\}$.

The following is an immediate corollary of Theorem 1.7:

THEOREM 1.13. *Suppose $f = u + iv$ is real differentiable as a map $U \to \mathbb{R}^2$. Then $f \in \mathscr{O}(U)$ if and only if*

$$(1.8) \qquad\qquad f_{\bar{z}} = 0,$$

or equivalently,

$$(1.9) \qquad\qquad u_x = v_y \qquad and \qquad u_y = -v_x$$

throughout U. In this case,

$$f' = f_z = f_x = -i f_y.$$

The Cauchy-Riemann equations had been studied earlier in the 18th century by d'Alembert and Euler.

The pair of equations (1.9) are classically known as the *Cauchy-Riemann equations*. The equivalent form (1.8) is called the *complex Cauchy-Riemann equation*.

EXAMPLE 1.14. The *exponential function* $\exp : \mathbb{C} \to \mathbb{C}$ defined by

$$\exp(z) = e^z = e^x \, e^{iy} = e^x (\cos y + i \sin y)$$

is entire. In fact,

$$\frac{\partial}{\partial x} \exp = e^x \, e^{iy} \quad \text{and} \quad \frac{\partial}{\partial y} \exp = i e^x \, e^{iy}, \quad \text{so} \quad \frac{\partial}{\partial \bar{z}} \exp = 0.$$

It follows that

$$\exp' = \frac{\partial}{\partial z} \exp = \exp.$$

The basic identity

$$\exp(z + w) = \exp(z) \, \exp(w) \qquad \text{for } z, w \in \mathbb{C}$$

can be proved as follows: Fix w and set $f(z) = \exp(z + w)$. By the chain rule, $f'(z) = \exp(z + w) = f(z)$. Since $\exp \neq 0$, the ratio $g = f / \exp$ is entire and $g' = 0$ everywhere by the quotient rule. It follows that g is a constant function (this can be seen, for example, by noting that $g' = 0$ implies that the real and imaginary parts of g have vanishing partial derivatives in the plane, hence are constant). Since $g(0) = \exp(w)$, we conclude that $g(z) = \exp(z + w) / \exp(z) = \exp(w)$ for all z, as required.

EXAMPLE 1.15. Let $\varphi : [0, 1] \to [0, 1]$ be a *Cantor function*, i.e., a continuous non-decreasing function which satisfies $\varphi(0) = 0$, $\varphi(1) = 1$, and $\varphi' = 0$ a.e. (the graph of such φ is often called a "devil's staircase"). Extend φ to a map $\mathbb{R} \to \mathbb{R}$ by setting $\varphi(x + n) = \varphi(x) + n$ for $0 \leq x \leq 1$ and $n \in \mathbb{Z}$. Define $f : \mathbb{C} \to \mathbb{C}$ by

$$f(x + iy) = x + i(y + \varphi(x)).$$

As is customary, "a.e." is short for "almost everywhere," that is, outside a set of Lebesgue measure zero.

Then f is a homeomorphism, with $f_x = 1$ and $f_y = i$; hence $f_z = 1$ and $f_{\bar{z}} = 0$ a.e. on $\mathbb{C}$. However, f is not holomorphic since φ', and hence f_z, fails to exist everywhere. This does not contradict Theorem 1.13 because f is not real differentiable.

REMARK 1.16. The implication $f_{\bar{z}} = 0 \Longrightarrow f \in \mathscr{O}(U)$ holds under much weaker conditions than real differentiability in Theorem 1.13. For example, a generalization of a classical theorem of Looman and Menshov asserts that if $f : U \to \mathbb{C}$ is continuous, f_z and $f_{\bar{z}}$ exist outside a countable set in U, and $f_{\bar{z}} = 0$ a.e. in U, then $f \in \mathscr{O}(U)$ [**GM**]. Another well-known result of the same flavor is "Weyl's lemma" which is important in the theory of quasiconformal maps (see [**A3**]).

We end this section with a brief justification for computing f_z and $f_{\bar{z}}$ as partial derivatives. Suppose $F(z, w)$ is holomorphic in each variable near a point $(p, \bar{p}) \in \mathbb{C} \times \mathbb{C}$. This means there is an $r > 0$ such that $z \mapsto F(z, w)$ is holomorphic in $\mathbb{D}(p, r)$ for each fixed $w \in \mathbb{D}(\bar{p}, r)$ and $w \mapsto F(z, w)$ is holomorphic in $\mathbb{D}(\bar{p}, r)$ for each fixed $z \in \mathbb{D}(p, r)$. Set $z = x + iy$ and $w = s + it$. Then F, viewed as a function of the four real variables (x, y, s, t), can be shown to be real differentiable. Consider the function $f(z) = F(z, \bar{z}) = F(x, y, x, -y)$ which is defined in some neighborhood of p. Using the symbol D_j for partial differentiation with respect to the j-th variable, we apply the chain rule to obtain

$$f_x = D_1F + D_3F \qquad \text{and} \qquad f_y = D_2F - D_4F,$$

where the left sides of these equations are evaluated at z and the right sides are evaluated at $(z, \bar{z}) = (x, y, x, -y)$. This gives

$$f_z = \tfrac{1}{2}(f_x - if_y) = \tfrac{1}{2}(D_1F - iD_2F) + \tfrac{1}{2}(D_3F + iD_4F)$$

$$f_{\bar{z}} = \tfrac{1}{2}(f_x + if_y) = \tfrac{1}{2}(D_1F + iD_2F) + \tfrac{1}{2}(D_3F - iD_4F).$$

Denote by D_zF and D_wF the complex derivatives of F with respect to each variable when the other is kept fixed. Then $(D_1F - iD_2F)/2 = D_zF$ and $D_3F + iD_4F = 0$ since F is holomorphic in w. Similarly, $(D_3F - iD_4F)/2 = D_wF$ and $D_1F + iD_2F = 0$ since F is holomorphic in z. It follows that

$$f_z = D_zF \qquad \text{and} \qquad f_{\bar{z}} = D_wF.$$

This means f_z and $f_{\bar{z}}$ are obtained by taking the partial derivatives of $F(z, w)$ with respect to z and w, respectively, and then substituting $w = \bar{z}$.

In Example 1.5, this result can be applied to the polynomial function $F(z, w) = \sum_{j,k} a_{jk} z^j w^k$ to justify the given formulas. In Example 1.6, it can be applied to $F(z, w) = \log(zw)$ which, as we will see in chapter 2, has well-defined holomorphic branches in each variable in a small neighborhood of $(p, \bar{p})$ provided that $p \neq 0$.

1.2 Complex analytic functions

For every $p \in \mathbb{C}$ and every sequence $\{a_n\}_{n=0}^{\infty}$ of complex numbers, we can form the **power series**

$$(1.10) \qquad \sum_{n=0}^{\infty} a_n (z - p)^n$$

in the complex variable z. Such series provide an abundance of examples of holomorphic functions and play a central role in complex analysis, especially the classical function theory according to Weierstrass. For now, the basic fact that we need to know (or remember) is that each power series (1.10) has a **disk of convergence** about p characterized by the property that it converges within this disk and diverges outside of it. Moreover, we can effectively compute the radius of this disk once we know the coefficients a_n or merely their asymptotic behavior as $n \to \infty$. This fact is stated more precisely in the following

THEOREM 1.17 (Cauchy, 1821). *Consider the power series* (1.10) *and define*

$$(1.11) \qquad R = \frac{1}{\limsup_{n \to \infty} \sqrt[n]{|a_n|}} \in [0, +\infty].$$

Then (1.10) converges absolutely and uniformly in the disk $\mathbb{D}(p, r)$ for every $r < R$ and diverges at every point z with $|z - p| > R$.

This result is also attributed to Hadamard who rediscovered it in 1888.

The number R is called the **radius of convergence** of the power series (1.10). Observe that the possibilities $R = 0$ or $R = +\infty$ have not been excluded.

PROOF. First consider the power series inside the disk of radius R. If $R = 0$ there is nothing to prove, so assume $R > 0$ and let $0 < r < s < R$. The definition of R shows that there is an integer $N \geq 1$ such that $|a_n| s^n < 1$ for all $n \geq N$. If $|z - p| < r$, then

$$\sum_{n=N}^{\infty} |a_n| \, |z - p|^n \leq \sum_{n=N}^{\infty} |a_n| \, r^n = \sum_{n=N}^{\infty} |a_n| \, s^n \left(\frac{r}{s}\right)^n \leq \sum_{n=N}^{\infty} \left(\frac{r}{s}\right)^n.$$

Since $r/s < 1$, the far right geometric series converges, which proves that the power series converges absolutely and uniformly in $\mathbb{D}(p, r)$.

Now consider the power series outside the disk of radius R. If $R = +\infty$ there is nothing to prove, so assume $R < +\infty$ and let $r > s > R$. The definition of R shows that there is an increasing sequence S of positive integers such that $|a_n| s^n > 1$ for all $n \in S$. If $|z - p| = r$, then

$$|a_n| \, |z - p|^n = |a_n| \, s^n \left(\frac{r}{s}\right)^n > \left(\frac{r}{s}\right)^n$$

whenever $n \in S$. Since $r/s > 1$, it follows that the power series diverges since its general term fails to converge to zero. $\square$

The behavior of power series on the circle of convergence $|z - p| = R$ is much more subtle. In fact, no general statement similar to Theorem 1.17 can be made for what should be happening on this circle.

EXAMPLE 1.18. The power series

$$\sum_{n=0}^{\infty} z^n \qquad \sum_{n=1}^{\infty} \frac{z^n}{n^2} \qquad \sum_{n=1}^{\infty} \frac{z^n}{n}$$

all have a radius of convergence of 1. The first diverges everywhere on the unit circle $\partial \mathbb{D}$ since its general term z^n does not tend to zero when $|z| = 1$. The second converges uniformly on $\partial \mathbb{D}$ since it is dominated by the convergent series $\sum 1/n^2$. The third converges at every point of $\partial \mathbb{D}$ other than 1 (see problem 13).

More on the behavior of power series on the circle of convergence will be discussed in chapter 10.

DEFINITION 1.19. Let $U \subset \mathbb{C}$ be non-empty and open. We call a function $f : U \to \mathbb{C}$ **complex analytic** if for every disk $\mathbb{D}(p, r) \subset U$ there exists a power series $\sum_{n=0}^{\infty} a_n \, (z - p)^n$ which converges to $f(z)$ whenever $z \in \mathbb{D}(p, r)$.

It is a fundamental fact that a function is complex analytic in U if and only if it is holomorphic in U. The following theorem proves the "only if" part of this statement. The "if" part, which is more difficult, will be proved in Theorem 1.37.

THEOREM 1.20. *Let $f : U \to \mathbb{C}$ be complex analytic. Then*

(i) *$f \in \mathcal{O}(U)$ and f' is also complex analytic in U.*

(ii) *All higher derivatives $f^{(k)}$ for $k \geq 1$ exist and are complex analytic in U. Moreover, the power series representation of the higher derivatives are obtained by term-by-term differentiation of that of f, that is, if $\mathbb{D}(p, r) \subset U$ and*

$$(1.12) \qquad f(z) = \sum_{n=0}^{\infty} a_n \, (z - p)^n \qquad for \ z \in \mathbb{D}(p, r),$$

then the representation

$$f^{(k)}(z) = \sum_{n=k}^{\infty} n(n-1) \cdots (n - k + 1) \, a_n \, (z - p)^{n-k}$$

holds in $\mathbb{D}(p, r)$.

(iii) *The coefficients $\{a_n\}$ of the power series (1.12) are given by*

$$(1.13) \qquad a_n = \frac{f^{(n)}(p)}{n!} \qquad (n \geq 0).$$

In particular, $\{a_n\}$ is uniquely determined by f, so any power series in $z - p$ which converges to f in some disk in U centered at p must coincide with (1.12).

PROOF. Define $g(z) = \sum_{n=1}^{\infty} n a_n \, (z - p)^{n-1}$. Note that by $\lim_{n \to \infty} \sqrt[n]{n} = 1$ and the formula (1.11), the power series with coefficients $\{n a_n\}$ has the same radius of convergence as the power series with coefficients $\{a_n\}$, so g converges in $\mathbb{D}(p, r)$. We will show that for every $z_0 \in \mathbb{D}(p, r)$, $f'(z_0)$ exists and is equal to $g(z_0)$. This will prove (i). Evidently, (ii) follows by induction from (i), and (iii) follows from (ii).

After replacing $z - p$ by z, we may assume $p = 0$. Fix $z_0 \in \mathbb{D}(0, r)$ and take any $\varepsilon > 0$. Choose s such that $|z_0| < s < r$. Since the power series of g converges absolutely in $\mathbb{D}(0, r)$, we can find an integer $N \geq 2$ such that

$$\sum_{n=N+1}^{\infty} n |a_n| \, s^{n-1} < \varepsilon.$$

For $z \neq z_0$ in $\mathbb{D}(0, r)$, write

$$\frac{f(z) - f(z_0)}{z - z_0} - g(z_0) = \sum_{n=2}^{\infty} a_n \left(\frac{z^n - z_0^n}{z - z_0} - n z_0^{n-1} \right)$$

$$= \left(\sum_{n=2}^{N} + \sum_{n=N+1}^{\infty} \right) a_n \, (z^{n-1} + z^{n-2} z_0 + \cdots + z_0^{n-1} - n z_0^{n-1}).$$

The first (finite) sum tends to 0 as $z \to z_0$. The second sum has its absolute value bounded above by $\sum_{n=N+1}^{\infty} 2n|a_n| s^{n-1} < 2\varepsilon$ if $|z| < s$. Hence,

$$\limsup_{z \to z_0} \left| \frac{f(z) - f(z_0)}{z - z_0} - g(z_0) \right| \le 2\varepsilon.$$

Since ε was arbitrary, it follows that $\lim_{z \to z_0} (f(z) - f(z_0))/(z - z_0)$ exists and is equal to $g(z_0)$. $\square$

EXAMPLE 1.21. The radius of convergence of the power series $f(z) = \sum_{n=0}^{\infty} z^n$ is 1, hence $f \in \mathscr{O}(\mathbb{D})$. The formula for the sum of a geometric series shows that in fact $f(z) = 1/(1-z)$. Term-by-term differentiation of this power series, which is legitimate by Theorem 1.20, yields other useful formulas. For example, it follows that

$$\sum_{n=1}^{\infty} n z^{n-1} = \left(\frac{1}{1-z} \right)' = \frac{1}{(1-z)^2},$$

so

$$\sum_{n=1}^{\infty} n z^n = \frac{z}{(1-z)^2} \qquad \text{for } |z| < 1.$$

Differentiating once more, we obtain

$$\sum_{n=1}^{\infty} n^2 z^{n-1} = \left(\frac{z}{(1-z)^2} \right)' = \frac{1+z}{(1-z)^3},$$

so

$$\sum_{n=1}^{\infty} n^2 z^n = \frac{z(1+z)}{(1-z)^3} \qquad \text{for } |z| < 1.$$

Continuing inductively, we can find closed expressions (as rational functions in z) for the power series $\sum_{n=1}^{\infty} n^p z^n$ in the unit disk for every positive integer p.

EXAMPLE 1.22. Since $\lim_{n \to \infty} 1/\sqrt[n]{n!}$ is easily seen to be 0, the radius of convergence of the power series $f(z) = \sum_{n=0}^{\infty} z^n/n!$ is $+\infty$. Hence, by Theorem 1.20, f is an entire function with $f(0) = 1$, and term-by-term differentiation gives $f' = f$. The exponential function exp defined in example 1.14 also satisfies $\exp(0) = 1$ and $\exp' = \exp$. It follows that the ratio $g = f/\exp$ is entire, $g(0) = 1$, and $g' = 0$ everywhere by the quotient rule, which shows g is the constant function 1. Thus, we arrive at the following alternative formula for the exponential function:

$$\exp(z) = \sum_{n=0}^{\infty} \frac{z^n}{n!} \qquad \text{for all } z \in \mathbb{C}.$$

1.3 Complex integration

We now turn to integration of complex-valued functions along curves. Our standing assumption in this section is that all curves are piecewise smooth. This regularity assumption greatly simplifies the arguments but it is not essential, as one can fashion a definition to allow more general curves such as those that are merely rectifiable. However, such generalizations are not worth the extra effort: The integration theory we are about to develop will be applied almost exclusively to holomorphic functions and such integrals, as we will see in chapter 2, depend only on the "homology class" of the curve. This means the integral along an arbitrary curve (rectifiable or not) can be defined as the integral along any piecewise smooth curve in the same homology class. Thus, one can ultimately arrive at the most general definition using only the special case treated here.

Let $U \subset \mathbb{C}$ be a non-empty open set. A **curve** in U is a continuous map $\gamma : [a, b] \to U$, where $[a, b] = \{t \in \mathbb{R} : a \leq t \leq b\}$. We call $\gamma(a)$ the **initial point** and $\gamma(b)$ the **end point** of γ. For simplicity we often say that γ is a curve **from** $\gamma(a)$ **to** $\gamma(b)$. γ is a **closed** curve if $\gamma(a) = \gamma(b)$. We denote by $|\gamma|$ the **image** of γ as a subset of $\mathbb{C}$, that is, $|\gamma| = \{\gamma(t) : t \in [a, b]\}$. We say γ is **piecewise** C^1 if there is a partition $a = t_0 < t_1 < \cdots < t_n = b$ such that for each $1 \leq k \leq n$, γ is continuously differentiable in the open interval (t_{k-1}, t_k), and the one-sided limits $\lim_{t \to t_{k-1}^+} \gamma'(t)$ and $\lim_{t \to t_k^-} \gamma'(t)$ exist.

Throughout this section we will assume that all curves are piecewise C^1, even if it is not explicitly mentioned.

DEFINITION 1.23. Let $\gamma : [a, b] \to \mathbb{C}$ be a curve and $f : |\gamma| \to \mathbb{C}$ be a continuous function. The **integral of f along γ** is the complex number defined by

$$(1.14) \qquad \int_\gamma f(z)\, dz = \int_a^b f(\gamma(t)) \gamma'(t)\, dt.$$

Here $\gamma' = d\gamma / dt$ is defined at all but finitely many points of $[a, b]$.

By writing $f = u + iv$ and $\gamma(t) = x(t) + iy(t)$, and separating the real and imaginary parts, we see that the integral in (1.14) can be written in terms of a pair of classical "line integrals" along γ:

$$(1.15) \qquad \int_\gamma f(z)\, dz = \int_\gamma (u\, dx - v\, dy) + i \int_\gamma (v\, dx + u\, dy).$$

It is easy to see that the right side of (1.14) remains unchanged if we reparametrize γ. In fact, if $\varphi : [c, d] \to [a, b]$ is a C^1 orientation-preserving homeomorphism and $\eta = \gamma \circ \varphi$, then by the change of variable formula in calculus,

$$\int_c^d f(\eta(t))\, \eta'(t)\, dt = \int_c^d f(\gamma(\varphi(t)))\, \gamma'(\varphi(t))\, \varphi'(t)\, dt$$

$$= \int_a^b f(\gamma(s))\, \gamma'(s)\, ds.$$

In particular, the domain of γ can always be chosen to be the unit interval $[0, 1]$ by precomposing it with an affine map, namely by considering the reparametrized curve $t \mapsto \gamma((1-t)a + tb)$ instead.

Given $\gamma : [0, 1] \to \mathbb{C}$, the **reverse curve** $\gamma^- : [0, 1] \to \mathbb{C}$ is defined by $\gamma^-(t) = \gamma(1-t)$. Since

$$\int_0^1 f(\gamma^-(t)) \, (\gamma^-)'(t) \, dt = - \int_0^1 f(\gamma(1-t)) \, \gamma'(1-t) \, dt$$

$$= - \int_0^1 f(\gamma(t)) \, \gamma'(t) \, dt,$$

we see that

$$\int_{\gamma^-} f(z) \, dz = - \int_\gamma f(z) \, dz.$$

There is an obvious way of combining two curves whenever the end point of one is the initial point of the other: If $\gamma, \eta : [0, 1] \to U$ are curves such that $\gamma(1) = \eta(0)$, we can define the **product** $\gamma \cdot \eta : [0, 1] \to U$ by

$$(\gamma \cdot \eta)(t) = \begin{cases} \gamma(2t) & t \in [0, 1/2] \\ \eta(2t - 1) & t \in [1/2, 1]. \end{cases}$$

This amounts to first traveling along γ and then along η, both with twice the usual speed so as to finish the journey in unit time. The additivity property of the integral shows that the relation

$$(1.16) \qquad \int_{\gamma \cdot \eta} f(z) \, dz = \int_\gamma f(z) \, dz + \int_\eta f(z) \, dz$$

holds for every continuous function $f : |\gamma| \cup |\eta| \to \mathbb{C}$.

Closely related to the complex integral is the notion of the line integral of $f : U \to \mathbb{C}$, viewed as a scalar function, as one learns in calculus:

$$(1.17) \qquad \int_\gamma f(z) \, |dz| = \int_0^1 f(\gamma(t)) \, |\gamma'(t)| \, dt.$$

For example, the case $f = 1$ gives

$$\int_\gamma |dz| = \int_0^1 |\gamma'(t)| \, dt.$$

This is by definition the **length** of γ for which we use the notation length(γ). It is evident that

$$\left| \int_\gamma f(z) \, dz \right| \leq \int_\gamma |f(z)| \, |dz|.$$

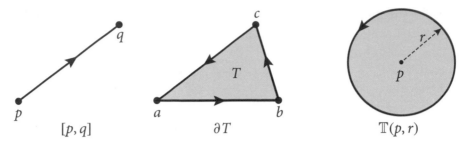

Figure 1.1. Three basic curves that frequently arise in complex integration. From left to right: an oriented segment, the oriented boundary of a triangle, and an oriented circle.

This proves the following useful inequality which we will frequently invoke when estimating integrals:

$$(1.18) \qquad \left| \int_\gamma f(z)\, dz \right| \leq \sup_{z \in |\gamma|} |f(z)| \cdot \text{length}(\gamma).$$

If we denote the supremum of $|f|$ on $|\gamma|$ by M and the length of γ by L, then (1.18) reads $|\int_\gamma f(z)\, dz| \leq ML$. This is why (1.18) is informally referred to as **the ML-inequality**.

EXAMPLE 1.24 (Oriented segments). For $p, q \in \mathbb{C}$, let $[p, q]$ denote the oriented line segment traversed once from p to q (see Fig. 1.1 left). We take $\gamma : [0, 1] \to \mathbb{C}$ defined by $\gamma(t) = (1 - t) p + tq$ as the standard parametrization of $[p, q]$. In the interest of simplifying our notations, we often denote the image $|\gamma|$ by $[p, q]$ as well. Thus, for any continuous function $f : [p, q] \to \mathbb{C}$,

$$\int_{[p,q]} f(z)\, dz = (q - p) \int_0^1 f((1 - t)p + tq)\, dt.$$

Note that $\int_{[q,p]} f(z)\, dz = - \int_{[p,q]} f(z)\, dz$ since $[q, p]$ is the reverse of $[p, q]$.

EXAMPLE 1.25 (Oriented triangle boundaries). Let T be the closed triangle with vertices a, b, c labeled counterclockwise. We use the notation $T = \triangle abc$, so $\triangle abc = \triangle bca = \triangle cab$. By definition, the oriented boundary ∂T is the product $([a, b] \cdot [b, c]) \cdot [c, a]$, that is, the closed curve obtained from the segment $[a, b]$, followed by $[b, c]$, followed by $[c, a]$ (see Fig. 1.1 middle). Again, to simplify notations, we denote the corresponding subset of the plane by ∂T as well. By (1.16), for any continuous function $f : \partial T \to \mathbb{C}$,

$$\int_{\partial T} f(z)\, dz = \int_{[a,b]} f(z)\, dz + \int_{[b,c]} f(z)\, dz + \int_{[c,a]} f(z)\, dz.$$

EXAMPLE 1.26 (Oriented circles). For $p \in \mathbb{C}$ and $r > 0$, let $\mathbb{T}(p, r)$ denote the oriented circle $|z - p| = r$ traversed once in the counterclockwise direction (see Fig. 1.1 right). We take $\gamma : [0, 2\pi] \to \mathbb{C}$ defined by $\gamma(t) = p + re^{it}$ as the standard parametrization of $\mathbb{T}(p, r)$, and often denote the image $|\gamma| = \{z \in \mathbb{C} : |z - p| = r\}$ by $\mathbb{T}(p, r)$ as well (the distinction is easily understood from the context).

Evidently, for any continuous function $f : \mathbb{T}(p, r) \to \mathbb{C}$,

$$\int_{\mathbb{T}(p,r)} f(z)\, dz = ir \int_0^{2\pi} f(p + re^{it}) e^{it}\, dt.$$

As a special case, consider a continuous complex-valued function f defined on the **unit circle** $\mathbb{T} = \mathbb{T}(0, 1)$. The integral of f as a scalar function can be expressed as a complex integral:

$$\int_0^{2\pi} f(e^{it})\, dt = \int_{\mathbb{T}} f(z)\, \frac{dz}{iz}.$$

More generally, the **Fourier coefficients** of f, defined by

$$\hat{f}(n) = \frac{1}{2\pi} \int_0^{2\pi} f(e^{it})\, e^{-int}\, dt \qquad (n \in \mathbb{Z}),$$

can be expressed as the complex integrals

$$\hat{f}(n) = \frac{1}{2\pi i} \int_{\mathbb{T}} \frac{f(z)}{z^{n+1}}\, dz.$$

The following elementary observation will be useful:

LEMMA 1.27 (Continuous dependence on vertices). *Let $f : U \to \mathbb{C}$ be continuous and $T = \triangle abc$ be a closed triangle in U. Then, for every $\varepsilon > 0$ there exists a $\delta > 0$ such that if $|a - a'|$, $|b - b'|$, and $|c - c'|$ are all less than δ, then $T' = \triangle a'b'c' \subset U$ and*

$$\left| \int_{\partial T} f(z)\, dz - \int_{\partial T'} f(z)\, dz \right| < \varepsilon.$$

PROOF. Since the integral along the oriented boundary of a triangle is the sum of three integrals along oriented segments, it suffices to prove continuous dependence for oriented segments. Fix $[a, b] \subset U$ and let V be any open neighborhood of $[a, b]$ whose closure $\overline{V}$ is a compact subset of U. Given $\varepsilon > 0$, use uniform continuity of f on V to find $0 < \delta < \varepsilon$ such that $|f(z) - f(w)| < \varepsilon$ whenever $z, w \in V$ and $|z - w| < \delta$. We can also arrange that $[a', b'] \subset V$ whenever $|a - a'| < \delta$ and $|b - b'| < \delta$. Let $[a', b']$ be any such segment and note that if $\gamma(t) = (1 - t)a + tb$ and $\eta(t) = (1 - t)a' + tb'$, then

$$|\gamma(t) - \eta(t)| \leq (1 - t)|a - a'| + t|b - b'| < \delta,$$

$$|\gamma'(t) - \eta'(t)| = |(b - a) - (b' - a')| \leq |b - b'| + |a - a'| < 2\delta$$

for all $0 \leq t \leq 1$. Hence,

$$\left| \int_{[a,b]} f(z)\, dz - \int_{[a',b']} f(z)\, dz \right| = \left| \int_0^1 \left[f(\gamma(t))\, \gamma'(t) - f(\eta(t))\, \eta'(t) \right] dt \right|$$

$$= \left| \int_0^1 \left[\bigl(f(\gamma(t)) - f(\eta(t))\bigr)\gamma'(t) + f(\eta(t))\bigl(\gamma'(t) - \eta'(t)\bigr) \right] dt \right|$$

$$\leq \int_0^1 |f(\gamma(t)) - f(\eta(t))| \, |\gamma'(t)| \, dt + \int_0^1 |f(\eta(t))| \, |\gamma'(t) - \eta'(t)| \, dt$$

$$\leq |b-a| \int_0^1 |f(\gamma(t)) - f(\eta(t))| \, dt + 2\delta \int_0^1 |f(\eta(t))| \, dt$$

$$\leq |b-a| \, \varepsilon + 2\delta \sup_{z \in V} |f(z)| \leq (|b-a| + 2 \sup_{z \in V} |f(z)|) \, \varepsilon,$$

which proves the asserted continuity. □

A primitive is what students of calculus call "antidetrivative."

> **DEFINITION 1.28.** A function $F \in \mathcal{O}(U)$ is called a **primitive** of a continuous function $f : U \to \mathbb{C}$ if $F'(z) = f(z)$ for all $z \in U$.

Suppose F is a primitive of f and $\gamma : [0,1] \to U$ is a piecewise C^1 curve. By the chain rule, the relation $(F \circ \gamma)'(t) = F'(\gamma(t))\gamma'(t)$ holds for all but finitely many $t \in [0,1]$ (see problem 6). Since $F'(\gamma(t))\gamma'(t)$ is piecewise continuous on $[0,1]$ with at worst jump discontinuities, the fundamental theorem of calculus shows that

$$\int_\gamma f(z) \, dz = \int_0^1 f(\gamma(t)) \, \gamma'(t) \, dt = \int_0^1 F'(\gamma(t)) \, \gamma'(t) \, dt$$

$$= \int_0^1 (F \circ \gamma)'(t) \, dt = F(\gamma(1)) - F(\gamma(0)).$$

THEOREM 1.29. *A continuous function $f : U \to \mathbb{C}$ has a primitive in U if and only if $\int_\gamma f(z) \, dz = 0$ for every closed curve γ in U.*

PROOF. First suppose f has a primitive F. If $\gamma : [0,1] \to U$ is a closed curve, then $\gamma(0) = \gamma(1)$, so

$$\int_\gamma f(z) \, dz = F(\gamma(1)) - F(\gamma(0)) = 0.$$

Conversely, suppose f integrates to zero along every closed curve in U. To show f has a primitive, it suffices to consider the case when U is connected (and therefore path-connected); the general case follows by applying this case to each connected component of U. If γ, η are two curves in U with the same initial and end points, then the product $\gamma \, . \, \eta^-$ is a closed curve. Hence, by additivity (1.16) and our assumption,

$$\int_\gamma f(\zeta) \, d\zeta - \int_\eta f(\zeta) \, d\zeta = \int_\gamma f(\zeta) \, d\zeta + \int_{\eta^-} f(\zeta) \, d\zeta = \int_{\gamma . \eta^-} f(\zeta) \, d\zeta = 0.$$

Now fix a point $p \in U$. For any $z \in U$ use path-connectivity of U to find a curve γ in U from p to z and define

$$F(z) = \int_\gamma f(\zeta) \, d\zeta.$$

By the above remark, the right-hand side is independent of the choice of γ and yields a well-defined function $F : U \to \mathbb{C}$. Let us show that F is a primitive of f. Fix $z_0 \in U$ and choose $r > 0$ small enough so that $\mathbb{D}(z_0, r) \subset U$. Let $z \in \mathbb{D}(z_0, r)$ and let γ be any curve in U from p to z_0. The product $\gamma \cdot [z_0, z]$ is then a curve in U from p to z. By additivity,

$$F(z) - F(z_0) = \int_{\gamma \cdot [z_0, z]} f(\zeta)\, d\zeta - \int_\gamma f(\zeta)\, d\zeta = \int_{[z_0, z]} f(\zeta)\, d\zeta,$$

so if $z \neq z_0$,

(1.19) $$\frac{F(z) - F(z_0)}{z - z_0} - f(z_0) = \frac{1}{z - z_0} \int_{[z_0, z]} (f(\zeta) - f(z_0))\, d\zeta.$$

Since f is continuous at z_0, for each $\varepsilon > 0$ we can find a $0 < \delta < r$ such that $|f(\zeta) - f(z_0)| < \varepsilon$ whenever $|\zeta - z_0| < \delta$. Since $|z - z_0| < \delta$ implies $|\zeta - z_0| < \delta$ for every $\zeta \in [z_0, z]$, the ML-inequality (1.18) applied to the right side of (1.19) gives

$$\left| \frac{F(z) - F(z_0)}{z - z_0} - f(z_0) \right| \leq \frac{1}{|z - z_0|} \cdot \varepsilon \cdot \text{length}([z_0, z]) = \varepsilon$$

whenever $0 < |z - z_0| < \delta$. Thus, $F'(z_0)$ exists and is equal to $f(z_0)$. Since $z_0 \in U$ was arbitrary, we conclude that F is a primitive of f in U.   □

EXAMPLE 1.30. For every integer $n \neq -1$, the power function $f(z) = z^n$ has a primitive $F(z) = z^{n+1}/(n+1)$. It follows from Theorem 1.29 that $\int_\gamma z^n\, dz = 0$ if γ is any closed curve in the punctured plane $\mathbb{C} \smallsetminus \{0\}$ and $n \neq -1$, or if γ is any closed curve in $\mathbb{C}$ and $n \geq 0$.

The case $n = -1$ is completely different: For any $r > 0$,

$$\int_{\mathbb{T}(0,r)} \frac{1}{z}\, dz = \int_0^{2\pi} \frac{1}{re^{it}}\, rie^{it}\, dt = 2\pi i \neq 0.$$

Note that the result is independent of the radius r. It follows from Theorem 1.29 that *the function $z \mapsto 1/z$ does not have a primitive in any punctured neighborhood of 0.*

1.4 Cauchy's theory in a disk

Our primary goal in this section is to prove that every holomorphic function in a disk has a primitive. Somewhat surprisingly, all the local properties of holomorphic functions are consequences of this central fact of Cauchy's theory. The special case of a disk will be enough for our purposes here; general domains and global issues will be dealt with in chapter 2.

According to Theorem 1.29, the existence of a primitive is equivalent to having vanishing integrals along all closed curves. Convexity of the disk allows us to replace the latter with something far simpler in terms of triangles.

THEOREM 1.31. *Let $D \subset \mathbb{C}$ be an open disk and $f : D \to \mathbb{C}$ be continuous. Suppose $\int_{\partial T} f(z)\, dz = 0$ for every closed triangle $T \subset D$. Then f has a primitive in D.*

PROOF. Let p be the center of D and define

$$F(z) = \int_{[p,z]} f(\zeta)\, d\zeta \qquad \text{for } z \in D.$$

We show that F is a primitive of f. Take distinct points $z_0, z \in D$ and apply the condition $\int_{\partial T} f(\zeta)\, d\zeta = 0$ to the closed triangle T with vertices p, z, z_0 to obtain

$$F(z) - F(z_0) = \int_{[p,z]} f(\zeta)\, d\zeta - \int_{[p,z_0]} f(\zeta)\, d\zeta = \int_{[z_0,z]} f(\zeta)\, d\zeta.$$

The rest of the argument, that is, dividing by $z - z_0$ and letting $z \to z_0$ to show that $F'(z_0) = f(z_0)$, is identical to the proof of Theorem 1.29. $\qquad \square$

The problem of constructing primitives in D is thus reduced to showing that every $f \in \mathscr{O}(D)$ satisfies the triangle condition of Theorem 1.31. If we knew that the derivative f' is continuous (which is true but we have not yet proved it), this would be an easy consequence of Green's theorem. To see this, suppose $f \in \mathscr{O}(D)$ and *assume* f' is continuous in D. Then the partial derivatives of $u = \operatorname{Re}(f)$ and $v = \operatorname{Im}(f)$ are continuous in D and Green's theorem together with the Cauchy-Riemann equations $u_x = v_y, u_y = -v_x$ shows that for every closed triangle $T \subset D$,

$$\int_{\partial T} (u\, dx - v\, dy) = \iint_T (-v_x - u_y)\, dx\, dy = 0$$

and

$$\int_{\partial T} (v\, dx + u\, dy) = \iint_T (u_x - v_y)\, dx\, dy = 0.$$

Hence, by (1.15), $\int_{\partial T} f(z)\, dz = 0$.

It was Goursat's key observation that the triangle condition for a holomorphic function can be proved directly without any reference to Green's theorem and continuity of the derivative.

Goursat's formulation of Theorem 1.32 was in fact more complicated. It was A. Pringsheim who in 1901 realized it suffices to consider triangles.

THEOREM 1.32 (Goursat, 1900). *If $f \in \mathscr{O}(U)$, then $\int_{\partial T} f(z)\, dz = 0$ for every closed triangle $T \subset U$.*

PROOF. Fix a closed triangle $T \subset U$ and set $I = \int_{\partial T} f(z)\, dz$. Connect the midpoints of the edges of T to form four congruent triangles, each having half the diameter of T. It is easy to see that I is the sum of the integrals of f along the oriented boundaries of these four triangles (see Fig. 1.2). Hence, one of these triangles, which we call T_1, satisfies

$$\left| \int_{\partial T_1} f(z)\, dz \right| \geq \frac{1}{4} |I|.$$

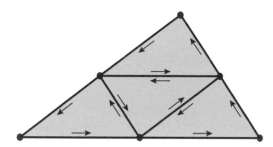

Figure 1.2. The integral along the oriented boundary of the large triangle is equal to the sum of the integrals along the oriented boundaries of the four smaller ones because each internal edge is traversed twice in opposite directions, so its net contribution to the integral is zero.

Replacing T by T_1 in the above construction and continuing inductively, we obtain a nested sequence $T \supset T_1 \supset T_2 \supset T_3 \supset \cdots$ of closed triangles with the properties

$$\operatorname{diam}(T_n) = 2^{-n}\operatorname{diam}(T) \qquad \text{and} \qquad \left| \int_{\partial T_n} f(z)\, dz \right| \geq 4^{-n}|I|.$$

Here "diam" denotes the Euclidean diameter.

The nested intersection $\bigcap_{n=1}^{\infty} T_n$ is a single point $p \in U$. By the assumption, $f'(p)$ exists, so given any $\varepsilon > 0$ there exists a $\delta > 0$ such that

$$|f(z) - f(p) - f'(p)(z-p)| \leq \varepsilon |z-p| \quad \text{whenever} \quad |z-p| < \delta.$$

Choose n large enough that $\operatorname{diam}(T_n) < \delta$. If $z \in \partial T_n$, then $|z-p| \leq \operatorname{diam}(T_n)$, so

$$|f(z) - f(p) - f'(p)(z-p)| \leq \varepsilon \operatorname{diam}(T_n).$$

Observe that by Theorem 1.29,

$$\int_{\partial T_n} (f(p) + f'(p)(z-p))\, dz = 0$$

since the integrand has a primitive $f(p)z + (1/2)f'(p)(z-p)^2$. Hence, by the *ML*-inequality (1.18),

$$4^{-n}|I| \leq \left| \int_{\partial T_n} f(z)\, dz \right| = \left| \int_{\partial T_n} (f(z) - f(p) - f'(p)(z-p))\, dz \right|$$

$$\leq \varepsilon \operatorname{diam}(T_n) \operatorname{length}(\partial T_n)$$

$$= \varepsilon \, 2^{-n}\operatorname{diam}(T) \cdot 2^{-n}\operatorname{length}(\partial T),$$

which implies

$$|I| \leq \varepsilon \operatorname{diam}(T) \operatorname{length}(\partial T).$$

Since this is true for every $\varepsilon > 0$, we must have $I = 0$.   □

Edouard Jean-Baptiste Goursat (1858–1936)

Theorems 1.31 and 1.32 put together now imply the following

THEOREM 1.33. *Let $D \subset \mathbb{C}$ be an open disk and $f \in \mathcal{O}(D)$. Then f has a primitive in D.*

Combining Theorem 1.29 and Theorem 1.33, we arrive at

THEOREM 1.34 (Cauchy's theorem in a disk, 1825). *Let $D \subset \mathbb{C}$ be an open disk and $f \in \mathcal{O}(D)$. Then for every closed curve γ in D,*

$$\int_\gamma f(z)\, dz = 0.$$

REMARK 1.35. Here is a minor technical point that will be exploited in the next result: Cauchy's Theorem 1.34 remains true under the apparently weaker assumption that f is continuous in D and holomorphic in $D \smallsetminus \{p\}$ for some $p \in D$. To see this, it suffices to show that $\int_{\partial T} f(z)\, dz = 0$ for every closed triangle $T \subset D$. If $T \subset D \smallsetminus \{p\}$, this follows from Theorem 1.32, so assume $p \in T$. First consider the case where p is on the boundary of T. By slightly moving a vertex of T, we can find a triangle T', arbitrarily close to T, for which $p \notin T'$. Since $\int_{\partial T'} f(z)\, dz = 0$ and since by Lemma 1.27 the integral along the boundary of a triangle depends continuously on vertices, we conclude that $\int_{\partial T} f(z)\, dz = 0$. If p belongs to the interior of $T = \triangle abc$, write $\int_{\partial T} f(z)\, dz$ as the sum of the integrals along the boundaries of $\triangle abp$, $\triangle bcp$, and $\triangle cap$, and reduce to the previous case.

Later we will see that such a point p is not really exceptional, so under the above assumptions $f \in \mathcal{O}(D)$ (compare Example 1.40 or Theorem 3.5).

THEOREM 1.36 (Cauchy's integral formula in a disk). *Let $D \subset \mathbb{C}$ be an open disk and $f \in \mathcal{O}(D)$. If $\overline{\mathbb{D}}(p, r) \subset D$, then*

$$f(z) = \frac{1}{2\pi i} \int_{\mathbb{T}(p,r)} \frac{f(\zeta)}{\zeta - z}\, d\zeta \qquad \textit{for } z \in \mathbb{D}(p, r).$$

In particular, the values of f on the circle $\mathbb{T}(p, r)$ uniquely determine the values of f inside the disk $\mathbb{D}(p, r)$.

PROOF. Fix $z \in \mathbb{D}(p, r)$ and define $g : D \to \mathbb{C}$ by

$$g(\zeta) = \begin{cases} \dfrac{f(\zeta) - f(z)}{\zeta - z} & \zeta \neq z \\ f'(z) & \zeta = z. \end{cases}$$

Evidently g is continuous in D and holomorphic in $D \smallsetminus \{z\}$. Hence by Remark 1.35, $\int_{\mathbb{T}(p,r)} g(\zeta)\, d\zeta = 0$. This gives

$$\frac{1}{2\pi i} \int_{\mathbb{T}(p,r)} \frac{f(\zeta)}{\zeta - z}\, d\zeta = f(z) \cdot \frac{1}{2\pi i} \int_{\mathbb{T}(p,r)} \frac{1}{\zeta - z}\, d\zeta.$$

To finish the proof, we need to show that the integral on the right is $2\pi i$. Take the parametrization of $\mathbb{T}(p, r)$ defined by $\gamma(t) = z + \rho(t) e^{it}$ for $t \in [0, 2\pi]$, where $\rho(t)$ is

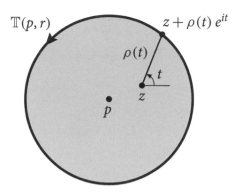

Figure 1.3. Parametrizing the oriented circle $\mathbb{T}(p,r)$ as seen from an off-center point z, used in the proof of Theorem 1.36.

the unique positive number which satisfies $|z + \rho(t)e^{it} - p| = r$ (see Fig. 1.3). It is easy to check that $t \mapsto \rho(t)$ is continuously differentiable. Hence

$$\int_{\mathbb{T}(p,r)} \frac{1}{\zeta - z} d\zeta = \int_0^{2\pi} \frac{\gamma'(t)}{\gamma(t) - z} dt = \int_0^{2\pi} \frac{(\rho'(t) + i\rho(t))\, e^{it}}{\rho(t)\, e^{it}} dt$$

$$= \int_0^{2\pi} \frac{\rho'(t)}{\rho(t)} dt + 2\pi i$$

$$= \log(\rho(2\pi)) - \log(\rho(0)) + 2\pi i = 2\pi i,$$

where the last equality holds since $\rho(2\pi) = \rho(0)$. □

More general versions of Theorems 1.34 and 1.36 will be proved in chapter 2. For now, let us collect some corollaries of these basic results. The first one is the converse of Theorem 1.20:

THEOREM 1.37 (Holomorphic implies complex analytic). *Every $f \in \mathscr{O}(U)$ is complex analytic in U: In every disk $\mathbb{D}(p,r) \subset U$ there is a power series representation*

$$f(z) = \sum_{n=0}^{\infty} a_n\, (z - p)^n$$

where the coefficients $\{a_n\}$ are given by

(1.20)
$$a_n = \frac{f^{(n)}(p)}{n!} = \frac{1}{2\pi i} \int_{\mathbb{T}(p,s)} \frac{f(\zeta)}{(\zeta - p)^{n+1}} d\zeta$$

for any $0 < s < r$.

PROOF. Fix $0 < s < r$ and a point $z \in \mathbb{D}(p, s)$. For any $\zeta \in \mathbb{T}(p, s)$,

$$\frac{1}{\zeta - z} = \frac{1}{(\zeta - p)\left[1 - \left(\frac{z - p}{\zeta - p}\right)\right]} = \frac{1}{\zeta - p} \sum_{n=0}^{\infty} \left(\frac{z - p}{\zeta - p}\right)^n.$$

Here the geometric series converges uniformly in ζ since its general term has absolute value $|z - p|/s < 1$ independent of ζ. Thus, we can integrate this series term-by-term on the circle $\mathbb{T}(p, s)$. By Theorem 1.36, we obtain

$$f(z) = \frac{1}{2\pi i} \int_{\mathbb{T}(p,s)} \sum_{n=0}^{\infty} \frac{f(\zeta)(z - p)^n}{(\zeta - p)^{n+1}} \, d\zeta = \sum_{n=0}^{\infty} a_n \, (z - p)^n,$$

where the a_n are given by (1.20). This proves that f can be represented by the power series $\sum_{n=0}^{\infty} a_n \, (z - p)^n$ in $\mathbb{D}(p, s)$. Since this holds for every $s < r$, Theorem 1.20(iii) shows that the power series with the same coefficients must converge to $f(z)$ for all $z \in \mathbb{D}(p, r)$. □

It follows from Theorem 1.20 that

COROLLARY 1.38. *If $f \in \mathscr{O}(U)$, then $f' \in \mathscr{O}(U)$. Therefore, the k-th derivative $f^{(k)}$ exists and belongs to $\mathscr{O}(U)$ for every $k \geq 1$.*

In particular, by Theorem 1.7, a differentiable map $f : U \to \mathbb{R}^2$ which satisfies the Cauchy-Riemann equation $f_{\bar{z}} = 0$ throughout U is automatically C^∞-smooth.

The following converse of Theorem 1.32 is a useful criterion for deciding when a continuous function is holomorphic:

THEOREM 1.39 (Morera, 1886). *Suppose $f : U \to \mathbb{C}$ is continuous and $\int_{\partial T} f(z) \, dz = 0$ for every closed triangle $T \subset U$. Then $f \in \mathscr{O}(U)$.*

PROOF. Let $D \subset U$ be a disk. By Theorem 1.31, f has a primitive F in D. Since $F \in \mathscr{O}(D)$ and since the derivative of a holomorphic function is holomorphic by Corollary 1.38, it follows that $f = F' \in \mathscr{O}(D)$. As this holds for every disk $D \subset U$, we conclude that $f \in \mathscr{O}(U)$. □

Giacinto Morera
(1856–1909)

EXAMPLE 1.40 (Lines are removable). Let $U \subset \mathbb{C}$ be open and L be a straight line which intersects U. Suppose $f : U \to \mathbb{C}$ is a continuous function which is holomorphic in $U \smallsetminus L$. We prove that f is holomorphic in U by showing that $\int_{\partial T} f(z) \, dz = 0$ for every triangle $T \subset U$. First assume that the interior of T is disjoint from L. Then, by moving the vertices of T slightly, we can find a triangle $T' \subset U \smallsetminus L$, arbitrarily close to T. By Goursat's Theorem 1.32, $\int_{\partial T'} f(z) \, dz = 0$. Since the integral along the boundary of a triangle depends continuously on vertices by Lemma 1.27, we must have $\int_{\partial T} f(z) \, dz = 0$. If the interior of T meets L, write T as the union of at most three triangles with pairwise disjoint interiors, each meeting L along a vertex or an edge, and reduce to the previous case.

This shows in particular that points are removable: If f is continuous in U and holomorphic in $U \smallsetminus \{p\}$, then $f \in \mathcal{O}(U)$. More general removability results are discussed in Theorem 3.5 and in chapter 10.

REMARK 1.41. Morera's theorem holds if we replace triangles with other special families of closed sets with nice boundaries. A typical example, which turns out to be more convenient in some situations, is the family of closed rectangles, or even squares. See problem 25.

THEOREM 1.42 (Cauchy's estimates, 1835). *Suppose f is continuous on $\overline{\mathbb{D}}(p, r)$ and holomorphic in $\mathbb{D}(p, r)$. Then,*

$$(1.21) \qquad |f^{(n)}(p)| \le \frac{n!}{r^n} \sup_{|z-p|=r} |f(z)| \qquad (n \ge 0).$$

The example $f(z) = z^n$ in the unit disk $\mathbb{D}$ shows that the bound in (1.21) is optimal for each n.

PROOF. Take $0 < s < r$ and represent f by a power series $\sum_{n=0}^{\infty} a_n (z-p)^n$ in $\mathbb{D}(p, s)$. By (1.20),

$$|f^{(n)}(p)| = n!\, |a_n| = \frac{n!}{2\pi} \left| \int_{\mathbb{T}(p,s)} \frac{f(z)}{(z-p)^{n+1}}\, dz \right|,$$

which by the *ML*-inequality implies

$$|f^{(n)}(p)| \le \frac{n!}{2\pi} \cdot \sup_{|z-p|=s} \frac{|f(z)|}{|z-p|^{n+1}} \cdot 2\pi s = \frac{n!}{s^n} \sup_{|z-p|=s} |f(z)|.$$

Letting $s \to r$, we obtain (1.21). $\qquad \square$

Cauchy's estimates lead to various quantitative results on holomorphic functions which have no counterpart in the smooth category. Here we prove two basic but important statements of this type.

THEOREM 1.43. *If a holomorphic function f maps the disk $\mathbb{D}(p, r)$ into the disk $\mathbb{D}(q, R)$, then $|f'(p)| \le R/r$.*

Note that we have *not* assumed $q = f(p)$. In particular, if $f : \mathbb{D} \to \mathbb{D}$ is holomorphic, then $|f'(0)| \le 1$. This is a basic version of the so-called "Schwarz lemma" which has deep applications and will be discussed at length in chapters 4, 11, and 13.

PROOF. Take $0 < s < r$ and apply (1.21) to the function $g = f - q$:

$$|f'(p)| = |g'(p)| \le \frac{1}{s} \sup_{|z-p|=s} |g(z)| \le \frac{R}{s}.$$

Letting $s \to r$ proves the result. $\qquad \square$

Liouville's theorem was
known to Cauchy in 1844.

THEOREM 1.44 (Liouville, 1847). *Every bounded entire function is constant.*

PROOF. Let $f \in \mathcal{O}(\mathbb{C})$ and $|f(z)| < M$ for all $z \in \mathbb{C}$. Then f maps any disk $\mathbb{D}(p, r)$ into $\mathbb{D}(0, M)$, so by Theorem 1.43, $|f'(p)| \leq M/r$. Letting $r \to +\infty$, we obtain $f'(p) = 0$. Since this holds for every $p \in \mathbb{C}$, f must be constant. □

Joseph Liouville
(1809–1882)

EXAMPLE 1.45 (The fundamental theorem of algebra). Let $P : \mathbb{C} \to \mathbb{C}$ be a polynomial of degree $d \geq 1$, so $\lim_{z \to \infty} P(z) = \infty$. If $P(z) \neq 0$ for all z, then $f(z) = 1/P(z)$ is entire and $\lim_{z \to \infty} f(z) = 0$. Hence there is an $R > 0$ such that $|f(z)| \leq 1$ whenever $|z| \geq R$. Since by continuity f is bounded on the closed disk $\overline{\mathbb{D}}(0, R)$, it follows that f is bounded on the plane. Liouville's theorem then implies that f is constant, which is a contradiction. Thus, P has at least one root z_1 and we can write $P(z) = (z - z_1)P_1(z)$ for some polynomial P_1 of degree $d - 1$. If $d - 1 = 0$ so P_1 is constant, stop. Otherwise repeat the argument with P_1 in place of P to find a root z_2 of P_1, and so on. This process stops after d steps and shows that P factors as $P(z) = a(z - z_1)(z - z_2) \cdots (z - z_d)$ for some $a, z_1, \ldots, z_d \in \mathbb{C}$. Thus, *every complex polynomial of degree $d \geq 1$ has precisely d roots counting multiplicities.*

We end this section with a useful theorem which, roughly speaking, says that the integral of a function which depends holomorphically on a parameter is a holomorphic function of that parameter, and differentiation under the integral sign is legitimate. We formulate a simple version of the theorem which will be sufficient for our purposes. One should note, however, that the result holds in much more general settings (see problem 27).

THEOREM 1.46. *Let $U \subset \mathbb{C}$ be open and $\varphi : U \times [a, b] \to \mathbb{C}$ be a continuous function such that for each $t \in [a, b]$, $z \mapsto \varphi(z, t)$ is holomorphic in U with derivative $\varphi'(z, t)$. Then, the function $f : U \to \mathbb{C}$ defined by*

$$f(z) = \int_a^b \varphi(z, t) \, dt$$

is holomorphic and we can differentiate under the integral sign:

$$f'(z) = \int_a^b \varphi'(z, t) \, dt \qquad \text{for all } z \in U.$$

PROOF. Fix $p \in U$ and take $r > 0$ such that $\overline{\mathbb{D}}(p, r) \subset U$. Let $0 < |z - p| < r/2$. By Theorem 1.36,

$$\varphi(z, t) - \varphi(p, t) = \frac{1}{2\pi i} \int_{\mathbb{T}(p, r)} \varphi(\zeta, t) \left(\frac{1}{\zeta - z} - \frac{1}{\zeta - p} \right) d\zeta,$$

so

$$\frac{\varphi(z, t) - \varphi(p, t)}{z - p} = \frac{1}{2\pi i} \int_{\mathbb{T}(p, r)} \frac{\varphi(\zeta, t)}{(\zeta - z)(\zeta - p)} \, d\zeta.$$

Since $|\zeta - z| > r/2$ whenever $|\zeta - p| = r$, we obtain the following estimate using the *ML*-inequality:

$$\left| \frac{\varphi(z,t) - \varphi(p,t)}{z - p} \right| \leq \frac{1}{2\pi} \cdot M \cdot \frac{2}{r^2} \cdot 2\pi r = \frac{2M}{r}.$$

Here M is the supremum of $|\varphi|$ on the compact set $\overline{\mathbb{D}}(p,r) \times [a,b]$. If $\{z_n\}$ is any sequence in $U \setminus \{p\}$ which tends to p, then

$$g_n(t) = \frac{\varphi(z_n, t) - \varphi(p, t)}{z_n - p}$$

is a sequence of continuous functions on $[a, b]$ which converges pointwise to $\varphi'(p, t)$ and is bounded by $2M/r$ for all large n. Hence, by Lebesgue's dominated convergence theorem, the function $t \mapsto \varphi'(p, t)$ is integrable on $[a, b]$ and

$$\lim_{n \to \infty} \frac{f(z_n) - f(p)}{z_n - p} = \lim_{n \to \infty} \int_a^b g_n(t)\, dt = \int_a^b \varphi'(p, t)\, dt.$$

Since this holds for every sequence $z_n \to p$, we conclude that $f'(p)$ exists and equals $\int_a^b \varphi'(p, t)\, dt$. $\square$

REMARK 1.47. Under the assumptions of the above theorem, the derivative $(z, t) \mapsto \varphi'(z, t)$ is in fact continuous on $U \times [a, b]$ (see problem 26). Thus, the result holds when $\varphi(z, t)$ is replaced with $\varphi'(z, t)$, and a simple induction proves the formula

$$f^{(n)}(z) = \int_a^b \varphi^{(n)}(z, t)\, dt \qquad \text{for all } z \in U,$$

where $\varphi^{(n)}(z, t)$ is the n-th derivative of $\varphi(z, t)$ with respect to z.

The following corollary of the above theorem will be used repeatedly:

COROLLARY 1.48. *Let $\gamma : [0, 1] \to \mathbb{C}$ be a piecewise C^1 curve and $g : |\gamma| \to \mathbb{C}$ be a continuous function. Then, for each integer $n \geq 1$, the function*

$$f(z) = \int_\gamma \frac{g(\zeta)}{(\zeta - z)^n}\, d\zeta$$

is holomorphic in $\mathbb{C} \setminus |\gamma|$, and

$$f'(z) = n \int_\gamma \frac{g(\zeta)}{(\zeta - z)^{n+1}}\, d\zeta \qquad \text{for all } z \in \mathbb{C} \setminus |\gamma|.$$

PROOF. This follows from Theorem 1.46 applied to $\varphi : (\mathbb{C} \setminus |\gamma|) \times [0, 1] \to \mathbb{C}$ defined by

$$\varphi(z, t) = \frac{g(\gamma(t))\, \gamma'(t)}{(\gamma(t) - z)^n}.$$

(Technically, we need to break up $[0, 1]$ into finitely many intervals in which γ' is continuous and add up the corresponding integrals, but that is a trivial matter.) ☐

EXAMPLE 1.49 (Cauchy's integral formula for higher derivatives). A special case of the above corollary is Cauchy's integral formula. If $f \in \mathcal{O}(U)$ and $\overline{\mathbb{D}}(p, r) \subset U$, then

$$\frac{1}{2\pi i} \int_{\mathbb{T}(p,r)} \frac{f(\zeta)}{\zeta - z} \, d\zeta$$

defines a holomorphic function in $\mathbb{C} \smallsetminus \mathbb{T}(p, r)$. By Theorem 1.36, this function coincides with f inside the disk $\mathbb{D}(p, r)$. Differentiation under the integral sign then gives

$$f'(z) = \frac{1}{2\pi i} \int_{\mathbb{T}(p,r)} \frac{f(\zeta)}{(\zeta - z)^2} \, d\zeta \qquad \text{for } z \in \mathbb{D}(p, r).$$

It follows by induction that for every $n \geq 0$,

$$f^{(n)}(z) = \frac{n!}{2\pi i} \int_{\mathbb{T}(p,r)} \frac{f(\zeta)}{(\zeta - z)^{n+1}} \, d\zeta \qquad \text{for } z \in \mathbb{D}(p, r).$$

Observe that for $z = p$ this is the formula (1.20) that we derived earlier.

1.5 Mapping properties of holomorphic functions

DEFINITION 1.50. Suppose $f \in \mathcal{O}(U)$ and f is not identically zero in the disk $\mathbb{D}(p, r) \subset U$. Let $f(z) = \sum_{n=0}^{\infty} a_n (z - p)^n$ be the power series representation of f in $\mathbb{D}(p, r)$. The smallest integer m with the property $a_m \neq 0$ is called the **order** of p and is denoted by $\mathrm{ord}(f, p)$. Thus, $\mathrm{ord}(f, p) \geq 1$ if and only if $f(p) = 0$. We call p a **simple zero** of f if $\mathrm{ord}(f, p) = 1$.

Alternatively, $\mathrm{ord}(f, p)$ can be described as the unique integer $m \geq 0$ for which f can be factored as

$$f(z) = (z - p)^m f_1(z)$$

with $f_1 \in \mathcal{O}(U)$ and $f_1(p) \neq 0$. The function f_1 is given by $(z - p)^{-m} f(z)$ in $U \smallsetminus \{p\}$. It is holomorphic in U since it is represented by the power series $\sum_{n=m}^{\infty} a_n (z - p)^{n-m}$ in $\mathbb{D}(p, r)$.

EXAMPLE 1.51 (Holomorphic L'Hôpital's rule). Suppose f and g are holomorphic in some neighborhood of p, with $\mathrm{ord}(f, p) = \mathrm{ord}(g, p) = m \geq 1$. Write $f(z) = (z - p)^m f_1(z)$ and $g(z) = (z - p)^m g_1(z)$, where f_1 and g_1 are non-zero and holomorphic near p. Since

$$f_1(p) = \frac{f^{(m)}(p)}{m!} \quad \text{and} \quad g_1(p) = \frac{g^{(m)}(p)}{m!},$$

it follows that

$$\lim_{z \to p} \frac{f(z)}{g(z)} = \frac{f_1(p)}{g_1(p)} = \frac{f^{(m)}(p)}{g^{(m)}(p)}.$$

Let us call $U \subset \mathbb{C}$ a **domain** if U is non-empty, open, and connected.

LEMMA 1.52. *Suppose $U \subset \mathbb{C}$ is a domain and $f \in \mathcal{O}(U)$. If the zero-set $f^{-1}(0) = \{z \in U : f(z) = 0\}$ has an accumulation point in U, then $f = 0$ everywhere in U.*

Connectivity of U is essential here: If U is the disjoint union of non-empty open sets U_1 and U_2, and if $f = 0$ in U_1 and $f = 1$ in U_2, then $f \in \mathcal{O}(U)$ and $f^{-1}(0) = U_1$ has accumulation points in U, but f is not identically zero in U.

PROOF. Let E be the non-empty set of accumulation points of $f^{-1}(0)$ in U. Then E is closed in U, and $E \subset f^{-1}(0)$ by continuity of f. Suppose $p \in E$ and there is a disk $\mathbb{D}(p, r) \subset U$ in which f is not identically zero. Then we can write $f(z) = (z - p)^m f_1(z)$, where $m = \mathrm{ord}(f, p) \geq 1$, $f_1 \in \mathcal{O}(U)$, and $f_1(p) \neq 0$. By continuity, f_1 does not vanish in some neighborhood of p. It follows that p is the only zero of f in this neighborhood, contradicting the fact that $p \in E$. Thus, if $\mathbb{D}(p, r) \subset U$, then f is identically zero in $\mathbb{D}(p, r)$ and therefore $\mathbb{D}(p, r) \subset E$. This shows that E is an open set. Since U is connected, we must have $E = f^{-1}(0) = U$. $\square$

Since every domain $U \subset \mathbb{C}$ is a countable union of open disks, it is clear that every uncountable subset of U must have an accumulation point in U. It follows from the above lemma that *a non-constant holomorphic function in a domain has at most countably many zeros, all of which are isolated.* Another immediate corollary is

THEOREM 1.53 (The identity theorem). *Suppose $U \subset \mathbb{C}$ is a domain, $f, g \in \mathcal{O}(U)$, and the set $\{z \in U : f(z) = g(z)\}$ has an accumulation point in U. Then $f = g$ everywhere in U.*

EXAMPLE 1.54. The complex cosine and sine are the entire functions defined by

$$\cos z = \frac{1}{2}(e^{iz} + e^{-iz}) = \sum_{n=0}^{\infty} \frac{(-1)^n}{(2n)!} z^{2n}$$

$$\sin z = \frac{1}{2i}(e^{iz} - e^{-iz}) = \sum_{n=0}^{\infty} \frac{(-1)^n}{(2n+1)!} z^{2n+1}.$$

They extend the usual cosine and sine functions defined on the real line. It follows from Theorem 1.53 that any trigonometric identity between cosine and sine that holds on $\mathbb{R}$ must continue to hold in $\mathbb{C}$. For example, the identities $\cos^2 z + \sin^2 z = 1$, $\sin(2z) = 2\sin z \cos z$, and $\cos(2z) = \cos^2 z - \sin^2 z$ remain valid for all $z \in \mathbb{C}$.

EXAMPLE 1.55. Suppose $f \in \mathcal{O}(\mathbb{C})$ has the power series representation $f(z) = \sum_{n=0}^{\infty} a_n z^n$. If every coefficient a_n is real, then clearly $f(\mathbb{R}) \subset \mathbb{R}$. Conversely, suppose $f(\mathbb{R}) \subset \mathbb{R}$ and consider the entire function

$$g(z) = \overline{f(\bar{z})} = \sum_{n=0}^{\infty} \overline{a_n}\, z^n.$$

Since $f(z)$ is real when z is real, we have $g = f$ on the real line. By Theorem 1.53, $g = f$ everywhere in $\mathbb{C}$. Uniqueness of power series then shows that every a_n is real.

Our next goal is to prove the fundamental fact that the image of a domain under a non-constant holomorphic function is open (Theorem 1.62). This will follow from a much stronger result on the local behavior of holomorphic functions (Theorem 1.59).

LEMMA 1.56. *If $f \in \mathcal{O}(U)$, the function $g : U \times U \to \mathbb{C}$ defined by*

$$g(\zeta, z) = \begin{cases} \dfrac{f(\zeta) - f(z)}{\zeta - z} & \zeta \neq z \\[2mm] f'(z) & \zeta = z \end{cases}$$

is continuous.

PROOF. Clearly g is continuous off the diagonal $\{(z,z) : z \in U\}$, so it is enough to check continuity of g at a diagonal point (p, p). Let $\varepsilon > 0$ be given. Since f' is continuous at p, there is an $r > 0$ such that

$$(1.22) \qquad |f'(z) - f'(p)| < \varepsilon \qquad \text{whenever } z \in \mathbb{D}(p, r).$$

Let $\zeta, z \in \mathbb{D}(p, r)$. If $\zeta = z$, then $|g(\zeta, z) - g(p, p)| = |f'(z) - f'(p)| < \varepsilon$. If $\zeta \neq z$, then

$$\frac{f(\zeta) - f(z)}{\zeta - z} = \frac{1}{\zeta - z} \int_{[z, \zeta]} f'(w)\, dw = \int_0^1 f'(\gamma(t))\, dt,$$

where $\gamma(t) = (1 - t)z + t\zeta$. Hence,

$$|g(\zeta, z) - g(p, p)| = \left| \frac{f(\zeta) - f(z)}{\zeta - z} - f'(p) \right| = \left| \int_0^1 [\, f'(\gamma(t)) - f'(p)\,]\, dt \right|$$

$$\leq \int_0^1 |f'(\gamma(t)) - f'(p)|\, dt \leq \varepsilon,$$

where the last inequality holds since by (1.22), $|f'(\gamma(t)) - f'(p)| < \varepsilon$ for every $t \in [0, 1]$.   □

THEOREM 1.57 (Holomorphic inverse function theorem). *Suppose $f \in \mathcal{O}(U)$, $p \in U$, and $f'(p) \neq 0$. Then, there exist open neighborhoods $V \subset U$ of p and $W \subset \mathbb{C}$ of $f(p)$ such* *that $f : V \to W$ is a bijection. Moreover, the local inverse $f^{-1} : W \to V$ is holomorphic and*

$$(f^{-1})'(w) = \frac{1}{f'(f^{-1}(w))} \qquad \text{for every } w \in W.$$

PROOF. Let g be the function defined in Lemma 1.56. By continuity of g at (p, p) there is an $r > 0$ such that

$$|g(\zeta, z)| \geq \frac{1}{2} |g(p, p)| \qquad \text{if } \zeta, z \in \mathbb{D}(p, r),$$

or

(1.23) $$|f(\zeta) - f(z)| \geq \frac{1}{2} |f'(p)| |\zeta - z| \qquad \text{if } \zeta, z \in \mathbb{D}(p, r).$$

Since by the assumption $f'(p) \neq 0$, this shows that f is injective in $V = \mathbb{D}(p, r)$.

We claim that $W = f(V)$ is open. Take any $z_0 \in V$, take $s > 0$ such that $\overline{\mathbb{D}}(z_0, s) \subset V$, and set

$$\varepsilon = \frac{1}{2} \min_{|z - z_0| = s} |f(z) - f(z_0)|.$$

By injectivity, $\varepsilon > 0$. Suppose there is a point $q \in \mathbb{D}(f(z_0), \varepsilon)$ that does not belong to W. Then $h = 1/(f - q)$ is holomorphic in V, and $\sup_{|z - z_0| = s} |h(z)| \leq 1/\varepsilon$. By the case $n = 0$ of Cauchy's estimates (1.21), $|h(z_0)| \leq 1/\varepsilon$. This contradicts the fact that $|h(z_0)| = 1/|f(z_0) - q| > 1/\varepsilon$. Thus, $\mathbb{D}(f(z_0), \varepsilon) \subset W$. Since z_0 was arbitrary, this proves that W is open.

To see that the local inverse $f^{-1} : W \to V$ is holomorphic, let $z, z_0 \in V$, $w = f(z)$, and $w_0 = f(z_0)$. Note that $f'(z_0) \neq 0$ since $|f'(z_0)| \geq |f'(p)|/2 > 0$ by (1.23). If $z \neq z_0$, then

$$\frac{f^{-1}(w) - f^{-1}(w_0)}{w - w_0} = \frac{z - z_0}{f(z) - f(z_0)}.$$

As $w \to w_0$, (1.23) shows that $z \to z_0$ and the right side tends to $1/f'(z_0)$. $\qquad \square$

REMARK 1.58. Despite what the above argument may suggest, the fact that injectivity of f in V implies $f(V)$ is open does not depend on f being holomorphic. According to Brouwer's ***invariance of domain theorem***, the image of an open set in $\mathbb{R}^2$ under any continuous injective map is open. He proved this theorem (which holds in every $\mathbb{R}^n$) in 1912 as a corollary to his famous fixed-point theorem. In dimension 2 it can be deduced from the Jordan curve theorem (see problem 19 of chapter 2).

The preceding result suggests a useful terminology: Let V and W be open sets in $\mathbb{C}$. We say that $f : V \to W$ is a **biholomorphism** if it is bijective and holomorphic, and the inverse map $f^{-1} : W \to V$ is also holomorphic (the last condition follows automatically from the first two; compare Corollary 1.63 below). Thus, we can restate Theorem 1.57 as saying that *every holomorphic function is a local biholomorphism near the points where its derivative is non-zero.*

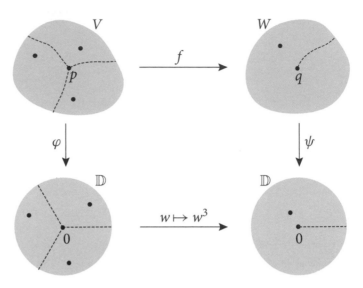

Figure 1.4. Illustration of Theorem 1.59. Here $m = \deg(f, p) = 3$.

The following is a generalization of Theorem 1.57 in which the assumption of having a non-zero derivative is dropped. It gives a complete description of the *local* behavior of holomorphic functions.

THEOREM 1.59 (Local normal form). *Suppose $f \in \mathcal{O}(U)$, $p \in U$, and f is non-constant near p. Then, there exist a positive integer m, neighborhoods $V \subset U$ of p and $W \subset \mathbb{C}$ of $q = f(p)$, and biholomorphisms $\varphi : V \to \mathbb{D}$ and $\psi : W \to \mathbb{D}$, with $\varphi(p) = \psi(q) = 0$, such that*

$$(\psi \circ f \circ \varphi^{-1})(w) = w^m$$

for all $w \in \mathbb{D}$.

The theorem says that the action of f, when viewed under suitable change of coordinates φ and ψ, is that of the power map $w \mapsto w^m$. In particular, f is locally m-to-1 near p. More precisely, for every w sufficiently close but not equal to q, there are m distinct solutions of $f(z) = w$ near p. The equation $f(z) = q$ has a unique solution $z = p$ in a neighborhood of p, but this solution should be counted with multiplicity m (see Fig. 1.4).

PROOF. Write $f(z) - q = (z - p)^m f_1(z)$, where $m = \mathrm{ord}(f - q, p)$, $f_1 \in \mathcal{O}(U)$ and $f_1(p) \neq 0$. The proof consists essentially of showing that f_1, and hence $f - q$, has a "holomorphic m-th root" in a neighborhood of p. Choose a small disk D centered at p such that $f_1 \neq 0$ in D. The function f_1'/f_1 is holomorphic in D, so it has a primitive $g \in \mathcal{O}(D)$ by Theorem 1.33. The computation

$$(f_1 e^{-g})' = (f_1' - f_1 g') e^{-g} = 0,$$

which holds in D, shows that $f_1 = c\,e^g$ for some non-zero constant c. By adding a suitable constant to g if necessary, we may therefore arrange $f_1 = e^g$ in D.

Define

$$\varphi_1(z) = (z - p) \exp\left(\frac{g(z)}{m}\right).$$

Then $\varphi_1 \in \mathcal{O}(D)$, $\varphi_1(p) = 0$, and $\varphi_1'(p) = \exp(g(p)/m) \neq 0$. Theorem 1.57 applied to φ_1 then gives a neighborhood $V \subset D$ of p and $\varepsilon > 0$ such that $\varphi_1 : V \to \mathbb{D}(0, \varepsilon)$ is a biholomorphism. Define

$$\varphi(z) = \frac{1}{\varepsilon}\,\varphi_1(z) \qquad \text{and} \qquad \psi(z) = \frac{1}{\varepsilon^m}(z - q).$$

Then $\varphi : V \to \mathbb{D}$ and $\psi : W = \mathbb{D}(q, \varepsilon^m) \to \mathbb{D}$ are biholomorphic, and

$$\psi(f(z)) = \frac{f(z) - q}{\varepsilon^m} = \frac{(\varphi_1(z))^m}{\varepsilon^m} = (\varphi(z))^m$$

for all $z \in V$. $\qquad\square$

The integer m in Theorem 1.59, which has both topological and analytic meanings, is important enough to deserve a name:

> **DEFINITION 1.60.** Let f be non-constant and holomorphic in a neighborhood of p. The order of p as a zero of the function $f - f(p)$ is called the **local degree** of f at p and is denoted by $\deg(f, p)$:
>
> $$\deg(f, p) = \mathrm{ord}(f - f(p), p).$$
>
> When $\deg(f, p) > 1$, equivalently when $f'(p) = 0$, we call p a **critical point** and $f(p)$ the corresponding **critical value** of f.

It is easily seen that

$$\deg(f, p) = \mathrm{ord}(f', p) + 1 \qquad \text{if } p \text{ is a critical point of } f.$$

EXAMPLE 1.61. The degree 4 polynomial $P(z) = 3z^4 - 4z^3 + 4$ with derivative $P'(z) = 12z^2$ $(z - 1)$ has critical points at 0 and 1, with critical values $P(0) = 4$ and $P(1) = 3$. We have $\deg(P, 0) = 3$ and $\deg(P, 1) = 2$. Since $P^{-1}(4) = \{0, 4/3\}$, Theorem 1.59 shows that for every q sufficiently close but not equal to 4 the equation $P(z) = q$ has three distinct solutions close to 0 and a unique solution close to $4/3$. Similarly, since $P^{-1}(3) = \{1, (-1 \pm \sqrt{2}i)/3\}$, for every q sufficiently close but not equal to 3 the equation $P(z) = q$ has two distinct solutions close to 1 and a pair of solutions close to $(-1 \pm \sqrt{2}i)/3$.

THEOREM 1.62 (The open mapping theorem). *Suppose $U \subset \mathbb{C}$ is a domain and $f \in \mathcal{O}(U)$ is non-constant. Then $f(U)$ is open.*

It follows that f is an **open map** in the sense that it sends every open subset of U to an open set in $\mathbb{C}$.

Proof. By Theorem 1.59, every $p \in U$ has an open neighborhood $V_p \subset U$ such that $f(V_p)$ is open. Hence

$$f(U) = f(\bigcup_{p \in U} V_p) = \bigcup_{p \in U} f(V_p)$$

is also open. $\qquad\square$

COROLLARY 1.63. *Suppose $U \subset \mathbb{C}$ is a domain and $f \in \mathscr{O}(U)$ is injective. Then $f'(z) \neq 0$ for all $z \in U$ and $f : U \to f(U)$ is a biholomorphism.*

Proof. If $f'(p) = 0$ for some $p \in U$, then $m = \deg(f, p) > 1$, which implies f is locally m-to-1 near p, contradicting injectivity of f. The image $f(U)$ is open by Theorem 1.62 and the inverse $f^{-1} : f(U) \to U$ is holomorphic by Theorem 1.57, so $f : U \to f(U)$ is a biholomorphism. $\qquad\square$

REMARK 1.64. The converse of Corollary 1.63 is false. For example, the derivative of $f(z) = e^z$ is nowhere zero yet f is infinite-to-one.

Let us also point out that Corollary 1.63 exhibits a complex analytic phenomenon which has no counterpart for smooth maps. For example, $f : \mathbb{R}^2 \to \mathbb{R}^2$ given by $f(x, y) = (x^3, y)$ is a C^∞-smooth homeomorphism, but it has critical points everywhere along the y-axis, and the inverse map $f^{-1}(x, y) = (\sqrt[3]{x}, y)$ is not differentiable there.

THEOREM 1.65 (The maximum principle for open maps). *Suppose $f : U \to \mathbb{C}$ is an open map. Then*

(i) *$|f|$ cannot have a local maximum at any point of U.*
(ii) *$|f|$ cannot have a local minimum at any point of U provided that $f(z) \neq 0$ for all $z \in U$.*

Proof. Let $p \in U$ and take any disk $\mathbb{D}(p, r) \subset U$. Since $f(\mathbb{D}(p, r))$ is open and contains $f(p)$, there is a point $z \in \mathbb{D}(p, r)$ for which $|f(z)| > |f(p)|$. It follows that $|f|$ does not have a local maximum at p. Similarly, if $f(p) \neq 0$, there is a point $z \in \mathbb{D}(p, r)$ for which $|f(z)| < |f(p)|$, so $|f|$ does not have a local minimum at p. $\qquad\square$

For holomorphic functions we often use the following version of the above result:

THEOREM 1.66 (The maximum principle for holomorphic functions). *Suppose $U \subset \mathbb{C}$ is a bounded domain, $f : \overline{U} \to \mathbb{C}$ is continuous, and $f \in \mathscr{O}(U)$. Then*

(i) *$|f(z)| \leq \sup_{\zeta \in \partial U} |f(\zeta)|$ for all $z \in U$.*
(ii) *$|f(z)| \geq \inf_{\zeta \in \partial U} |f(\zeta)|$ for all $z \in U$ provided that $f \neq 0$ in U.*

In either case, if equality holds at some $z \in U$, then f is constant.

PROOF. To see (i), assume $|f(z)| \geq \sup_{\zeta \in \partial U} |f(\zeta)|$ for some $z \in U$. Then the supremum of $|f|$ on the compact set $\overline{U}$ is attained at some point of U, so $|f|$ has a local maximum there. By Theorem 1.65(i), f cannot be an open map in U, so by Theorem 1.62 f must be constant.

For (ii), note that the inequality is trivial if f has a zero on the boundary ∂U. Otherwise, $f \neq 0$ in $\overline{U}$ and the statement follows from (i) applied to the holomorphic function $1/f$.   □

EXAMPLE 1.67. Here is an alternative route to the fundamental theorem of algebra discussed in Example 1.45. Every polynomial $P : \mathbb{C} \to \mathbb{C}$ of degree $d \geq 1$ has the property that $\lim_{z \to \infty} P(z) = \infty$. Hence

$$|P(0)| < \inf_{|z|=r} |P(z)| \qquad \text{if } r \text{ is large enough.}$$

It follows from Theorem 1.66(ii) that P has a zero in the disk $\mathbb{D}(0, r)$. The same inductive argument as in Example 1.45 now shows that P has precisely d zeros counting multiplicities.

EXAMPLE 1.68. Recall that for a set $X \subset \mathbb{C}$, the Euclidean diameter $\operatorname{diam}(X)$ is the supremum of the distance $|p - q|$ as p, q vary over X. For any domain $U \subsetneq \mathbb{C}$, the inequality $\operatorname{diam}(\partial U) \leq \operatorname{diam}(U)$ trivially holds, and it may well be strict as the example $U = \{z \in \mathbb{C} : |z| > 1\}$ shows. However, when U is a bounded domain, one has the equality $\operatorname{diam}(\partial U) = \operatorname{diam}(U)$. This simple fact can of course be proved directly (for a given pair of points in U, look at the straight line passing through them), but can also be deduced from the maximum principle: Given $p, q \in U$, let $a \in \partial U$ be a point where $|z - q|$ reaches its maximum on ∂U and let $b \in \partial U$ be a point where $|z - a|$ reaches its maximum on ∂U. Applying Theorem 1.66(i) twice, we obtain

$$|p - q| < |a - q| < |a - b| \leq \operatorname{diam}(\partial U).$$

Taking the supremum over all $p, q \in U$ now proves $\operatorname{diam}(U) \leq \operatorname{diam}(\partial U)$.

More general versions of the maximum principle (for the larger class of harmonic functions) will be discussed in chapter 7.

Problems

(1) Show that all roots of the monic polynomial $P(z) = z^n + a_{n-1}z^{n-1} + \cdots + a_1 z + a_0$ lie in the disk $\mathbb{D}(0, R)$, where

$$R = \sqrt{1 + |a_{n-1}|^2 + \cdots + |a_1|^2 + |a_0|^2}.$$

(Hint: If $P(z) = 0$, then $-z^n = \sum_{k=0}^{n-1} a_k z^k$. Use the Cauchy-Schwarz inequality to prove $|z| < R$.)

(2) (Gauss-Lucas) Recall that the **convex hull** of a set $\{p_1, \ldots, p_n\}$ in $\mathbb{C}$ consists of all points $z \in \mathbb{C}$ which can be written as $z = \sum_{k=1}^{n} t_k p_k$ for some real numbers $0 \leq t_k \leq 1$ with $\sum_{k=1}^{n} t_k = 1$. If P is a complex polynomial, show that the roots of the derivative P' belong to the convex

hull of the roots of P. (Hint: Write $P(z) = c \prod_{k=1}^{n}(z - p_k)^{m_k}$, so

$$\frac{P'(z)}{P(z)} = \sum_{k=1}^{n} \frac{m_k}{z - p_k} = \sum_{k=1}^{n} \frac{m_k(\bar{z} - \bar{p}_k)}{|z - p_k|^2}.$$

If $P'(z) = 0$ but $P(z) \neq 0$, then

$$z \sum_{k=1}^{n} \frac{m_k}{|z - p_k|^2} = \sum_{k=1}^{n} \frac{m_k p_k}{|z - p_k|^2}.)$$

(3)   This exercise develops several useful formulas in the complex-variable notation. Use subscripts for the operators $\partial/\partial z$ and $\partial/\partial \bar{z}$ acting on smooth functions $U \to \mathbb{C}$, so $f_z = \partial f/\partial z$ and $f_{\bar{z}} = \partial f/\partial \bar{z}$. Recall that for $f = u + iv$, the notation $\bar{f}$ is used for the complex conjugate function $u - iv$.

(i)   Verify the product rules

$$(fg)_z = f_z g + f g_z \qquad (fg)_{\bar{z}} = f_{\bar{z}} g + f g_{\bar{z}}.$$

(ii)   Show that $\bar{f}_{\bar{z}} = \overline{(f_z)}$ and $\bar{f}_z = \overline{(f_{\bar{z}})}$.

(iii) Verify the chain rule

$$(f \circ g)_z = (f_z \circ g)\, g_z + (f_{\bar{z}} \circ g)\, \bar{g}_z$$

$$(f \circ g)_{\bar{z}} = (f_z \circ g)\, g_{\bar{z}} + (f_{\bar{z}} \circ g)\, \bar{g}_{\bar{z}}.$$

(iv)   Show that $\Delta f = 4f_{z\bar{z}} = 4f_{\bar{z}z}$, where Δ is the Laplace operator defined by $\Delta f = f_{xx} + f_{yy}$. In particular, if $f \in \mathcal{O}(U)$, then $\Delta f = 0$.

(v)   The **Jacobian** of f, viewed as a map $U \to \mathbb{R}^2$, is defined by $J_f = \det(Df)$. Show that $J_f = |f_z|^2 - |f_{\bar{z}}|^2$. In particular, if $f \in \mathcal{O}(U)$, then $J_f = |f'|^2$.

(vi)   Show that if $f \in \mathcal{O}(U)$, then $\Delta|f|^2 = 4|f'|^2$.

(4)   Let (r, θ) denote the polar coordinates in $\mathbb{R}^2$. Show that the partial differentiation operators $f_r = \partial f/\partial r$ and $f_\theta = \partial f/\partial \theta$ satisfy

$$r f_r = \; x f_x + y f_y = z f_z + \bar{z} f_{\bar{z}}$$

$$f_\theta = -y f_x + x f_y = iz f_z - i\bar{z} f_{\bar{z}}.$$

Conclude that a smooth function f is holomorphic if and only if it satisfies the polar form of the Cauchy-Riemann equation

$$r f_r = -i f_\theta,$$

in which case

$$f' = e^{-i\theta} f_r = -\frac{ie^{-i\theta}}{r} f_\theta.$$

(5)   Let κ denote complex conjugation $z \mapsto \bar{z}$. Show that the following conditions on a function $f : U \to \mathbb{C}$ are equivalent:

(i) $\kappa \circ f \in \mathcal{O}(U)$.

(ii) $f \circ \kappa \in \mathcal{O}(\kappa(U))$.

(iii) In every disk $\mathbb{D}(p, r) \subset U$, $f(z)$ has a power series representation of the form $\sum_{n=0}^{\infty} a_n (\bar{z} - \bar{p})^n$.

(iv) f is differentiable as a map $U \to \mathbb{R}^2$ and $\partial f/\partial z = 0$ throughout U.

A function which satisfies any (hence all) of these conditions is called **anti-holomorphic**.

(6)   Let $\gamma : [0, 1] \to U$ be a piecewise C^1 curve and $f \in \mathcal{O}(U)$.

(i)   Show that $f \circ \gamma : [0, 1] \to \mathbb{C}$ is piecewise C^1 and

$$(f \circ \gamma)'(t) = f'(\gamma(t))\, \gamma'(t)$$

for all but finitely many $t \in [0, 1]$.

(ii) Verify that the change of variable formulas

$$\int_{f \circ \gamma} g(w)\, dw = \int_{\gamma} (g \circ f)(z) f'(z)\, dz$$

$$\int_{f \circ \gamma} g(w)\, |dw| = \int_{\gamma} (g \circ f)(z)\, |f'(z)|\, |dz|$$

hold for every continuous function $g : |f \circ \gamma| \to \mathbb{C}$.

(7)   Let $f = u + iv \in \mathcal{O}(U)$. Show that

$$|f'| = \|\nabla u\| = \|\nabla v\| \qquad \text{in } U,$$

where ∇ is the gradient and $\| \cdot \|$ is the Euclidean norm in $\mathbb{R}^2$. At every point where $f' \neq 0$, find a simple geometric interpretation for the Cauchy-Riemann equations in terms of the vectors ∇u and ∇v at that point.

(8)   Verify that the conditions in Theorem 1.7 are also equivalent to the following: The real derivative $Df(p)$ exists and commutes with the 90° rotation $J : \mathbb{R}^2 \to \mathbb{R}^2$ defined by $J(x, y) = (-y, x)$:
$$Df(p) \circ J = J \circ Df(p).$$

(9)   Let $\langle \cdot, \cdot \rangle$ be an inner product on a 2-dimensional vector space V over $\mathbb{R}$. Let $\|u\| = \sqrt{\langle u, u \rangle}$ be the induced norm and
$$\measuredangle(u, v) = \arccos \left(\frac{\langle u, v \rangle}{\|u\|\, \|v\|} \right)$$

be the angle between two non-zero vectors u, v. A linear isomorphism $T : V \to V$ is called **conformal** with respect to $\langle \cdot, \cdot \rangle$ if $\langle T(u), T(v) \rangle = |\det T|\, \langle u, v \rangle$ for all $u, v \in V$. Show that the following conditions are equivalent:

(i) T is conformal.

(ii) T preserves angles in the sense that $\measuredangle(T(u), T(v)) = \measuredangle(u, v)$ for all non-zero vectors $u, v \in V$.

(iii) The matrix of T in any orthonormal basis has the form $\begin{bmatrix} \alpha & -\beta \\ \beta & \alpha \end{bmatrix}$ or $\begin{bmatrix} \alpha & \beta \\ \beta & -\alpha \end{bmatrix}$ depending on whether $\det T > 0$ or $\det T < 0$.

(iv) If $\{e_1, e_2\}$ is any orthonormal basis and $J : V \to V$ is the linear isomorphism defined by $J(e_1) = e_2$ and $J(e_2) = -e_1$, then $T \circ J = J \circ T$ or $T \circ J = -J \circ T$ depending on whether $\det T > 0$ or $\det T < 0$.

(v) T maps every "circle" $\{u \in V : \|u\| = R\}$ onto the "circle" $\{u \in V : \|u\| = R\, |\det T|\}$.

(10) (i)   For any sequence $\{a_n\}$ of non-zero complex numbers, show that

$$\liminf_{n \to \infty} \left| \frac{a_{n+1}}{a_n} \right| \leq \liminf_{n \to \infty} \sqrt[n]{|a_n|} \leq \limsup_{n \to \infty} \sqrt[n]{|a_n|} \leq \limsup_{n \to \infty} \left| \frac{a_{n+1}}{a_n} \right|.$$

(ii) Conclude that if $\{a_n\}$ is a sequence of non-zero complex numbers for which $L = \lim_{n \to \infty} |a_{n+1}/a_n|$ exists, then the radius of convergence of the power series $\sum_{n=0}^{\infty} a_n z^n$ is $1/L$.

(iii)   Give an example of a sequence $\{a_n\}$ of positive numbers such that

$$\liminf_{n\to\infty} \frac{a_{n+1}}{a_n} = 0 \qquad\qquad \liminf_{n\to\infty} \sqrt[n]{a_n} = 1$$

$$\limsup_{n\to\infty} \sqrt[n]{a_n} = 2 \qquad\qquad \limsup_{n\to\infty} \frac{a_{n+1}}{a_n} = +\infty.$$

This shows that in general, the radius of convergence of a power series cannot be determined by looking at limsup or liminf of the sequence $|a_{n+1}/a_n|$ as $n \to \infty$.

(11)   Show that $f(z) = 1/(1 - z^6)^3$ is complex analytic in the unit disk by finding the power series representation for f about 0 explicitly. Using the coefficients of this series, verify that the radius of convergence is 1. (Hint: Look at the higher derivatives of $1/(1 - z)$.)

(12)   Prove the following theorem of Dirichlet: Suppose $\{a_n\}$ is a decreasing sequence of positive numbers with $\lim_{n\to\infty} a_n = 0$ and $\{b_n\}$ is a sequence of complex numbers for which the partial sums $s_n = \sum_{k=1}^{n} b_k$ form a bounded sequence. Then the series $\sum_{n=1}^{\infty} a_n b_n$ converges. (Hint: Use $b_n = s_n - s_{n-1}$ to write

$$\sum_{n=m}^{N} a_n b_n = \sum_{n=m}^{N-1} (a_n - a_{n+1})s_n + a_N s_N - a_m s_{m-1}.)$$

(13)   Use the result of the previous problem to show that for $0 < p \leq 1$ the power series $\sum_{n=1}^{\infty} z^n/n^p$ converges if $|z| = 1$ but $z \neq 1$.

(14)   A sequence $\{a_n\}$ of complex numbers is called **recursive** if there are complex numbers $b_0, \ldots, b_m$ such that

$$a_{n+1} = b_0 a_n + b_1 a_{n-1} + \cdots + b_m a_{n-m}$$

for all large n.
  (i)   Show that a power series $f(z) = \sum_{n=0}^{\infty} a_n z^n$ with positive radius of convergence represents a rational function if and only if the sequence $\{a_n\}$ is recursive.
  (ii)   Find the rational function corresponding to the Fibonacci sequence $\{a_n\}$ defined by $a_0 = a_1 = 1$ and $a_{n+1} = a_n + a_{n-1}$ for $n \geq 1$.
(Hint: For the "if" part in (i), set $P(z) = b_0 + b_1 z + \cdots + b_m z^m$ and consider the power series representation of $f(z)P(z)$. This should give you an idea to deal with the other parts as well.)

(15)   Suppose $U \subset \mathbb{C}$ is a domain and $f, g \in \mathcal{O}(U)$. If $|f(z)|^2 + |g(z)|^2 = 1$ for all $z \in U$, show that f and g are constant.

(16)   Verify that the function

$$f(z) = \frac{1}{2\pi} \int_0^{2\pi} \exp(2z \cos t) \, dt$$

is entire, with the power series representation

$$f(z) = \sum_{n=0}^{\infty} \left(\frac{z^n}{n!}\right)^2 \qquad \text{for } z \in \mathbb{C}.$$

(Hint: Use the power series of the exponential map. You will need the value of $\int_0^{2\pi} \cos^n t \, dt$, which you can find by repeated applications of integration by parts.)

(17)   Suppose $f = u + iv \in \mathcal{O}(\mathbb{C})$. If u or v is bounded from above or below in the whole plane, show that f is constant.

(18)   A function $f : \mathbb{C} \to \mathbb{C}$ is called **doubly periodic** if there are non-zero complex numbers α, β, with $\alpha/\beta \notin \mathbb{R}$, such that

$$f(z + \alpha) = f(z + \beta) = f(z) \qquad \text{for all } z \in \mathbb{C}.$$

Show that every doubly periodic entire function is constant.

(19)   Let $f \in \mathcal{O}(\mathbb{C})$. Suppose there is a *non*-integer $\beta > 0$ and a constant $C > 0$ such that

$$\int_0^{2\pi} |f(re^{it})| \, dt \le C r^\beta \qquad \text{for all } r > 0.$$

Prove that f is identically zero. (Hint: Use Cauchy's integral formula to show that $f^{(n)}(0) = 0$ for all $n \ge 0$.)

(20)   Suppose $f \in \mathcal{O}(\mathbb{D}(0, R))$ has the power series representation $f(z) = \sum_{n=0}^\infty a_n z^n$. Prove **Parseval's formula**

$$\frac{1}{2\pi} \int_0^{2\pi} |f(re^{it})|^2 \, dt = \sum_{n=0}^\infty |a_n|^2 r^{2n} \qquad \text{for } 0 \le r < R.$$

(21)   Use Cauchy's integral formula (Theorem 1.36 and Example 1.49) to compute the integrals

$$\int_\mathbb{T} \frac{dz}{4z^2 + 1} \qquad \int_\mathbb{T} \frac{e^z - e^{-z}}{z^4} \, dz \qquad \int_\mathbb{T} z^n e^z \, dz \ \ (n \in \mathbb{Z}).$$

(22)   Compute the integral

$$\int_0^{2\pi} \frac{dt}{|e^{it} - a|^2},$$

where $|a| \neq 1$. (Hint: Use $|e^{it} - a|^2 = (e^{it} - a)(e^{-it} - \bar{a})$ to turn it into a complex integral along the unit circle $\mathbb{T}$, which can then be computed by applying Cauchy's integral formula. Compare Example 1.26.)

(23)   If $\varphi : [0, 1] \to \mathbb{C}$ is integrable, verify that the function f defined by

$$f(z) = \int_0^1 \varphi(t) \, e^{izt} \, dt$$

is entire. Show that there are positive constants A, B such that

$$|f(z)| \le A \, e^{B|y|} \qquad \text{for all } z = x + iy \in \mathbb{C}.$$

(24)   Let $f : (0, +\infty) \to \mathbb{R}$ be continuous and

$$\sigma = \limsup_{t \to +\infty} \frac{\log |f(t)|}{t} < +\infty,$$

which means $|f(t)|$ grows at most exponentially as $t \to +\infty$. Let U be the half-plane $\{z \in \mathbb{C} : \mathrm{Re}(z) > \sigma\}$. Define the **Laplace transform** of f by

$$\mathcal{L}(f)(z) = \int_0^\infty f(t) \, e^{-zt} \, dt \qquad \text{for } z \in U.$$

Show that $\mathcal{L}(f)$ is defined and holomorphic in U, and $(\mathcal{L}(f))' = -\mathcal{L}(tf)$.

(25)   This is a variant of Morera's theorem: Suppose $f : U \to \mathbb{C}$ is continuous.

(i) If $\int_{\partial R} f(z)\,dz = 0$ for every closed rectangle $R \subset U$ with sides parallel to the coordinate axes, show that $f \in \mathscr{O}(U)$.

(ii) If $\int_{\partial S} f(z)\,dz = 0$ for every closed square $S \subset U$ with sides parallel to the coordinate axes, show that $f \in \mathscr{O}(U)$.

(Hint: We may assume $U = \mathbb{D}(p, r)$. For (i), construct a primitive of f similar to that in the proof of Theorem 1.31 by integrating along a horizontal followed by a vertical segment. For (ii), show that the assumption implies $\int_{\partial R} f(z)\,dz = 0$ whenever the height-to-width ratio of R is rational, and use continuous dependence of integral on vertices to conclude the same for arbitrary R.)

(26) Under the assumptions of Theorem 1.46, show that the derivative $(z, t) \mapsto \varphi'(z, t)$ is continuous on $U \times [a, b]$. (Hint: Fix $(z_0, t_0) \in U \times [a, b]$ and take $r > 0$ so that $\overline{\mathbb{D}}(z_0, r) \subset U$. Apply Lebesgue's dominated convergence theorem to the formula

$$\varphi'(z, t) = \frac{1}{2\pi i} \int_{\mathbb{T}(z_0, r)} \frac{\varphi(\zeta, t)}{(\zeta - z)^2}\,d\zeta \qquad \text{for } (z, t) \in \mathbb{D}(z_0, r) \times [a, b]$$

to deduce that $\varphi'(z_n, t_n) \to \varphi'(z_0, t_0)$ whenever $z_n \to z_0$ and $t_n \to t_0$.)

(27) Prove the following generalization of Theorem 1.46 by slightly modifying the given argument: Let $U \subset \mathbb{C}$ be open, X be a measure space, μ be a positive measure on X, and $\varphi : U \times X \to \mathbb{C}$ be a function such that

(i) For every $z \in U$, $t \mapsto \varphi(z, t)$ is measurable on X;

(ii) For almost every $t \in X$, $z \mapsto \varphi(z, t)$ is holomorphic in U with derivative $\varphi'(z, t)$;

(iii) φ is locally bounded in the following sense: For every $z \in U$ there is a disk $\mathbb{D}(z, r) \subset U$ and a positive function $M \in L^1(X, \mu)$ such that $|\varphi(\zeta, t)| \leq M(t)$ for every $\zeta \in \mathbb{D}(z, r)$ and almost every $t \in X$.

Then, the function $f(z) = \int_X \varphi(z, t)\,d\mu(t)$ is defined and holomorphic in U and $f'(z) = \int_X \varphi'(z, t)\,d\mu(t)$ for every $z \in U$.

The same holds if μ is a complex measure, provided that in (iii) we require $M \in L^1(X, |\mu|)$. Note that (iii) holds automatically if X is a compact topological space, μ is a finite measure on X and φ is continuous on $U \times X$.

(28) Suppose μ is a complex measure with compact support K in $\mathbb{C}$. Define the **Cauchy transform** of μ by

$$\hat{\mu}(z) = \int_K \frac{d\mu(\zeta)}{\zeta - z} \qquad \text{for } z \in \mathbb{C} \smallsetminus K.$$

Show that $\hat{\mu}$ is holomorphic in $\mathbb{C} \smallsetminus K$ and find a formula for its derivative.

(29) Use Morera's Theorem 1.39 to find another proof for the claim $f \in \mathscr{O}(U)$ in Theorem 1.46. (Hint: Continuity of f follows from Lebesgue's dominated convergence theorem. To use Morera, you need to invoke Fubini's theorem on interchanging the order of an iterated integral.)

(30) What can be said about a function $f \in \mathscr{O}(\mathbb{D})$ which satisfies $|f(1/n)| \leq e^{-n}$ for all large positive integers n?

(31) Let $f \in \mathscr{O}(\mathbb{C})$, $f(z) \in \mathbb{R}$ if $z \in \mathbb{R}$, and $\text{Im}(f(z)) > 0$ if $\text{Im}(z) > 0$. Show that $f'(z) > 0$ for all $z \in \mathbb{R}$.

(32) Suppose $f \in \mathscr{O}(\mathbb{C})$ maps the imaginary axis to the imaginary axis, that is, $\text{Re}(f(z)) = 0$ whenever $\text{Re}(z) = 0$. What can you say about the coefficients of the power series representation $f(z) = \sum_{n=0}^{\infty} a_n z^n$?

(33)   Suppose $U \subset \mathbb{C}$ is a domain and $f, g \in \mathcal{O}(U)$. Assume $g(z) = g(w)$ implies $f(z) = f(w)$. Show that there is a holomorphic function h such that $f = h \circ g$ in U.

(34)   Let $p_1, \ldots, p_n$ be on the unit circle $\mathbb{T}$. Show that there is a point $z \in \mathbb{T}$ such that the product of the distances from z to the p_k is greater than 1.

(35)   Suppose $f \in \mathcal{O}(U)$ and $\overline{\mathbb{D}}(p, r) \subset U$. Show that there is a point z with $|z - p| = r$ such that

$$\left| f(z) - \frac{1}{z - p} \right| \geq \frac{1}{r}.$$

(36)   Suppose $U \subset \mathbb{C}$ is a bounded domain, $f : \overline{U} \to \mathbb{C}$ is continuous, $f \in \mathcal{O}(U)$, and $|f|$ is constant on ∂U. If $f \neq 0$ in U, show that f must be constant in U. Show by an example that the assumption $f \neq 0$ cannot be dispensed with.

(37)   Let $U \subset \mathbb{C}$ be open. A map $\varphi : \mathcal{O}(U) \to \mathbb{C}$ is called a $\mathbb{C}$-*algebra homomorphism* if

$$\varphi(f + g) = \varphi(f) + \varphi(g) \qquad \varphi(fg) = \varphi(f)\varphi(g) \qquad \varphi(\lambda f) = \lambda \varphi(f)$$

for every $f, g \in \mathcal{O}(U)$ and every $\lambda \in \mathbb{C}$. Similarly we can define $\mathbb{C}$-algebra homomorphisms $\mathcal{O}(U) \to \mathcal{O}(V)$ for open sets $U, V \subset \mathbb{C}$.

   (i)   Show that every $\mathbb{C}$-algebra homomorphism $\varphi : \mathcal{O}(U) \to \mathbb{C}$ which is not identically zero is the *point evaluation* $\varphi_p(f) = f(p)$ for some $p \in U$.

   (ii)   (Bers) Show that every $\mathbb{C}$-algebra homomorphism $\varphi : \mathcal{O}(U) \to \mathcal{O}(V)$ which is not identically zero is a composition, i.e., there is a unique holomorphic function $h : V \to U$ such that $\varphi(f) = f \circ h$ for all $f \in \mathcal{O}(U)$. In particular, U and V are biholomorphic if and only if $\mathcal{O}(U)$ and $\mathcal{O}(V)$ are isomorphic $\mathbb{C}$-algebras [**Re**].

(Hint: Let id denote the identity map of U. For (i), show that $p = \varphi(\text{id})$ is in U. If $f \in \mathcal{O}(U)$, use the expansion $f(z) = f(p) + (z - p)f_1(z)$, with $f_1 \in \mathcal{O}(U)$, to show that $\varphi = \varphi_p$. For (ii), let $h = \varphi(\text{id})$ and look at $\varphi_q \circ \varphi$ for $q \in V$.)

2

Topological aspects of Cauchy's theory

This chapter introduces the basic topological apparatus needed in order to develop a global version of Cauchy's theory which deals with holomorphic functions in arbitrary open sets and allows paths of integration that are not necessarily circles or even closed curves. The centerpiece of our brief topological excursion, the idea of the exponential map as a covering space, will be further explored in a general setting in chapter 12.

2.1 Homotopy of curves

A **curve** in a topological space X is a continuous map $\gamma : [a, b] \to X$, where $[a, b] = \{t \in \mathbb{R} : a \le t \le b\}$. We call $\gamma(a)$ the **initial point** and $\gamma(b)$ the **end point** of γ. We often say that γ is a curve **from** $\gamma(a)$ **to** $\gamma(b)$. The compact set $\gamma([a, b]) \subset X$ is called the **image** of γ and is denoted by $|\gamma|$. We say γ is a **closed curve** if $\gamma(a) = \gamma(b)$. Evidently, a closed curve in X can be identified with a continuous map of the circle into X. For simplicity we often assume $[a, b]$ is the unit interval $[0, 1]$. This can always be achieved by an affine change of parameter.

> **DEFINITION 2.1.** Let $p, q \in X$ and $\gamma_0, \gamma_1 : [0, 1] \to X$ be curves from p to q. A **homotopy** between γ_0 and γ_1 is a continuous map $H : [0, 1] \times [0, 1] \to X$ such that
>
> - $H(t, 0) = \gamma_0(t)$ and $H(t, 1) = \gamma_1(t)$ for all $t \in [0, 1]$.
> - $H(0, s) = p$ and $H(1, s) = q$ for all $s \in [0, 1]$.
>
> When such a homotopy exists, we say γ_0 and γ_1 are **homotopic** in X and we write $\gamma_0 \simeq \gamma_1$.

Roughly speaking, $\gamma_0 \simeq \gamma_1$ if γ_0 can be continuously deformed to γ_1 through a family $\{\gamma_s : t \mapsto H(t, s)\}_{s \in [0,1]}$ of curves in X from p to q. Note that the extremities $\gamma_s(0)$ and $\gamma_s(1)$ are required to be kept fixed during this deformation.

EXAMPLE 2.2. Let $X \subset \mathbb{R}^n$ be a convex set. For $p, q \in X$, any two curves $\gamma_0, \gamma_1 : [0, 1] \to X$ from p to q are homotopic in X. The simplest homotopy between γ_0 and γ_1 is given by the **convex combination** $H(t, s) = (1 - s)\gamma_0(t) + s\gamma_1(t)$.

If $\gamma, \eta : [0, 1] \to X$ are curves such that $\gamma(1) = \eta(0)$, the **product** $\gamma \cdot \eta : [0, 1] \to X$ is defined by

$$(\gamma \cdot \eta)(t) = \begin{cases} \gamma(2t) & t \in [0, 1/2] \\ \eta(2t - 1) & t \in [1/2, 1]. \end{cases}$$

This product is *not* associative, so there is no canonical way of defining $\gamma \cdot \eta \cdot \xi$ whenever $\gamma(1) = \eta(0)$ and $\eta(1) = \xi(0)$ (but we will see in Theorem 2.4(iv) that the two curves $(\gamma \cdot \eta) \cdot \xi$ and $\gamma \cdot (\eta \cdot \xi)$ are the same up to reparametrization, and therefore are homotopic). To have a specific choice to work with, we define $\gamma \cdot \eta \cdot \xi = (\gamma \cdot \eta) \cdot \xi$. Inductively, if $\gamma_1, \ldots, \gamma_n : [0, 1] \to X$ are given curves such that $\gamma_k(1) = \gamma_{k+1}(0)$ for $1 \le k \le n - 1$, we define

$$\gamma_1 \cdot \ldots \cdot \gamma_{n-1} \cdot \gamma_n = (\gamma_1 \cdot \ldots \cdot \gamma_{n-1}) \cdot \gamma_n.$$

Given a curve $\gamma : [0, 1] \to X$, the **reverse curve** $\gamma^- : [0, 1] \to X$ is defined by $\gamma^-(t) = \gamma(1 - t)$. Clearly, $(\gamma \cdot \eta)^- = \eta^- \cdot \gamma^-$. We denote by $\varepsilon_p : [0, 1] \to X$ the **constant curve at** $p \in X$ defined by $\varepsilon_p(t) = p$.

LEMMA 2.3. *Let* $\gamma : [0, 1] \to X$ *be a curve and* $\varphi : [0, 1] \to [0, 1]$ *be a continuous map.*

(i) If $\varphi(0) = 0$ *and* $\varphi(1) = 1$, *then* $\gamma \circ \varphi \simeq \gamma$.
(ii) If $\varphi(0) = \varphi(1)$, *then* $\gamma \circ \varphi \simeq \varepsilon_p$, *where* $p = \gamma(\varphi(0))$.

PROOF. For (i), consider the homotopy $\{\varphi_s : t \mapsto (1 - s)\varphi(t) + st\}_{s \in [0,1]}$ between φ and the identity map of $[0, 1]$ (compare Example 2.2). The composition $\{\gamma_s = \gamma \circ \varphi_s\}_{s \in [0,1]}$ will then be a homotopy between $\gamma_0 = \gamma \circ \varphi$ and $\gamma_1 = \gamma$. For (ii), consider the homotopy $\{\varphi_s : t \mapsto (1 - s)\varphi(t) + s\varphi(0)\}_{s \in [0,1]}$ between φ and the constant function $\varphi(0)$. The composition $\{\gamma_s = \gamma \circ \varphi_s\}_{s \in [0,1]}$ will then be a homotopy between $\gamma_0 = \gamma \circ \varphi$ and $\gamma_1 = \gamma \circ \varphi_1 = \varepsilon_p$. □

THEOREM 2.4. *The homotopy* $\simeq$ *is an equivalence relation on the set of all curves in* X *with given initial and end points. Moreover,*

(i) If $\gamma \simeq \eta$, *then* $\gamma^- \simeq \eta^-$.
(ii) If $\gamma(1) = \eta(0)$, $\gamma \simeq \hat{\gamma}$ *and* $\eta \simeq \hat{\eta}$, *then* $\gamma \cdot \eta \simeq \hat{\gamma} \cdot \hat{\eta}$.
(iii) If $\gamma(0) = p$ *and* $\gamma(1) = q$, *then*

$$\varepsilon_p \cdot \gamma \simeq \gamma \cdot \varepsilon_q \simeq \gamma \qquad and \qquad \gamma \cdot \gamma^- \simeq \varepsilon_p.$$

(iv) If $\gamma(1) = \eta(0)$ *and* $\eta(1) = \xi(0)$, *then*

$$(\gamma \cdot \eta) \cdot \xi \simeq \gamma \cdot (\eta \cdot \xi).$$

PROOF. To prove $\simeq$ is an equivalence relation, only transitivity needs a comment: If $\{\gamma_s\}_{s\in[0,1]}$ is a homotopy between γ_0 and γ_1 and $\{\eta_s\}_{s\in[0,1]}$ is a homotopy between γ_1 and γ_2, then $\{\xi_s\}_{s\in[0,1]}$ defined by

$$\xi_s = \begin{cases} \gamma_{2s} & s\in[0,1/2] \\ \eta_{2s-1} & s\in[1/2,1] \end{cases}$$

will be a homotopy between γ_0 and γ_2. The proofs of parts (i)–(iv) go as follows:

(i) If $\{\gamma_s\}_{s\in[0,1]}$ is a homotopy between γ and η, then $\{\gamma_s^-\}_{s\in[0,1]}$ is a homotopy between γ^- and η^-.

(ii) Suppose $\{\gamma_s\}_{s\in[0,1]}$ is a homotopy between γ and $\hat{\gamma}$ and $\{\eta_s\}_{s\in[0,1]}$ is a homotopy between η and $\hat{\eta}$. Then it is easy to check that $\{\gamma_s \cdot \eta_s\}_{s\in[0,1]}$ is a homotopy between $\gamma \cdot \eta$ and $\hat{\gamma} \cdot \hat{\eta}$.

(iii) By definition, $\varepsilon_p \cdot \gamma = \gamma \circ \varphi$, where $\varphi : [0,1] \to [0,1]$ is the continuous function defined by

$$\varphi(t) = \begin{cases} 0 & t\in[0,1/2] \\ 2t-1 & t\in[1/2,1]. \end{cases}$$

By Lemma 2.3(i), $\varepsilon_p \cdot \gamma \simeq \gamma$. Similarly, $\gamma \cdot \varepsilon_q = \gamma \circ \varphi$, where

$$\varphi(t) = \begin{cases} 2t & t\in[0,1/2] \\ 1 & t\in[1/2,1], \end{cases}$$

and it follows from Lemma 2.3(i) that $\gamma \cdot \varepsilon_q \simeq \gamma$. Finally, note that $\gamma \cdot \gamma^- = \gamma \circ \varphi$, where

$$\varphi(t) = \begin{cases} 2t & t\in[0,1/2] \\ 2-2t & t\in[1/2,1]. \end{cases}$$

By Lemma 2.3(ii), $\gamma \cdot \gamma^- \simeq \varepsilon_p$.

(iv) It is easy to check that $(\gamma \cdot \eta) \cdot \xi = (\gamma \cdot (\eta \cdot \xi)) \circ \varphi$, where

$$\varphi(t) = \begin{cases} 2t & t\in[0,1/4] \\ t+1/4 & t\in[1/4,1/2] \\ (t+1)/2 & t\in[1/2,1], \end{cases}$$

and the result follows from Lemma 2.3(i).   □

The equivalence class of a curve γ under $\simeq$ is called the **homotopy class** of γ and is denoted by $[\gamma]$. By Theorem 2.4(ii), if $\gamma, \eta : [0,1] \to X$ are curves with $\gamma(1) = \eta(0)$, the product

(2.1) $$[\gamma] \cdot [\eta] = [\gamma \cdot \eta]$$

of their homotopy classes is well defined.

DEFINITION 2.5. Let X be a topological space and $p \in X$. The set of all homotopy classes of closed curves $\gamma : [0, 1] \to X$ with $\gamma(0) = \gamma(1) = p$ forms a group under the product (2.1), which is called the ***fundamental group*** of X with the ***base point*** p and is denoted by $\pi_1(X, p)$.

The idea of the fundamental group was pioneered by Poincaré circa 1895.

Note that by Theorem 2.4(iv) the product (2.1) is associative. By Theorem 2.4(iii) the homotopy class $[\varepsilon_p]$ acts as the identity element of $\pi_1(X, p)$, and the inverse $[\gamma]^{-1}$ is given by $[\gamma^-]$ since $[\gamma] \cdot [\gamma^-] = [\varepsilon_p]$.

Suppose $f : X \to Y$ is a continuous map between topological spaces. If $H : [0, 1] \times [0, 1] \to X$ is a homotopy between the curves $\gamma_0, \gamma_1 : [0, 1] \to X$, then $f \circ H : [0, 1] \times [0, 1] \to Y$ is a homotopy between the image curves $f \circ \gamma_0$ and $f \circ \gamma_1$ in Y. It follows that the induced map $f_* : \pi_1(X, p) \to \pi_1(Y, f(p))$ given by

$$f_*([\gamma]) = [f \circ \gamma]$$

is a well-defined group homomorphism. If $g : Y \to Z$ is another continuous map, then

$$(g \circ f)_* = g_* \circ f_*.$$

Jules Henri Poincaré (1854–1912)

In particular, when $f : X \to Y$ is a homeomorphism, the induced map f_* is a group isomorphism with the inverse $(f^{-1})_*$. Thus, *homeomorphic spaces have isomorphic fundamental groups.*

The dependence of the fundamental group on the base point is easy to understand when X is ***path-connected***, i.e., when each pair of points in X can be joined by a curve. Assuming this, let $p, q \in X$ and η be a curve in X from p to q. It is not hard to verify that the map $\Phi_\eta : \pi_1(X, p) \to \pi_1(X, q)$ defined by $\Phi_\eta([\gamma]) = [\eta^- \cdot \gamma \cdot \eta]$ is a group isomorphism. If ξ is another curve in X from p to q, then

$$\Phi_\xi([\gamma]) = [\xi^- \cdot \eta] \cdot \Phi_\eta([\gamma]) \cdot [\eta^- \cdot \xi].$$

Setting $y = [\xi^- \cdot \eta] \in \pi_1(X, q)$, it follows that Φ_η and Φ_ξ differ by the automorphism $x \mapsto y \cdot x \cdot y^{-1}$ of $\pi_1(X, q)$. In particular, when $\pi_1(X, q)$ is abelian, $\Phi_\eta = \Phi_\xi$ and one obtains a canonical isomorphism between $\pi_1(X, p)$ and $\pi_1(X, q)$.

DEFINITION 2.6. A path-connected topological space X is called ***simply connected*** if the fundamental group $\pi_1(X, p)$ is trivial for some (hence for every) $p \in X$.

The concept of simple connectivity (for surfaces) was introduced by Riemann in 1851.

Thus, X is simply connected if every closed curve γ in X with the initial and end point p is ***null-homotopic*** in the sense that $\gamma \simeq \varepsilon_p$. Here is an alternative characterization:

THEOREM 2.7. *A path-connected topological space X is simply connected if and only if any two curves in X with the same initial and end points are homotopic.*

PROOF. Suppose X is simply connected and $\gamma, \eta : [0,1] \to X$ are two curves from p to q. The closed curve $\gamma \centerdot \eta^-$ is null-homotopic, so $[\gamma] \centerdot [\eta]^{-1} = [\gamma \centerdot \eta^-] = [\varepsilon_p] = 1$. It follows that $[\gamma] = [\eta]$.

The converse statement is trivial. □

EXAMPLE 2.8. Let $U \subset \mathbb{C}$ be a ***star-shaped*** domain. This means there is a $p \in U$ such that for every $z \in U$ the segment $[p, z]$ is contained in U. Suppose $\gamma : [0,1] \to U$ is a closed curve with $\gamma(0) = \gamma(1) = p$. The map $H : [0,1] \times [0,1] \to U$ defined by $H(t, s) = (1-s)\gamma(t) + sp$ is a homotopy between γ and the constant curve ε_p. It follows that $\pi_1(U, p)$ is trivial, so U is simply connected.

EXAMPLE 2.9. Since by the previous example the open unit disk $\mathbb{D} \subset \mathbb{C}$ is simply connected, so is any subset of the plane homeomorphic to $\mathbb{D}$. Later in chapter 6 we will prove the converse statement: *Every simply connected domain in the plane is homeomorphic to $\mathbb{D}$.* The analogous statement in higher dimensions is false.

We end this section with a more general notion of homotopy for *closed* curves in which no base point is required to be kept fixed during deformation.

DEFINITION 2.10. Two closed curves $\gamma_0, \gamma_1 : [0,1] \to X$ are said to be ***freely homotopic*** in X if there exists a continuous map $H : [0,1] \times [0,1] \to X$ such that

- $H(t, 0) = \gamma_0(t)$ and $H(t, 1) = \gamma_1(t)$ for all $t \in [0,1]$.
- $H(0, s) = H(1, s)$ for all $s \in [0,1]$.

Thus, γ_0 and γ_1 are freely homotopic in X if γ_0 can be continuously deformed to γ_1 through a family of closed curves $\{\gamma_s : t \mapsto H(t, s)\}_{s \in [0,1]}$ in X. It is easy to check that a free homotopy is an equivalence relation on the set of all closed curves.

EXAMPLE 2.11. Let $X = \mathbb{C} \smallsetminus \{-1, 1\}$ be the twice punctured plane, choose $p = 0$ as the base point of X, and let $\gamma, \eta : [0,1] \to X$ be the closed curves defined by

$$\gamma(t) = e^{2\pi it} - 1 \qquad \text{and} \qquad \eta(t) = -\gamma(t) = 1 - e^{2\pi it}.$$

Thus, each curve winds around a puncture once in the counterclockwise direction (see Fig. 2.1). It is not hard to see that the closed curves $\gamma \centerdot \eta$ and $\eta \centerdot \gamma$ are freely homotopic in X. In fact, they both are freely homotopic to a loop that winds around both punctures once in the counterclockwise direction. On the other hand, there is no homotopy between $\gamma \centerdot \eta$ and

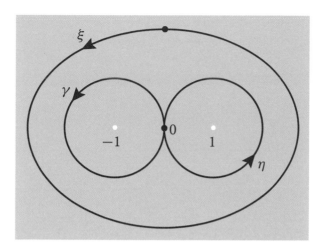

Figure 2.1. Illustration of Example 2.11. The closed curves $\gamma \cdot \eta$ and $\eta \cdot \gamma$ are freely homotopic in the twice punctured plane $\mathbb{C} \smallsetminus \{-1, 1\}$ since each is freely homotopic to ξ.

$\eta \cdot \gamma$ that keeps the base point 0 fixed. In other words, $[\gamma \cdot \eta] \neq [\eta \cdot \gamma]$ in $\pi_1(X, 0)$ (see problem 3).

REMARK 2.12. It is not hard to show that if a closed curve is freely homotopic to a constant curve, then it is homotopic to a constant curve. In other words, being "freely null-homotopic" and "null-homotopic" are identical notions for closed curves (see problem 2).

2.2 Covering properties of the exponential map

In this section we will briefly study the topology of the exponential map $\exp \colon \mathbb{C} \to \mathbb{C} \smallsetminus \{0\}$ as a "covering space" and the question of lifting under exp. Since a more detailed account of general covering spaces will be presented in chapter 12, the discussion here will be limited to a few fundamental facts that are relevant to the global version of Cauchy's theory. While the special case of the exponential map does not fully reflect the power of the covering space theory, it beautifully captures many essential features. Ironically, the fact that exp is holomorphic will not play a role in the following discussion, until Theorem 2.29.

Recall that for $z = x + iy \in \mathbb{C}$, $\exp(z) = e^z = e^x e^{iy} = e^x(\cos y + i \sin y)$. We have $\exp(z) = \exp(\zeta)$ if and only if $z - \zeta$ is an integer multiple of $2\pi i$. The image $\exp(\mathbb{C})$ is the punctured plane $\mathbb{C}^* = \mathbb{C} \smallsetminus \{0\}$, so given $w_0 \in \mathbb{C}^*$ one can find $z_0 \in \mathbb{C}$ such that $\exp(z_0) = w_0$. Thus, the preimage of w_0 under exp is

$$\exp^{-1}(w_0) = z_0 + 2\pi i \mathbb{Z} = \{z_0 + 2\pi i k : k \in \mathbb{Z}\}.$$

More generally, w_0 has a neighborhood W which is **evenly covered** by the exponential map in the sense that $\exp^{-1}(W)$ is the disjoint union of open sets $O_k = O_0 + 2\pi i k$

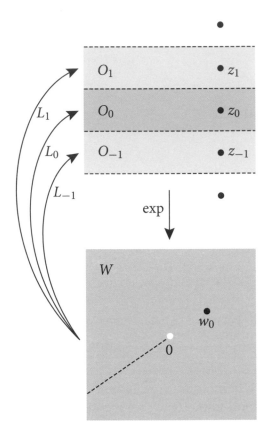

Figure 2.2. An evenly covered neighborhood W of $w_0 \in \mathbb{C}^*$ under the exponential map, the components O_k of $\exp^{-1}(W)$, and the local inverses $L_k : W \to O_k$.

containing $z_k = z_0 + 2\pi i k$ such that $\exp : O_k \to W$ is a homeomorphism for each $k \in \mathbb{Z}$. It follows that for each k there is a well-defined **local inverse** $L_k : W \to O_k$ of the exponential map (the letter L should remind you of "logarithm"). One convenient choice for the evenly covered neighborhood W is the slit plane $W = \mathbb{C} \setminus \{tw_0 : t \leq 0\}$, in which case O_0 is the horizontal strip $\{z \in \mathbb{C} : \operatorname{Im}(z_0) - \pi < \operatorname{Im}(z) < \operatorname{Im}(z_0) + \pi\}$ (see Fig. 2.2). Notice that since W is connected, the strips O_k form the connected components of $\exp^{-1}(W)$. The local inverse L_k is given by the formula

$$(2.2) \qquad L_k(w) = \log |w| + i\theta_k,$$

where θ_k is the unique argument of w in the interval $(\operatorname{Im}(z_k) - \pi, \operatorname{Im}(z_k) + \pi)$.

DEFINITION 2.13. Let X be a topological space and $f : X \to \mathbb{C}^*$ be a continuous map. A **lift** (under exp) of f is a continuous map $\tilde{f} : X \to \mathbb{C}$ which satisfies $\exp(\tilde{f}) = f$.

In other words, a lift makes the following diagram commute:

$$
\begin{array}{ccc}
 & & \mathbb{C} \\
 & \tilde{f} \nearrow & \downarrow \text{exp} \\
X & \xrightarrow{\quad f \quad} & \mathbb{C}^*
\end{array}
$$

For obvious reasons, $\tilde{f}$ is also called a **branch of the logarithm of** f.

EXAMPLE 2.14. The curve $\gamma : [0, 1] \to \mathbb{C}^*$ defined by $\gamma(t) = e^{2\pi i n t}$ $(n \in \mathbb{Z})$ has a lift $\tilde{\gamma} : [0, 1] \to \mathbb{C}$ given by $\tilde{\gamma}(t) = 2\pi i n t$. This shows that *a lift of a closed curve may not be closed.*

As another example, if W is evenly covered by the exponential map and $\exp^{-1}(W) = \bigcup_k O_k$ as in the definition, then each local inverse $L_k : W \to O_k$ is a lift of the identity map $W \to W$.

The issue of uniqueness of lifts is easy to address:

THEOREM 2.15 (Uniqueness of lifts). *If X is a connected topological space, any two lifts under $\exp$ of a continuous map $X \to \mathbb{C}^*$ differ by an integer multiple of $2\pi i$.*

PROOF. Suppose $\tilde{f}_1, \tilde{f}_2 : X \to \mathbb{C}$ are two lifts of the same map $f : X \to \mathbb{C}^*$. For each $p \in X$, $\exp(\tilde{f}_1(p)) = f(p) = \exp(\tilde{f}_2(p))$, so there exists an integer $n(p)$ such that $\tilde{f}_1(p) = \tilde{f}_2(p) + 2\pi i\, n(p)$. The function

$$
p \mapsto n(p) = \frac{1}{2\pi i}(\tilde{f}_1(p) - \tilde{f}_2(p))
$$

is continuous and integer-valued. Since X is assumed connected, this function must be constant. ◻

The question of existence of lifts is more delicate and will be addressed in Theorem 2.21. It will be convenient to treat the cases $X = [0, 1]$ and $X = [0, 1] \times [0, 1]$ first.

THEOREM 2.16.

(i) *Given a curve $\gamma : [0, 1] \to \mathbb{C}^*$ and a point $p \in \mathbb{C}$ such that $\exp(p) = \gamma(0)$, there is a unique lift $\tilde{\gamma} : [0, 1] \to \mathbb{C}$ of γ such that $\tilde{\gamma}(0) = p$.*

(ii) *Given a continuous map $H : [0, 1] \times [0, 1] \to \mathbb{C}^*$ and a point $p \in \mathbb{C}$ such that $\exp(p) = H(0, 0)$, there is a unique lift $\tilde{H} : [0, 1] \times [0, 1] \to \mathbb{C}$ of H such that $\tilde{H}(0, 0) = p$.*

Part (i) is often referred to as the **curve lifting property** of the exponential map. The proof of both parts will make use of an elementary result in topology which we now recall for convenience:

LEMMA 2.17 (Lebesgue's cover lemma). *For every open cover $\{U_\alpha\}$ of a compact metric space X there is a $\delta > 0$ with the property that if $E \subset X$ has $\operatorname{diam} E < \delta$, then $E \subset U_\alpha$ for some α.*

Such δ is called a **Lebesgue number** for the cover $\{U_\alpha\}$.

PROOF. Let $B(x, r)$ denote the open ball of radius r centered at x. For each $x \in X$ we can find a radius $r(x) > 0$ and some index $\alpha(x)$ such that $B(x, 2r(x)) \subset U_{\alpha(x)}$. By compactness, there are finitely many points $x_1, \ldots, x_n$ in X such that $X = \bigcup_{j=1}^n B(x_j, r(x_j))$. Set $\delta = \min_{1 \le j \le n} r(x_j)$. If $E \subset X$ has diameter $< \delta$, since E meets some $B(x_j, r(x_j))$, it will be contained in $B(x_j, 2r(x_j))$, which proves $E \subset U_{\alpha(x_j)}$. $\qquad \square$

PROOF OF THEOREM 2.16. (i) For each $t \in [0, 1]$ the point $\gamma(t)$ has a neighborhood that is evenly covered by exp. By continuity, the preimages under γ of these neighborhoods form an open cover of $[0, 1]$. Let $\delta > 0$ be a Lebesgue number for this cover and take a partition $0 = t_0 < t_1 < \cdots < t_n = 1$ for which $t_k - t_{k-1} < \delta$ for all $1 \le k \le n$. It follows that each image $\gamma([t_{k-1}, t_k])$ is contained in an evenly covered neighborhood W_k. Let O_1 be the component of $\exp^{-1}(W_1)$ containing p and $L_1 : W_1 \to O_1$ be the corresponding local inverse of exp. Define $\tilde{\gamma} = L_1 \circ \gamma$ on $[t_0, t_1]$. Let O_2 be the component of $\exp^{-1}(W_2)$ containing $\tilde{\gamma}(t_1)$ and $L_2 : W_2 \to O_2$ be the corresponding local inverse of exp, and define $\tilde{\gamma} = L_2 \circ \gamma$ on $[t_1, t_2]$. Repeating this process n times, we construct the desired lift $\tilde{\gamma} : [0, 1] \to \mathbb{C}$. Uniqueness of $\tilde{\gamma}$ follows from Theorem 2.15.

(ii) The proof is very similar to (i). For each $(t, s) \in [0, 1] \times [0, 1]$ the point $H(t, s)$ has a neighborhood that is evenly covered by exp. By continuity, the preimages under H of these neighborhoods form an open cover of $[0, 1] \times [0, 1]$. Let $\delta > 0$ be a Lebesgue number for this cover and take partitions $0 = t_0 < t_1 < \cdots < t_n = 1$ and $0 = s_0 < s_1 < \cdots < s_n = 1$ for which the rectangles $R_{jk} = [t_{j-1}, t_j] \times [s_{k-1}, s_k]$ have diameter less than δ for every $1 \le j, k \le n$. It follows that each image $H(R_{jk})$ is contained in an evenly covered neighborhood W_{jk}. Note that by Theorem 2.15 a lift $\tilde{H}_{jk} : R_{jk} \to \mathbb{C}$ of $H : R_{jk} \to \mathbb{C}^*$ is uniquely determined once its value p_{jk} at the lower left corner (t_{j-1}, s_{k-1}) is known. In fact, let O_{jk} be the component of $\exp^{-1}(W_{jk})$ containing p_{jk} and $L_{jk} : W_{jk} \to O_{jk}$ be the corresponding local inverse of exp. Then $\tilde{H}_{jk} = L_{jk} \circ H$ is the desired lift. So, starting with $p_{11} = p$ we define a lift $\tilde{H}_{11}$ on R_{11}, then use $p_{21} = \tilde{H}_{11}(t_1, s_0)$ to define $\tilde{H}_{21}$ on R_{21}, then use $p_{31} = \tilde{H}_{21}(t_2, s_0)$ to define $\tilde{H}_{31}$ on R_{31}, and so on. We repeat this process in every row from left to right, each time using the boundary value of the lift on a rectangle to define the lift on the next one, until all the rectangles are exhausted. Clearly each $\tilde{H}_{jk}$ is continuous on R_{jk} and satisfies $\exp \circ \tilde{H}_{jk} = H$. Moreover, it is easy to check that if R_{jk} and $R_{h\ell}$ have a common edge E, then $\tilde{H}_{jk}, \tilde{H}_{h\ell} : E \to \mathbb{C}$ are two lifts of the same curve $H : E \to \mathbb{C}^*$ which agree at one point of E (the left end if E is horizontal and the lower end if E is vertical). Hence, by uniqueness of lifts in (i), they must agree on E. Thus, the local lifts $\{\tilde{H}_{jk}\}$ patch together consistently and yield a lift $\tilde{H}$ on $[0, 1] \times [0, 1]$. Uniqueness of $\tilde{H}$ follows again from Theorem 2.15. $\qquad \square$

COROLLARY 2.18. *Suppose* $\gamma_0, \gamma_1 : [0, 1] \to \mathbb{C}^*$ *are homotopic curves. Take* $p \in \mathbb{C}$ *such that* $\exp(p) = \gamma_0(0) = \gamma_1(0)$ *and let* $\tilde{\gamma}_0, \tilde{\gamma}_1 : [0, 1] \to \mathbb{C}$ *be the unique lifts of* γ_0, γ_1 *with* $\tilde{\gamma}_0(0) = \tilde{\gamma}_1(0) = p$. *Then* $\tilde{\gamma}_0, \tilde{\gamma}_1$ *are homotopic and in particular* $\tilde{\gamma}_0(1) = \tilde{\gamma}_1(1)$.

PROOF. Let $H : [0, 1] \times [0, 1] \to \mathbb{C}^*$ be a homotopy between γ_0 and γ_1, as in Definition 2.1. By Theorem 2.16(ii), there is a lift $\tilde{H} : [0, 1] \times [0, 1] \to \mathbb{C}$ of H with $\tilde{H}(0, 0) = p$. By uniqueness of lifts, $\tilde{\gamma}_0 = \tilde{H}(\cdot, 0)$ and $\tilde{\gamma}_1 = \tilde{H}(\cdot, 1)$. To show $\tilde{H}$ is a homotopy between $\tilde{\gamma}_0, \tilde{\gamma}_1$, we only need to check that $s \mapsto \tilde{H}(1, s)$ is constant. But this is easy because the exponential map sends $\tilde{H}(1, s)$ down to $H(1, s) = \gamma_0(1)$, so the continuous map $s \mapsto \tilde{H}(1, s)$ takes values in the totally disconnected set $\exp^{-1}(\gamma_0(1)) = \tilde{\gamma}_0(1) + 2\pi i \mathbb{Z}$. □

COROLLARY 2.19. *A closed curve in $\mathbb{C}^*$ is null-homotopic if and only if every lift of it under* exp *is a closed curve.*

PROOF. Suppose γ_0 is null-homotopic in $\mathbb{C}^*$ and $\tilde{\gamma}_0$ is any lift of it. Let γ_1 be the constant curve at $\gamma_0(0)$. Then the lift $\tilde{\gamma}_1$ with the initial point $\tilde{\gamma}_0(0)$ is just the constant curve at $\tilde{\gamma}_0(0)$. By Corollary 2.18, $\gamma_0 \simeq \gamma_1$ implies $\tilde{\gamma}_0(1) = \tilde{\gamma}_1(1) = \tilde{\gamma}_0(0)$, which means $\tilde{\gamma}_0$ is a closed curve. Conversely, suppose some (hence every) lift $\tilde{\gamma}_0$ of γ_0 is a closed curve. Since the plane is simply connected, there is a homotopy between $\tilde{\gamma}_0$ and the constant curve at $\tilde{\gamma}_0(0)$. Postcompositng this homotopy with the exponential map, we obtain a homotopy in $\mathbb{C}^*$ between γ_0 and the constant curve at $\gamma_0(0)$. □

EXAMPLE 2.20 (The fundamental group of $\mathbb{C}^*$). We can use the preceding results to show that the fundamental group of the punctured plane $\mathbb{C}^*$ is infinite cyclic. Define a map $\Phi : \pi_1(\mathbb{C}^*, 1) \to \mathbb{Z}$ as follows. Take a homotopy class $[\gamma] \in \pi_1(\mathbb{C}^*, 1)$ and take the lift $\tilde{\gamma}$ of γ under exp with $\tilde{\gamma}(0) = 0$. The end point $\tilde{\gamma}(1)$ is of the form $2\pi i n$ for some integer n which, by Corollary 2.18, does not depend on the choice of the representative of $[\gamma]$. Set $\Phi([\gamma]) = n$. It is easily seen that Φ is a group homomorphism. Φ is surjective because it sends the homotopy class represented by $\gamma(t) = e^{2\pi i n t}$ to n. It is injective since $\Phi([\gamma]) = 0$ means $\tilde{\gamma}$ is a closed curve, which by Corollary 2.19 implies γ is null-homotopic. It follows that Φ is a group isomorphism.

We now address the existence of lifts on more general spaces. Recall that a topological space is ***locally path-connected*** if every neighborhood of every point contains a path-connected neighborhood of that point.

THEOREM 2.21 (Existence of lifts). *Suppose X is a simply connected and locally path-connected topological space and $f : X \to \mathbb{C}^*$ is continuous. Then $f = \exp(g)$ for some continuous function $g : X \to \mathbb{C}$. Moreover, g is unique up to addition of an integer multiple of $2\pi i$.*

Local path-connectivity is a necessary assumption in general (see problem 1) but always holds for open subsets of the plane.

PROOF. It suffices to construct one lift since the uniqueness issue has already been dealt with in Theorem 2.15. Fix a base point $p \in X$ and a point $z_0 \in \mathbb{C}$ such that $\exp(z_0) = f(p)$. Take any $q \in X$ and let $\gamma : [0, 1] \to X$ be any curve from p to q. By Theorem 2.16(i), the curve $f \circ \gamma : [0, 1] \to \mathbb{C}^*$ has a unique lift $\xi : [0, 1] \to \mathbb{C}$ with

$\xi(0) = z_0$. Define $g(q) = \xi(1)$. This does not depend on the choice of γ since any other curve $\hat{\gamma} : [0,1] \to X$ from p to q is homotopic to γ by Theorem 2.7, so $f \circ \hat{\gamma}$ is homotopic to $f \circ \gamma$ in $\mathbb{C}^*$, so their unique lifts with the same initial point z_0 have the same end point by Corollary 2.18. Evidently, $\exp(g(q)) = \exp(\xi(1)) = f(\gamma(1)) = f(q)$. It remains to check that $g : X \to \mathbb{C}$ defined this way is continuous.

Fix $q \in X$ and take an arbitrary neighborhood O of $g(q)$ which is small enough so it maps homeomorphically to a neighborhood $W = \exp(O)$ of $f(q)$. By continuity of f and local path-connectivity of X, there is a path-connected neighborhood V of q such that $f(V) \subset W$. Take any $x \in V$ and let η be a curve in V from q to x, so $f \circ \eta$ is a curve in W from $f(q)$ to $f(x)$. Let γ and ξ be as above. Then $\gamma \cdot \eta$ is a curve from p to x and the lift of $f \circ (\gamma \cdot \eta) = (f \circ \gamma) \cdot (f \circ \eta)$ with the initial point z_0 is given by $\xi \cdot (L \circ f \circ \eta)$, where $L : W \to O$ is the local inverse of exp. It follows that $g(x)$ is the end point of $L \circ f \circ \eta$, which belongs to O. This shows $g(V) \subset O$ and proves continuity of g at q. □

COROLLARY 2.22 (Existence of holomorphic logarithms and *n*-th roots). *Suppose $U \subset \mathbb{C}$ is a simply connected domain and $f \in \mathcal{O}(U)$ has no zeros in U.*

(i) *There exists a $g \in \mathcal{O}(U)$ such that $f = \exp(g)$, and g is unique up to addition of an integer multiple of $2\pi i$.*

(ii) *For each positive integer n, there exists an $h \in \mathcal{O}(U)$ such that $f = h^n$, and h is unique up to multiplication by an n-th root of unity.*

PROOF. (i) Since every open subset of $\mathbb{C}$ is locally path-connected, Theorem 2.21 gives a continuous function $g : U \to \mathbb{C}$ which satisfies $f = \exp(g)$. To see g is holomorphic, take any $p \in U$, let W be an evenly covered neighborhood of $f(p)$, O be the connected component of $\exp^{-1}(W)$ containing $g(p)$, and V be the connected component of $f^{-1}(W)$ containing p. The local inverse $L : W \to O$ of exp is holomorphic since exp is holomorphic in O. The composition $L \circ f : V \to O$ is a lift of f in V which sends p to $g(p)$. By Theorem 2.15, $g = L \circ f$ holds in V, so $g \in \mathcal{O}(V)$.

(ii) Write $f = \exp(g)$ for some $g \in \mathcal{O}(U)$. The function $h = \exp(g/n)$ is clearly holomorphic in U and satisfies $f = h^n$. If h_1, h_2 satisfy $h_1^n = h_2^n = f$, the continuous function h_1/h_2 takes its values in the finite set of the n-th roots of unity. Since U is connected, this ratio must be constant. □

EXAMPLE 2.23 (Principal branch of the logarithm). According to Corollary 2.22(i), there is a unique holomorphic function $g : \mathbb{C} \setminus (-\infty, 0] \to \mathbb{C}$ which satisfies $\exp(g(z)) = z$ and $g(1) = 0$. Traditionally, g is called the **principal branch of the logarithm function** and is denoted by Log. Each $z \in \mathbb{C} \setminus (-\infty, 0]$ has a unique representation of the form re^{it} where $r > 0$ and $-\pi < t < \pi$, and it is easily checked that $\text{Log}\, z = \log r + it$. Observe that Log is just the local inverse L_0 in (2.2) determined by the choice $w_0 = 1, z_0 = 0$.

If z, w have positive real parts so their arguments are between $-\pi/2$ and $\pi/2$ in the above polar representation, then $\text{Log}(zw) = \text{Log}\, z + \text{Log}\, w$.

Let us find the power series representation of Log in the disk $\mathbb{D}(1,1)$. Differentiation of the equation $\exp(\mathrm{Log}(z)) = z$ gives $z\,\mathrm{Log}'(z) = 1$, or $\mathrm{Log}'(z) = 1/z$. The geometric series formula

$$\frac{1}{z} = \frac{1}{1 + (z-1)} = \sum_{n=0}^{\infty} (-1)^n (z-1)^n,$$

which is valid as long as $|z - 1| < 1$, then shows that

$$\mathrm{Log}\,z = \sum_{n=1}^{\infty} \frac{(-1)^{n-1}}{n}\,(z-1)^n \qquad \text{for } |z-1| < 1.$$

This is usually written in the more convenient form

$$(2.3) \qquad\qquad \mathrm{Log}(1+z) = \sum_{n=1}^{\infty} \frac{(-1)^{n-1}}{n}\,z^n \qquad \text{for } |z| < 1.$$

The construction of holomorphic n-th roots via logarithms can be imitated to produce arbitrary powers. Let f be a non-vanishing holomorphic function in a simply connected domain U and take any $g \in \mathcal{O}(U)$ such that $f = \exp(g)$. For $\alpha \in \mathbb{C}$ the function $\exp(\alpha g) \in \mathcal{O}(U)$ is called a **holomorphic branch of the power f^α**. Since g is unique up to addition of an integer multiple of $2\pi i$, it follows that any two holomorphic branches of f^α differ by a multiplicative constant of the form $e^{2\pi i n\alpha}$ for some $n \in \mathbb{Z}$. Not surprisingly, this implies that f^α is uniquely defined when α is an integer. The computation

$$\big(\exp(\alpha g)\big)' = \exp(\alpha g)\,\alpha g' = \alpha \exp(\alpha g) \exp(-g) f' = \alpha \exp((\alpha - 1)g) f'$$

proves that the familiar differentiation formula

$$(f^\alpha)' = \alpha f^{\alpha-1} f'$$

remains valid for suitable choices of power branches on each side.

2.3 The winding number

Our next goal is to introduce the concept of the "winding number" of a planar curve with respect to a point. In the case of a closed curve, the winding number is an integer which, roughly speaking, counts how many times the curve goes around the given point in the counterclockwise direction. We will make use of the fact that for each $p \in \mathbb{C}$, all the lifting results of the last section hold if we replace $\mathbb{C}^* = \mathbb{C} \smallsetminus \{0\}$ with $\mathbb{C} \smallsetminus \{p\}$ and $z \mapsto \exp(z)$ with the translated exponential map $z \mapsto \exp(z) + p$.

Let $\gamma : [0,1] \to \mathbb{C}$ be a curve and $p \in \mathbb{C} \smallsetminus |\gamma|$. View γ as a curve in the punctured plane $\mathbb{C} \smallsetminus \{p\}$ and lift it under $z \mapsto \exp(z) + p$ to a curve $\tilde{\gamma} : [0,1] \to \mathbb{C}$, so $\exp(\tilde{\gamma}) + p = \gamma$. Any other lift of γ is of the form $\tilde{\gamma} + 2\pi i k$ for some integer k, so the difference $\tilde{\gamma}(1) - \tilde{\gamma}(0)$ is independent of the choice of the lift. Note that this difference is an integer multiple of $2\pi i$ if and only if $\exp(\tilde{\gamma}(1)) = \exp(\tilde{\gamma}(0))$, which happens precisely when γ is a closed curve.

DEFINITION 2.24. Let $\gamma : [0, 1] \to \mathbb{C}$ be a curve, $p \in \mathbb{C} \smallsetminus |\gamma|$, and $\tilde{\gamma} : [0, 1] \to \mathbb{C}$ be any lift of γ under $z \mapsto \exp(z) + p$. The quantity

$$W(\gamma, p) = \frac{1}{2\pi i}\big(\tilde{\gamma}(1) - \tilde{\gamma}(0)\big)$$

is called the **winding number** of γ with respect to p. It is an integer if and only if γ is a closed curve.

The winding number is a spin-off of the general concept of the degree of a map introduced by Brouwer in 1910.

It is easy to justify the terminology. Suppose for example that γ is a closed curve and $\tilde{\gamma}(t) = x(t) + iy(t)$ is a lift of γ under $z \mapsto \exp(z) + p$ so $\gamma(t) = e^{x(t)+iy(t)} + p$. The imaginary part $y(t)$ represents a continuous choice of the argument of $\gamma(t)$ as seen from p. Every time γ winds around p the value of $y(t)$ increases or decreases by 2π depending on whether the winding is counterclockwise or clockwise. Since $x(0) = x(1)$, the quantity

The term "index" is also widely used in the literature for what we call the winding number.

$$\frac{1}{2\pi i}\big(\tilde{\gamma}(1) - \tilde{\gamma}(0)\big) = \frac{1}{2\pi}\big(y(1) - y(0)\big)$$

measures the algebraic number of full-turns that γ makes around p in the counterclockwise direction (compare Fig. 2.3).

EXAMPLE 2.25. For each $n \in \mathbb{Z}$, the closed curve $\gamma : [0, 1] \to \mathbb{C}^*$ defined by $\gamma(t) = e^{2\pi int}$ lifts to $\tilde{\gamma}(t) = 2\pi int$, so $W(\gamma, 0) = n$.

THEOREM 2.26 (Properties of the winding number). *Let $\gamma : [0, 1] \to \mathbb{C}$ be a curve and $p \in \mathbb{C} \smallsetminus |\gamma|$.*

(i) *For any $w \in \mathbb{C}$,*
$$W(\gamma, p) = W(\gamma + w, p + w),$$
   where $\gamma + w$ denotes the translated curve $t \mapsto \gamma(t) + w$.

(ii) $W(\gamma^-, p) = -W(\gamma, p).$

(iii) *If $\eta : [0, 1] \to \mathbb{C} \smallsetminus \{p\}$ is a curve with $\eta(0) = \gamma(1)$, then*
$$W(\gamma \cdot \eta, p) = W(\gamma, p) + W(\eta, p).$$

(iv) *If γ and η are homotopic in $\mathbb{C} \smallsetminus \{p\}$, then*
$$W(\gamma, p) = W(\eta, p).$$

(v) *Suppose γ is a closed curve. Then γ and η are freely homotopic in $\mathbb{C} \smallsetminus \{p\}$ if and only if $W(\gamma, p) = W(\eta, p)$. In particular, γ is null-homotopic in $\mathbb{C} \smallsetminus \{p\}$ if and only if $W(\gamma, p) = 0$.*

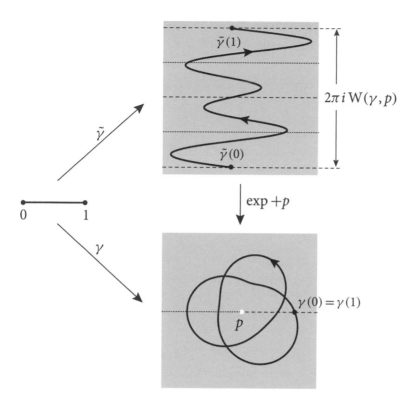

Figure 2.3. Definition of the winding number $W(\gamma, p)$. In this example γ is a closed curve and $W(\gamma, p) = 2$ is an integer.

(vi) *Suppose γ is a closed curve. Then the function $z \mapsto W(\gamma, z)$ is constant in each connected component of $\mathbb{C} \setminus |\gamma|$, and vanishes in the unbounded connected component of $\mathbb{C} \setminus |\gamma|$.*

PROOF. (i) follows from the fact that any lift of γ under $z \mapsto \exp(z) + p$ is a lift of $\gamma + w$ under $z \mapsto \exp(z) + p + w$.

For the rest of the proof, all lifts are meant to be under $z \mapsto \exp(z) + p$.

(ii) follows from the fact that if $\tilde{\gamma}$ is a lift of γ, then $\tilde{\gamma}^-$ is a lift of γ^- and $\tilde{\gamma}^-(1) - \tilde{\gamma}^-(0) = \tilde{\gamma}(0) - \tilde{\gamma}(1)$.

(iii) Let $\tilde{\gamma}$ be any lift of γ and $\tilde{\eta}$ be the unique lift of η with $\tilde{\eta}(0) = \tilde{\gamma}(1)$. Then $\tilde{\gamma} \cdot \tilde{\eta}$ is a lift of $\gamma \cdot \eta$, so

$$W(\gamma \cdot \eta, p) = \frac{1}{2\pi i}((\tilde{\gamma} \cdot \tilde{\eta})(1) - (\tilde{\gamma} \cdot \tilde{\eta})(0)) = \frac{1}{2\pi i}(\tilde{\eta}(1) - \tilde{\gamma}(0))$$

$$= \frac{1}{2\pi i}(\tilde{\gamma}(1) - \tilde{\gamma}(0)) + \frac{1}{2\pi i}(\tilde{\eta}(1) - \tilde{\eta}(0)) = W(\gamma, p) + W(\eta, p).$$

(iv) follows from Corollary 2.18 since a lift of γ and a lift of η with the same initial points have the same end points.

(v) First suppose $H : [0, 1] \times [0, 1] \to \mathbb{C} \smallsetminus \{p\}$ is a free homotopy between γ and η, as in Definition 2.10. Use Theorem 2.16(ii) to lift H to a map $\tilde{H} : [0, 1] \times [0, 1] \to \mathbb{C}$. Then, for every $s \in [0, 1]$,

$$\exp(\tilde{H}(0, s)) + p = H(0, s) = H(1, s) = \exp(\tilde{H}(1, s)) + p.$$

This shows that the continuous map

$$s \mapsto n(s) = \frac{1}{2\pi i}(\tilde{H}(1, s) - \tilde{H}(0, s))$$

is integer-valued, therefore constant. Hence, $W(\gamma, p) = n(0) = n(1) = W(\eta, p)$.

Conversely, suppose γ and η are closed curves with $W(\gamma, p) = W(\eta, p)$. Take arbitrary lifts $\tilde{\gamma}$ and $\tilde{\eta}$, so $\tilde{\gamma}(1) - \tilde{\gamma}(0) = \tilde{\eta}(1) - \tilde{\eta}(0)$. Consider the family of translated curves $\tilde{\gamma}_s = \tilde{\gamma} + s(\tilde{\eta}(0) - \tilde{\gamma}(0))$ for $s \in [0, 1]$. The corresponding family $\gamma_s = \exp(\tilde{\gamma}_s) + p$ defines a free homotopy in $\mathbb{C} \smallsetminus \{p\}$ between the closed curves $\gamma = \gamma_0$ and γ_1. The proof will be complete once we check that γ_1 and η are (freely) homotopic in $\mathbb{C} \smallsetminus \{p\}$. But this is trivial since $\tilde{\gamma}_1$ and $\tilde{\eta}$, having the same initial and end points, are homotpic in $\mathbb{C}$, and postcomposing such homotopy with $\exp + p$ will give a homotopy between γ_1 and η.

(vi) Let U be a component of $\mathbb{C} \smallsetminus |\gamma|$. Take $z_0 \in U$ and a disk $D \subset U$ centered at z_0. Let $z \in D$ and let $\eta : [0, 1] \to \mathbb{C} \smallsetminus \{z_0\}$ be the closed curve defined by $\eta(t) = \gamma(t) + z_0 - z$. Then γ and η are freely homotopic in $\mathbb{C} \smallsetminus \{z_0\}$ by the map $H(t, s) = \gamma(t) + s(z_0 - z)$. Applying (i) and (v), we obtain

$$W(\gamma, z) = W(\eta, z_0) = W(\gamma, z_0).$$

Thus, $z \mapsto W(\gamma, z)$ is constant in D. Since z_0 was arbitrary, it follows that this function is locally constant in U. Connectedness of U now shows that it must be constant throughout U.

Finally, let U be the unbounded component of $\mathbb{C} \smallsetminus |\gamma|$ and choose $r > 0$ large enough that $|z| \geq r$ implies $z \in U$. Then $|\gamma| \subset \mathbb{D}(0, r) \subset \mathbb{C} \smallsetminus \{r\}$, so there is a radial free homotopy $H(t, s) = s \gamma(t)$ in $\mathbb{C} \smallsetminus \{r\}$ between the constant curve $\varepsilon_0 = H(\cdot, 0)$ and $\gamma = H(\cdot, 1)$. By (v), we have $W(\gamma, r) = 0$. Hence $z \mapsto W(\gamma, z)$ vanishes identically in U. $\qquad \square$

EXAMPLE 2.27. The oriented circle $\gamma : [0, 1] \to \mathbb{C} \smallsetminus \{p\}$ defined by $\gamma(t) = p + re^{2\pi i t}$ lifts under $z \mapsto \exp(z) + p$ to the curve $\tilde{\gamma} : [0, 1] \to \mathbb{C}$ given by $\tilde{\gamma}(t) = \log r + 2\pi i t$, so $W(\gamma, p) = 1$. By Theorem 2.26(vi),

$$W(\gamma, z) = \begin{cases} 1 & \text{if } |z - p| < r \\ 0 & \text{if } |z - p| > r. \end{cases}$$

See Theorem 2.28 for a generalization.

It will be useful to have an efficient method for computing the winding number of a closed curve γ on each component of $\mathbb{C} \smallsetminus |\gamma|$. This is provided by a principle which can be informally stated as follows:

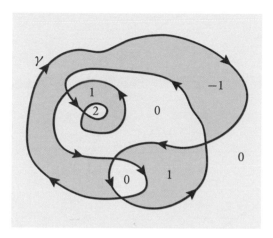

Figure 2.4. Illustration of the jump principle for the winding number. Each integer indicates the value of $W(\gamma, \cdot)$ in the corresponding connected component of $\mathbb{C} \setminus |\gamma|$.

The jump principle for the winding number: *Each time γ is crossed from right to left, the winding number increases by 1.*

In practice, we can always start with the unbounded component of $\mathbb{C} \setminus |\gamma|$ in which the winding number is zero, and determine $W(\gamma, \cdot)$ in each "adjacent" component using the jump principle, and continue until all the complementary components are exhausted. An example is shown in Fig. 2.4.

Before we give a precise formulation, recall that a closed curve $\gamma : [0, 1] \to \mathbb{C}$ is a **Jordan curve** (or a **simple closed curve**) if it is injective on $[0, 1)$. Alternatively, if γ induces a homeomorphism of the circle into the complex plane. According to the **Jordan curve theorem**, in this case $\mathbb{C} \setminus |\gamma|$ has precisely two connected components: a bounded component called the **interior** of γ and an unbounded component called the **exterior** of γ, denoted by $\mathrm{int}(\gamma)$ and $\mathrm{ext}(\gamma)$, respectively. Moreover, $|\gamma| = \partial \, \mathrm{int}(\gamma) = \partial \, \mathrm{ext}(\gamma)$. The proof can be found in many books, such as [**Fu**] or [**W**].

Now let us formulate the jump principle rigorously. Suppose $\gamma : [0, 1] \to \mathbb{C}$ is a closed curve and D is an open disk for which there exist $0 \le a < b < 1$ such that

- $\gamma(t) \in D$ if and only if $a < t < b$,
- $\gamma(t) \in \partial D$ if and only if $t = a$ and $t = b$,
- γ is injective on $[a, b]$.

If we consider the closed disk $\overline{D} = D \cup \partial D$ and collapse the boundary ∂D to a point, we obtain a topological 2-sphere in which the image $\gamma([a, b])$ is homeomorphic to a circle and therefore separates the 2-sphere into two connected components by the Jordan curve theorem. It follows that $\gamma([a, b])$ separates D into two connected components which we label as follows: As we go counterclockwise around ∂D, the closure of the "right" component R meets ∂D along the arc from $\gamma(a)$ to $\gamma(b)$ and the closure of the "left" component L meets ∂D along the arc from $\gamma(b)$ to $\gamma(a)$ (see Fig. 2.5). The jump

Jordan's 1887 proof of this theorem was incomplete, but the fact that he recognized a proof was needed deserves much credit.

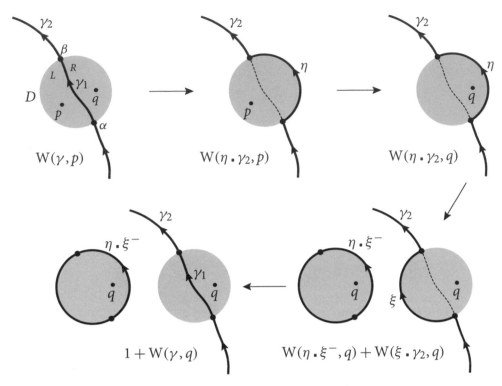

Figure 2.5. Proof of the jump principle.

principle asserts that

$$(2.4) \qquad \mathrm{W}(\gamma, p) = 1 + \mathrm{W}(\gamma, q) \qquad \text{whenever } p \in L \text{ and } q \in R.$$

To see this, choose for simplicity a parametrization for γ in which $a = 0, b = 1/2$ (this will not affect the winding numbers). Set $\alpha = \gamma(0), \beta = \gamma(1/2)$ and consider the following curves $[0, 1] \to \mathbb{C}$, as illustrated in Fig. 2.5:

- $\gamma_1(t) = \gamma(t/2)$, parametrizing the part of $|\gamma|$ in $\overline{D}$ from α to β.
- $\gamma_2(t) = \gamma((t+1)/2)$, parametrizing the part of $|\gamma|$ outside D from β back to α.

Thus, $\gamma = \gamma_1 \cdot \gamma_2$.

- η, parametrizing the counterclockwise arc on ∂D from α to β.
- ξ, parametrizing the clockwise arc on ∂D from α to β.

Fix $p \in L$ and $q \in R$. The product $\gamma_1 \cdot \eta^-$ is a Jordan curve with $L \subset \mathrm{ext}(\gamma_1 \cdot \eta^-)$. Hence $\mathrm{W}(\gamma_1 \cdot \eta^-, p) = 0$ by Theorem 2.26(vi). It follows from Theorem 2.26(ii) and (iii) that

$$(2.5) \qquad \mathrm{W}(\gamma, p) = \mathrm{W}(\gamma_1 \cdot \gamma_2, p) = \mathrm{W}(\gamma_1 \cdot \eta^-, p) + \mathrm{W}(\eta \cdot \gamma_2, p) = \mathrm{W}(\eta \cdot \gamma_2, p).$$

The image of the closed curve $\eta \cdot \gamma_2$ is disjoint from the connected set D which has both p and q in it. By Theorem 2.26(vi),

(2.6) $W(\eta \cdot \gamma_2, p) = W(\eta \cdot \gamma_2, q).$

The closed curve $\eta \cdot \xi^-$ parametrizes the oriented boundary ∂D, so $W(\eta \cdot \xi^-, q) = 1$ (compare Example 2.27). It follows that

(2.7) $W(\eta \cdot \gamma_2, q) = W(\eta \cdot \xi^-, q) + W(\xi \cdot \gamma_2, q) = 1 + W(\xi \cdot \gamma_2, q).$

Finally, $\xi \cdot \gamma_1^-$ is a Jordan curve with $R \subset \text{ext}(\xi \cdot \gamma_1^-)$, so $W(\xi \cdot \gamma_1^-, q) = 0$. This gives

(2.8) $W(\xi \cdot \gamma_2, q) = W(\xi \cdot \gamma_1^-, q) + W(\gamma_1 \cdot \gamma_2, q) = W(\gamma, q).$

Now (2.4) follows by combining the equalities (2.5)–(2.8).

The following theorem serves to illustrate an application of the jump principle:

THEOREM 2.28. *For any Jordan curve* $\gamma : [0, 1] \to \mathbb{C}$,

$$W(\gamma, p) = \begin{cases} \pm 1 & \text{if } p \in \text{int}(\gamma) \\ 0 & \text{if } p \in \text{ext}(\gamma) \end{cases}$$

We say that γ is **positively** or **negatively oriented** according as $W(\gamma, p) = 1$ or -1 for every $p \in \text{int}(\gamma)$.

PROOF. Since $W(\gamma, \cdot)$ vanishes in $\text{ext}(\gamma)$ by Theorem 2.26(vi), we need to look at $\text{int}(\gamma)$ only. First consider the special case where γ is affine on some interval $[a, b] \subset [0, 1]$ so the image $|\gamma|$ contains the line segment $[\gamma(a), \gamma(b)] \subset \mathbb{C}$. Let D be a disk centered at the midpoint of $[\gamma(a), \gamma(b)]$. If the radius of D is sufficiently small, $D \setminus |\gamma|$ has two connected components R, L and the jump principle (2.4) applies. Since by the Jordan curve theorem each point of $|\gamma|$ belongs to the boundary of both $\text{int}(\gamma)$ and $\text{ext}(\gamma)$, exactly one of R and L must be contained in $\text{ext}(\gamma)$ where the winding number vanishes. In the first case, $W(\gamma, \cdot) = 1$ in L and $L \subset \text{int}(\gamma)$, so we have $W(\gamma, \cdot) = 1$ in $\text{int}(\gamma)$. In the second case, $W(\gamma, \cdot) = -1$ in R and $R \subset \text{int}(\gamma)$, so we have $W(\gamma, \cdot) = -1$ in $\text{int}(\gamma)$.

Now let γ be an arbitrary Jordan curve. The idea is to find a homotopy between γ and a Jordan curve of the special type considered above and reduce to the previous case. Let $q = \gamma(1/2)$, and for $r > 0$ define a_r and b_r to be the infimum and supremum of the set $\{t \in [0, 1] : \gamma(t) \in \mathbb{D}(q, r)\}$, respectively. Then $0 \le a_r < 1/2 < b_r \le 1$. Moreover, $a_r, b_r \to 1/2$ as $r \to 0$ because γ is injective on $[0, 1)$. Take $0 < \varepsilon < |\gamma(0) - q|$ and find an $0 < r < \varepsilon$ such that $|\gamma(t) - q| < \varepsilon$ for all $t \in [a_r, b_r]$. Fix this r and, to simplify the notation, let $a = a_r$ and $b = b_r$. Let η be the curve obtained from γ by replacing the arc $\gamma([a, b])$ with the straight line segment $[\gamma(a), \gamma(b)]$. Explicitly, define $\eta : [0, 1] \to \mathbb{C}$ by

$$\eta(t) = \begin{cases} \left(\dfrac{b-t}{b-a}\right) \gamma(a) + \left(\dfrac{t-a}{b-a}\right) \gamma(b) & t \in [a, b] \\ \gamma(t) & t \in [0, a] \cup [b, 1]. \end{cases}$$

Marie Ennemond Camille Jordan (1838–1922)

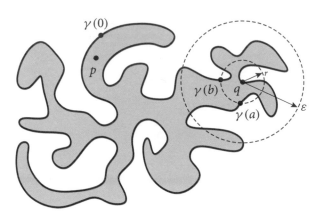

Figure 2.6. Proof of Theorem 2.28.

The definition of a and b shows that η is a Jordan curve. Since η is affine on the interval $[a, b]$, it follows from the special case treated above that $\mathrm{W}(\eta, \cdot) = \pm 1$ in $\mathrm{int}(\eta)$. Note that $\eta(0) = \gamma(0)$ is outside the closed disk $\overline{\mathbb{D}}(q, \varepsilon)$. Since $\eta(0) \in |\eta| = \partial\,\mathrm{int}(\eta)$, we can choose a point p, close to $\eta(0)$, such that $p \in \mathrm{int}(\eta) \smallsetminus \overline{\mathbb{D}}(q, \varepsilon)$ (see Fig. 2.6). Consider the homotopy $H : [0, 1] \times [0, 1] \to \mathbb{C}$ between γ and η defined by

$$H(t, s) = \begin{cases} (1 - s)\gamma(t) + s\,\eta(t) & t \in [a, b] \\ \gamma(t) & t \in [0, a] \cup [b, 1]. \end{cases}$$

For $(t, s) \in [a, b] \times [0, 1]$,

$$|H(t, s) - q| \le (1 - s)|\gamma(t) - q| + s|\eta(t) - q| \le (1 - s)\varepsilon + sr < \varepsilon.$$

This shows that the homotopy H takes values in $\mathbb{C} \smallsetminus \{p\}$. By Theorem 2.26(iv),

$$\mathrm{W}(\gamma, p) = \mathrm{W}(\eta, p) = \pm 1.$$

Since $\mathrm{W}(\gamma, \cdot)$ vanishes in $\mathrm{ext}(\gamma)$, we must have $p \in \mathrm{int}(\gamma)$. It follows that $\mathrm{W}(\gamma, \cdot)$ is identically 1 or -1 in $\mathrm{int}(\gamma)$. □

Complex integration provides an alternative approach to the topological notion of winding number when the curve is piecewise C^1. It is here that analyticity of the exponential map becomes essential for the first time.

THEOREM 2.29 (Analytic description of the winding number). *Let $\gamma : [0, 1] \to \mathbb{C}$ be a piecewise C^1 curve and $p \in \mathbb{C} \smallsetminus |\gamma|$. Then*

$$(2.9) \qquad \mathrm{W}(\gamma, p) = \frac{1}{2\pi i} \int_\gamma \frac{dz}{z - p}.$$

PROOF. Lift γ under $z \mapsto \exp(z) + p$ to a curve $\tilde{\gamma} : [0, 1] \to \mathbb{C}$. Then $\tilde{\gamma}$ is piecewise C^1 since near each parameter t it is given by the composition of γ and a local inverse of exp $+p$. Differentiating $\gamma(t) = \exp(\tilde{\gamma}(t)) + p$, we obtain

$$\gamma'(t) = \exp(\tilde{\gamma}(t))\, \tilde{\gamma}'(t) = (\gamma(t) - p)\, \tilde{\gamma}'(t)$$

for all but finitely many $t \in [0, 1]$. Hence

$$\frac{1}{2\pi i} \int_\gamma \frac{dz}{z-p} = \frac{1}{2\pi i} \int_0^1 \frac{\gamma'(t)}{\gamma(t) - p}\, dt = \frac{1}{2\pi i} \int_0^1 \tilde{\gamma}'(t)\, dt = \frac{1}{2\pi i}\big(\tilde{\gamma}(1) - \tilde{\gamma}(0)\big),$$

which is the definition of $\mathrm{W}(\gamma, p)$.   □

2.4 Cycles and homology

A subtlety in the definition of the fundamental group is the requirement that a base point should be kept fixed during deformation. This enriches homotopy theory but complicates it at the same time by introducing non-commutativity. On the other hand, integration along curves is inherently a commutative operation. For example, if γ and η are piecewise C^1 closed curves in an open set $U \subset \mathbb{C}$ with the initial and end point p, then

$$\int_{\gamma \cdot \eta} f(z)\, dz = \int_\gamma f(z)\, dz + \int_\eta f(z)\, dz = \int_{\eta \cdot \gamma} f(z)\, dz$$

for any continuous function $f : U \to \mathbb{C}$, whether or not the homotopy classes $[\gamma]$ and $[\eta]$ commute in $\pi_1(U, p)$. This suggests that as far as integration is concerned, it would be useful, and easier, to work with an abelianized version of the fundamental group. This leads to the idea of cycles and homology which we now discuss.

For any set S, there is a simple algebraic way of constructing an abelian group whose elements are finite linear combinations of elements of S with integer coefficients. To see this, consider those functions $\varphi : S \to \mathbb{Z}$ with the property that $\varphi(x) = 0$ for all but finitely many $x \in S$. The collection of all such φ is an abelian group $\mathcal{G}(S)$ under addition of functions which is called the *free abelian group generated by* S. Let $\varphi_x \in \mathcal{G}(S)$ denote the function that takes the value 1 at $x \in S$ and 0 everywhere else. If $\varphi \in \mathcal{G}(S)$ vanishes outside $\{x_1, \ldots, x_m\}$ and if $\varphi(x_k) = n_k$, then

(2.10)
$$\varphi = n_1\, \varphi_{x_1} + \cdots + n_m\, \varphi_{x_m}.$$

The coefficients n_k in this expression can take any integer value, but we naturally suppress a term if the corresponding coefficient is zero. Adopting this convention, it is easy to see that every $\varphi \in \mathcal{G}(S)$ has a *unique* representation of the form (2.10). It is customary to identify φ_x with the point x itself and simplify the notation to the formal sum

$$\varphi = n_1\, x_1 + \cdots + n_m\, x_m.$$

In this notation, the zero element of $\mathcal{G}(S)$ corresponds to the empty sum (or the one in which all the coefficients n_k are zero). The additive inverse of the above φ is given by $-\varphi = -n_1\, x_1 - \cdots - n_m\, x_m$.

> **DEFINITION 2.30.** Let $U \subset \mathbb{C}$ be non-empty and open. A **chain** in U is an element of the free abelian group generated by the set of all curves in U. In other words, a chain is a formal sum
>
> $$\gamma = n_1\, \gamma_1 + \cdots + n_m\, \gamma_m,$$
>
> where each $\gamma_k : [0,1] \to U$ is a curve and each n_k is an integer which we think of as the **multiplicity** of γ_k. The **image** of γ is defined as the union
>
> $$|\gamma| = |\gamma_1| \cup \cdots \cup |\gamma_m|.$$
>
> The chain γ is called a **cycle** if for every $p \in U$,
>
> $$\sum_{\{k\,:\,\gamma_k(0)=p\}} n_k = \sum_{\{k\,:\,\gamma_k(1)=p\}} n_k.$$
>
> In other words, if each point occurs as an initial point the same number of times as it occurs as an end point, both counting multiplicities.

The defining condition of a cycle is reminiscent of Kirchhoff's law for electric currents: What goes in a node must come out. Here the "current" flowing in each curve is its multiplicity.

Alternatively, consider the **boundary map** ∂ defined by

$$\partial\left(\sum_{k=1}^{m} n_k\, \gamma_k\right) = \sum_{k=1}^{m} n_k\, \gamma_k(1) - \sum_{k=1}^{m} n_k\, \gamma_k(0),$$

where the formal sum on the right is an element of the free abelian group generated by all points in U. Then γ is a cycle if and only if $\partial\gamma = 0$. Since it is easy to see that ∂ is a group homomorphism, it follows that the collection of cycles, the kernel of ∂, is an additive group.

EXAMPLE 2.31. A chain $\gamma = \sum_{k=1}^{m} n_k\, \gamma_k$ in which every γ_k is a closed curve is a cycle. However, γ may well be a cycle without the γ_k being closed. For example, whenever the curves γ_1, γ_2 have the same initial and end points, the chain $\gamma_1 - \gamma_2$ is a cycle.

Although the curves that appear in the formal expression of a cycle need not be closed, they can be suitably grouped such that the product of the curves in each group forms a closed curve:

LEMMA 2.32 (Cycles decompose into closed curves). *Let $\gamma = \sum_{k=1}^{m} n_k\, \gamma_k$ be a cycle and $\mathcal{C}$ be the finite collection of curves in which γ_k appears n_k times when $n_k > 0$ and the reverse curve γ_k^- appears $-n_k$ times when $n_k < 0$. Then $\mathcal{C}$ can be partitioned into*

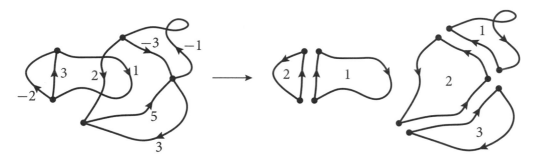

Figure 2.7. Left: The formal sum of these eight curves, each with the multiplicity shown, defines a cycle. Right: Decomposing this cycle into five closed curves with multiplicities, as suggested by Lemma 2.32.

subsets $C_1, \ldots, C_N$ such that the product of the curves in each C_k, when put in appropriate order, is defined and forms a closed curve.

Fig. 2.7 illustrates this lemma.

PROOF. Denote by $i(\xi)$ and $e(\xi)$ the initial and end point of $\xi \in C$. Note that since γ is a cycle, we have

$$(2.11) \qquad \#\{\xi \in C : i(\xi) = p\} = \#\{\xi \in C : e(\xi) = p\}$$

for every $p \in U$.

Start with any $\xi_1 \in C$. If $e(\xi_1) = i(\xi_1)$, stop. Otherwise by (2.11) there must be some $\xi_2 \in C$ with $e(\xi_1) = i(\xi_2)$. If $e(\xi_2) = i(\xi_1)$, stop. Otherwise by (2.11) there must be some $\xi_3 \in C$ with $e(\xi_2) = i(\xi_3)$. Since C is finite, this process must eventually stop, i.e., we must have $e(\xi_n) = i(\xi_1)$ for some n. Set $C_1 = \{\xi_1, \ldots, \xi_n\}$ and note that the product $\eta_1 = \xi_1 \cdot \ldots \cdot \xi_n$ is a closed curve.

If $C_1 = C$, we are done. Otherwise, we repeat the same argument for the collection $C \setminus C_1$, which also satisfies (2.11), to find a collection C_2 whose elements join together to form a closed curve η_2. Continuing this process finitely many times, we eventually exhaust all elements of C and obtain the desired decomposition $\{C_1, \ldots, C_N\}$ and closed curves $\{\eta_1, \ldots, \eta_N\}$. □

The notion of the winding number introduced in the previous section extends to chains and cycles by linearity: If $\gamma = \sum_{k=1}^m n_k \gamma_k$ is a chain and $p \in \mathbb{C} \setminus |\gamma|$, define

$$(2.12) \qquad W(\gamma, p) = \sum_{k=1}^m n_k \, W(\gamma_k, p).$$

THEOREM 2.33 (Winding numbers of chains and cycles).

(i) For any chain γ,

$$W(-\gamma, p) = - W(\gamma, p) \qquad if \, p \in \mathbb{C} \setminus |\gamma|.$$

(ii) For any two chains γ and η,

$$W(\gamma + \eta, p) = W(\gamma, p) + W(\eta, p) \qquad if\ p \in \mathbb{C} \smallsetminus (|\gamma| \cup |\eta|).$$

(iii) If γ is a cycle, the function $W(\gamma, \cdot)$ is integer-valued and constant on each connected component of $\mathbb{C} \smallsetminus |\gamma|$, and it vanishes on the unbounded connected component of $\mathbb{C} \smallsetminus |\gamma|$.

PROOF. Only (iii) requires justification. Let $\gamma = \sum_{k=1}^{m} n_k\,\gamma_k$ and consider the collection $\mathcal{C}$ and the decomposition of γ into (not necessarily distinct) closed curves $\eta_1, \ldots, \eta_N$ as in Lemma 2.32. Then, for $p \in \mathbb{C} \smallsetminus |\gamma|$,

$$W(\gamma, p) = \sum_{k=1}^{m} n_k\, W(\gamma_k, p) = \sum_{\xi \in \mathcal{C}} W(\xi, p) = \sum_{k=1}^{N} W(\eta_k, p).$$

The second equality uses Theorem 2.26(ii), while the third equality follows from Theorem 2.26(iii). Since each η_k is a closed curve, each term in the last sum is an integer, so $W(\gamma, p)$ is also an integer.

Every component U of $\mathbb{C} \smallsetminus |\gamma|$ is contained in a unique component of $\mathbb{C} \smallsetminus |\eta_k|$ for $1 \le k \le N$. By Theorem 2.26(vi), $W(\eta_k, \cdot)$ is constant in U, so the same must be true of $W(\gamma, \cdot)$. Similarly, the unbounded component U of $\mathbb{C} \smallsetminus |\gamma|$ is contained in the unbounded component of $\mathbb{C} \smallsetminus |\eta_k|$ for $1 \le k \le N$. By Theorem 2.26(vi), $W(\eta_k, \cdot)$ vanishes in U, so $W(\gamma, \cdot)$ too must vanish in U. $\square$

According to Ahlfors, "it was Emil Artin who discovered that the characterization of homology by vanishing winding numbers ties in precisely with what is needed for the general version of Cauchy's theorem" [A2].

> **DEFINITION 2.34.** Let $U \subset \mathbb{C}$ be non-empty and open. We say that a cycle γ in U is **null-homologous** in U, and write $\gamma \sim 0$, if
>
> $$W(\gamma, p) = 0 \qquad \text{for every } p \in \mathbb{C} \smallsetminus U.$$
>
> More generally, we say that two cycles γ, η in U are **homologous** in U, and write $\gamma \sim \eta$, if $\gamma - \eta \sim 0$; in other words, if
>
> $$W(\gamma, p) = W(\eta, p) \qquad \text{for every } p \in \mathbb{C} \smallsetminus U.$$

EXAMPLE 2.35. Let γ be a Jordan curve in a domain $U \subset \mathbb{C}$. Then $\gamma \sim 0$ in U if and only if $\mathrm{int}(\gamma)$ is contained in U. Intuitively, this means γ does not enclose any "hole" of U.

It is easy to check that $\sim$ is an equivalence relation on the additive group of all cycles in U. The equivalence class of a cycle γ under $\sim$ is called the **homology class** of γ and is denoted by $\langle \gamma \rangle$. Since $\gamma \sim \hat{\gamma}$ and $\eta \sim \hat{\eta}$ imply $\gamma + \eta \sim \hat{\gamma} + \hat{\eta}$, the addition

$$(2.13) \qquad\qquad \langle \gamma \rangle + \langle \eta \rangle = \langle \gamma + \eta \rangle$$

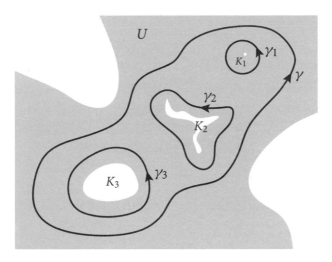

Figure 2.8. Illustration of Example 2.36. Here the complement of U has three bounded components K_1, K_2, K_3, and $\gamma \sim \gamma_1 + \gamma_2 + \gamma_3$.

between homology classes is well defined. Thus, the set of all homology classes of cycles in U has the structure of an abelian group which is called the ***first homology group*** of U and is denoted by $H_1(U)$. Alternatively, $H_1(U)$ can be identified with the quotient of the group of all cycles in U by the subgroup of all null-homologous cycles.

Emil Artin (1898–1962)

EXAMPLE 2.36. Let $U \subset \mathbb{C}$ be a domain whose complement $\mathbb{C} \smallsetminus U$ has finitely many bounded connected components $K_1, \ldots, K_m$. Thus, the K_j are disjoint compact connected subsets of $\mathbb{C}$. For $1 \leq j \leq m$, let γ_j be a cycle in U with the property that $\mathrm{W}(\gamma_j, p) = 1$ if $p \in K_j$ and $\mathrm{W}(\gamma_j, p) = 0$ if $p \in K_n$ for some $n \neq j$. Intuitively, γ_j winds around K_j once in the counterclockwise direction without winding around other complementary components (see Fig. 2.8). Suppose also that γ is a cycle in U such that $\mathrm{W}(\gamma, p) = 1$ whenever $p \in \bigcup_{j=1}^{m} K_j$. Then

$$\gamma \sim \gamma_1 + \cdots + \gamma_m \qquad \text{in } U.$$

In fact, if $p \in K_j$ for some j, then $\mathrm{W}(\gamma, p) = \mathrm{W}(\gamma_1 + \cdots + \gamma_m, p) = 1$. On the other hand, if p belongs to an unbounded component E of $\mathbb{C} \smallsetminus U$, then $\mathrm{W}(\gamma, p) = \mathrm{W}(\gamma_1 + \cdots + \gamma_m, p) = 0$ since E is contained in the unbounded component of $\mathbb{C} \smallsetminus |\gamma|$ and every $\mathbb{C} \smallsetminus |\gamma_j|$.

One can show that such cycles γ_j always exist and the homology group $H_1(U)$ is the free abelian group generated by $\langle \gamma_j \rangle$ for $1 \leq j \leq m$ (see Theorem 9.31).

The following statements are easy consequences of Definition 2.34:

COROLLARY 2.37.

(i) *Suppose the product $\gamma_1 \cdot \gamma_2 \cdot \ldots \cdot \gamma_n$ of curves in U defines a closed curve. Then, as cycles,*

$$\gamma_1 \cdot \gamma_2 \cdot \ldots \cdot \gamma_n \sim \gamma_1 + \gamma_2 + \cdots + \gamma_n \qquad \text{in } U.$$

(ii) *Every cycle in U is homologous to a finite sum of closed curves in U.*

PROOF. (i) simply follows from Theorem 2.26 since for every $p \in \mathbb{C} \smallsetminus U$,

$$W(\gamma_1 \cdot \gamma_2 \cdot \ldots \cdot \gamma_n, p) = \sum_{k=1}^{n} W(\gamma_k, p) = W\left(\sum_{k=1}^{n} \gamma_k, p\right).$$

For (ii), let γ be a cycle in U and decompose it into closed curves $\eta_1, \ldots, \eta_N$ in the sense of Lemma 2.32. Then, as in the proof of Theorem 2.33(iii), $W(\gamma, p) = \sum_{k=1}^{N} W(\eta_k, p)$ for every $p \in \mathbb{C} \smallsetminus |\gamma|$ and in particular for every $\mathbb{C} \smallsetminus U$. It follows that $\gamma \sim \sum_{k=1}^{N} \eta_k$ in U.   □

THEOREM 2.38 (Homotopic implies homologous). *Let γ, η be curves in an open set $U \subset \mathbb{C}$.*

(i) If γ, η are homotopic in U, then the cycle $\gamma - \eta$ is null-homologous in U.

(ii) If γ, η are freely homotopic closed curves in U, then γ, η are homologous cycles in U.

A trivial case of (ii) occurs when we reparametrize a closed curve: By Lemma 2.3(i), reparametrizing a closed curve results in a homotopic curve and hence a homologous cycle. The converse of Theorem 2.38 is false. For example, there are null-homologous closed curves in the twice punctured plane that are not null-homotopic (see problem 3).

PROOF. (i) For any $p \in \mathbb{C} \smallsetminus U$ the curves γ, η are homotopic in $\mathbb{C} \smallsetminus \{p\}$ since they are homotopic in U. Hence, by Theorem 2.26(iv),

$$W(\gamma - \eta, p) = W(\gamma, p) - W(\eta, p) = 0.$$

It follows that $\gamma - \eta \sim 0$ in U.

(ii) Since γ, η are freely homotopic in U, they are freely homotopic in $\mathbb{C} \smallsetminus \{p\}$ for every choice of $p \in \mathbb{C} \smallsetminus U$. By Theorem 2.26(v), $W(\gamma, p) = W(\eta, p)$, which proves $\gamma \sim \eta$ in U.   □

REMARK 2.39. Let $U \subset \mathbb{C}$ be a domain and fix a base point $p \in U$. Theorem 2.38 shows that the map $\varphi : \pi_1(U, p) \to H_1(U)$ which sends each homotopy class $[\gamma]$ to the homology class $\langle \gamma \rangle$ is well defined. It follows from Corollary 2.37(i) that φ is a group homomorphism. Since $H_1(U)$ is an abelian group, φ must vanish on the commutator subgroup $C \subset \pi_1(U, p)$ generated by all elements of the form $[\gamma] \cdot [\eta] \cdot [\gamma]^{-1} \cdot [\eta]^{-1}$. This gives an induced homomorphism $\Phi : \pi_1(U, p)/C \to H_1(U)$. A classical theorem of Poincaré and Hurewicz asserts that Φ is indeed an isomorphism. For a proof, see for example [**Fu**] or [**H**].

COROLLARY 2.40. *Every cycle in a simply connected domain $U \subset \mathbb{C}$ is null-homologous, so $H_1(U) = 0$.*

The converse statement is also true, that is, $H_1(U) = 0$ for a domain U implies that U is simply connected (see Theorem 9.27).

PROOF. Let γ be a cycle in U and use Corollary 2.37(ii) to find closed curves $\eta_1, \ldots, \eta_N$ such that $\gamma \sim \sum_{k=1}^{N} \eta_k$. Since each η_k is null-homotopic in U, we have $\eta_k \sim 0$ by Theorem 2.38. It follows that $\gamma \sim 0$.   □

2.5 The homology version of Cauchy's theorem

As we remarked in chapter 1, Cauchy's theorem in a disk is just about the only ingredient one needs to establish the most basic properties of holomorphic functions. However, this local version is too confining for many practical purposes. With minimal effort, one can drop all the restrictions and prove a general version of Cauchy's theorem which applies to arbitrary open sets and deals with integration over cycles.

To state the theorem, we need to define what it means to integrate a function over a chain. Take a chain $\gamma = \sum_{k=1}^{m} n_k \gamma_k$ in the plane and suppose γ is piecewise C^1, which means each γ_k is a piecewise C^1 curve. If $f : |\gamma| \to \mathbb{C}$ is continuous, we define the integral of f along γ by

$$\int_\gamma f(z)\, dz = \sum_{k=1}^{m} n_k \int_{\gamma_k} f(z)\, dz.$$

Observe that the special case of $f(z) = 1/(z-p)$ with $p \in \mathbb{C} \setminus |\gamma|$ gives

$$(2.14) \qquad \frac{1}{2\pi i} \int_\gamma \frac{dz}{z-p} = \sum_{k=1}^{m} n_k \, \mathrm{W}(\gamma_k, p) = \mathrm{W}(\gamma, p),$$

where the first equality follows from Theorem 2.29 and the second from the definition (2.12). This generalizes formula (2.9) to all piecewise C^1 chains.

THEOREM 2.41 (Homology version of Cauchy's theorem). *Suppose γ is a piecewise C^1 null-homologous cycle in an open set $U \subset \mathbb{C}$. Then, for every $f \in \mathcal{O}(U)$,*

$$(2.15) \qquad \int_\gamma f(z)\, dz = 0$$

and

$$(2.16) \qquad f(z) \cdot \mathrm{W}(\gamma, z) = \frac{1}{2\pi i} \int_\gamma \frac{f(\zeta)}{\zeta - z}\, d\zeta \qquad \textit{if } z \in U \setminus |\gamma|.$$

Augustin-Louis Cauchy
(1789–1857)

The assumption $\gamma \sim 0$ in U is necessary in order for (2.15) and (2.16) to hold for every $f \in \mathcal{O}(U)$. This can be seen by applying (2.15) to the holomorphic function

$f(z) = 1/(z - p)$, where $p \in \mathbb{C} \setminus U$:

$$W(\gamma, p) = \frac{1}{2\pi i} \int_\gamma \frac{1}{z - p} \, dz = \frac{1}{2\pi i} \int_\gamma f(z) \, dz = 0.$$

The short elegant proof presented here is due to the group theorist J. Dixon [Dx].

PROOF. Assume (2.16) holds for every $f \in \mathcal{O}(U)$. Take any $p \in U \setminus |\gamma|$ and apply (2.16) to $F(z) = (z - p)f(z)$:

$$\int_\gamma f(z) \, dz = \int_\gamma \frac{F(z)}{z - p} \, dz = 2\pi i \, F(p) \cdot W(\gamma, p) = 0.$$

This proves (2.15). Thus, it suffices to verify (2.16) only.

By (2.14), the formula (2.16) is equivalent to

(2.17)
$$\int_\gamma \frac{f(\zeta) - f(z)}{\zeta - z} \, d\zeta = 0 \qquad \text{for } z \in U \setminus |\gamma|.$$

Consider the function $g : U \times U \to \mathbb{C}$ defined by

$$g(\zeta, z) = \begin{cases} \dfrac{f(\zeta) - f(z)}{\zeta - z} & \zeta \neq z \\ f'(z) & \zeta = z, \end{cases}$$

which is continuous in $U \times U$ by Lemma 1.56. For each fixed $\zeta \in U$, the function $z \mapsto g(\zeta, z)$ is holomorphic in U since it is trivially holomorphic in $U \setminus \{\zeta\}$ and it is continuous at $z = \zeta$ (compare Example 1.40). Theorem 1.46 now shows that $G : U \to \mathbb{C}$ defined by

$$G(z) = \frac{1}{2\pi i} \int_\gamma g(\zeta, z) \, d\zeta$$

is holomorphic in U. Equation (2.17) and therefore formula (2.16) will follow once we prove that G is identically zero in U.

Let V be the open set consisting of all points $z \in \mathbb{C} \setminus |\gamma|$ for which $W(\gamma, z) = 0$. Since γ is null-homologous in U, we see that $\mathbb{C} \setminus U \subset V$, or $U \cup V = \mathbb{C}$. Define

$$\tilde{G}(z) = \frac{1}{2\pi i} \int_\gamma \frac{f(\zeta)}{\zeta - z} \, d\zeta \qquad \text{for } z \in V.$$

By Theorem 1.46 (or Corollary 1.48), $\tilde{G} \in \mathcal{O}(V)$. If $z \in U \cap V$, then

$$G(z) = \frac{1}{2\pi i} \int_\gamma \frac{f(\zeta)}{\zeta - z} \, d\zeta - f(z) \cdot W(\gamma, z) = \tilde{G}(z).$$

This shows that the function $\varphi : \mathbb{C} \to \mathbb{C}$ defined by

$$\varphi(z) = \begin{cases} G(z) & z \in U \\ \tilde{G}(z) & z \in V \end{cases}$$

is entire. Choose $\tilde{R} > 0$ so that $|\gamma| \subset \mathbb{D}(0, R)$. Theorem 2.33(iii) implies that $W(\gamma, z) = 0$ whenever $|z| > R$, so $\{z \in \mathbb{C} : |z| > R\} \subset V$. Since $\zeta \in |\gamma|$ and $|z| > 2R$ imply $|\zeta - z| > R$, the ML-inequality gives the bound

$$|\tilde{G}(z)| \leq \frac{1}{2\pi R} \cdot \sup_{\zeta \in |\gamma|} |f(\zeta)| \cdot \text{length}(\gamma)$$

whenever $|z| > 2R$. The definition of φ now shows that

$$\lim_{z \to \infty} \varphi(z) = \lim_{z \to \infty} \tilde{G}(z) = 0.$$

It follows from Liouville's Theorem 1.44 that φ is identically zero. In particular, $G = 0$ in U, as required.   □

Here are a few immediate corollaries:

COROLLARY 2.42. *If γ and η are piecewise C^1 homotopic curves or homologous cycles in U and $f \in \mathcal{O}(U)$, then*

$$\int_\gamma f(z)\, dz = \int_\eta f(z)\, dz.$$

PROOF. This simply follows from Theorem 2.41 applied to the cycle $\gamma - \eta$, which is null-homologous by Theorem 2.38.   □

EXAMPLE 2.43. Consider the domain U and the closed curves $\gamma, \gamma_1, \gamma_2, \gamma_3$ in Fig. 2.8. Since $\gamma \sim \gamma_1 + \gamma_2 + \gamma_3$ in U, for every $f \in \mathcal{O}(U)$ we have

$$\int_\gamma f(z)\, dz = \int_{\gamma_1} f(z)\, dz + \int_{\gamma_2} f(z)\, dz + \int_{\gamma_3} f(z)\, dz.$$

EXAMPLE 2.44. Let $U \subset \mathbb{C}$ be a bounded domain whose bounded complementary components consist of m points $p_1, \ldots, p_m$. Choose $r > 0$ small enough so the oriented circles $\gamma_k = \mathbb{T}(p_k, r)$ for $1 \leq k \leq m$ are all in U. Then, any closed curve γ in U is homologous to the cycle $\sum_{k=1}^m n_k \gamma_k$, where $n_k = W(\gamma, p_k)$. Hence, for every $f \in \mathcal{O}(U)$,

$$\int_\gamma f(z)\, dz = \sum_{k=1}^m n_k \int_{\gamma_k} f(z)\, dz$$

(compare Fig. 2.9).

COROLLARY 2.45. *Let $U \subset \mathbb{C}$ be a simply connected domain and $f \in \mathcal{O}(U)$. Then Cauchy's theorem (2.15) and the integral representation formula (2.16) hold for every cycle γ in U.*

PROOF. Every cycle in U is null-homologous by Corollary 2.40.   □

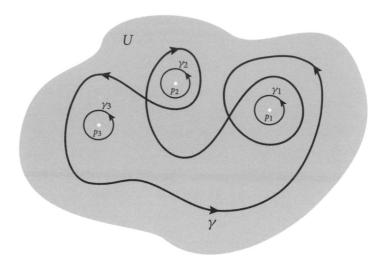

Figure 2.9. Illustration of Example 2.44 for $m = 3$. Here $\gamma \sim 2\gamma_1 - \gamma_2 + \gamma_3$.

Combining the above corollary with Theorem 1.29, we obtain the following important generalization of Theorem 1.33:

THEOREM 2.46. *Every holomorphic function in a simply connected domain has a primitive.*

REMARK 2.47. This theorem can be used to find an alternative proof for Corollary 2.22: If U is simply connected and $f \in \mathscr{O}(U)$ is non-vanishing, let g be a primitive of $f'/f \in \mathscr{O}(U)$. Then $(fe^{-g})' = 0$ in U, so $f = ce^{g}$ for a non-zero constant c. After adding a suitable constant to g, we can arrange $c = 1$ and $f = e^{g}$.

COROLLARY 2.48 (Cauchy's integral formula for higher derivatives). *Suppose $f \in \mathscr{O}(U)$ and γ is a piecewise C^1 null-homologous cycle in U. Then, for every integer $n \geq 1$,*

$$(2.18) \qquad f^{(n)}(z) \cdot \mathrm{W}(\gamma, z) = \frac{n!}{2\pi i} \int_{\gamma} \frac{f(\zeta)}{(\zeta - z)^{n+1}} \, d\zeta \qquad \text{if } z \in U \smallsetminus |\gamma|.$$

PROOF. This follows by taking n successive derivatives of each side of Cauchy's integral formula (2.16). The left side becomes $f^{(n)}(z) \cdot \mathrm{W}(\gamma, z)$ since $\mathrm{W}(\gamma, z)$ is a locally constant function of z in $U \smallsetminus |\gamma|$. On the right side, one can appeal to Corollary 1.48 to differentiate under the integral sign, which would give the right side of (2.18). □

We conclude by showing how to define the integral of a *holomorphic* function along an arbitrary curve whether or not it is piecewise C^1. Let us first show that every curve $\gamma : [0, 1] \to U$ has a piecewise C^1 curve $\eta : [0, 1] \to U$ homotopic to it. One simple way to construct such an η is as follows: Since $|\gamma|$ is a compact subset of U, we can find an $\varepsilon > 0$ such that $\mathbb{D}(p, \varepsilon) \subset U$ for every $p \in |\gamma|$. By uniform continuity of

γ on $[0, 1]$, there is an integer n such that $|\gamma(t) - \gamma(t')| < \varepsilon$ whenever $|t - t'| \leq 1/n$. Set $t_k = k/n$ for $0 \leq k \leq n$. By the choice of ε, each line segment $[\gamma(t_{k-1}), \gamma(t_k)]$ lies in U. Define $\eta : [0, 1] \to U$ by setting $\eta(t_k) = \gamma(t_k)$ and interpolating affinely on each interval $[t_{k-1}, t_k]$. Explicitly,

$$\eta(t) = (k - nt)\gamma(t_{k-1}) + (nt - k + 1)\gamma(t_k) \qquad \text{if } t \in [t_{k-1}, t_k].$$

The curve η is clearly piecewise C^1, and by the choice of ε the convex combination $H(t, s) = (1 - s)\gamma(t) + s\eta(t)$ is a homotopy in U between γ and η. (A more careful argument shows that η can be taken C^∞-smooth; see problem 5.)

Now suppose $f \in \mathcal{O}(U)$ and $\gamma : [0, 1] \to U$ is any curve. Define

$$\int_\gamma f(z)\, dz = \int_\eta f(z)\, dz,$$

where $\eta : [0, 1] \to U$ is any piecewise C^1 curve homotopic to γ in U. By Corollary 2.42, the right side is independent of the particular choice of such η. The definition extends to arbitrary cycles by linearity, namely, if $\gamma = \sum_{k=1}^m n_k \gamma_k$ is any cycle in U, we define

$$\int_\gamma f(z)\, dz = \sum_{k=1}^m n_k \int_{\gamma_k} f(z)\, dz.$$

This, in effect, amounts to replacing γ by a homologous cycle $\eta = \sum_{k=1}^m n_k \eta_k$, where each η_k is piecewise C^1 and homotopic to γ_k.

Thus, as far as integration of holomorphic functions is concerned, we can always assume that all paths of integration are piecewise C^1 by replacing an arbitrary curve (resp. cycle) with a piecewise C^1 curve (resp. cycle) in the same homotopy (resp. homology) class.

Problems

(1)   Let X be the subset of the plane shown in Fig. 2.10 (where the top part is an ever oscillating curve which accumulates on the vertical segment, much like the graph of $x \mapsto \sin(1/x)$). Consider the continuous map from X to $\mathbb{C}^*$ which is the identity on the union

$$[-1 + i, -1 - i] \cup [-1 - i, 1 - i] \cup [1 - i, 1 + i]$$

and projects the rest of X vertically on the segment $[-1 + i, 1 + i]$. Show that this map has no continuous lift to $\mathbb{C}$ under the exponential map, even though X is simply connected. Contrast this example with Theorem 2.21.

(2)   Suppose $H : [0, 1] \times [0, 1] \to X$ is a free homotopy between closed curves $\gamma_0 = H(\cdot, 0)$, $\gamma_1 = H(\cdot, 1)$, as in Definition 2.10. Define $\eta : [0, 1] \to X$ by $\eta(s) = H(0, s) = H(1, s)$. Show that γ_0 is homotopic to $\eta \cdot \gamma_1 \cdot \eta^-$. Conclude that if γ_0 is freely homotopic to a constant curve, then it is null-homotopic.

(3)   Show that the closed curves $\gamma \cdot \eta$ and $\eta \cdot \gamma$ of Example 2.11 are freely homotopic in $X = \mathbb{C} \setminus \{-1, 1\}$, but there is no homotopy between the two which fixes the base point 0. Conclude that the closed curve $\gamma \cdot \eta \cdot \gamma^- \cdot \eta^-$ is not null-homotopic in X even though it is

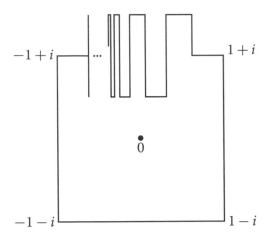

Figure 2.10. The simply connected set X of problem 1. There is a continuous map $X \to \mathbb{C}^*$ which does not lift under exp.

null-homologous. (Hint: Look at the unique lifts of $\gamma \cdot \eta$ and $\eta \cdot \gamma$ under $z \mapsto \exp(z) - 1$ whose initial points are 0. A homotopy in X between $\gamma \cdot \eta$ and $\eta \cdot \gamma$ would give rise to a homotopy in $\mathbb{C} \smallsetminus \exp^{-1}(2)$ between their lifts.)

(4) Can you find a domain $U \subset \mathbb{C}$ and two closed curves in U which are homologous but not freely homotopic in U?

(5) Prove that every curve in a planar domain is homotopic to a C^∞ curve. Use this to show that every cycle in a planar domain is homologous to a C^∞ closed curve. (Hint: Consider the piecewise affine approximation constructed at the end of §2.5. Round the corners.)

(6) Let Log denote the principal branch of the logarithm defined in Example 2.23. Prove the inequality

$$|\mathrm{Log}(1 + z)| \leq -\log(1 - |z|) \qquad \text{for } |z| < 1,$$

where the right side is the natural logarithm of the positive number $1 - |z|$.

(7) Show that the identity function has no holomorphic square root in $\mathbb{D}$, i.e., there is no $f \in \mathcal{O}(\mathbb{D})$ such that $f^2(z) = z$ for $|z| < 1$. By contrast, show that there is a unique $f \in \mathcal{O}(\mathbb{D}(1,1))$ which satisfies $f^2(z) = z$ and $f(1) = 1$. Writing $\sqrt{z}$ for this $f(z)$, verify the expansion

$$\sqrt{z} = 1 + \frac{1}{2}(z - 1) - \frac{1}{8}(z - 1)^2 + \frac{1}{16}(z - 1)^3 - \frac{5}{128}(z - 1)^4 + \cdots$$

whenever $|z - 1| < 1$.

(8) Let $U \subset \mathbb{C}$ be a simply connected domain. Suppose $f \in \mathcal{O}(U)$, $f^{-1}(0) = \{p_1, \ldots, p_n\}$, and $\mathrm{ord}(f, p_k)$ is an even integer for every $1 \leq k \leq n$. Show that f has a holomorphic square root in U.

(9) Suppose $U \subset \mathbb{C}$ is a domain and $f : U \to \mathbb{C}$ is continuous. If there exists a positive integer n such that $f^n \in \mathcal{O}(U)$, show that $f \in \mathcal{O}(U)$.

(10) (A different product on planar curves) For given curves $\gamma, \eta : [0, 1] \to \mathbb{C}^*$, define $\gamma\eta : [0, 1] \to \mathbb{C}^*$ by $(\gamma\eta)(t) = \gamma(t)\eta(t)$ (i.e., multiply pointwise as complex numbers). Show that

$$W(\gamma\eta,0)=W(\gamma,0)+W(\eta,0).$$

(11)   Is there a closed curve γ in the plane for which the winding number $p\mapsto W(\gamma,p)$ defined in $\mathbb{C}\smallsetminus|\gamma|$ takes on
    (i)   only the values $0,1,2,3$?
    (ii) only the values $0,1,3$?
    (iii) every integer value?

(12)   Suppose $g:\mathbb{T}\to\mathbb{C}$ is continuous. Show that

$$\int_{\mathbb{T}}\frac{g(\zeta)}{\zeta-z}\,d\zeta=0\qquad\text{for all }z\in\mathbb{D}$$

if and only if

$$\int_{\mathbb{T}}\zeta^{n}g(\zeta)\,d\zeta=0\qquad\text{for all integers }n\le-1.$$

(13)   Let $U\subsetneqq\mathbb{C}$ be a domain and $f\in\mathcal{O}(U)$. If $\iint_{U}|f(z)|^{\alpha}\,dx\,dy<+\infty$ for some $\alpha\ge1$, show that there is a constant $C>0$ such that

$$|f(z)|\le\frac{C}{\operatorname{dist}(z,\partial U)^{2/\alpha}}\qquad\text{for }z\in U.$$

Here $\operatorname{dist}(z,\partial U)=\inf_{\zeta\in\partial U}|z-\zeta|$. (Hint: Use Cauchy's integral formula on $\mathbb{T}(z,r)$ for every $0<r<\operatorname{dist}(z,\partial U)$ and integrate over r.)

(14)   Prove the complex form of Green's theorem: Let $U\subset\mathbb{C}$ be a bounded domain whose boundary ∂U is a finite union of piecewise C^{1} Jordan curves, each positively or negatively oriented depending whether or not its interior meets U (intuitively, moving along each boundary component, the domain U is seen on the left). Suppose f is a complex-valued function which is C^{1}-smooth in some neighborhood of $\overline{U}$. Then,

$$\int_{\partial U}f(z)\,dz=2i\iint_{U}f_{\bar{z}}(z)\,dx\,dy.$$

(Hint: Write the left side as a pair of classical line integrals and apply the real version of Green's theorem to each.)

(15)   Suppose γ is a piecewise C^{1} positively oriented Jordan curve in the plane. What quantity does the complex integral

$$\frac{1}{2i}\int_{\gamma}\bar{z}\,dz$$

measure? How would the answer change if γ has self-intersections like a figure eight?

(16)   Prove the following version of Cauchy's integral formula for smooth functions, often known as the **Cauchy-Pompeiu formula**: Suppose $f:U\to\mathbb{C}$ is C^{1}-smooth and γ is a piecewise C^{1} positively oriented Jordan curve in U such that $D=\operatorname{int}(\gamma)\subset U$. Then

$$f(z)=\frac{1}{2\pi i}\int_{\gamma}\frac{f(\zeta)}{\zeta-z}\,d\zeta-\frac{1}{\pi}\iint_{D}\frac{f_{\bar{z}}(\zeta)}{\zeta-z}\,dx\,dy\qquad\text{if }z\in D.$$

(Hint: Fix $z\in D$ and apply the complex form of Green's theorem (problem 14) to the smooth function $\zeta\mapsto f(\zeta)/(\zeta-z)$ in the domain $D\smallsetminus\overline{\mathbb{D}}(z,\varepsilon)$. Then let $\varepsilon\to0$.)

(17)   (Jump principle for Cauchy transforms) This is a generalization of the jump principle for the winding number formulated in §2.3. Let $\gamma:[0,1]\to\mathbb{C}$ be a curve, $g:|\gamma|\to\mathbb{C}$ be

continuous, and

$$f(z) = \frac{1}{2\pi i} \int_\gamma \frac{g(\zeta)}{\zeta - z} \, d\zeta \qquad \text{for } z \in \mathbb{C} \setminus |\gamma|.$$

Then f is holomorphic off $|\gamma|$ (Corollary 1.48). Suppose D is a disk centered at some $z_0 \in |\gamma|$ for which there exist $0 \le a < b < 1$ such that

- $\gamma(t) \in D$ if and only if $a < t < b$,
- $\gamma(t) \in \partial D$ if and only if $t = a$ and $t = b$,
- γ is injective on $[a, b]$.

As in the discussion leading to the equation (2.4) in this chapter, label the left and right components of $D \setminus |\gamma|$ by L and R. Show that $f(p) - f(q) \to g(z_0)$ as $p \in L$ and $q \in R$ tend to z_0. The jump principle for the winding number corresponds to the case where γ is a closed curve and g is the constant function 1.

(18)   Suppose $g : [-1, 1] \to \mathbb{C}$ is continuous and

$$f(z) = \frac{1}{2\pi i} \int_{[-1,1]} \frac{g(\zeta)}{\zeta - z} \, d\zeta \qquad \text{for } z \in \mathbb{C} \setminus [-1, 1].$$

Compute

$$\lim_{y \to 0^+} \left(f(x + iy) - f(x - iy) \right)$$

for $x \in [-1, 1]$.

(19)   Assuming the Jordan curve theorem, prove Brouwer's *invariance of domain theorem* in the plane: "If $U \subset \mathbb{C}$ is open and $f : U \to \mathbb{C}$ is continuous and injective, then $f(U)$ is open, hence $f : U \to f(U)$ is a homeomorphism." Setting $D = \mathbb{D}(p, \varepsilon)$, it suffices to show that $f(D)$ is open whenever $p \in U$ and $\varepsilon > 0$ is small. Consider the Jordan curve $\gamma : [0, 1] \to \mathbb{C}$ defined by $\gamma(t) = f(p + \varepsilon e^{2\pi i t})$. Show that $f(D) = \text{int}(\gamma)$ by completing the following steps:

(i) $f(D) \subset \text{int}(\gamma)$ or $f(D) \subset \text{ext}(\gamma)$.

(ii) If $f(D) \subset \text{ext}(\gamma)$ and $q \in \text{int}(\gamma)$ then $W(\gamma_r, q) = \pm 1$ for all $0 < r < \varepsilon$, where $\gamma_r(t) = f(p + r e^{2\pi i t})$. This is impossible since $\lim_{r \to 0} W(\gamma_r, q) = 0$. Thus, $f(D) \subset \text{int}(\gamma)$.

(iii) $f(D)$ is closed in $\text{int}(\gamma)$, so $\text{int}(\gamma) \setminus f(D)$ is open. If there is a $q \in \text{int}(\gamma) \setminus f(D)$, then $W(\gamma_r, q) = \pm 1$ for all $0 < r < \varepsilon$. Again impossible since $\lim_{r \to 0} W(\gamma_r, q) = 0$.

3

Meromorphic functions

3.1 Isolated singularities

> **DEFINITION 3.1.** Suppose $U \subset \mathbb{C}$ is open and $p \in U$. If f is defined and holomorphic in $U \smallsetminus \{p\}$, we call p an ***isolated singularity*** of f. We say that p is ***removable*** if f can be extended to a holomorphic function in U, that is, if there exists a function $g \in \mathcal{O}(U)$ such that $f(z) = g(z)$ for all $z \in U \smallsetminus \{p\}$.

Here are a few examples:

EXAMPLE 3.2. Every rational function $f = P/Q$ has isolated singularities at the roots of the polynomial Q. When P and Q have no common factors, these singularities are not removable: If $Q(p) = 0$, then $P(p) \neq 0$ and $\lim_{z \to p} f(z) = \infty$, so clearly p cannot be removable.

On the other hand, the singularity at $z = 0$ of $f \in \mathcal{O}(\mathbb{C}^*)$ defined by $f(z) = (e^z - 1)/z$ is removable: Since $e^z - 1 = \sum_{n=1}^{\infty} z^n/n!$, the power series

$$\sum_{n=1}^{\infty} \frac{z^{n-1}}{n!} = 1 + \frac{z}{2!} + \frac{z^2}{3!} + \frac{z^3}{4!} + \cdots$$

extends f to an entire function.

EXAMPLE 3.3. Suppose $f, g \in \mathcal{O}(U)$ have a common zero at $p \in U$ and assume g is not identically zero. Let $n = \mathrm{ord}(f, p)$ and $m = \mathrm{ord}(g, p)$. If $n \geq m$, the ratio $h = f/g$ has a removable singularity at p. This follows by writing $f(z) = (z - p)^n f_1(z)$ and $g(z) = (z - p)^m g_1(z)$, where $f_1, g_1 \in \mathcal{O}(U)$ with $f_1(p) \neq 0$ and $g_1(p) \neq 0$, and noting that $(z - p)^{n-m} f_1(z)/g_1(z)$ is a holomorphic extension of h in a neighborhood of p. In particular, if $n = m$, it follows that $h = f/g$ extends to a non-vanishing holomorphic function in some neighborhood of p.

EXAMPLE 3.4. The function $f(z) = 1/(\sin(1/z))$ has an isolated singularity at $1/(n\pi)$ for every $n \in \mathbb{Z} \smallsetminus \{0\}$. However, the origin is not an isolated singularity of f.

It follows from Morera's Theorem 1.39 that an isolated singularity p of f for which $\lim_{z \to p} f(z)$ exists must be removable (compare Example 1.40). It turns out that a much weaker assumption guarantees removability:

The theorem appeared in Riemann's dissertation, but was known to Weierstrass earlier.

THEOREM 3.5 (Riemann's removable singularity theorem, 1851). *If a holomorphic function is bounded in a neighborhood of an isolated singularity, then the singularity is removable.*

PROOF. Let f be bounded and holomorphic in the punctured disk $\mathbb{D}^*(p, r) = \{z \in \mathbb{C} : 0 < |z - p| < r\}$. The function $g(z) = (z - p)^2 f(z)$ is holomorphic in $\mathbb{D}^*(p, r)$ and $\lim_{z \to p} g(z) = 0$. Hence g extends continuously to $\mathbb{D}(p, r)$ if we set $g(p) = 0$. Using boundedness of f once again gives

$$\lim_{z \to p} \frac{g(z) - g(p)}{z - p} = \lim_{z \to p} (z - p) f(z) = 0,$$

which shows $g'(p)$ exists and is 0. Thus g is holomorphic in $\mathbb{D}(p, r)$ and is represented by a power series of the form $\sum_{n=2}^{\infty} a_n (z - p)^n$. The power series $\sum_{n=2}^{\infty} a_n (z - p)^{n-2}$ is then a holomorphic extension of f to $\mathbb{D}(p, r)$. $\square$

REMARK 3.6. The above proof shows that boundedness of f can be replaced by the weaker assumption $\lim_{z \to p} (z - p) f(z) = 0$. For example, if $f \in \mathscr{O}(\mathbb{D}^*(0, r))$ and

$$|f(z)| \leq \frac{C}{|z|^{\alpha}}$$

for some constants $C > 0$ and $0 < \alpha < 1$, then 0 is a removable singularity of f.

It is easy to classify isolated singularities of holomorphic functions in terms of their behavior in a vicinity of such points. The following theorem describes the three mutually exclusive possibilities and establishes the relevant terminology:

THEOREM 3.7 (Classification of isolated singularities). *An isolated singularity p of a holomorphic function f is of one of the following types:*

(i) *A removable singularity for which $\lim_{z \to p} f(z)$ exists as a complex number.*
(ii) *A **pole** for which $\lim_{z \to p} f(z) = \infty$. In this case, there exists a unique positive integer m such that $f_1(z) = (z - p)^m f(z)$ has a removable singularity at p, with $f_1(p) \neq 0$.*
(iii) *An **essential singularity** for which $\lim_{z \to p} f(z)$ does not exist. In this case, for every small $r > 0$, the image $f(\mathbb{D}^*(p, r))$ is dense in $\mathbb{C}$.*

The density statement in part (iii) is known as the **Casorati-Weierstrass theorem, 1868-76**. A substantially deeper statement, due to Picard, asserts that $f(\mathbb{D}^*(p, r))$ misses at most one value in $\mathbb{C}$ (see Theorem 11.15).

PROOF. Suppose (iii) fails. Then, we can find disks $\mathbb{D}(p,r)$ and $\mathbb{D}(q,s)$ such that $f(\mathbb{D}^*(p,r)) \cap \mathbb{D}(q,s) = \emptyset$. This implies that the function $g = 1/(f-q)$ is holomorphic and bounded by $1/s$ in $\mathbb{D}^*(p,r)$. By Theorem 3.5, p is a removable singularity of g, so $g \in \mathscr{O}(\mathbb{D}(p,r))$. If $g(p) \neq 0$, then f is bounded in $\mathbb{D}^*(p,r)$, hence p is a removable singularity of f by another application of Theorem 3.5, and we are in case (i). Otherwise, $g(p) = 0$ but g is not identically zero. Write $g(z) = (z-p)^m g_1(z)$, where $m = \mathrm{ord}(g,p)$, $g_1 \in \mathscr{O}(\mathbb{D}(p,r))$, and $g_1(p) \neq 0$. In fact, g_1 has no zeros in $\mathbb{D}(p,r)$ because g does not vanish in $\mathbb{D}^*(p,r)$. It follows that $1/g_1 \in \mathscr{O}(\mathbb{D}(p,r))$. Setting $f_1(z) = (z-p)^m f(z)$, we see that

$$f_1(z) = q(z-p)^m + \frac{(z-p)^m}{g(z)} = q(z-p)^m + \frac{1}{g_1(z)}$$

Felice Casorati
(1835–1890)

has a removable singularity at p, with $f_1(p) = 1/g_1(p) \neq 0$, so we are in case (ii). Since $\lim_{z \to p}(z-p)^n f(z)$ is 0 if $n > m$ and is ∞ if $n < m$, it is evident that the integer m in (ii) is unique. $\square$

EXAMPLE 3.8. The function $\exp(1/z)$ has an isolated singularity at $z = 0$ and $\lim_{z \to 0} \exp(1/z)$ does not exist since $\exp(1/z) \to \infty$ or 0 according as $z \to 0$ along the positive or negative real axis. By Theorem 3.7, $z = 0$ must be an essential singularity. Let us verify the density property of Theorem 3.7(iii) directly: For every $w \in \mathbb{C}$ there exists a sequence $z_n \to 0$ such that $\exp(1/z_n) \to w$. If $w = 0$, simply take $z_n = -1/n$. If $w \neq 0$, find some $z \neq 0$ such that $\exp(1/z) = w$, and set $z_n = z/(1 + 2\pi i n z)$. Evidently $z_n \to 0$ as $n \to \infty$ and $\exp(1/z_n) = \exp(1/z) = w$.

DEFINITION 3.9. Suppose f has a pole at p. The positive integer m in Theorem 3.7(ii) is called the **order** of p and, in analogy with Definition 1.50, is denoted by $\mathrm{ord}(f,p)$. We call p a **simple pole** if $\mathrm{ord}(f,p) = 1$. The function $f_1(z) = (z-p)^m f(z)$ has a power series $\sum_{n=0}^{\infty} b_n (z-p)^n$ in some disk $\mathbb{D}(p,r)$, with $b_0 = f_1(p) \neq 0$. Hence

(3.1) $$f(z) = \sum_{n=0}^{\infty} b_n (z-p)^{n-m}$$

in the punctured disk $\mathbb{D}^*(p,r)$. The rational function

$$\frac{b_0}{(z-p)^m} + \frac{b_1}{(z-p)^{m-1}} + \cdots + \frac{b_{m-1}}{(z-p)}$$

is called the **principal part** of f at p.

The principal part is characterized as the unique polynomial P in $(z-p)^{-1}$ such that $P(0) = 0$ and $f(z) - P((z-p)^{-1})$ has a removable singularity at p. Alternatively, it is the unique rational function R with p as the only pole, such that $\lim_{z \to \infty} R(z) = 0$ and $f(z) - R(z)$ has a removable singularity at p.

EXAMPLE 3.10. The function $f \in \mathscr{O}(\mathbb{C}^*)$ defined by $f(z) = (\sin z)/z^5$ has a pole of order 4 at the origin. This is immediate from the fact that

$$f_1(z) = z^4 f(z) = \frac{\sin z}{z} = \sum_{n=0}^{\infty} \frac{(-1)^n}{(2n+1)!} z^{2n}$$

has a removable singularity at $z = 0$, with $f_1(0) = 1 \neq 0$. The expansion

$$f(z) = \frac{1}{z^4} - \frac{1}{6z^2} + \sum_{n=2}^{\infty} \frac{(-1)^n}{(2n+1)!} z^{2n-4} \qquad (z \in \mathbb{C}^*)$$

shows that the rational function $1/z^4 - 1/(6z^2)$ is the principal part of f at $z = 0$.

EXAMPLE 3.11 (Local degree at a pole). If f has a zero of order m at p, then $1/f$ has a pole of order m at p. Conversely, if f has a pole of order m at p, then $g = 1/f$ has a removable singularity at p and its extension has a zero of order m at p. It follows from Theorem 1.59 that g, hence f, is locally m-to-1 near p. Thus, we can extend Definition 1.60 by setting $\deg(f, p) = \mathrm{ord}(f, p)$ when p is a pole of f.

3.2 The Riemann sphere

"Meromorphic" comes from the Greek $\mu\acute{\epsilon}\rho o\varsigma$ (part) and $\mu o\rho\phi\acute{\eta}$ (form).

DEFINITION 3.12. A function f is said to be ***meromorphic*** in an open set $U \subset \mathbb{C}$ if there is a set $E \subset U$ (possibly empty) with no accumulation point in U such that $f \in \mathscr{O}(U \smallsetminus E)$ and f has a pole at every point of E. The set of all meromorphic functions in U is denoted by $\mathscr{M}(U)$.

Thus, $\mathscr{O}(U) \subset \mathscr{M}(U)$.

EXAMPLE 3.13. Every rational function is meromorphic in the plane. More generally, it is not hard to check that if $U \subset \mathbb{C}$ is a domain (that is, open and connected), if $f, g \in \mathscr{M}(U)$, and if g is not identically zero, then $f/g \in \mathscr{M}(U)$. A sharper converse statement is also true but harder to prove: Every meromorphic function in U is the ratio of two holomorphic functions in U (see Corollary 8.26).

When U is a domain, $\mathscr{M}(U)$ is a field under pointwise addition, multiplication, and division of functions. In fact, $\mathscr{M}(U)$ is just the "field of fractions" constructed from the commutative ring $\mathscr{O}(U)$.

EXAMPLE 3.14. If f is meromorphic in some domain, so is the derivative f'. This follows from the fact that every pole of f is also a pole of f', as can be easily seen by differentiating the representation (3.1).

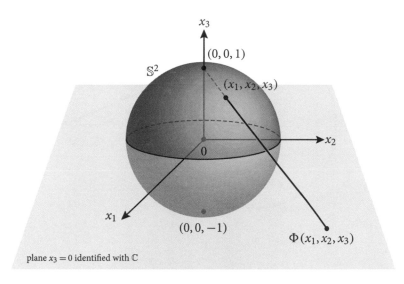

Figure 3.1. The stereographic projection $\Phi : \mathbb{S}^2 \to \hat{\mathbb{C}}$.

It is often convenient to compactify $\mathbb{C}$ by adding a "point at infinity" to form a new topological space $\hat{\mathbb{C}}$ homeomorphic to the 2-sphere. This process makes it possible to treat infinity as any point in the plane, but what makes $\hat{\mathbb{C}}$ particularly important in complex analysis is the fact that it can be given a "complex structure" that allows us to talk about holomorphic maps between open subsets of $\hat{\mathbb{C}}$. Rather than trying to explain this in the proper language of "Riemann surfaces," and to preserve the elementary nature of our presentation, we only touch upon the core idea by introducing as little formalism as possible.

DEFINITION 3.15. The *Riemann sphere* $\hat{\mathbb{C}} = \mathbb{C} \cup \{\infty\}$ is the one-point compactification of the complex plane. The topology of $\hat{\mathbb{C}}$ is generated by the open disks $\mathbb{D}(p, r)$ for $p \in \mathbb{C}$ and $r > 0$, as well as the neighborhoods of ∞ of the form $\{z \in \mathbb{C} : |z| > r\} \cup \{\infty\}$.

It is easy to find explicit homeomorphisms between $\hat{\mathbb{C}}$ and the 2-sphere

$$\mathbb{S}^2 = \{(x_1, x_2, x_3) \in \mathbb{R}^3 : x_1^2 + x_2^2 + x_3^2 = 1\}.$$

A particularly nice one is provided by the *stereographic projection* $\Phi : \mathbb{S}^2 \to \hat{\mathbb{C}}$ from the north pole $(0, 0, 1)$ to the plane $x_3 = 0$ in $\mathbb{R}^3$ identified with the complex plane in the usual way, $(x_1, x_2, 0) \leftrightarrow z = x_1 + ix_2$. In other words, $\Phi(x_1, x_2, x_3)$ is the intersection of the line through the north pole and $(x_1, x_2, x_3) \in \mathbb{S}^2$ with the plane $x_3 = 0$, with the convention that the north pole itself maps to ∞ (see Fig. 3.1). One readily checks

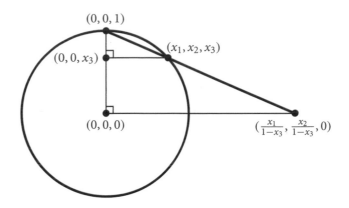

Figure 3.2. The two similar triangles have a scale ratio of $1 - x_3$ to 1.

by looking at Fig. 3.2 that

$$(3.2) \qquad \Phi(x_1, x_2, x_3) = \begin{cases} \dfrac{x_1 + ix_2}{1 - x_3} & x_3 \neq 1 \\ \infty & x_3 = 1. \end{cases}$$

Φ is a homeomorphism since it is clearly continuous and has a continuous inverse given by

$$(3.3) \qquad \Phi^{-1}(z) = \begin{cases} \left(\dfrac{z + \bar{z}}{|z|^2 + 1}, \dfrac{1}{i}\dfrac{z - \bar{z}}{|z|^2 + 1}, \dfrac{|z|^2 - 1}{|z|^2 + 1} \right) & z \neq \infty \\ (0, 0, 1) & z = \infty. \end{cases}$$

For a marvelous animated demonstration of the geometry of the stereographic projection, watch the first episode of the series *Dimensions* at www.dimensions-math.org

The stereographic projection has remarkable geometric properties, some of which will be discussed in chapters 4 and 5.

EXAMPLE 3.16. Under Φ, the $180°$ rotation $R: (x_1, x_2, x_3) \mapsto (-x_1, -x_2, x_3)$ of $\mathbb{S}^2$ corresponds to the rotation $\Phi \circ R \circ \Phi^{-1} : z \mapsto -z$ of $\hat{\mathbb{C}}$ which fixes $0, \infty$ and swaps ± 1. Similarly, the $180°$ rotation $(x_1, x_2, x_3) \mapsto (x_1, -x_2, -x_3)$ of $\mathbb{S}^2$ corresponds to the inversion $z \mapsto 1/z$ of $\hat{\mathbb{C}}$ which fixes ± 1 and swaps $0, \infty$. The reflection $(x_1, x_2, x_3) \mapsto (x_1, -x_2, x_3)$ on $\mathbb{S}^2$ corresponds to the complex conjugation $z \mapsto \bar{z}$.

REMARK 3.17. It is geometrically evident that the stereographic projection Φ as defined here is orientation-*reversing*: An ordinary map of the world drawn on $\mathbb{S}^2$ will project to a map on the plane $x_3 = 0$ in which all countries appear as their mirror images. For this reason, perhaps it would be more natural to modify the definition of Φ, for example by sending (x_1, x_2, x_3) to $(x_1 - ix_2)/(1 - x_3)$.

Now let $U \subset \hat{\mathbb{C}}$ be open. A continuous map $f : U \to \hat{\mathbb{C}}$ is said to be holomorphic if every $p \in U$ has a neighborhood in which f is holomorphic. To define what the latter means, it suffices to consider the case where either p or $q = f(p)$ is ∞; otherwise by continuity a neighborhood (in $\mathbb{C}$) of p is mapped into a neighborhood (in $\mathbb{C}$) of q and the usual definition of holomorphic applies. The idea is to precompose or postcompose f with the homeomorphism $\iota : \hat{\mathbb{C}} \to \hat{\mathbb{C}}$ given by

$$\iota(z) = \frac{1}{z}$$

in order to bring ∞ to 0, near which being holomorphic makes sense (notice that by definition $\iota(0) = \infty$ and $\iota(\infty) = 0$). Specifically, we distinguish three cases:

- $p \neq \infty$ and $q = \infty$. Then f is holomorphic in a neighborhood of p if $\iota \circ f$ is holomorphic (in the usual sense) in a neighborhood of p.
- $p = \infty$ and $q \neq \infty$. Then f is holomorphic in a neighborhood of p if $f \circ \iota$ is holomorphic in a neighborhood of 0.
- $p = \infty$ and $q = \infty$. Then f is holomorphic in a neighborhood of p if $\iota \circ f \circ \iota$ is holomorphic in a neighborhood of 0.

EXAMPLE 3.18. Suppose f has a pole of order m at $p \in \mathbb{C}$. By Theorem 3.7, $\lim_{z \to p} f(z) = \infty$, so setting $f(p) = \infty$ extends f continuously to a map from a neighborhood of p to $\hat{\mathbb{C}}$. This extension is actually holomorphic: In fact, the function $f_1(z) = (z - p)^m f(z)$ is non-zero and holomorphic in some neighborhood of p, so $1/f(z) = (z - p)^m / f_1(z)$ is holomorphic in that neighborhood.

This shows that every meromorphic function f in a domain $U \subset \mathbb{C}$ can be viewed as a holomorphic map $f : U \to \hat{\mathbb{C}}$ by setting $f(p) = \infty$ for every pole p of f. Conversely, if $f : U \to \hat{\mathbb{C}}$ is holomorphic and not identically ∞ in U, then $f \in \mathscr{M}(U)$ and every point of $f^{-1}(\infty)$ is a pole.

Justified by the last example, we will often use the notation $f^{-1}(\infty)$ for the set of poles of a meromorphic function f.

EXAMPLE 3.19. Suppose f is holomorphic in a punctured neighborhood of ∞ of the form $\{z \in \mathbb{C} : |z| > r\}$. Then the function $g : \mathbb{D}^*(0, 1/r) \to \mathbb{C}$ defined by $g(z) = f(1/z)$ has an isolated singularity at 0. If this singularity is removable, f extends to a holomorphic map $\{z \in \mathbb{C} : |z| > r\} \cup \{\infty\} \to \mathbb{C}$ by setting $f(\infty) = g(0)$. If g has a pole of order m at 0, the function $g_1(z) = z^m g(z)$ is non-zero and holomorphic in a neighborhood of 0, so $1/f(1/z) = 1/g(z) = z^m / g_1(z)$ has a zero of order m at 0. It follows that f extends to a holomorphic map $\{z \in \mathbb{C} : |z| > r\} \cup \{\infty\} \to \hat{\mathbb{C}}$ by setting $f(\infty) = \infty$.

Let P and Q be polynomials in z with no common factors, and assume Q is not identically zero. Using the preceding examples, we see that the rational function $f = P/Q$ extends to a holomorphic map $\hat{\mathbb{C}} \to \hat{\mathbb{C}}$ if we set $f(p) = \infty$ at every root p of Q and $f(\infty) = \lim_{z \to \infty} f(z)$. Here $\lim_{z \to \infty} f(z)$ is 0 if $\deg P < \deg Q$, is ∞ if $\deg P > \deg Q$, and is the ratio of the leading coefficients of P and Q if $\deg P = \deg Q$.

It turns out that rational functions are the only non-trivial examples of holomorphic self-maps of the sphere.

THEOREM 3.20. *Every holomorphic map* $f : \hat{\mathbb{C}} \to \hat{\mathbb{C}}$ *which is not identically* ∞ *is a rational function.*

PROOF. The set $f^{-1}(\infty)$ is finite since otherwise it would have an accumulation point in $\hat{\mathbb{C}}$ and f would be identically ∞ by Theorem 1.53. Let $f^{-1}(\infty) = \{p_1, \ldots, p_n\}$. By replacing f with $1/f$ if necessary, we may assume that $f(\infty) \neq \infty$ so $p_k \neq \infty$ for every $1 \leq k \leq n$. The restriction of f to $\mathbb{C}$ is meromorphic with poles at the p_k. Let $P_k((z - p_k)^{-1})$ be the principal part of f at p_k, where P_k is a polynomial of degree $\mathrm{ord}(f, p_k)$ with $P_k(0) = 0$. The function

$$g(z) = f(z) - \sum_{k=1}^{n} P_k \left(\frac{1}{z - p_k} \right)$$

has removable singularities at every p_k, so it extends to an entire function. Moreover,

$$\lim_{z \to \infty} g(z) = \lim_{z \to \infty} f(z) - \sum_{k=1}^{n} P_k(0) = f(\infty)$$

belongs to $\mathbb{C}$ since by our assumption $f(\infty) \neq \infty$. In particular, g stays bounded outside a disk $\mathbb{D}(0, r)$ for large enough r. Since g is certainly bounded in $\mathbb{D}(0, r)$, it follows that g is bounded in $\mathbb{C}$, hence constant by Liouville's Theorem 1.44. Thus, up to a constant, f is the sum of the principal parts at its poles, clearly a rational function. □

3.3 Laurent series

Near a pole p, a holomorphic function can be represented by a power series with finitely many negative powers of $z - p$ (compare (3.1)). A similar statement is true even when p is an essential singularity, provided that we allow infinitely many negative powers of $z - p$. In fact, we can use Cauchy's integral formula to prove more generally that every holomorphic function in a round annulus has a power series representation containing both positive and negative powers. Here is the precise statement:

THEOREM 3.21 (Laurent, 1843). *Let* $U = \{z \in \mathbb{C} : R_1 < |z - p| < R_2\}$, *where* $0 \leq R_1 < R_2 \leq +\infty$. *Every* $f \in \mathcal{O}(U)$ *has a power series representation of the form*

$$(3.4) \qquad f(z) = \sum_{n=-\infty}^{\infty} a_n (z - p)^n \qquad \text{for } z \in U,$$

where the coefficients $\{a_n\}$ *are given by*

$$(3.5) \qquad a_n = \frac{1}{2\pi i} \int_{\mathbb{T}(p,r)} \frac{f(\zeta)}{(\zeta - p)^{n+1}} \, d\zeta \qquad \text{for any } R_1 < r < R_2.$$

The series converges uniformly in every sub-annulus $\{z \in \mathbb{C} : a < |z - p| < b\}$ *with* $R_1 < a < b < R_2$.

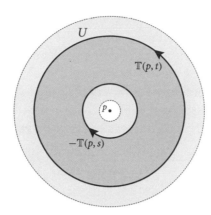

Figure 3.3. Any holomorphic function f in the round annulus U can be written as $f = g + h$ in the darker sub-annulus, where g is holomorphic inside the disk of radius t and h is holomorphic outside the disk of radius s. Combining the power series of g and h, one obtains the Laurent series of f.

We call (3.4) the ***Laurent series*** representation of f in U.

PROOF. Take any s, t such that $R_1 < s < t < R_2$. The cycle $\mathbb{T}(p, t) - \mathbb{T}(p, s)$ is null-homologous in U and has winding number 1 with respect to any point in the annulus $\{z : s < |z - p| < t\}$ (see Fig. 3.3). Hence, by Cauchy's Theorem 2.41, in this annulus we can write $f = g + h$, where

$$g(z) = \frac{1}{2\pi i} \int_{\mathbb{T}(p,t)} \frac{f(\zeta)}{\zeta - z}\, d\zeta \quad \text{and} \quad h(z) = -\frac{1}{2\pi i} \int_{\mathbb{T}(p,s)} \frac{f(\zeta)}{\zeta - z}\, d\zeta.$$

Pierre Alphonse Laurent
(1813–1854)

By Corollary 1.48, g is holomorphic off $\mathbb{T}(p, t)$, so it has a power series representation $g(z) = \sum_{n=0}^{\infty} \alpha_n (z - p)^n$ in $\mathbb{D}(p, t)$. Similarly, h is holomorphic off $\mathbb{T}(p, s)$, so the function $\hat{h}(z) = h(p + 1/z)$ is holomorphic in the punctured disk $\mathbb{D}^*(0, 1/s)$. Moreover, it is easy to see using the *ML*-inequality that $\lim_{z \to 0} \hat{h}(z) = \lim_{z \to \infty} h(z) = 0$. Hence the singularity of $\hat{h}$ at 0 is removable and $\hat{h}$ has a power series representation $\hat{h}(z) = \sum_{n=1}^{\infty} \beta_n z^n$ in $\mathbb{D}(0, 1/s)$. It follows that $h(z) = \hat{h}(1/(z - p))$ has the power series representation $h(z) = \sum_{n=1}^{\infty} \beta_n (z - p)^{-n}$ for $|z - p| > s$. Setting $a_n = \alpha_n$ for $n \geq 0$ and $a_n = \beta_{-n}$ for $n \leq -1$, we see that f has a power series representation of the form (3.4) in the annulus $\{z : s < |z - p| < t\}$. This series converges uniformly in $\{z : a < |z - p| < b\}$ whenever $s < a < b < t$ since the series of g converges uniformly for $|z - p| < b$ and that of h converges uniformly for $|z - p| > a$.

To verify (3.5), choose any radius r with $s < r < t$ and use the uniform convergence of (3.4) on $\mathbb{T}(p, r)$ to integrate term-by-term:

$$\int_{\mathbb{T}(p,r)} \frac{f(\zeta)}{(\zeta - p)^{n+1}}\, d\zeta = \sum_{k=-\infty}^{\infty} a_k \left(\int_{\mathbb{T}(p,r)} (\zeta - p)^{k-n-1}\, d\zeta \right).$$

The integral on the right vanishes unless $k = n$, in which case its value is $2\pi i$ (compare Example 1.30). Hence the right side is $2\pi i a_n$.

Finally, note that the integrals in (3.5) are independent of r since the oriented circles $\mathbb{T}(p, r)$ are all homologous in U as long as $R_1 < r < R_2$. Hence the coefficients $\{a_n\}$ depend only on f and not on the choice of the radius r. Since s and t can be chosen arbitrarily close to R_1 and R_2, respectively, it follows that the Laurent series (3.4) with the same coefficients $\{a_n\}$ should converge to $f(z)$ for every $z \in U$. $\qquad \square$

Although (3.5) provides an explicit formula for the coefficients of the Laurent series, in practice it is often easier to find these coefficients by relating the function to the ones whose expansions are already known. Here are two examples:

EXAMPLE 3.22. The function $\exp(1/z)$ is holomorphic in $\mathbb{C}^*$ and has an essential singularity at $z = 0$. The Laurent series of f in $\mathbb{C}^*$ is readily obtained from the power series of the exponential function about $z = 0$: Since $\exp(z) = \sum_{n=0}^{\infty} z^n/n!$ for all $z \in \mathbb{C}$, we have

$$\exp\left(\frac{1}{z}\right) = \sum_{n=0}^{\infty} \frac{z^{-n}}{n!} = 1 + \frac{1}{z} + \frac{1}{2!z^2} + \frac{1}{3!z^3} + \cdots \qquad \text{for } z \in \mathbb{C}^*.$$

EXAMPLE 3.23. The rational function

$$f(z) = \frac{1}{1 - z^2} + \frac{1}{3 - z}$$

has poles at ± 1 and 3, hence it is holomorphic in the disk $|z| < 1$ and in the annuli $1 < |z| < 3$ and $|z| > 3$. Let us find the Laurent series of f in each of these domains. In all three cases the computation will depend on the geometric series formula $1/(1 - z) = \sum_{n=0}^{\infty} z^n$ for $z \in \mathbb{D}$.

- $|z| < 1$. Then

$$\frac{1}{1 - z^2} = \sum_{n=0}^{\infty} z^{2n}$$

and

$$(3.6) \qquad \frac{1}{3 - z} = \frac{1}{3(1 - z/3)} = \frac{1}{3} \sum_{n=0}^{\infty} \frac{z^n}{3^n} = \sum_{n=0}^{\infty} 3^{-(n+1)} z^n.$$

Hence

$$f(z) = \sum_{n=0}^{\infty} a_n z^n, \qquad \text{where} \quad a_n = \begin{cases} 3^{-(n+1)} + 1 & n \text{ even} \\ 3^{-(n+1)} & n \text{ odd.} \end{cases}$$

- $1 < |z| < 3$. Then

$$(3.7) \qquad \frac{1}{1 - z^2} = \frac{-1}{z^2(1 - 1/z^2)} = \frac{-1}{z^2} \sum_{n=0}^{\infty} \frac{1}{z^{2n}} = -\sum_{n=1}^{\infty} z^{-2n}$$

and the expansion (3.6) for $1/(3 - z)$ is still valid. Hence

$$f(z) = \sum_{n=-\infty}^{\infty} a_n z^n, \qquad \text{where} \quad a_n = \begin{cases} -1 & n < 0 \text{ even} \\ 0 & n < 0 \text{ odd} \\ 3^{-(n+1)} & n \geq 0. \end{cases}$$

● $|z| > 3$. Then

$$\frac{1}{3-z} = \frac{-1}{z(1-3/z)} = \frac{-1}{z}\sum_{n=0}^{\infty}\frac{3^n}{z^n} = -\sum_{n=1}^{\infty}3^{n-1}z^{-n}$$

and the expansion (3.7) for $1/(1-z^2)$ is still valid. Hence

$$f(z) = \sum_{n=1}^{\infty}a_n\,z^{-n}, \qquad \text{where} \quad a_n = \begin{cases} -3^{n-1}-1 & n \text{ even} \\ -3^{n-1} & n \text{ odd.} \end{cases}$$

As an application of Laurent series, we consider the problem of representing periodic holomorphic functions as a superposition of exponentials. Suppose for simplicity that f is a 2π-periodic holomorphic function in the horizontal strip $S = \{x+iy \in \mathbb{C} : a < y < b\}$. Thus, $f(z+2\pi) = f(z)$ for all $z \in S$. Let $\zeta = \varphi(z) = \exp(iz)$. Under φ, the strip S maps to the annulus $U = \{\zeta \in \mathbb{C} : e^{-b} < |\zeta| < e^{-a}\}$. By periodicity of f, there is a unique $g \in \mathscr{O}(U)$ which satisfies

$$f = g \circ \varphi \qquad \text{in } S.$$

If $g(\zeta) = \sum_{n=-\infty}^{\infty}a_n\,\zeta^n$ is the Laurent series of g in U, then f has the representation

(3.8)
$$f(z) = \sum_{n=-\infty}^{\infty}a_n\,e^{inz} \qquad \text{for } z \in S.$$

This is called the **Fourier series** of f in the strip U. The **Fourier coefficients** a_n, often denoted by $\hat{f}(n)$, can be expressed as complex integrals in terms of f. To see this, choose any s such that $a < s < b$. Then $e^{-b} < r = e^{-s} < e^{-a}$ and by (3.5)

$$\hat{f}(n) = \frac{1}{2\pi i}\int_{\mathbb{T}(0,r)}\frac{g(\zeta)}{\zeta^{n+1}}\,d\zeta.$$

Consider the parametrization $z(t) = t + is$ $(0 \le t \le 2\pi)$ of the oriented segment $[is, is + 2\pi] \subset S$. Then $\zeta(t) = \varphi(z(t)) = e^{iz(t)} = re^{it}$ is the standard parametrization of the oriented circle $\mathbb{T}(0,r)$ (see Fig. 3.4). Hence,

$$\hat{f}(n) = \frac{1}{2\pi i}\int_0^{2\pi}\frac{g(\zeta(t))}{\zeta(t)^{n+1}}\,\zeta'(t)\,dt = \frac{1}{2\pi}\int_0^{2\pi}\frac{f(z(t))}{e^{i(n+1)z(t)}}\,e^{iz(t)}\,z'(t)\,dt$$

$$= \frac{1}{2\pi}\int_{[is,is+2\pi]}f(z)\,e^{-inz}\,dz.$$

EXAMPLE 3.24. The complex cosine and sine functions are entire and 2π-periodic, so they have Fourier series representations of the form (3.8) whose coefficients can be computed from the above formula by integrating along the segment $[0,2\pi]$ on the real line. However, such computation

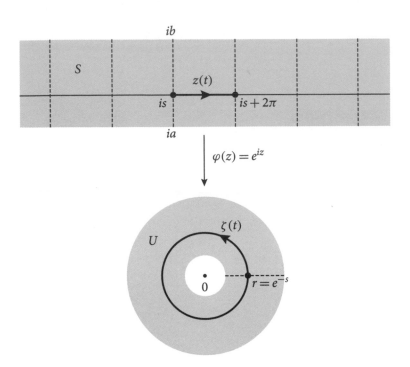

Figure 3.4. Relating the Fourier series of a 2π-periodic holomorphic function in the strip S to the Laurent series of the induced holomorphic function in the annulus $U = \varphi(S)$.

would be unnecessary since the definitions of cosine and sine already give us these coefficients: For $f(z) = \cos z, \hat{f}(1) = \hat{f}(-1) = 1/2$, and $\hat{f}(n) = 0$ otherwise; for $f(z) = \sin z, \hat{f}(1) = -\hat{f}(-1) = 1/(2i)$, and $\hat{f}(n) = 0$ otherwise.

A similar reasoning applies if f is an ω-periodic holomorphic function, that is, if there is a non-zero $\omega \in \mathbb{C}$ so that $f(z + \omega) = f(z)$ for all z:

THEOREM 3.25 (Fourier series of periodic holomorphic functions). *Suppose f is ω-periodic and holomorphic in a strip S which is invariant under the translation $z \mapsto z + \omega$. Then f has a Fourier series representation of the form*

$$f(z) = \sum_{n=-\infty}^{\infty} \hat{f}(n) \, e^{2\pi i n z/\omega} \qquad \text{for } z \in S$$

which converges uniformly in every closed sub-strip of S. The Fourier coefficients of f are given by

$$\hat{f}(n) = \frac{1}{\omega} \int_{[p,p+\omega]} f(z) \, e^{-2\pi i n z/\omega} \, dz,$$

where p is an arbitrary point of S.

PROOF. This simply follows from the 2π-periodic case treated above, by considering the function $f(p + \omega z/2\pi)$. □

3.4 Residues

The idea of residue goes back to an 1814 paper of Cauchy. But he first used the term explicitly in a paper that appeared in 1826.

DEFINITION 3.26. Let p be an isolated singularity of a holomorphic function f, so $f \in \mathcal{O}(U \smallsetminus \{p\})$ for some neighborhood U of p. The **residue** of f at p is defined by

(3.9)
$$\text{res}(f, p) = \frac{1}{2\pi i} \int_{\mathbb{T}(p,r)} f(z)\, dz,$$

where $r > 0$ is any radius such that $\overline{\mathbb{D}}(p, r) \subset U$.

The definition is of course independent of the choice of the radius r since by Corollary 2.42 the oriented circles $\mathbb{T}(p, r_1)$ and $\mathbb{T}(p, r_2)$ are homologous in $U \smallsetminus \{p\}$ as long as $\overline{\mathbb{D}}(p, r_1)$ and $\overline{\mathbb{D}}(p, r_2)$ are both contained in U. More generally, we can replace these circles by any positively oriented Jordan curve γ in U such that $p \in \text{int}(\gamma) \subset U$.

It is easy to compute the residue using the Laurent series representation in a punctured neighborhood of the singular point. In fact, from the formula (3.5) one immediately obtains the following

THEOREM 3.27 (Computing the residue). *Suppose $f \in \mathcal{O}(\mathbb{D}^*(p, r))$ has the Laurent series $\sum_{n=-\infty}^{\infty} a_n (z - p)^n$. Then*

$$\text{res}(f, p) = a_{-1}.$$

EXAMPLE 3.28. The function $f(z) = e^z/z^3$ has a pole of order 3 at $z = 0$. To compute the residue of f at 0, use $e^z = \sum_{n=0}^{\infty} z^n/n!$ to obtain

$$f(z) = \frac{1}{z^3} + \frac{1}{z^2} + \frac{1}{2!z} + \frac{1}{3!} + \frac{z}{4!} + \cdots,$$

which shows $\text{res}(f, 0) = 1/2$.

More generally, suppose f has a pole of order m at p so the function $f_1(z) = (z - p)^m f(z)$ has a removable singularity at p, with $f_1(p) \neq 0$. Writing $f_1(z) = \sum_{n=0}^{\infty} b_n(z - p)^n$ gives the Laurent series $f(z) = \sum_{n=0}^{\infty} b_n(z - p)^{n-m}$, hence

$$\text{res}(f, p) = b_{m-1} = \frac{f_1^{(m-1)}(p)}{(m - 1)!}.$$

This can be written in the form

(3.10)
$$\text{res}(f, p) = \frac{1}{(m - 1)!} \lim_{z \to p} \left[(z - p)^m f(z) \right]^{(m-1)}.$$

EXAMPLE 3.29. The rational function $f(z) = 1/(z^2(z+1))$ has a simple pole at $z = -1$ and a pole of order 2 at $z = 0$. By (3.10),

$$\text{res}(f, -1) = \lim_{z \to -1} \left[(z+1)f(z)\right] = 1$$

and

$$\text{res}(f, 0) = \lim_{z \to 0} \left[z^2 f(z)\right]' = \lim_{z \to 0} \frac{-1}{(z+1)^2} = -1.$$

Residues can be used to compute complex integrals. Take for instance the thrice punctured domain U and the closed curves $\gamma, \gamma_1, \gamma_2, \gamma_3$ of Example 2.44 as illustrated in Fig. 2.9. We have $\gamma \sim 2\gamma_1 - \gamma_2 + \gamma_3$ in U, so for every $f \in \mathcal{O}(U)$,

$$\int_\gamma f(z)\, dz = 2 \int_{\gamma_1} f(z)\, dz - \int_{\gamma_2} f(z)\, dz + \int_{\gamma_3} f(z)\, dz$$

$$= 2\pi i \big(2\, \text{res}(f, p_1) - \text{res}(f, p_2) + \text{res}(f, p_3)\big).$$

This type of computation is the idea behind the following

THEOREM 3.30 (The residue theorem). *Suppose $U \subset \mathbb{C}$ is open, $E \subset U$ has no accumulation point in U, and $f \in \mathcal{O}(U \smallsetminus E)$. Let γ be a null-homologous cycle in U such that $|\gamma| \cap E = \emptyset$. Then*

$$(3.11) \qquad \int_\gamma f(z)\, dz = 2\pi i \sum_{p \in E} W(\gamma, p)\, \text{res}(f, p).$$

The case $E = \emptyset$ is just the homology version of Cauchy's Theorem 2.41. Note also that the theorem applies in particular to the case where $f \in \mathcal{M}(U)$ and $E = f^{-1}(\infty)$ is the set of poles of f.

PROOF. Set $A = \{p \in E : W(\gamma, p) \neq 0\}$. Evidently A is a bounded set since it does not meet the unbounded component of $\mathbb{C} \smallsetminus |\gamma|$ where $W(\gamma, \cdot) = 0$. We claim that A is a finite set. Otherwise, it must have an accumulation point $q \in \overline{U}$. Since by the assumption E has no accumulation point in U, the point q must lie on the boundary ∂U. Take a small disk D centered at q that is disjoint from $|\gamma|$ so $W(\gamma, \cdot)$ is constant in D. Since D meets $\mathbb{C} \smallsetminus U$, the assumption $\gamma \sim 0$ in U implies that $W(\gamma, \cdot)$ vanishes in D, so $A \cap D = \emptyset$. This contradicts the fact that q is an accumulation point of A and proves our claim. Thus, A is a finite set $\{p_1, \ldots, p_n\}$ and the right side of (3.11) is actually a finite sum.

Choose $r > 0$ small enough so that the closed disks $\overline{\mathbb{D}}(p_k, r)$ are disjoint and contained in $U \smallsetminus |\gamma|$ and satisfy $\overline{\mathbb{D}}(p_k, r) \cap E = \{p_k\}$ for every $1 \le k \le n$. The cycles γ and

$$\eta = \sum_{k=1}^n W(\gamma, p_k)\, \mathbb{T}(p_k, r)$$

are then homologous in $U \smallsetminus E$. In fact, if $p \in \mathbb{C} \smallsetminus U$ or $p \in E \smallsetminus A$, then each $\mathbb{T}(p_k, r)$ has winding number 0 with respect to p, so $\mathrm{W}(\eta, p) = 0 = \mathrm{W}(\gamma, p)$. On the other hand, if $p = p_j \in A$, then each $\mathbb{T}(p_k, r)$ has winding number 0 with respect to p except when $k = j$, in which case the winding number is 1. Thus, again, $\mathrm{W}(\eta, p) = \mathrm{W}(\gamma, p)$. It follows from Corollary 2.42 that

$$\int_\gamma f(z)\,dz = \int_\eta f(z)\,dz = \sum_{k=1}^n \mathrm{W}(\gamma, p_k) \int_{\mathbb{T}(p_k, r)} f(z)\,dz$$

$$= 2\pi i \sum_{k=1}^n \mathrm{W}(\gamma, p_k)\ \mathrm{res}(f, p_k). \qquad \square$$

The following special case is the most typical situation where the residue theorem is invoked:

COROLLARY 3.31. *Let γ be a positively oriented Jordan curve in a domain $U \subset \mathbb{C}$ and suppose $\{p_1, \ldots, p_n\} \subset \mathrm{int}(\gamma) \subset U$. If f is holomorphic in $U \smallsetminus \{p_1, \ldots, p_n\}$, then*

$$\int_\gamma f(z)\,dz = 2\pi i \sum_{k=1}^n \mathrm{res}(f, p_k).$$

EXAMPLE 3.32. Let us use the residue theorem to evaluate the integral

$$I = \int_{\mathbb{T}(0,2)} \frac{dz}{(z-3)(z^4 - 1)}.$$

The rational function $f(z) = 1/((z-3)(z^4 - 1))$ has simple poles at $z = \pm 1, \pm i$, and 3. By (3.10),

$$\mathrm{res}(f, 1) = \lim_{z \to 1}\left[(z-1)f(z)\right] = \lim_{z \to 1} \frac{1}{(z-3)(z+1)(z^2+1)} = -\frac{1}{8},$$

$$\mathrm{res}(f, -1) = \lim_{z \to -1}\left[(z+1)f(z)\right] = \lim_{z \to -1} \frac{1}{(z-3)(z-1)(z^2+1)} = \frac{1}{16},$$

$$\mathrm{res}(f, i) = \lim_{z \to i}\left[(z-i)f(z)\right] = \lim_{z \to i} \frac{1}{(z-3)(z+i)(z^2-1)} = \frac{1}{4(1+3i)},$$

$$\mathrm{res}(f, -i) = \lim_{z \to -i}\left[(z+i)f(z)\right] = \lim_{z \to -i} \frac{1}{(z-3)(z-i)(z^2-1)} = \frac{1}{4(1-3i)}.$$

Corollary 3.31 now shows that

$$I = 2\pi i \left(-\frac{1}{8} + \frac{1}{16} + \frac{1}{4(1+3i)} + \frac{1}{4(1-3i)}\right) = -\frac{\pi i}{40}.$$

We will soon return to this example.

It is often useful to treat the point at infinity as an isolated singularity and talk about the residue there. Suppose f is holomorphic in a punctured neighborhood of

∞ of the form $\{z \in \mathbb{C} : |z| > R\}$. Define the ***residue of f at infinity*** by

$$(3.12) \qquad \operatorname{res}(f, \infty) = -\frac{1}{2\pi i} \int_{\mathbb{T}(0,r)} f(z)\, dz$$

where r is any radius $> R$. A negative sign is included since to an observer on the Riemann sphere, the oriented circle $\mathbb{T}(0, r)$ appears to wind around ∞ clockwise.

LEMMA 3.33. *Suppose f is holomorphic in* $\{z \in \mathbb{C} : |z| > R\}$.

(i) *If* $\sum_{n=-\infty}^{\infty} a_n z^n$ *is the Laurent series of* f *for* $|z| > R$, *then* $\operatorname{res}(f, \infty) = -a_{-1}$.

(ii) $\operatorname{res}(f, \infty) = \operatorname{res}(g, 0)$, *where* $g(z) = -1/z^2 f(1/z)$.

Part (ii) is a special example of the invariance of residue under a local biholomorphic change of coordinates (see problem 13).

PROOF. Part (i) follows from formula (3.5). For (ii), note that g has the Laurent series

$$g(z) = -\frac{1}{z^2} f\left(\frac{1}{z}\right) = -\sum_{n=-\infty}^{\infty} a_n z^{-(n+2)}$$

in the punctured disk $\mathbb{D}^*(0, 1/R)$. Hence, by Theorem 3.27,

$$\operatorname{res}(g, 0) = \text{ the coefficient of } 1/z \text{ in the Laurent series of } g$$

$$= -a_{-1} = \operatorname{res}(f, \infty). \qquad \square$$

THEOREM 3.34 (Residues add up to zero). *Suppose* $p_1, \ldots, p_n$ *are distinct points in* $\mathbb{C}$ *and f is holomorphic in* $\mathbb{C} \smallsetminus \{p_1, \ldots, p_n\}$. *Then,*

$$\operatorname{res}(f, \infty) + \sum_{k=1}^{n} \operatorname{res}(f, p_k) = 0.$$

If we count ∞ as an isolated singularity of f, the theorem asserts that the residues of f at its isolated singularities in $\hat{\mathbb{C}}$ sum to zero. The most typical application of this result is when $f \in \mathscr{M}(\mathbb{C})$ has finitely many poles, such as a rational function.

PROOF. Choose an r larger than $|p_k|$ for every $1 \le k \le n$. Then f is holomorphic in $\{z \in \mathbb{C} : |z| > r\}$, and by (3.12) and Corollary 3.31,

$$\operatorname{res}(f, \infty) = -\frac{1}{2\pi i} \int_{\mathbb{T}(0,r)} f(z)\, dz = -\sum_{k=1}^{n} \operatorname{res}(f, p_k). \qquad \square$$

EXAMPLE 3.35. Let us revisit Example 3.32. Since the value of the integral I is $2\pi i$ times the sum of the residues at $\pm 1, \pm i$, Theorem 3.34 implies that

$$I = -2\pi i \big(\operatorname{res}(f, 3) + \operatorname{res}(f, \infty)\big).$$

We have
$$\mathrm{res}(f,3) = \lim_{z\to 3}\left[(z-3)f(z)\right] = \lim_{z\to 3}\frac{1}{z^4-1} = \frac{1}{80}$$
and since $g(z) = -1/z^2 f(1/z) = -z^3/((1-3z)(1-z^4))$ has a removable singularity at 0, Lemma 3.33(ii) shows that
$$\mathrm{res}(f,\infty) = \mathrm{res}(g,0) = 0.$$
It follows that $I = -\pi i/40$, matching our previous answer.

The advantage of using the residue at infinity over the standard method of Example 3.32 may not be fully appreciated in this particular example. But suppose we were to evaluate the same integral for the rational function
$$f(z) = \frac{1}{(z-3)(z^n-1)}$$
instead. The method of Example 3.32 would require computing the residues of f at the n-th roots of unity, while the above method would require only two residues, at 3 and ∞. The latter is far more efficient if n is large, and easily gives the answer
$$\int_{\mathbb{T}(0,2)} \frac{dz}{(z-3)(z^n-1)} = -\frac{2\pi i}{3^n-1}.$$

As another application of the notion of residue, we consider the behavior of a holomorphic function f near a fixed point $p = f(p) \in \mathbb{C}$. Let us suppose that p is an isolated fixed point, that is, an isolated zero of the function $z - f(z)$. By the **multiplicity** of p we mean the order of p as a zero of this function, or equivalently, as a pole of the function $1/(z-f(z))$. In analogy with the case of zeros and poles, we call p a **simple** fixed point if it has multiplicity 1. Evidently, p is simple if and only if $f'(p) \neq 1$.

DEFINITION 3.36. The **index** of f at an isolated fixed point $p \in \mathbb{C}$ is defined by
$$\mathrm{ind}(f,p) = \mathrm{res}\left(\frac{1}{z-f(z)}, p\right).$$

Take the power series representation $f(z) = \sum_{n=0}^{\infty} a_n(z-p)^n$ near p, where $a_0 = f(p) = p$. Then
$$z - f(z) = (z-p)\left(1 - a_1 - \sum_{n=2}^{\infty} a_n(z-p)^{n-1}\right).$$
If $a_1 = f'(p) \neq 1$, then p is a simple pole of $1/(z-f(z))$, so by (3.10),
$$\mathrm{res}\left(\frac{1}{z-f(z)}, p\right) = \lim_{z\to p}\frac{z-p}{z-f(z)} = \frac{1}{1-a_1}.$$
This prove the formula
$$(3.13) \qquad \mathrm{ind}(f,p) = \frac{1}{1-f'(p)} \qquad \text{if } p \text{ is simple.}$$

The definition of index generalizes to the case where the isolated fixed point is at ∞. In this case the function $\hat{f}(z) = 1/f(1/z)$ has an isolated fixed point at 0 and we define

$$(3.14) \qquad \operatorname{ind}(f,\infty) = \operatorname{ind}(\hat{f},0).$$

If we set $f'(\infty) = \hat{f}'(0)$, it follows that (3.13) holds even when $p = \infty$. Word of caution: $\operatorname{ind}(f,\infty)$ is *not* the same as $\operatorname{res}(1/(z-f(z)),\infty)$ (compare problem 17).

EXAMPLE 3.37 (Index of polynomials at ∞). Let $f(z) = a_d z^d + \cdots + a_1 z + a_0$ be a polynomial of degree $d \geq 2$, so ∞ is a fixed point of f. The polynomial $P(z) = z^d f(1/z) = a_0 z^d + \cdots + a_{d-1}z + a_d$ has a non-zero constant term $P(0) = a_d$. Since

$$\frac{1}{z-\hat{f}(z)} = \frac{1}{z - z^d/P(z)} = \frac{P(z)}{z(P(z) - z^{d-1})},$$

it follows that $1/(z - \hat{f}(z))$ has a simple pole at $z = 0$ and

$$\operatorname{res}\left(\frac{1}{z-\hat{f}(z)}, 0\right) = \lim_{z\to 0}\frac{z}{z-\hat{f}(z)} = \lim_{z\to 0}\frac{P(z)}{P(z)-z^{d-1}} = 1.$$

This proves

$$\operatorname{ind}(f,\infty) = \operatorname{ind}(\hat{f},0) = 1.$$

THEOREM 3.38 (Rational fixed point formula). *For every rational map $f:\hat{\mathbb{C}} \to \hat{\mathbb{C}}$ other than the identity,*

$$\sum_{p=f(p)} \operatorname{ind}(f,p) = 1.$$

Applying this to a polynomial map of degree ≥ 2 which has a fixed point at ∞ with index 1 (Example 3.37), we obtain the following

COROLLARY 3.39. *If $p_1,\ldots,p_n \in \mathbb{C}$ are the distinct fixed points of a polynomial map $f:\mathbb{C} \to \mathbb{C}$ of degree ≥ 2, then*

$$\sum_{k=1}^{n} \operatorname{ind}(f,p_k) = 0.$$

In fact, a non-identity rational function in reduced form P/Q has $d+1$ fixed points counting multiplicities, where $d = \max\{\deg(P),\deg(Q)\}$.

PROOF OF THEOREM 3.38. Evidently f has finitely many fixed points. Let E be the set of fixed points of f in $\mathbb{C}$ (the case $E = \emptyset$ can occur, as the example $f(z) = z+1$ shows). Then, by Theorem 3.34 and Lemma 3.33(ii),

$$\sum_{p\in E} \operatorname{ind}(f,p) = \sum_{p\in E} \operatorname{res}\left(\frac{1}{z-f(z)}, p\right)$$

$$(3.15) \qquad = -\operatorname{res}\left(\frac{1}{z-f(z)}, \infty\right) = -\operatorname{res}(g,0),$$

where

$$g(z) = -\frac{1}{z^2}\frac{1}{1/z - f(1/z)} = \frac{1}{z(zf(1/z) - 1)}.$$

If $f(\infty) \neq \infty$, we have $\lim_{z\to 0} zf(1/z) = 0$. The above formula then shows that g has a simple pole at $z = 0$ with $\mathrm{res}(g, 0) = \lim_{z\to 0} zg(z) = -1$. Thus, $\sum_{p=f(p)} \mathrm{ind}(f, p) = \sum_{p\in E} \mathrm{ind}(f, p) = 1$, which proves the theorem in this case.

Now assume $f(\infty) = \infty$. Then, by (3.14) and (3.15),

$$\sum_{p=f(p)} \mathrm{ind}(f, p) = \sum_{p\in E} \mathrm{ind}(f, p) + \mathrm{ind}(f, \infty)$$

$$= -\mathrm{res}(g, 0) + \mathrm{res}(h, 0) = \mathrm{res}(h - g, 0),$$

where $h(z) = 1/(z - \hat{f}(z))$. But

$$h(z) - g(z) = \frac{f(1/z)}{zf(1/z) - 1} - \frac{1}{z(zf(1/z) - 1)} = \frac{1}{z},$$

which shows $\mathrm{res}(h - g, 0) = 1$, and completes the proof. $\qquad\square$

Further applications of residues, including their use in computing real and improper integrals, are outlined in the problems at the end of this chapter. For applications of Theorem 3.38 in the dynamics of rational maps, see [**Mi**].

3.5 The argument principle

The number of times a non-constant meromorphic function takes on a certain value in a given domain has both topological and analytic descriptions. This dual interpretation, often known as the "argument principle," is another beautiful example of the interplay between complex analysis and topology and turns out to be very useful in a variety of situations.

To get a taste of what is at work here, consider the behavior of a holomorphic function f which has a zero of order 3 at p and so, by Theorem 1.59, behaves locally as the power map $w \mapsto w^3$. Let γ be a *small* positively oriented Jordan curve that winds around p. Counting multiplicity, the number of solutions of the equation $f(z) = 0$ in $\mathrm{int}(\gamma)$ is clearly 3. This is the same as the number of times the image curve $f \circ \gamma$ winds around the origin in the counterclockwise direction (see Fig. 3.5). At the same time, it is the value of the "logarithmic integral" $1/(2\pi i) \int_\gamma f'(z)/f(z)\, dz$: Write $f(z) = (z - p)^3 f_1(z)$, with f_1 non-zero and holomorphic near p, and note that

$$\frac{1}{2\pi i}\int_\gamma \frac{f'(z)}{f(z)}\, dz = \frac{1}{2\pi i}\int_\gamma \left(\frac{3}{z - p} + \frac{f_1'(z)}{f_1(z)}\right) dz = 3,$$

where the second term integrates to zero by Cauchy's theorem since f_1'/f_1 is holomorphic.

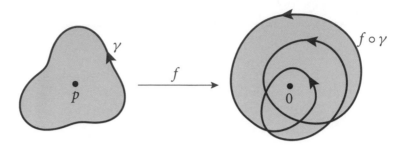

Figure 3.5. A holomorphic function f with a zero of order 3 at p. Counting multiplicity, the number of times f takes the value 0 in the interior of the small loop γ is the same as the number of times the image curve $f \circ \gamma$ winds around the origin.

We now show how the local statement

"number of roots = winding number = logarithmic integral"

can be globalized. The equality "winding number = logarithmic integral" is straightforward from the definition (Theorem 3.40 below), while the equality "logarithmic integral = number of roots" follows from the residue theorem (Theorem 3.42 below). In both cases the proof works more generally for meromorphic functions.

Take a domain $U \subset \mathbb{C}$ and a piecewise C^1 curve $\gamma : [0, 1] \to U$. Suppose $f \in \mathcal{M}(U)$ has no zeros or poles on $|\gamma|$. Then,

$$\frac{1}{2\pi i} \int_\gamma \frac{f'(z)}{f(z)} \, dz = \frac{1}{2\pi i} \int_0^1 \frac{f'(\gamma(t))}{f(\gamma(t))} \, \gamma'(t) \, dt = \frac{1}{2\pi i} \int_0^1 \frac{(f \circ \gamma)'(t)}{(f \circ \gamma)(t)} \, dt$$

$$= \frac{1}{2\pi i} \int_{f \circ \gamma} \frac{d\zeta}{\zeta} = W(f \circ \gamma, 0),$$

where the last equality follows from Theorem 2.29. The computation extends by linearity to the case where $\gamma = \sum_{k=1}^m n_k \gamma_k$ is a chain in U, once we define

$$f \circ \gamma = \sum_{k=1}^m n_k \, (f \circ \gamma_k).$$

Thus, we arrive at the following

THEOREM 3.40. *For any chain γ in a domain U and any $f \in \mathcal{M}(U)$ without zeros or poles on $|\gamma|$,*

$$\frac{1}{2\pi i} \int_\gamma \frac{f'(z)}{f(z)} \, dz = W(f \circ \gamma, 0).$$

When γ is a null-homologous cycle in U, the residue theorem can be used to compute the logarithmic integral $\int_\gamma f'/f$. This will be the content of Theorem 3.42 below.

NOTATION 3.41. Suppose f is a non-constant meromorphic function in a domain U and D is an open set whose closure $\overline{D}$ is a compact subset of U. For $q \in \hat{\mathbb{C}}$, we set

$$\mathrm{N}_f(D,q) = \sum_{p \in f^{-1}(q) \cap D} \deg(f,p).$$

In other words, $\mathrm{N}_f(D,q)$ is the number of solutions of the equation $f(z) = q$ in D, counting multiplicities. Note that the set $f^{-1}(q) \cap D$ is finite (possibly empty) by the identity theorem, so the above sum is well defined.

THEOREM 3.42 (The argument principle). *Let γ be a null-homologous cycle in a domain $U \subset \mathbb{C}$. If $f \in \mathscr{M}(U)$ has no zeros or poles on $|\gamma|$, then*

(3.16)
$$\frac{1}{2\pi i} \int_\gamma \frac{f'(z)}{f(z)}\, dz = \sum_{p \in f^{-1}(0)} \mathrm{W}(\gamma,p)\, \deg(f,p)$$
$$- \sum_{p \in f^{-1}(\infty)} \mathrm{W}(\gamma,p)\, \deg(f,p).$$

In particular, if γ is a positively oriented Jordan curve in U such that $D = \mathrm{int}(\gamma) \subset U$, then

(3.17)
$$\frac{1}{2\pi i} \int_\gamma \frac{f'(z)}{f(z)}\, dz = \mathrm{N}_f(D,0) - \mathrm{N}_f(D,\infty).$$

Why is it called the argument principle? By Theorem 3.40 the left side of (3.16) is the net change in the argument of $f(z)$ as z traverses $|\gamma|$.

The right side of (3.17) is the number of zeros minus the number of poles of f in D, both counting multiplicities.

PROOF. First note that f does not vanish identically in U since $f \neq 0$ on $|\gamma|$ and U is connected. Hence the function $g = f'/f$ is meromorphic in U. Each pole p of g is either a zero or a pole of f. First consider the case where $p \in f^{-1}(0)$. Write $f(z) = (z-p)^m f_1(z)$, where $m = \deg(f,p)$ and f_1 is holomorphic near p with $f_1(p) \neq 0$. Then

$$g(z) = \frac{m}{z-p} + \frac{f_1'(z)}{f_1(z)}.$$

Since the function f_1'/f_1 is holomorphic near p, the definition of residue gives

(3.18)
$$\mathrm{res}(g,p) = m = \deg(f,p).$$

Next, consider the case where $p \in f^{-1}(\infty)$. Write $f(z) = (z-p)^{-m} f_1(z)$, where $m = \deg(f,p)$ and again f_1 is holomorphic near p with $f_1(p) \neq 0$. This time

$$g(z) = \frac{-m}{z-p} + \frac{f_1'(z)}{f_1(z)},$$

so

(3.19)
$$\mathrm{res}(g,p) = -\deg(f,p).$$

Now the residue theorem together with (3.18) and (3.19) shows that

$$\frac{1}{2\pi i}\int_\gamma g(z)\,dz = \sum_{p\in f^{-1}(\{0,\infty\})} W(\gamma,p)\ \mathrm{res}(g,p)$$

$$= \sum_{p\in f^{-1}(0)} W(\gamma,p)\ \deg(f,p) - \sum_{p\in f^{-1}(\infty)} W(\gamma,p)\ \deg(f,p).$$

This proves (3.16). The formula (3.17) follows from (3.16) since by Theorem 2.28, $W(\gamma,p)$ is 1 if $p \in D$ and is 0 otherwise. □

Combining Theorems 3.40 and 3.42, we arrive at the following

COROLLARY 3.43. *Let γ be a positively oriented Jordan curve in a domain U such that $\mathrm{int}(\gamma) \subset U$. Suppose $f \in \mathcal{M}(U)$ has no zeros or poles on $|\gamma|$. Then the number of zeros minus the number of poles of f in $\mathrm{int}(\gamma)$ equals the number of times the closed curve $f \circ \gamma$ winds around the origin in the counterclockwise direction.*

EXAMPLE 3.44. Consider the rational function

$$f(z) = \frac{z^2}{(z-1)(z+2i)}$$

For an animated demonstration of this example, visit http://qcpages.qc.cuny.edu/~zakeri/CAbook/AP.html

with a zero of order 2 at 0 and simple poles at 1 and $-2i$. Let $\gamma_r : [0, 2\pi] \to \mathbb{C}$ be the closed curve defined by $\gamma_r(t) = f(re^{it})$. By Corollary 3.43, the curve $\gamma_{0.5}$ winds around the origin twice, while $\gamma_{1.5}$ winds around the origin only once, and $\gamma_{2.5}$ does not wind around the origin at all (see Fig. 3.6).

As a good exercise, try to visualize what happens to γ_r qualitatively as r crosses the values 1 and 2. This should also explain why the curve $\gamma_{2.5}$ is negatively oriented.

Theorem 3.42 shows how to count the number of zeros of a given holomorphic function f in a domain without knowing their exact location. This is a theoretically important result but has limited computational value since to evaluate the logarithmic integral $\int_\gamma f'/f$, say, using residues, we need to know where the zeros of f are, and if we had that knowledge we could count them directly! One possible way out of this dilemma is to compare f with a simpler function for which the location of zeros is easier to find out. This is the motivating idea behind Theorem 3.46 below.

First we need a topological lemma:

This is the *dog-on-a-leash lemma*: A man walks his dog (on a retractable leash) around a tree, always keeping the leash shorter than his distance to the tree. No matter how he and his dog move, they wind around the tree the same number of times.

LEMMA 3.45. *Let $\eta, \xi : [0, 1] \to \mathbb{C}^*$ be closed curves which satisfy*

$$|\eta(t) - \xi(t)| < |\xi(t)| \qquad \text{for all } t \in [0, 1].$$

Then

$$W(\eta, 0) = W(\xi, 0).$$

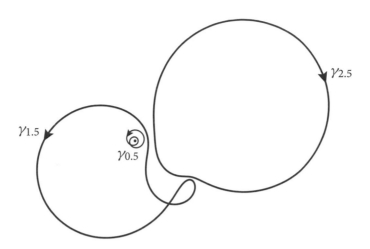

Figure 3.6. The closed curves $\gamma_r(t) = f(re^{it})$ of Example 3.44. The dot shows the location of the origin 0.

PROOF. By Theorem 2.26(v) it suffices to show that η and ξ are freely homotopic in $\mathbb{C}^*$. Consider the free homotopy $H : [0,1] \times [0,1] \to \mathbb{C}$ between η and ξ defined by

$$H(t,s) = (1-s)\eta(t) + s\xi(t).$$

Suppose $H(t,s) = 0$ for some t and s. Then $(1-s)(\eta(t) - \xi(t)) = -\xi(t)$, which implies

$$|\eta(t) - \xi(t)| \geq (1-s)|\eta(t) - \xi(t)| = |\xi(t)|,$$

which contradicts the assumption on η, ξ. Thus, the free homotopy H takes values in $\mathbb{C}^*$, as required. □

THEOREM 3.46 (Rouché, 1862). *Let γ be a Jordan curve in a domain U such that $D = \text{int}(\gamma) \subset U$. Suppose $f, g \in \mathscr{O}(U)$ satisfy*

$$|f(z) - g(z)| < |g(z)| \qquad \text{for all } z \in |\gamma|.$$

Then, counting multiplicities, f and g have the same number of zeros in D:

$$N_f(D,0) = N_g(D,0).$$

PROOF. The assumed inequality shows that neither f nor g can vanish on $|\gamma|$. Hence $\eta = f \circ \gamma$ and $\xi = g \circ \gamma$ are closed curves in $\mathbb{C}^*$, and $|\eta(t) - \xi(t)| < |\xi(t)|$ for all t. After replacing γ by its reverse γ^- if necessary, we may assume that γ is positively oriented. Corollary 3.43 and Lemma 3.45 then show that

Eugène Rouché
(1832–1910)

$$N_f(D,0) = W(\eta,0) = W(\xi,0) = N_g(D,0). \qquad □$$

For a rather different proof as well as a more general version of Rouché's theorem, see problems 36 and 37.

EXAMPLE 3.47. The polynomial $f(z) = z^5 + 15z + 1$ has all its roots in the disk $\mathbb{D}(0,2)$ but only one of these roots lies in the disk $\mathbb{D}(0,3/2)$. To verify the first claim, let $g(z) = z^5$ and observe that if $|z| = 2$, then

$$|f(z) - g(z)| = |15z + 1| \leq 15|z| + 1 = 31 < 32 = |g(z)|.$$

Hence, by Theorem 3.46,

$$N_f(\mathbb{D}(0,2), 0) = N_g(\mathbb{D}(0,2), 0) = 5.$$

To verify the second claim, let $g(z) = 15z + 1$ and note that if $|z| = 3/2$, then

$$|f(z) - g(z)| = |z^5| = \frac{243}{32} < \frac{43}{2} = 15|z| - 1 \leq |g(z)|.$$

Hence,

$$N_f(\mathbb{D}(0,3/2), 0) = N_g(\mathbb{D}(0,3/2), 0) = 1.$$

The five roots of f have approximate absolute values of 0.066666578, 1.956251223, and 1.979799860.

Problems

(1) Recall that $\cos z = (e^{iz} + e^{-iz})/2$ and $\sin z = (e^{iz} - e^{-iz})/(2i)$. Find the isolated singularities of the functions

$$\frac{z}{\sin z} \qquad \frac{1 - \cos z}{z^2} \qquad \sin\left(\frac{1}{z}\right)$$

and determine their type.

(2) What can you say about a meromorphic function f in the plane which satisfies the inequality $|f(z)| \leq 1/|z - 3|^2$ for all $z \neq 3$?

(3) Suppose $f, g \in \mathcal{M}(\mathbb{C})$ and $|f(z)| \leq |g(z)|$ for all $z \in \mathbb{C}$ that is not a pole of f or g. Show that $f = cg$ for some constant $c \in \mathbb{C}$.

(4) Let P and Q be complex polynomials without common roots. Suppose $Q(p) = 0$ and $m = \operatorname{ord}(Q, p)$. Find the principal part of the rational function P/Q at the pole p if $m = 1$ and if $m = 2$.

(5) Let p be an isolated singularity of f. If p is not removable, show that the function $\exp(f)$ has an essential singularity at p.

(6) Suppose $f \in \mathcal{O}(\mathbb{C})$ and $\lim_{z \to \infty} f(z) = \infty$. Show that f is a polynomial.

(7) Represent the rational function

$$\frac{z}{z^2 - 4} + \frac{1}{z - 1}$$

by a Laurent series $\sum_{n=-\infty}^{\infty} a_n z^n$. How many such representations are there? In which annulus is each of them valid? (Compare Example 3.23.)

(8) Prove that if f is bounded and holomorphic in $\{z : |z - p| > R\}$, then the Laurent series of f has the form $\sum_{n \leq 0} a_n (z - p)^n$ (all positive powers missing) and the convergence is uniform in $\{z : |z - p| > R + \varepsilon\}$ for every small $\varepsilon > 0$. (Hint: Use the ML-inequality to estimate the coefficients of the Laurent series in (3.5).)

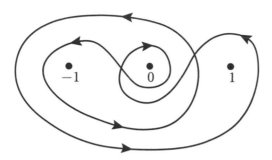

Figure 3.7. The closed curve γ of problem 10.

(9)   Use Laurent series to give another proof of Riemann's removable singularity theorem.

(10)   Compute the integrals

$$\int_{\mathbb{T}} \frac{e^{3z} - e^{-2z}}{z^4}\, dz \quad \text{and} \quad \int_{\gamma} \frac{dz}{z^3(z^2 - 1)},$$

where γ is the closed curve shown in Fig. 3.7.

(11)   Show that Cauchy's integral formula for higher derivatives (Corollary 2.48) follows from the residue Theorem 3.30.

(12)   Let $U \subset \mathbb{C}$ be a simply connected domain and $f \in \mathcal{M}(U)$. Show that f has a primitive (that is, a function $F \in \mathcal{M}(U)$ such that $F' = f$ away from the poles) if and only if $\mathrm{res}(f, p) = 0$ for every pole p of f. (Hint: For the "if" part, imitate the proof of Theorem 1.31. Let $E = f^{-1}(\infty)$, fix a base point $z_0 \in U \smallsetminus E$, and define $F(z) = \int_{\gamma} f(\zeta)\, d\zeta$, where γ is any curve in $U \smallsetminus E$ from z_0 to z.)

(13)   Suppose φ maps a neighborhood of p biholomorphically to a neighborhood of $q = \varphi(p)$. Show that for every holomorphic function f with an isolated singularity at q,

$$\mathrm{res}(f, q) = \mathrm{res}((f \circ \varphi)\, \varphi', p).$$

This shows that although the residue of the function $f(z)$ is not invariant under local change of coordinates, the residue of the differential $f(z)dz$ is.

(14)   Show that if P and Q are polynomials with $\deg Q \geq \deg P + 2$, then $\mathrm{res}(P/Q, \infty) = 0$.

(15)   Let P be a quadratic polynomial with distinct fixed points $z_1, z_2 \in \mathbb{C}$. Show that $P'(z_1) + P'(z_2) = 2$.

(16)   Let $P(z) = (z - z_1) \cdots (z - z_n)$, where the $z_1, \ldots, z_n$ are distinct complex numbers and $n \geq 2$. Show that $\sum_{k=1}^{n} 1/P'(z_k) = 0$. (Hint: Use (3.13) and Corollary 3.39.)

(17)   If ∞ is an isolated fixed point of a holomorphic map f, show that

$$\mathrm{ind}(f, \infty) = \mathrm{res}\left(\frac{f(z)}{z(z - f(z))}, \infty\right).$$

(18)   Compute $\mathrm{res}(1/(z - f(z)), \infty)$ when f is a polynomial of degree $d \geq 1$ other than the identity map. You will need to consider the cases $d = 1$ and $d > 1$ separately.

(19)   (Fractional residue) Let $f \in \mathscr{O}(\mathbb{D}^*)$ have a simple pole at 0 (here $\mathbb{D}^* = \mathbb{D}^*(0, 1) = \mathbb{D} \smallsetminus \{0\}$). For $0 \leq \alpha \leq 2\pi$ and $0 < r < 1$, let $\gamma_r : [0, \alpha] \to \mathbb{C}$ be the curve defined by $\gamma_r(t) = re^{it}$. Show that

$$\lim_{r \to 0} \int_{\gamma_r} f(z)\, dz = \alpha\, i\ \operatorname{res}(f, 0).$$

(20)   Let P and Q be polynomials with real coefficients. Suppose $Q(x) \neq 0$ for all $x \in \mathbb{R}$ and $\deg Q \geq \deg P + 2$. Show that

$$\int_{-\infty}^{+\infty} \frac{P(x)}{Q(x)}\, dx = 2\pi i \sum_k \operatorname{res}\left(\frac{P}{Q}, p_k\right),$$

where the sum is taken over all roots p_k of Q in the upper half-plane. (Hint: For $a > 0$, let γ_a be the positively oriented Jordan curve formed by the real segment from $-a$ to a, followed by the semicircle of radius a in the upper half-plane from a back to $-a$. Apply the residue theorem to $\int_{\gamma_a} P(z)/Q(z)\, dz$ for large a, and use the ML-inequality to show that the integral along the semicircle part tends to zero as $a \to +\infty$.)

(21)   Evaluate $\displaystyle\int_{-\infty}^{+\infty} \frac{x^2}{x^4 + 1}\, dx$.

(22)   Show that for every integer $n \geq 2$,

$$\int_0^\infty \frac{dx}{x^n + 1} = \frac{\pi/n}{\sin(\pi/n)}.$$

(Hint: For $a > 0$, let γ_a be the positively oriented Jordan curve formed by the real segment from 0 to a, followed by the circular arc of radius a in the upper half-plane from a to $ae^{2\pi i/n}$, followed by the radial segment from $ae^{2\pi i/n}$ back to 0. Apply the residue theorem to $\int_{\gamma_a} dz/(z^n + 1)$ for large a, relate the integrals along the two segments and show that the integral along the circular arc tends to zero as $a \to +\infty$.)

(23)   Residues can be used to evaluate integrals of the form

$$\int_0^{2\pi} f(\cos t, \sin t)\, dt,$$

where f is a rational function of two variables. Parametrize the unit circle $\mathbb{T}$ by $z = e^{it}$ for $0 \leq t \leq 2\pi$, so

$$\cos t = \frac{e^{it} + e^{-it}}{2} = \frac{z + z^{-1}}{2}, \qquad \sin t = \frac{e^{it} - e^{-it}}{2i} = \frac{z - z^{-1}}{2i}.$$

This turns the above integral into

$$\int_{\mathbb{T}} f\left(\frac{z + z^{-1}}{2}, \frac{z - z^{-1}}{2i}\right) \frac{dz}{iz}.$$

Assuming the integrand, now a rational function of z, has no poles on the unit circle, the integral can be evaluated by applying the residue theorem. Use this method to show that

$$\int_0^{2\pi} \frac{dt}{a + \sin t} = \frac{2\pi}{\sqrt{a^2 - 1}} \qquad \text{for every } a > 1.$$

(24)   Deduce the fundamental theorem of algebra from the argument principle.

(25) Prove the following statements:
  (i) Let γ be a Jordan curve, $D = \text{int}(\gamma)$, and f be holomorphic in a neighborhood of $\overline{D}$. If f is injective on $|\gamma|$, then it is injective in D.
  (ii) If $f \in \mathcal{O}(\mathbb{D})$ is not injective, there exist distinct points $z, w \in \mathbb{D}$ with $|z| = |w|$ such that $f(z) = f(w)$.

(26) Let $U \subset \mathbb{C}$ be a domain. Show that a non-vanishing $f \in \mathcal{O}(U)$ has a holomorphic logarithm in U if and only if it has a holomorphic n-th root in U for every $n \geq 2$. (Hint: The "only if" part is trivial. For the "if" part, it suffices to show that f'/f has a primitive in U, or equivalently $I = \int_\gamma (f'(z)/f(z))\, dz = 0$ for every closed curve γ in U. Verify that f having a holomorphic n-th root implies that I is an integer multiple of $2\pi i n$.)

(27) (Generalized argument principle) Let γ be a positively oriented Jordan curve in a domain U such that $\text{int}(\gamma) \subset U$. Suppose $f \in \mathcal{M}(U)$ has no zeros or poles on $|\gamma|$. Let $\{z_k\}_{k=1}^m$ and $\{p_k\}_{k=1}^n$ denote the zeros and poles of f in $\text{int}(\gamma)$. If $\varphi \in \mathcal{O}(U)$, show that

$$\frac{1}{2\pi i} \int_\gamma \varphi(z) \frac{f'(z)}{f(z)}\, dz = \sum_{k=1}^m \deg(f, z_k)\, \varphi(z_k) - \sum_{k=1}^n \deg(f, p_k)\, \varphi(p_k).$$

The standard argument principle corresponds to the case $\varphi = 1$.

(28) Suppose f is holomorphic in a neighborhood of the closed unit disk $\overline{\mathbb{D}}$, and

$$\int_{\mathbb{T}} \frac{f'(z)}{f(z)}\, dz = 4\pi i, \qquad \int_{\mathbb{T}} z \frac{f'(z)}{f(z)}\, dz = -2\pi, \qquad \int_{\mathbb{T}} z^2 \frac{f'(z)}{f(z)}\, dz = -\pi i.$$

Locate the zeros of f in $\mathbb{D}$.

(29) (A formula for the inverse) Let γ be a positively oriented Jordan curve in a domain U such that $D = \text{int}(\gamma) \subset U$. Suppose $f \in \mathcal{O}(U)$ is injective in D, so $f : D \to f(D)$ is a biholomorphism. Show that

$$f^{-1}(w) = \frac{1}{2\pi i} \int_\gamma \frac{\zeta f'(\zeta)}{f(\zeta) - w}\, d\zeta$$

for every $w \in f(D)$.

(30) (Holomorphic implicit function theorem) Let U and V be domains in $\mathbb{C}$ and $\varphi : U \times V \to \mathbb{C}$ be a continuous function such that $z \mapsto \varphi(z, t)$ is holomorphic in U for each $t \in V$, with complex derivative $\varphi_z(z, t)$. Suppose $\varphi(z_0, t_0) = 0$ and $\varphi_z(z_0, t_0) \neq 0$ for some $(z_0, t_0) \in U \times V$. Choose $r > 0$ small enough so that $\varphi(z, t_0) \neq 0$ if $0 < |z - z_0| \leq r$.
  (i) Show that for small $\varepsilon > 0$, $\varphi(z, t) \neq 0$ if $(z, t) \in \mathbb{T}(z_0, r) \times \mathbb{D}(t_0, \varepsilon)$, and

$$\frac{1}{2\pi i} \int_{\mathbb{T}(z_0, r)} \frac{\varphi_z(\zeta, t)}{\varphi(\zeta, t)}\, d\zeta = 1 \qquad \text{for } t \in \mathbb{D}(t_0, \varepsilon).$$

  Conclude that for each $t \in \mathbb{D}(t_0, \varepsilon)$ the equation $\varphi(z, t) = 0$ has a unique solution $z = z(t)$ in $\mathbb{D}(z_0, r)$.
  (ii) Prove the formula

$$z(t) = \frac{1}{2\pi i} \int_{\mathbb{T}(z_0, r)} \zeta \frac{\varphi_z(\zeta, t)}{\varphi(\zeta, t)}\, d\zeta \qquad \text{for } t \in \mathbb{D}(t_0, \varepsilon).$$

(iii)   Show that

$$\varphi(z,t) = (z - z(t))\,\psi(z,t) \qquad \text{for } (z,t) \in \mathbb{D}(z_0, r) \times \mathbb{D}(t_0, \varepsilon),$$

where ψ is continuous and non-zero, with $z \mapsto \psi(z,t)$ holomorphic for each t.

(iv)   Assume further that $t \mapsto \varphi(z,t)$ is holomorphic in V for each $z \in U$, with complex derivative $\varphi_t(z,t)$. Show that $z(t)$ is holomorphic and

$$z'(t) = -\frac{\varphi_t(z(t),t)}{\varphi_z(z(t),t)} \qquad \text{for } t \in \mathbb{D}(t_0, \varepsilon).$$

(31)   Show that the equation $z^4 - 5z + 1 = 0$ has all its roots in the disk $\mathbb{D}(0,2)$. How many of these roots are in the unit disk $\mathbb{D}$?

(32)   Let $a > 1$. Use Rouché's theorem to show that the equation $z\exp(a - z) = 1$ has exactly one root in $\mathbb{D}$. Verify that this root is real and positive.

(33)   Let $\varepsilon > 0$ and $f : \mathbb{D}(0, 1 + \varepsilon) \to \mathbb{D}$ be holomorphic. Use Rouché's theorem to show that there exists a unique $p \in \mathbb{D}$ such that $f(p) = p$.

(34)   Redo problem 1 of chapter 1 using Rouché's theorem.

(35)   Suppose $f \in \mathcal{O}(\mathbb{D})$ has the power series $z + \sum_{n=2}^{\infty} a_n z^n$. If $\rho = \sum_{n=2}^{\infty} |a_n| < 1$, show that the inverse function f^{-1} is defined in the disk $\mathbb{D}(0, 1 - \rho)$. (Hint: First assume f is holomorphic in a neighborhood of $\overline{\mathbb{D}}$ and use Rouché's theorem to show that for each $w \in \mathbb{D}(0, 1 - \rho)$ the equation $f(z) = w$ has a unique solution in $\mathbb{D}$.)

(36)   Complete the following outline of an alternative proof of Rouché's theorem: Under the assumptions of Theorem 3.46, the values of the meromorphic function $h = f/g$ on $|\gamma|$ belong to the open disk $\mathbb{D}(1, 1)$. Any closed curve in this disk must have winding number zero with respect to the origin, hence by the argument principle, $N_h(D, 0) = N_h(D, \infty)$. This implies the equality $N_f(D, 0) = N_g(D, 0)$ (be cautious about when f and g have common zeros).

(37)   Prove the following symmetric (and sharper) version of Rouché's theorem [**Gl**]: If γ is a Jordan curve in U with $D = \text{int}(\gamma) \subset U$ and if $f, g \in \mathcal{O}(U)$ satisfy

$$|f(z) - g(z)| < |f(z)| + |g(z)| \qquad \text{for all } z \in |\gamma|,$$

then $N_f(D, 0) = N_g(D, 0)$. (Hint: The inequality shows that the values of the meromorphic function $h = f/g$ on $|\gamma|$ belong to the slit plane $\mathbb{C} \smallsetminus (-\infty, 0]$. Now proceed as in the previous problem.)

4

Möbius maps and the Schwarz lemma

4.1 The Möbius group

Rational functions of degree 1 are ubiquitous objects in complex analysis. They exhibit rich algebraic, geometric, dynamical, and arithmetical structures.

> **DEFINITION 4.1.** A *Möbius map* is a rational function of the form
> $$z \mapsto \frac{az+b}{cz+d},$$
> where $a, b, c, d \in \mathbb{C}$ and $ad - bc \neq 0$.

Möbius maps are also called "fractional linear transformations."

Every Möbius map $f(z) = (az+b)/(cz+d)$ is a one-to-one holomorphic map $\hat{\mathbb{C}} \to \hat{\mathbb{C}}$ whose inverse $f^{-1}(z) = (dz-b)/(-cz+a)$ is also Möbius. Note that $f(\infty) = a/c$ or ∞ according as $c \neq 0$ or $c = 0$; similarly, $f^{-1}(\infty) = -d/c$ or ∞ according as $c \neq 0$ or $c = 0$. Thus, every Möbius map defines a biholomorphism $\hat{\mathbb{C}} \to \hat{\mathbb{C}}$.

The collection of all Möbius maps is a group under composition which we call the *Möbius group* and denote by Möb.

Algebraically, Möb can be identified with a matrix group. There is an obvious way of associating a Möbius map to every non-singular 2×2 matrix over $\mathbb{C}$:

$$\text{to} \quad A = \begin{bmatrix} a & b \\ c & d \end{bmatrix} \quad \text{associate} \quad f_A(z) = \frac{az+b}{cz+d}.$$

Under this association, multiplication of matrices corresponds to composition of maps:

$$f_{AB} = f_A \circ f_B.$$

However, this association is not one-to-one because simultaneous multiplication of the entries of a matrix by a non-zero complex number would yield the same map:

$$f_{\alpha A} = f_A \qquad \text{for all } \alpha \in \mathbb{C}^*.$$

We can remedy this problem by normalizing our matrices, for example by choosing among all multiples of A a representative αA for which $\det(\alpha A) = \alpha^2 \det(A) = 1$. Since there are exactly two such multiples which only differ by a sign, it follows that every Möbius map is of the form f_A for a matrix A with $\det(A) = 1$, and A is unique up to multiplication by -1. Thus, Möb is isomorphic to the quotient group

PSL stands for "Projective Special Linear."

(4.1)
$$\mathrm{PSL}_2(\mathbb{C}) = \left\{ \begin{bmatrix} a & b \\ c & d \end{bmatrix} : a, b, c, d \in \mathbb{C}, ad - bc = 1 \right\} / \{\pm I\},$$

where $I = \begin{bmatrix} 1 & 0 \\ 0 & 1 \end{bmatrix}$ is the identity matrix.

We will use three special types of Möbius maps whose actions are easy to understand:

- **translations** $T_\beta : z \mapsto z + \beta$, where $\beta \in \mathbb{C}$;

(4.2)   - **linear maps** $L_\alpha : z \mapsto \alpha z$, where $\alpha \in \mathbb{C}^*$; and

- the **inversion** across the unit circle: $\iota : z \mapsto 1/z$.

THEOREM 4.2 (Basic properties of **Möb**).

  (i) *The group* Möb *is generated by translations, linear maps, and the inversion across the unit circle.*
 (ii) *The action of* Möb *on* $\hat{\mathbb{C}}$ *is "simply 3-transitive," that is, for any given triples* (p_1, p_2, p_3) *and* (q_1, q_2, q_3) *of distinct points in* $\hat{\mathbb{C}}$, *there exists a unique element of* Möb *which sends* p_k *to* q_k *for* $k = 1, 2, 3$.
(iii) *The action of* Möb *preserves the family of circles in* $\hat{\mathbb{C}}$.

By a **circle** in $\hat{\mathbb{C}}$ we mean a Euclidean circle in the plane or a straight line with the point ∞ added (the latter can be thought of as a circle passing through ∞). One can verify that these are precisely the images of Euclidean circles on $\mathbb{S}^2 \subset \mathbb{R}^3$ under the stereographic projection introduced in §3.2 (see problem 1).

PROOF. (i) Take any $f : z \mapsto (az + b)/(cz + d)$ in Möb and consider two cases: If $c = 0$, then $ad \neq 0$ and

$$f = T_\beta \circ L_\alpha, \quad \text{where} \quad \alpha = \frac{a}{d} \quad \text{and} \quad \beta = \frac{b}{d}.$$

If $c \neq 0$, a brief computation shows that

$$f(z) = \frac{a}{c} + \frac{bc - ad}{c^2} \frac{1}{z + d/c},$$

so

$$f = T_\beta \circ L_\alpha \circ \iota \circ T_\gamma,$$

where

$$\alpha = \frac{bc - ad}{c^2}, \quad \beta = \frac{a}{c}, \quad \text{and} \quad \gamma = \frac{d}{c}.$$

(ii) It suffices to show that every triple (p_1, p_2, p_3) can be sent to $(0, 1, \infty)$ by a unique Möbius map. The existence of such a map is easy to see; simply take

$$f(z) = \frac{(p_2 - p_3)}{(p_2 - p_1)} \frac{(z - p_1)}{(z - p_3)}.$$

(If one of the p_k is ∞, the formula for f must be interpreted as the limiting case when $p_k \to \infty$. Thus

$$f(z) = \frac{p_2 - p_3}{z - p_3} \quad \text{or} \quad \frac{z - p_1}{z - p_3} \quad \text{or} \quad \frac{z - p_1}{p_2 - p_1}$$

August Ferdinand
Möbius (1790–1868)

if p_1 or p_2 or p_3 is ∞, respectively.) To see uniqueness, suppose $g \in \text{Möb}$ also sends (p_1, p_2, p_3) to $(0, 1, \infty)$. Then $f \circ g^{-1}$ is a Möbius map which fixes 0, 1, and ∞. Writing $f \circ g^{-1}(z) = (az + b)/(cz + d)$ and imposing these three conditions, it follows that $b = c = 0$ and $a = d$, which gives $f \circ g^{-1}(z) = z$ or $f = g$.

(iii) We need to show that the image of a Euclidean circle or line in $\mathbb{C}$ under any Möbius map is again a circle or line. By (i), it suffices to verify this for the special types of maps in (4.2). Since the claim is clearly true for translations and linear maps, we need only consider the inversion ι. A line in $\mathbb{C}$ has an equation $ax + by + c = 0$, where a, b, c are real and a, b are not simultaneously zero. Switching to the complex-variable notation $z = x + iy$, this becomes

(4.3)
$$\bar{p}z + p\bar{z} + c = 0,$$

where $p = (a + ib)/2$ is a non-zero complex number and c is real. On the other hand, a Euclidean circle of radius $r > 0$ centered at $p \in \mathbb{C}$ has an equation $|z - p| = r$ or $(z - p)(\bar{z} - \bar{p}) = r^2$, which can be written as

(4.4)
$$|z|^2 - (\bar{p}z + p\bar{z}) + |p|^2 - r^2 = 0.$$

Comparing (4.3) and (4.4), we see that every circle in $\hat{\mathbb{C}}$ has an equation of the form

(4.5)
$$t|z|^2 - (\bar{q}z + q\bar{z}) + s = 0,$$

where t, s are real, q is complex, and $ts < |q|^2$. The case $t = 0$ gives a line as in (4.3). The case $t \neq 0$ gives a circle centered at $p = q/t$ of radius $r = \sqrt{|q|^2 - ts}/|t|$. Now, under the inversion $w = \iota(z) = 1/z$ the equation (4.5) transforms into

$$\frac{t}{|w|^2} - \left(\frac{\bar{q}}{w} + \frac{q}{\bar{w}}\right) + s = 0,$$

or

$$s|w|^2 - (qw + \bar{q}\bar{w}) + t = 0,$$

which is another equation of the form (4.5). This proves the assertion (iii). □

EXAMPLE 4.3 (Communicated to me by G. Yassiyevich). Suppose p_1, p_2, p_3, p_4 are four distinct points lying on a circle $\Gamma \subset \mathbb{C}$ in counterclockwise order. The Möbius map $f(z) = 1/(z - p_1)$ sends

p_1 to ∞, hence Γ to a straight line $L = f(\Gamma)$. Let $q_k = f(p_k)$. Looking at the cyclic order of the points on Γ, we see that q_3 lies between q_2 and q_4 on L. Hence,

$$|q_2 - q_4| = |q_2 - q_3| + |q_3 - q_4|.$$

Substituting $q_k = 1/(p_k - p_1)$ and simplifying, we obtain

(4.6) $$|p_1 - p_3|\,|p_2 - p_4| = |p_1 - p_2|\,|p_3 - p_4| + |p_1 - p_4|\,|p_2 - p_3|.$$

Running this argument backwards, it follows that (4.6) is in fact equivalent to the four points p_k lying on a circle. This is ***Ptolemy's theorem*** in Euclidean geometry: *A quadrilateral can be inscribed in a circle if and only if the sum of the products of its two pairs of opposite sides is equal to the product of its diagonals.*

Möbius maps enjoy the common property of all holomorphic functions with non-vanishing derivative in that they preserve angles. This follows from the fact that the real derivative of such maps is a composition of a rotation and a dilation (equivalently, it acts as multiplication by a non-zero complex number) and thus is an angle-preserving linear transformation (see the discussion in chapter 1 leading to Corollary 1.10). Let us explain this in a little more detail. To begin with, consider two smooth curves $\gamma_1, \gamma_2 : (-1, 1) \to \mathbb{C}$ with $\gamma_1(0) = \gamma_2(0) = p$ such that the derivatives $v_1 = \gamma_1'(0)$ and $v_2 = \gamma_2'(0)$ are non-zero. The complex numbers v_1 and v_2 can be identified with vectors that are tangent to γ_1 and γ_2 at p. Under this identification, the quantity $\mathrm{Re}(v_1 \bar{v}_2)$ is easily seen to correspond to the dot product $v_1 \cdot v_2$. It follows that the angle θ between γ_1 and γ_2 at p is the unique number in $[0, \pi]$ that satisfies

(4.7) $$\cos \theta = \frac{\mathrm{Re}(v_1 \bar{v}_2)}{|v_1|\,|v_2|}.$$

Note that θ is independent of how the curves are parametrized since changing the parametrizations will only multiply v_1 and v_2 by positive numbers.

Now suppose f is holomorphic in some neighborhood of p and $f'(p) = \alpha \neq 0$. Then the tangent vectors $\tilde{v}_1$ and $\tilde{v}_2$ to the image curves $f \circ \gamma_1$ and $f \circ \gamma_2$ at $f(p)$ can be found by the chain rule: For $k = 1, 2$,

$$(f \circ \gamma_k)'(0) = f'(p)\,\gamma_k'(0) \quad \text{so} \quad \tilde{v}_k = \alpha\,v_k$$

(see Fig. 4.1). The right side of (4.7) remains unchanged if we multiply both v_1 and v_2 by α. It follows that the angle between $f \circ \gamma_1$ and $f \circ \gamma_2$ at $f(p)$ equals θ.

THEOREM 4.4 (Angle preservation). *Suppose f is holomorphic in some neighborhood U of p, with $f'(p) \neq 0$, and γ_1 and γ_2 are smooth curves in U which meet at p. Then, the angle between γ_1 and γ_2 at p is the same as the angle between $f \circ \gamma_1$ and $f \circ \gamma_2$ at $f(p)$.*

The theorem holds in particular if $f \in \mathrm{M\ddot{o}b}$ and $U = \hat{\mathbb{C}}$ since f, being a biholomorphism of the sphere, has non-vanishing derivative everywhere. When p or $f(p)$ or both are ∞, the theorem is interpreted as saying that the Möbius map $f \circ \iota$ or $\iota \circ f$ or $\iota \circ f \circ \iota$ preserves angles at 0.

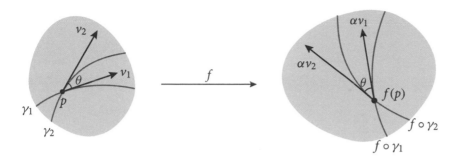

Figure 4.1. A holomorphic function f near a point p with $f'(p) = \alpha \neq 0$ acts infinitesimally as the complex affine map $z \mapsto f(p) + \alpha(z - p)$. As seen from p and $f(p)$, this is just multiplication by the complex number α which, geometrically, is a rotation followed by a dilation. Thus, f preserves the angle between any two smooth curves meeting at p.

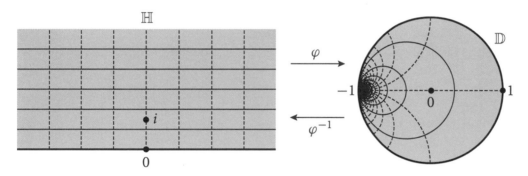

Figure 4.2. The Cayley map $w = \varphi(z) = (i - z)/(i + z)$ and its inverse $z = \varphi^{-1}(w) = i(1 - w)/(1 + w)$ define biholomorphisms between $\mathbb{H}$ and $\mathbb{D}$. Under φ the horizontal lines in $\mathbb{H}$ map to circles in $\mathbb{D}$ that are tangent to $\mathbb{T}$ at -1, while the vertical lines in $\mathbb{H}$ map to circular arcs in $\mathbb{D}$ that are orthogonal to $\mathbb{T}$ at -1.

EXAMPLE 4.5 (The Cayley map). Let us find a Möbius map which sends the ***upper half-plane*** $\mathbb{H} = \{z \in \mathbb{C} : \operatorname{Im}(z) > 0\}$ biholomorphically to $\mathbb{D}$ (of course there are infinitely many such maps). We claim that the unique $\varphi \in$ Möb for which

$$(4.8) \qquad \varphi(0) = 1 \qquad \varphi(i) = 0 \qquad \varphi(\infty) = -1,$$

called the ***Cayley map***, has this property. In fact, φ sends the imaginary axis (the unique circle passing through $0, i, \infty$) to the real axis (the unique circle passing through $1, 0, -1$). The real axis, i.e., the circle passing through $0, \infty$ which meets the imaginary axis orthogonally at 0, must map to a circle passing through $1, -1$ which meets the real axis orthogonally at 1. The only such circle is the unit circle centered at the origin, hence $\varphi(\mathbb{R} \cup \{\infty\}) = \mathbb{T}$. Since φ is a homeomorphism, the upper half-plane must map onto one of the two components of $\hat{\mathbb{C}} \smallsetminus \mathbb{T}$. Because $\varphi(i) = 0$, this component must be $\mathbb{D}$. This proves that φ maps $\mathbb{H}$ biholomorphically onto $\mathbb{D}$ (see Fig. 4.2).

Alternatively, it is easy to check that φ subject to the conditions (4.8) has the formula

(4.9)
$$\varphi(z) = \frac{i-z}{i+z}.$$

We have $\text{Im}(z) > 0$ if and only if z is closer to i than to $-i$. This is equivalent to the condition $|i-z| < |i+z|$ or $|\varphi(z)| < 1$.

We now introduce a geometric invariant associated with four points on the sphere which nicely ties in with the action of the Möbius group.

DEFINITION 4.6. The ***cross ratio*** of an ordered quadruple (p_1, p_2, p_3, p_4) of distinct points in $\hat{\mathbb{C}}$ is defined by

$$[p_1, p_2, p_3, p_4] = \frac{(p_3 - p_1)(p_4 - p_2)}{(p_2 - p_1)(p_4 - p_3)}.$$

If one of the points is ∞, the definition is interpreted as a limit; thus

$$[\infty, p_2, p_3, p_4] = \frac{p_4 - p_2}{p_4 - p_3}, \qquad [p_1, \infty, p_3, p_4] = \frac{p_3 - p_1}{p_3 - p_4},$$

and so on. Observe that $[0, 1, p, \infty] = p$, and $[p_1, p_2, p_3, p_4] > 1$ whenever p_1, p_2, p_3, p_4 all lie on the real line and $p_1 < p_2 < p_3 < p_4$.

THEOREM 4.7 (Properties of the cross ratio).

  (i) *If f is the unique element of Möb which maps (p_1, p_2, p_4) to $(0, 1, \infty)$, then $[p_1, p_2, p_3, p_4] = f(p_3)$. In particular, $[p_1, p_2, p_3, p_4]$ takes values in $\hat{\mathbb{C}} \smallsetminus \{0, 1, \infty\}$.*

  (ii) *Two quadruples (p_1, p_2, p_3, p_4) and (q_1, q_2, q_3, q_4) can be mapped to one another by a Möbius map if and only if*

$$[p_1, p_2, p_3, p_4] = [q_1, q_2, q_3, q_4].$$

  (iii) *If $f : \hat{\mathbb{C}} \to \hat{\mathbb{C}}$ is a homeomorphism which preserves cross ratios of all quadruples, then $f \in$ Möb.*

  (iv) *Four points p_1, p_2, p_3, p_4 lie on a circle in $\hat{\mathbb{C}}$ if and only if $[p_1, p_2, p_3, p_4]$ is real.*

PROOF. (i) The unique $f \in$ Möb which maps (p_1, p_2, p_4) to $(0, 1, \infty)$ has the formula

$$f(z) = \frac{(z - p_1)(p_2 - p_4)}{(z - p_4)(p_2 - p_1)}$$

and it is immediate that $f(p_3) = [p_1, p_2, p_3, p_4]$.

(ii) Take the $\varphi \in$ Möb which maps (p_1, p_2, p_4) to $(0, 1, \infty)$, and the $\psi \in$ Möb which maps (q_1, q_2, q_4) to $(0, 1, \infty)$. If there is an $f \in$ Möb which maps (p_1, p_2, p_3, p_4) to

(q_1, q_2, q_3, q_4), then $\psi \circ f \circ \varphi^{-1} \in$ Möb fixes $0, 1, \infty$, so it must be the identity map. Hence, by part (i),

$$[p_1, p_2, p_3, p_4] = \varphi(p_3) = \psi(f(p_3)) = \psi(q_3) = [q_1, q_2, q_3, q_4].$$

Conversely, if $[p_1, p_2, p_3, p_4] = [q_1, q_2, q_3, q_4]$, then $\varphi(p_3) = \psi(q_3)$, so $\psi^{-1} \circ \varphi \in$ Möb maps (p_1, p_2, p_3, p_4) to (q_1, q_2, q_3, q_4).

(iii) Set $p_1 = f^{-1}(0), p_2 = f^{-1}(1), p_4 = f^{-1}(\infty)$. Parts (i) and (ii) then show that for any $z \in \hat{\mathbb{C}}$ other than p_1, p_2, p_4,

$$f(z) = [0, 1, f(z), \infty] = [f(p_1), f(p_2), f(z), f(p_4)] = [p_1, p_2, z, p_4]$$

$$= \frac{(z - p_1)(p_4 - p_2)}{(p_2 - p_1)(p_4 - z)}.$$

This proves $f \in$ Möb.

(iv) Take the unique $f \in$ Möb that maps (p_1, p_2, p_4) to $(0, 1, \infty)$. Then f maps the circle Γ passing through p_1, p_2, p_4 to $\mathbb{R} \cup \{\infty\}$. Thus, $p_3 \in \Gamma$ if and only if $f(p_3) = [p_1, p_2, p_3, p_4] \in \mathbb{R}$. $\qquad \square$

EXAMPLE 4.8. Suppose p_1, p_2, p_3, p_4 are the vertices of a rectangle in the plane, labeled counterclockwise. Let

$$m = \frac{|p_1 - p_2|}{|p_2 - p_3|}.$$

There exists a complex affine map of the form $z \mapsto \alpha z + \beta$ sending the quadruple (p_1, p_2, p_3, p_4) to $(0, m, m + i, i)$. It follows from Theorem 4.7(ii) that

$$[p_1, p_2, p_3, p_4] = [0, m, m + i, i] = \frac{m^2 + 1}{m^2}.$$

In the special case of a square where $m = 1$, the cross ratio of the four vertices, labeled counterclockwise, is 2.

Let us point out that the choice of the cross ratio formula in Definition 4.6 is not unique, as any permutation of p_1, p_2, p_3, p_4 would lead to another cross ratio with similar properties. Our particular choice is motivated by its connection with the distance formula in the hyperbolic disk (problem 26). To explore the other choices, note that simultaneous swapping of any two pairs in a quadruple leaves the cross ratio unchanged:

$$[p_1, p_2, p_3, p_4] = [p_2, p_1, p_4, p_3] = [p_3, p_4, p_1, p_2] = [p_4, p_3, p_2, p_1].$$

It follows that only 6 out of $4! = 24$ possible permutations can produce new cross ratios. These correspond to pinning down, say, p_3 in the third place and permuting p_1, p_2, p_4 among themselves. The resulting formulas can of course be obtained by direct computation, but here is a more conceptual approach. Let $f \in$ Möb send (p_1, p_2, p_4) to $(0, 1, \infty)$, so $\chi = [p_1, p_2, p_3, p_4] = f(p_3)$. Every permutation

σ of $\{p_1, p_2, p_4\}$ is realized by a unique $\hat{\sigma} \in \text{Möb}$. The permutation $\nu = f \circ \sigma^{-1} \circ f^{-1}$ of $\{0, 1, \infty\}$ is then realized by the map $\hat{\nu} = f \circ \hat{\sigma}^{-1} \circ f^{-1} \in \text{Möb}$ whose formula can be easily derived. The Möbius invariance of the cross ratio now shows that

$$[\sigma(p_1), \sigma(p_2), p_3, \sigma(p_4)] = [p_1, p_2, \hat{\sigma}^{-1}(p_3), p_4]$$

$$= [f(p_1), f(p_2), f(\hat{\sigma}^{-1}(p_3)), f(p_4)]$$

$$= [0, 1, \hat{\nu}(\chi), \infty] = \hat{\nu}(\chi).$$

The result of this computation for various permutations is summarized in the following table:

permutation σ	permutation ν	$\hat{\nu}(z)$	cross ratio formula
identity	identity	z	$[p_1, p_2, p_3, p_4] = \chi$
$p_1 \rightleftarrows p_2, \ p_4 \circlearrowleft$	$0 \rightleftarrows 1, \ \infty \circlearrowleft$	$1 - z$	$[p_2, p_1, p_3, p_4] = 1 - \chi$
$p_1 \rightleftarrows p_4, \ p_2 \circlearrowleft$	$0 \rightleftarrows \infty, \ 1 \circlearrowleft$	$\dfrac{1}{z}$	$[p_4, p_2, p_3, p_1] = \dfrac{1}{\chi}$
$p_2 \rightleftarrows p_4, \ p_1 \circlearrowleft$	$1 \rightleftarrows \infty, \ 0 \circlearrowleft$	$\dfrac{z}{z-1}$	$[p_1, p_4, p_3, p_2] = \dfrac{\chi}{\chi - 1}$
$p_1 \to p_2 \to p_4 \to p_1$	$0 \to \infty \to 1 \to 0$	$\dfrac{z-1}{z}$	$[p_2, p_4, p_3, p_1] = \dfrac{\chi - 1}{\chi}$
$p_1 \to p_4 \to p_2 \to p_1$	$0 \to 1 \to \infty \to 0$	$\dfrac{1}{1-z}$	$[p_4, p_1, p_3, p_2] = \dfrac{1}{1 - \chi}$

4.2 Three automorphism groups

DEFINITION 4.9. Let U be a domain in $\hat{\mathbb{C}}$. The **automorphism group** $\text{Aut}(U)$ is the group of all biholomorphisms $U \to U$ under composition. Thus f belongs to $\text{Aut}(U)$ if and only if $f : U \to U$ is bijective and holomorphic.

Our main goal in this section is to determine $\text{Aut}(U)$ for the three simply connected domains $U = \hat{\mathbb{C}}, \mathbb{C},$ and $\mathbb{D}$. The importance of these cases will be clear when we discuss uniformization of spherical domains in chapter 13.

THEOREM 4.10. $\text{Aut}(\hat{\mathbb{C}}) = \text{Möb}$.

PROOF. Since Möb is a subgroup of $\text{Aut}(\hat{\mathbb{C}})$, it suffices to check that every $f \in \text{Aut}(\hat{\mathbb{C}})$ is a Möbius map. By transitivity, there is a $g \in \text{Möb}$ which maps $f(\infty)$ to ∞, so the composition $h = g \circ f \in \text{Aut}(\hat{\mathbb{C}})$ satisfies $h(\infty) = \infty$. Injectivity of h shows that in fact $h^{-1}(\infty) = \infty$. Since h is a rational map by Theorem 3.20, it must be a polynomial, necessarily of degree 1 since under a polynomial of higher degree each point has two or more preimages. It follows that $h \in$ Möb and therefore $f = g^{-1} \circ h \in$ Möb. □

THEOREM 4.11. Aut$(\mathbb{C})$ *is the subgroup of* Möb *consisting of all affine maps of the form* $z \mapsto az + b$ *with* $a \in \mathbb{C}^*$ *and* $b \in \mathbb{C}$.

Alternatively, Aut$(\mathbb{C})$ can be described as the subgroup of Möb consisting of all maps which fix the point ∞. Geometrically, they constitute all orientation-preserving similarities of the plane, that is, maps that preserve shapes but may change the scale.

PROOF. Clearly every affine map belongs to Aut$(\mathbb{C})$. Let $f \in$ Aut$(\mathbb{C})$ and extend it to a homeomorphism $\hat{\mathbb{C}} \to \hat{\mathbb{C}}$ by setting $f(\infty) = \infty$. By Riemann's removable singularity theorem the extended map is holomorphic in a neighborhood of ∞ since the map $z \mapsto 1/f(1/z)$ is holomorphic in a punctured neighborhood of the origin and tends to 0 as $z \to 0$. Thus $f \in$ Aut$(\hat{\mathbb{C}})$. It follows from Theorem 4.10 that f is a Möbius map. Writing $f(z) = (az + b)/(cz + d)$, the condition $f(\infty) = \infty$ implies $c = 0$, which proves f is affine. □

COROLLARY 4.12. *The action of* Aut$(\mathbb{C})$ *is "simply 2-transitive," that is, for any given pairs* (p_1, p_2) *and* (q_1, q_2) *of distinct points in* $\mathbb{C}$, *there exists a unique element of* Aut$(\mathbb{C})$ *which sends* p_k *to* q_k *for* $k = 1, 2$.

PROOF. This follows from Aut$(\mathbb{C}) \subset$ Aut$(\hat{\mathbb{C}})$ and Theorem 4.2(ii) since there is a unique Möbius map which sends (p_1, p_2, ∞) to (q_1, q_2, ∞). Alternatively, it suffices to show that every pair (p_1, p_2) can be sent to $(0, 1)$ by a unique affine map. The map $f(z) = (z - p_1)/(p_2 - p_1)$ certainly does this. If $g \in$ Aut$(\mathbb{C})$ also sends (p_1, p_2) to $(0, 1)$, then $f \circ g^{-1}$ is an affine map which fixes 0 and 1. Writing $f \circ g^{-1}(z) = az + b$ and imposing these conditions, it follows that $a = 1$ and $b = 0$, which shows $f \circ g^{-1}(z) = z$ or $f = g$. □

To determine the group Aut$(\mathbb{D})$, we first need the following deceptively simple but fundamental result; variations and generalizations of this lemma will be discussed later (see Theorem 4.40 and Theorem 11.31).

THEOREM 4.13 (The Schwarz lemma, 1869). *Suppose* $f : \mathbb{D} \to \mathbb{D}$ *is holomorphic and* $f(0) = 0$. *Then*

This version of the Schwarz lemma was formulated and proved by Carathéodory in 1912.

(4.10) $$|f(z)| \leq |z| \qquad \text{for all } z \in \mathbb{D}$$

and

(4.11) $$|f'(0)| \leq 1.$$

If $|f(z)| = |z|$ *for some* $z \in \mathbb{D}^*$, *or if* $|f'(0)| = 1$, *then* f *is a rigid rotation of the form* $f(z) = \alpha z$, *where* $|\alpha| = 1$.

The inequality $|f'(0)| \leq 1$, even without the assumption $f(0) = 0$, already follows from Theorem 1.43.

PROOF. Let $f(z) = \sum_{n=1}^{\infty} a_n z^n$ be the power series representation of f in $\mathbb{D}$, where the constant term is missing because $f(0) = 0$. The function $g(z) = f(z)/z = \sum_{n=0}^{\infty} a_{n+1} z^n$ has a removable singularity at the origin, so $g \in \mathcal{O}(\mathbb{D})$. If $0 < |z| = r < 1$, then $|g(z)| = |f(z)|/|z| < 1/r$. By the maximum principle, $|g(z)| < 1/r$ whenever $|z| \leq r$. Letting $r \to 1$, we obtain $|g(z)| \leq 1$ for all $z \in \mathbb{D}$. This proves (4.10) and (4.11) simultaneously since $f'(0) = g(0)$.

If $|f(z)| = |z|$ for some $z \in \mathbb{D}^*$, or if $|f'(0)| = 1$, then $|g(z)| = 1$ for some $z \in \mathbb{D}$. Hence, by the maximum principle, g is a constant α with $|\alpha| = 1$, which shows $f(z) = \alpha z$ for all $z \in \mathbb{D}$. □

EXAMPLE 4.14. Suppose $f \in \mathcal{O}(\mathbb{D})$ has positive real part and $f(0) = 1$. We can think of f as a holomorphic function into the right half-plane $U = \{z \in \mathbb{C} : \text{Re}(z) > 0\}$. The Möbius map $\varphi(z) = (1 - z)/(1 + z)$ maps U biholomorphically onto $\mathbb{D}$ (compare Example 4.5), so the composition $g = \varphi \circ f$ is a holomorphic function $\mathbb{D} \to \mathbb{D}$ which fixes the origin. Applying the Schwarz lemma to g, we obtain

$$|g'(0)| = |\varphi'(1)| \, |f'(0)| = \frac{1}{2}|f'(0)| \leq 1, \quad \text{or} \quad |f'(0)| \leq 2.$$

Equality holds if and only if f is of the form $f(z) = \varphi^{-1}(\alpha z) = (1 - \alpha z)/(1 + \alpha z)$ for some α of absolute value 1.

Karl Herman Amandus
Schwarz (1843–1921)

It will be convenient to introduce a special notation for a class of Möbius maps that occur frequently in complex function theory. If $p \in \mathbb{C}$ and $|p| \neq 1$, we set

$$(4.12) \qquad \qquad \varphi_p(z) = \frac{z - p}{1 - \bar{p} z}.$$

THEOREM 4.15. *The following properties hold:*

(i) $\varphi_p(0) = -p$ *and* $\varphi_p(p) = 0$;

(ii) $\varphi_p'(z) = (1 - |p|^2)/(1 - \bar{p} z)^2$, *hence*

$$\varphi_p'(0) = 1 - |p|^2 \qquad \text{and} \qquad \varphi_p'(p) = \frac{1}{1 - |p|^2};$$

(iii) $(\varphi_p)^{-1} = \varphi_{-p} : z \mapsto (z + p)/(1 + \bar{p} z)$;

(iv) *If* $|p| < 1$, *then* $\varphi_p \in \text{Aut}(\mathbb{D})$.

PROOF. Only (iv) requires a comment: If $|z| = 1$, then

$$|\varphi_p(z)| = \frac{|z - p|}{|1 - \bar{p} z|} = \frac{|z - p|}{|z| \, |\bar{z} - \bar{p}|} = 1.$$

In other words, φ_p maps the unit circle $\mathbb{T}$ onto itself. Since φ_p is a homeomorphism of the sphere, it should map $\mathbb{D}$ bijectively either to $\mathbb{D}$ or to $\hat{\mathbb{C}} \smallsetminus \overline{\mathbb{D}}$. The assumption $|p| < 1$ shows that $\varphi_p(0) = -p \in \mathbb{D}$. Hence $\varphi_p(\mathbb{D}) = \mathbb{D}$. □

THEOREM 4.16. Aut($\mathbb{D}$) *is the subgroup of* Möb *consisting of all maps of the form*

$$\alpha\,\varphi_p : z \mapsto \alpha\,\frac{z-p}{1-\bar{p}z},$$

where $|p| < 1$ *and* $|\alpha| = 1$. *In particular, the only automorphisms of* $\mathbb{D}$ *which fix the origin are the rigid rotations* $z \mapsto \alpha z$.

Alternatively, Aut($\mathbb{D}$) can be described as the subgroup of Möb consisting of all maps which send the unit circle to itself, preserving its orientation.

PROOF. Every Möbius map of the above form is an automorphism of the disk by Theorem 4.15(iv). Conversely, take any $f \in$ Aut($\mathbb{D}$) and let $p = f^{-1}(0) \in \mathbb{D}$. Then $g = f \circ \varphi_{-p} \in$ Aut($\mathbb{D}$) and $g(0) = 0$. By the Schwarz lemma, $|g'(0)| \leq 1$. But $g^{-1} \in$ Aut($\mathbb{D}$) as well, and the Schwarz lemma applied to g^{-1} gives $|(g^{-1})'(0)| = 1/|g'(0)| \leq 1$. It follows that $|g'(0)| = 1$, so $g(z) = \alpha z$ for some constant α with $|\alpha| = 1$. This proves $f = \alpha\varphi_p$, as required. $\square$

REMARK 4.17. It is easy to see that Aut($\mathbb{D}$) acts transitively: Given any two points p, q in $\mathbb{D}$, there are (infinitely many) automorphisms which send p to q; simply take the composition $\varphi_{-q} \circ \alpha\varphi_p$ for any constant α with $|\alpha| = 1$. However, the action of Aut($\mathbb{D}$) is *not* 2-transitive. For instance, if $f \in$ Aut($\mathbb{D}$) fixes the origin, it must be a rigid rotation by Theorem 4.16, so f cannot send the pair $(0, p)$ to the pair $(0, q)$ unless $|p| = |q|$. The question of which pairs can be mapped to each other by a disk automorphism will be addressed in Corollary 4.39.

Sometimes it is more convenient to work in the upper half-plane $\mathbb{H}$ instead of the unit disk $\mathbb{D}$. As the two domains are biholomorphic (Example 4.5), they have isomorphic automorphism groups. Nonetheless, it is desirable to have an explicit description for Aut($\mathbb{H}$):

THEOREM 4.18. Aut($\mathbb{H}$) *is the subgroup of* Möb *consisting of all maps of the form*

$$z \mapsto \frac{az+b}{cz+d},$$

where $a, b, c, d \in \mathbb{R}$ *and* $ad - bc > 0$.

Alternatively, Aut($\mathbb{H}$) can be described as the subgroup of Möb consisting of all elements which map the extended real line $\hat{\mathbb{R}} = \mathbb{R} \cup \{\infty\}$ to itself, preserving its orientation. It can also be identified with the matrix group

$$\mathrm{PSL}_2(\mathbb{R}) = \left\{ \begin{bmatrix} a & b \\ c & d \end{bmatrix} : a, b, c, d \in \mathbb{R}, ad - bc = 1 \right\} / \{\pm I\}$$

(see the discussion leading to (4.1)).

PROOF. Suppose $f \in$ Möb has the given form. Since the coefficients a, b, c, d are real, f maps $\hat{\mathbb{R}}$ onto itself. Being a homeomorphism of the sphere, it follows that f maps

$\mathbb{H}$ biholomorphically onto one of the two components of $\hat{\mathbb{C}} \smallsetminus \hat{\mathbb{R}}$. This component must be $\mathbb{H}$ since

$$(4.13) \qquad \operatorname{Im}(f(i)) = \operatorname{Im}\left(\frac{ai+b}{ci+d}\right) = \frac{ad-bc}{|ci+d|^2} > 0.$$

Thus, $f \in \operatorname{Aut}(\mathbb{H})$.

Conversely, suppose $f \in \operatorname{Aut}(\mathbb{H})$ and consider the map $g = \varphi \circ f \circ \varphi^{-1} \in \operatorname{Aut}(\mathbb{D})$, where $\varphi : \mathbb{H} \to \mathbb{D}$ is the Cayley map (4.9). By Theorem 4.16, g is the restriction of a Möbius map. Since φ is also Möbius, it follows that $f = \varphi^{-1} \circ g \circ \varphi \in \operatorname{Möb}$, say $f(z) = (az+b)/(cz+d)$. The assumption $f(\mathbb{H}) = \mathbb{H}$ shows that f maps $\hat{\mathbb{R}}$ to itself. Thus, the Möbius map $z \mapsto \overline{f(\bar{z})} = (\bar{a}z + \bar{b})/(\bar{c}z + \bar{d})$ coincides with f along $\hat{\mathbb{R}}$, hence everywhere by the identity theorem. This proves that a, b, c, d are real. That $ad - bc > 0$ follows from (4.13) since $f(i) \in \mathbb{H}$. $\qquad \square$

While the automorphism groups of the three simply connected examples $\hat{\mathbb{C}}, \mathbb{C}, \mathbb{D}$ are rich and easy to describe, finding the automorphism group of general domains is often a subtle problem. As the topology of U becomes more complicated, we typically expect the group $\operatorname{Aut}(U)$ to be smaller, often trivial.

EXAMPLE 4.19 (The automorphism group of punctured spheres). Consider an n-punctured sphere $U = \hat{\mathbb{C}} \smallsetminus \{p_1, \ldots, p_n\}$ where the p_k are distinct and $n \geq 1$. Every homeomorphism $U \to U$ extends to a homeomorphism $\hat{\mathbb{C}} \to \hat{\mathbb{C}}$ which permutes the points $p_1, \ldots, p_n$ (why?). If $f \in \operatorname{Aut}(U)$, it follows that the isolated singularities $p_1, \ldots, p_n$ of f are removable since f remains continuous near the punctures. The extension of f is thus a Möbius map which permutes the points $p_1, \ldots, p_n$. Conversely, any such Möbius map clearly restricts to an automorphism of U. We conclude that

$$\operatorname{Aut}(U) \cong \left\{ f \in \operatorname{Möb} : f(\{p_1, \ldots, p_n\}) = \{p_1, \ldots, p_n\} \right\}.$$

For $n = 1$, $\operatorname{Aut}(U)$ is an infinite non-abelian group isomorphic to the affine group $\operatorname{Aut}(\mathbb{C})$. If φ is any Möbius map which sends p_1 to ∞, the conjugation $f \mapsto \varphi \circ f \circ \varphi^{-1}$ defines an isomorphism $\operatorname{Aut}(U) \to \operatorname{Aut}(\mathbb{C})$.

For $n = 2$, $\operatorname{Aut}(U)$ is an infinite non-abelian group isomorphic to $\operatorname{Aut}(\mathbb{C}^*)$ which is generated by the linear maps $L_\alpha : z \mapsto \alpha z$ and the inversion $\iota : z \mapsto 1/z$ (problem 12). If φ is any Möbius map which sends (p_1, p_2) to $(0, \infty)$, the conjugation $f \mapsto \varphi \circ f \circ \varphi^{-1}$ defines an isomorphism $\operatorname{Aut}(U) \to \operatorname{Aut}(\mathbb{C}^*)$.

For $n = 3$, $\operatorname{Aut}(U)$ is a finite non-abelian group isomorphic to the group S_3 of all permutations of p_1, p_2, p_3. This is because the action of Möb is simply 3-transitive (Theorem 4.2(ii)), so every permutation of p_1, p_2, p_3 is realized by a *unique* Möbius map. If φ is the Möbius map which sends (p_1, p_2, p_3) to $(0, 1, \infty)$, the conjugation $f \mapsto \varphi \circ f \circ \varphi^{-1}$ defines an isomorphism $\operatorname{Aut}(U) \to \operatorname{Aut}(\hat{\mathbb{C}} \smallsetminus \{0, 1, \infty\})$. The latter group consists of the maps

$$z \mapsto z, \quad z \mapsto 1 - z, \quad z \mapsto \frac{1}{z}, \quad z \mapsto \frac{z}{z-1}, \quad z \mapsto \frac{z-1}{z}, \quad z \mapsto \frac{1}{1-z}.$$

This subgroup of Möb is often called the ***anharmonic group*** (compare the table of cross ratios at the end of §4.1).

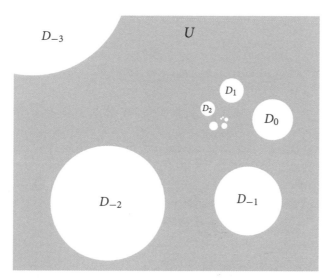

Figure 4.3. The invariant domain $U = \hat{\mathbb{C}} \smallsetminus \bigcup_{n=-\infty}^{\infty} \overline{D_n}$ of the linear map L_α, as in Example 4.20. Here α is non-real with $|\alpha| = 0.6$. The disks D_n spiral in toward 0 as $n \to \infty$ and out toward ∞ as $n \to -\infty$.

For $n = 4$, $\mathrm{Aut}(U)$ is a finite group, typically isomorphic to $\mathbb{Z}_2 \times \mathbb{Z}_2$, but isomorphic to the larger dihedral groups of order 8 or 12 in the exceptional cases where the cross ratio $[p_1, p_2, p_3, p_4]$ is $-1, 2, 1/2$, or $e^{\pm i\pi/3}$ (problem 13).

For $n \geq 5$, the group $\mathrm{Aut}(U)$ is typically trivial. To get a feel for the kind of restriction having five or more points imposes, consider a non-identity automorphism f of $\hat{\mathbb{C}} \smallsetminus \{0, 1, \infty, p, q\}$. If $f(q) = q$, then f induces an automorphism of $\hat{\mathbb{C}} \smallsetminus \{0, 1, p, \infty\}$ which fixes q. This gives finitely many choices for f and hence for q. On the other hand, if $f(q) \neq q$, then $q \in f(\{0, 1, p, \infty\})$ and the algebraic equation

$$p = [0, 1, p, \infty] = [f(0), f(1), f(p), f(\infty)]$$

can be solved uniquely for q in terms of p. In either case, it follows that for a given p there are only finitely many q for which $\hat{\mathbb{C}} \smallsetminus \{0, 1, \infty, p, q\}$ admits a non-trivial automorphism.

EXAMPLE 4.20. Here is a simple example of a domain in $\hat{\mathbb{C}}$ with infinitely many complementary components and non-trivial automorphism group. Consider the linear map $L_\alpha : z \mapsto \alpha z$ with $|\alpha| \neq 1$. The image of the disk $D_0 = \mathbb{D}(1, r)$ under the n-th iterate of L_α is the disk $D_n = \mathbb{D}(\alpha^n, |\alpha|^n r)$. If $r > 0$ is sufficiently small, the disks D_n for $n \in \mathbb{Z}$ have disjoint closures. Thus, the complement $U = \hat{\mathbb{C}} \smallsetminus \bigcup_{n=-\infty}^{\infty} \overline{D_n}$ is a domain that is invariant under L_α. Evidently $\mathrm{Aut}(U)$ contains the infinite cyclic group generated by L_α (see Fig. 4.3).

4.3 Dynamics of Möbius maps

In this section we investigate the action of a Möbius map on the Riemann sphere and the behavior of its orbits. We begin with some standard definitions. For $f \in \mathrm{M\ddot{o}b}$ and an integer $n \geq 1$, we use the notation $f^{\circ n}$ for the n-fold *iterate* of f:

The notation $f^{\circ n}$, introduced by J. Milnor, distinguishes the n-th iterate from the n-th power.

$$f^{\circ n} = \underbrace{f \circ \cdots \circ f}_{n \text{ times}}.$$

For $n < 0$, $f^{\circ n}$ will denote the $(-n)$-fold iterate of f^{-1}, and we set $f^{\circ 0} = \text{id}$. Thus, $\{f^{\circ n}\}_{n \in \mathbb{Z}}$ is the (finite or infinite) cyclic subgroup of Möb generated by f. The **orbit** of $p \in \hat{\mathbb{C}}$ under f is the set $\{f^{\circ n}(p)\}_{n \in \mathbb{Z}}$. We say that p is a **periodic point** of f if $f^{\circ n}(p) = p$ for some positive integer n. The smallest such n is called the **period** of p. When the period is 1, so $f(p) = p$, we call p a **fixed point** of f. The **multiplier** of f at a fixed point p is the derivative $f'(p)$, with the proviso that when $p = \infty$, the derivative $f'(\infty)$ is understood as the derivative of $1/f(1/z)$ at the origin. Since f is injective, the multiplier at a fixed point is always a *non-zero* complex number.

Two elements $f, g \in$ Möb are said to be **conjugate** if there exists a $\varphi \in$ Möb such that

(4.14) $$\varphi \circ f = g \circ \varphi.$$

In other words, the following diagram is commutative:

$$\begin{array}{ccc} \hat{\mathbb{C}} & \xrightarrow{f} & \hat{\mathbb{C}} \\ {\scriptstyle\varphi}\downarrow & & \downarrow{\scriptstyle\varphi} \\ \hat{\mathbb{C}} & \xrightarrow{g} & \hat{\mathbb{C}} \end{array}$$

Conjugate maps have similar orbit structures: From (4.14) it follows by an easy induction that $\varphi \circ f^{\circ n} = g^{\circ n} \circ \varphi$ for all $n \in \mathbb{Z}$, which shows φ maps the orbit of p under f to the orbit of $q = \varphi(p)$ under g. In particular, p is a fixed point of f if and only if q is a fixed point of g. When this is the case, differentiating each side of (4.14) at p gives

$$\varphi'(p)\, f'(p) = g'(q)\, \varphi'(p),$$

which implies $f'(p) = g'(q)$ since $\varphi'(p) \neq 0$ (again, if p or q is ∞, this argument needs a straightforward modification by applying a suitable change of coordinate). Thus, *the multiplier at a fixed point is a conjugacy invariant.*

As in §3.4, the multiplicity of a fixed point p of $f \in$ Möb is defined as $\text{ord}(z - f(z), p)$ if $p \neq \infty$, and $\text{ord}(z - 1/f(1/z), 0)$ if $p = \infty$. We call p a **simple** or **double** fixed point if its multiplicity is 1 or 2, respectively. Being simple is therefore equivalent to $(z - f(z))'\big|_{z=p} = 1 - f'(p) \neq 0$ if $p \neq \infty$, and to $(z - 1/f(1/z))'\big|_{z=0} = 1 - f'(\infty) \neq 0$ if $p = \infty$. It follows that

(4.15) $$p \text{ is a simple fixed point of } f \iff f'(p) \neq 1.$$

LEMMA 4.21. *Every non-identity $f \in$ Möb has two fixed points counting multiplicities. Moreover, there are only two possibilities:*

(i) *f has a pair of simple fixed points p, q with multipliers $f'(p) = \alpha$, $f'(q) = 1/\alpha$ for some $\alpha \in \mathbb{C} \setminus \{0, 1\}$. In this case, f is conjugate to the linear map $L_\alpha : z \mapsto \alpha z$.*

(ii) f has a double fixed point at p with multiplier $f'(p) = 1$. In this case, f is conjugate to the unit translation $T_1 : z \mapsto z + 1$.

In either case, it follows that the product of the multipliers at the fixed points is 1.

PROOF. First suppose $f(\infty) \neq \infty$ so $f(z) = (az + b)/(cz + d)$ with $c \neq 0$. In this case the fixed points of f are the roots of the equation $(az + b)/(cz + d) = z$, which is equivalent to the quadratic equation $cz^2 + (d - a)z - b = 0$. Next suppose $f(\infty) = \infty$, so f is an affine map $z \mapsto az + b$. If $a \neq 1$, there is a second fixed point at $b/(1 - a)$ and both fixed points are clearly simple. If $a = 1$, then ∞ is the only fixed point because $b \neq 0$. In this case, the function

$$z - \frac{1}{f(1/z)} = z - \frac{1}{(1/z) + b} = \frac{bz^2}{1 + bz}$$

has a zero of order 2 at 0, which shows ∞ is a double fixed point of f. This proves the first claim.

If f has a pair of simple fixed points at p, q, take any $\varphi \in$ Möb which sends p to 0 and q to ∞. The conjugate map $\varphi \circ f \circ \varphi^{-1} \in$ Möb has fixed points at 0 and ∞, so it must be the linear map L_α for some $\alpha \in \mathbb{C} \setminus \{0, 1\}$. We have $f'(p) = L'_\alpha(0) = \alpha$ and $f'(q) = L'_\alpha(\infty) = 1/\alpha$.

If f has a double fixed point at p, then $f'(p) = 1$ by (4.15). This time take any $\varphi \in$ Möb which sends p to ∞. The conjugate map $g = \varphi \circ f \circ \varphi^{-1} \in$ Möb has a double fixed point at ∞ and no other fixed point. It follows that g must be a translation $g(z) = z + b$ for some $b \neq 0$. The map g in turn is conjugate to T_1 since $T_1 = \psi \circ g \circ \psi^{-1}$ if $\psi(z) = z/b$. □

EXAMPLE 4.22. The inversion $\iota : z \mapsto 1/z$ has a pair of fixed points of multiplier -1 at ± 1. By the above lemma, ι must be conjugate to the 180° rotation $L_{-1} : z \mapsto -z$. This, of course, can be checked directly: $\varphi \circ \iota = L_{-1} \circ \varphi$, where $\varphi(z) = (z - 1)/(z + 1)$.

REMARK 4.23. The fact that the product of the multipliers at fixed points is 1 also follows from the notion of index that we introduced in §3.4. In fact, if the fixed points are simple and have multipliers α and β, then by (3.13) they have indices $1/(1 - \alpha)$ and $1/(1 - \beta)$. Hence, by the rational fixed point formula in Theorem 3.38,

$$\frac{1}{1 - \alpha} + \frac{1}{1 - \beta} = 1,$$

which proves $\alpha\beta = 1$.

In view of the above lemma, to understand the dynamics of a general $f \in$ Möb, it suffices to study the models T_1 and L_α. This is rather straightforward:

● If f has a double fixed point at p, it is conjugate to T_1 which has a double fixed point at ∞. For every $w \in \hat{\mathbb{C}} \setminus \{\infty\}$, the iterate $T_1^{\circ n}(w) = w + n$ tends to ∞ along a

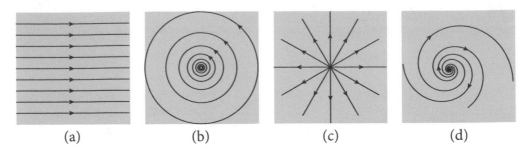

Figure 4.4. The invariant curves of four Möbius models: (a) the unit translation T_1; (b) a rotation L_α with $|\alpha| = 1$; (c) a dilation L_α with $\alpha > 1$; (d) a linear map L_α with α non-real and $|\alpha| > 1$.

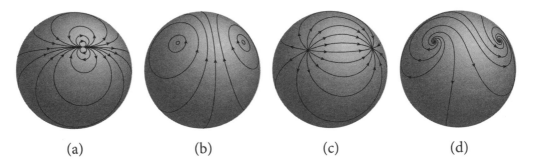

Figure 4.5. The invariant curves of Möbius maps that are conjugate to the models in Fig. 4.4. Their types are (a) parabolic; (b) elliptic; (c) hyperbolic; (d) loxodromic.

horizontal line in the plane as $n \to \pm\infty$. We say that these lines are ***invariant curves*** under the action of T_1 (Fig. 4.4(a)). It follows that for every $z \in \hat{\mathbb{C}} \smallsetminus \{p\}$, the sequence $f^{\circ n}(z)$ tends to p along a circle passing through p as $n \to \pm\infty$. These invariant circles are all tangent to each other at p (Fig. 4.5(a)).

When f has a pair of simple fixed points p, q with multipliers $f'(p) = \alpha, f'(q) = 1/\alpha$, it is conjugate to the linear map L_α which has simple fixed points at $0, \infty$ with the same multipliers $\alpha, 1/\alpha$. The behavior of the orbits in this case depends on α.

• If $|\alpha| = 1$, L_α is a rotation around $0, \infty$, and the Euclidean circles centered at 0 are invariant curves. Writing $\alpha = e^{2\pi i\theta}$, it is not hard to check that for every $w \in \hat{\mathbb{C}} \smallsetminus \{0, \infty\}$, the orbit $\{L_\alpha^{\circ n}(w) = \alpha^n w\}_{n \in \mathbb{Z}}$ is finite if θ is rational, and is dense on the corresponding invariant circle if θ is irrational (Fig. 4.4(b)). It follows that for every $z \in \hat{\mathbb{C}} \smallsetminus \{p, q\}$, the orbit $\{f^{\circ n}(z)\}_{n \in \mathbb{Z}}$ lies on a circle separating p from q and is either finite or dense on this circle. These invariant circles are disjoint and fill up the twice punctured sphere $\hat{\mathbb{C}} \smallsetminus \{p, q\}$ (Fig. 4.5(b)).

• If $|\alpha| \neq 1$, we may assume without loss of generality that $|\alpha| > 1$ (otherwise replace f with f^{-1}). For every $w \in \hat{\mathbb{C}} \smallsetminus \{0, \infty\}$ the iterate $L_\alpha^{\circ n}(w) = \alpha^n w$ tends to ∞ as $n \to \infty$ and to 0 as $n \to -\infty$. Writing $\alpha = re^{2\pi i\theta}$, it is easy to check that the curves

$$t \mapsto c\alpha^t = cr^t e^{2\pi i\theta t} \qquad \text{for } c \neq 0 \tag{4.16}$$

Sphere Spirals by
M. C. Escher, 1958.

are invariant under the action of L_α. They are straight lines if θ is an integer (Fig. 4.4(c)) and logarithmic spirals with ends approaching $0, \infty$ otherwise (Fig. 4.4(d)). It follows that for every $z \in \hat{\mathbb{C}} \setminus \{p, q\}$, the sequence $f^{\circ n}(z)$ tends to q as $n \to \infty$ and to p as $n \to -\infty$. The invariant curves for f are circles passing through p, q if α is real (Fig. 4.5(c)) and logarithmic spirals with ends approaching p, q otherwise (Fig. 4.5(d)).

DEFINITION 4.24. A non-identity Möbius map f is called *parabolic* if it has a double fixed point. If f has a pair of simple fixed points with multipliers $\alpha, 1/\alpha$ in $\mathbb{C} \setminus \{0, 1\}$, we call it *elliptic* when $|\alpha| = 1$, *hyperbolic* when $\alpha > 0$, and *loxodromic* when α is not positive and $|\alpha| \neq 1$.

"Loxodromic" comes from the Greek $\lambda o\xi\acute{o}\varsigma$ (oblique) and $\delta\rho o\mu\acute{o}\varsigma$ (running). In navigation, a loxodrome is a path that crosses all meridians at the same angle. Each spiral of the form (4.16) has this property.

A non-identity $f \in$ Möb has an invariant disk (that is, a Euclidean disk D with $f(D) = D$) except when it is loxodromic. This follows from the above analysis of the invariant curves of the models T_1 and L_α. In particular, *elements of* $\mathrm{Aut}(\mathbb{D})$ *can only be parabolic, elliptic or hyperbolic, with all three types occurring.*

To determine the type of a non-identity Möbius map, we can of course locate its fixed points by solving an algebraic equation and compute the multipliers. But there is an alternative and easier approach. Recall that Möb is isomorphic to the matrix group $\mathrm{PSL}_2(\mathbb{C})$ in (4.1), so every $f \in$ Möb is represented by a 2×2 matrix A over $\mathbb{C}$ with $\det(A) = 1$, which is unique up to multiplication by -1. Under this isomorphism, conjugate maps correspond to conjugate matrices.

DEFINITION 4.25. Let f be a non-identity Möbius map with an associated matrix $\pm A \in \mathrm{PSL}_2(\mathbb{C})$. We define the invariant $\tau(f)$ by

$$\tau(f) = (\mathrm{tr}(A))^2$$

where $\mathrm{tr}(A)$ stands for the trace of A.

Squaring the trace yields an invariant which only depends on f and not on its representative matrix. Since conjugate Möbius maps correspond to conjugate matrices in $\mathrm{PSL}_2(\mathbb{C})$, and since conjugate matrices have the same trace, it follows that $\tau(f) = \tau(g)$ *whenever* f *and* g *are conjugate Möbius maps.*

THEOREM 4.26. *Let* f *be a non-identity element of* Möb, *with fixed point multipliers* α, α^{-1} *(so* $\alpha = 1$ *iff* f *is parabolic). Then*

(4.17)
$$\tau(f) = \alpha + \alpha^{-1} + 2.$$

Moreover, f *is*

- *parabolic if* $\tau(f) = 4$;
- *elliptic if* $\tau(f) \in [0, 4)$;

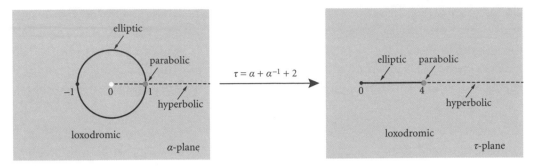

Figure 4.6. The loci of various types of Möbius maps in the multiplier plane $\alpha \in \mathbb{C}^*$ and the trace-squared plane $\tau \in \mathbb{C}$.

- *hyperbolic if $\tau(f) \in (4, +\infty)$;*
- *loxodromic if $\tau(f) \in \mathbb{C} \smallsetminus [0, +\infty)$.*

See Fig. 4.6.

PROOF. We have the association

$$T_1 \longleftrightarrow \pm \begin{bmatrix} 1 & 1 \\ 0 & 1 \end{bmatrix} \quad \text{so} \quad \tau(T_1) = 4.$$

Similarly, for $\alpha \in \mathbb{C} \smallsetminus \{0, 1\}$,

$$L_\alpha \longleftrightarrow \pm \begin{bmatrix} \sqrt{\alpha} & 0 \\ 0 & 1/\sqrt{\alpha} \end{bmatrix} \quad \text{so} \quad \tau(L_\alpha) = \alpha + \alpha^{-1} + 2.$$

By Lemma 4.21, every non-identity $f \in$ Möb with fixed point multipliers α, α^{-1} is conjugate to T_1 if $\alpha = 1$ and to L_α otherwise. This gives the required formula $\tau(f) = \alpha + \alpha^{-1} + 2$ since $\tau(f)$ is a conjugacy invariant.

The remaining assertions are now easy to prove. In fact, setting $\zeta = \alpha + \alpha^{-1}$, a brief computation shows that $\zeta = 2$ if and only if $\alpha = 1$, $\zeta \in [-2, 2]$ if and only if $|\alpha| = 1$, and $\zeta \in [2, +\infty)$ if and only if $\alpha > 0$. The details are straightforward and will be left to the reader. $\square$

COROLLARY 4.27. *Two non-identity elements $f, g \in$ Möb are conjugate if and only if $\tau(f) = \tau(g)$.*

PROOF. We have already noted that $\tau(f) = \tau(g)$ if f and g are conjugate. Suppose $\tau(f) = \tau(g) = \tau$. If $\tau = 2$, then f and g are parabolic, each having a double fixed point with multiplier 1. By Lemma 4.21, f and g are conjugate to T_1, hence conjugate to each other. If $\tau \neq 2$, then the formula (4.17) determines the multiplier pair $\alpha, \alpha^{-1} \in \mathbb{C} \smallsetminus \{0, 1\}$ uniquely. By Lemma 4.21, f and g are conjugate to L_α, hence conjugate to each other. $\square$

EXAMPLE 4.28. The inversion $\iota : z \mapsto 1/z$ is elliptic since it has a pair of fixed points of multiplier -1 at ± 1. Alternatively, note that ι is represented by the matrix $\pm \begin{bmatrix} 0 & i \\ i & 0 \end{bmatrix} \in \mathrm{PSL}_2(\mathbb{C})$, so $\tau(\iota) = 0$ and ellipticity follows from Theorem 4.26.

On the other hand, the disk automorphism φ_p defined by (4.12) is represented by the matrix

$$\frac{\pm 1}{\sqrt{1-|p|^2}} \cdot \begin{bmatrix} 1 & -p \\ -\bar{p} & 1 \end{bmatrix} \in \mathrm{PSL}_2(\mathbb{C}).$$

For $p = 0$, φ_p is the identity map. Otherwise, $0 < |p| < 1$ and

$$\tau(\varphi_p) = \frac{4}{1-|p|^2} \in (4, +\infty).$$

It follows from Theorem 4.26 that φ_p is hyperbolic.

4.4 Conformal metrics

Let us make a short digression to introduce conformal metrics and develop a convenient language to express and manipulate them in the complex-variable notation. This apparatus will be used in the next section where we construct the hyperbolic metric of the disk and arrive at a geometric interpretation of the classical Schwarz lemma. Other applications of conformal metrics will occur in §5.3, §11.4, and §13.3.

We begin with some standard definitions from differential geometry. Let $U \subset \mathbb{C}$ be a domain. A **(Riemannian) metric** g in U is a continuous way of measuring lengths of tangent vectors attached to points of U. More precisely, for each $p \in U$ consider the 2-dimensional **tangent plane** T_pU spanned by all vectors with the initial point p, which can be naturally identified with $\mathbb{R}^2$. The metric g assigns a **norm** $\|\cdot\|_g$ on each T_pU which by definition satisfies the following conditions:

- $\|v\|_g \geq 0$ for every $v \in T_pU$, and $\|v\|_g = 0$ if and only if $v = 0$;
- $\|tv\|_g = |t| \, \|v\|_g$ for every $v \in T_pU$ and $t \in \mathbb{R}$; and
- $\|v_1 + v_2\|_g \leq \|v_1\|_g + \|v_2\|_g$ for every $v_1, v_2 \in T_pU$.

Moreover, we require g to be continuous in the sense that for every continuous vector field $p \mapsto v(p)$ in U, the function $p \mapsto \|v(p)\|_g$ is continuous in U.

DEFINITION 4.29. A metric g in U is called a **conformal metric** if its assigned norm on each tangent plane is a positive multiple of the Euclidean norm. In other words, if there is a positive continuous function $\rho : U \to \mathbb{R}$ such that

$$\|v\|_g = \rho(p)|v| \qquad \text{for all } p \in U \text{ and } v \in T_pU.$$

We call ρ the **density** function of the metric g. We say that g is a C^r-smooth metric if its density function ρ is C^r-smooth in the sense that it has continuous partial derivatives of all orders up to r.

Thus, to specify a conformal metric it suffices to know its density. Note that the Euclidean metric itself is conformal with constant density $\rho = 1$.

Definition 4.29 suggests the following classical notation which is very useful in local computations: If g is a conformal metric in U with the density function ρ, we write

$$g = \rho(z)\, |dz|.$$

This notation is motivated by the way the length of a piecewise C^1 curve $\gamma : [a, b] \to U$ is computed in the metric g:

$$\text{length}_g(\gamma) = \int_a^b \|\gamma'(t)\|_g\, dt = \int_a^b \rho(\gamma(t))\, |\gamma'(t)|\, dt = \int_\gamma \rho(z)\, |dz|.$$

The notion of length leads to the notion of **distance** between points. For $p, q \in U$, we define

(4.18)
$$\text{dist}_g(p, q) = \inf_\gamma \ \text{length}_g(\gamma),$$

where the infimum is taken over all piecewise C^1 curves γ in U from p to q. It is not hard to verify that $\text{dist}_g(\cdot, \cdot)$ indeed has all the properties of a distance function and thus turns U into a metric space. A C^1 curve $\gamma : [a, b] \to U$ is called a **minimal geodesic** (or g-geodesic, to emphasize the dependence on g) if

$$\text{length}_g(\gamma) = \text{dist}_g(\gamma(a), \gamma(b)).$$

Thus, minimal geodesics play the role of "straight lines" in the geometry of U induced by the metric g. It can be shown that every pair of *nearby* points in U can be joined by a unique minimal geodesic, but we will not prove this classical result here (see for example [**doC**]). By contrast, minimal geodesics between arbitrary pairs may fail to exist, or they may fail to be unique.

Suppose $\tilde{U}, U \subset \mathbb{C}$ are domains and $f : \tilde{U} \to U$ is a holomorphic map with non-vanishing derivative. Given a conformal metric g in U, the **pull-back** f^*g is the conformal metric in $\tilde{U}$ defined as follows: For each $p \in \tilde{U}$ and each tangent vector $\tilde{v} \in T_p\tilde{U}$, the norm $\|\tilde{v}\|_{f^*g}$ is defined to be the norm $\|v\|_g$ of the image vector $v = Df(p)\tilde{v} = f'(p)\tilde{v} \in T_{f(p)}U$:

$$\|\tilde{v}\|_{f^*g} = \rho(f(p))|f'(p)|\, |\tilde{v}|.$$

In terms of densities, the pull-back f^*g is obtained by formally substituting $w = f(z)$ in the expression $g = \rho(w)\, |dw|$:

(4.19)   If $g = \rho(w)\, |dw|$,   then $f^*g = \rho(f(z))|f'(z)|\, |dz|$.

The definition can be extended to any non-constant holomorphic function f, but we must take special care at the critical points of f. The pull-back f^*g in this case is a well-defined conformal metric in the subdomain $\{p \in \tilde{U} : f'(p) \neq 0\}$. Alternatively, we can think of f^*g as a "singular metric" in $\tilde{U}$ which vanishes at every critical point of f.

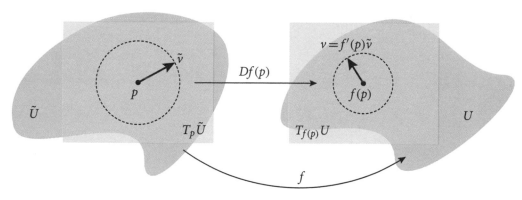

Figure 4.7. The derivative norm $\| f'(p) \|$ compares the length of a tangent vector $\tilde{v}$ at p with that of the image vector $v = Df(p)(\tilde{v}) = f'(p)\tilde{v}$ at $f(p)$, with respect to the given metrics in $\tilde{U}$ and U. The ratio of the two lengths is independent of $\tilde{v}$ because f is holomorphic and the metrics are conformal.

Now suppose $f : \tilde{U} \to U$ is holomorphic, where $\tilde{U}$ and U are equipped with their own conformal metrics $\tilde{g} = \tilde{\rho}(z)\, |dz|$ and $g = \rho(w)\, |dw|$, respectively. We express this situation conveniently by writing $f : (\tilde{U}, \tilde{g}) \to (U, g)$. For each $p \in \tilde{U}$, the **derivative norm** $\| f'(p) \|$ compares the norm of a tangent vector in $T_p \tilde{U}$ with respect to $\tilde{g}$ with the norm of its image vector in $T_{f(p)} U$ with respect to g. Equivalently, it compares the norms of tangent vectors in $T_p \tilde{U}$ with respect to $\tilde{g}$ and the pull-back $f^* g$: If $\tilde{v} \in T_p \tilde{U}$ is non-zero and $v = f'(p)\tilde{v} \in T_{f(p)} U$, then

$$\| f'(p) \| = \frac{\|v\|_g}{\|\tilde{v}\|_{\tilde{g}}} = \frac{\|\tilde{v}\|_{f^* g}}{\|\tilde{v}\|_{\tilde{g}}} = \frac{\rho(f(p))|f'(p)|\,|\tilde{v}|}{\tilde{\rho}(p)|\tilde{v}|} = \frac{\rho(f(p))|f'(p)|}{\tilde{\rho}(p)}$$

(see Fig. 4.7). Observe that the result is independent of $\tilde{v}$ precisely because the metrics are conformal. Thus, the derivative norm is the Euclidean norm of the derivative, modified by the target-to-domain ratio of the densities of the given metrics:

$$(4.20) \qquad \| f'(z) \| = \frac{\rho(f(z))}{\tilde{\rho}(z)}\, |f'(z)|.$$

Suggested by this definition, the map $f : (\tilde{U}, \tilde{g}) \to (U, g)$ is called a **contraction**, **expansion**, or **local isometry** according as $\| f'(z) \|$ is less than, greater than, or equal to 1 for every $z \in \tilde{U}$. Note that these attributes depend not only on f, but on the choice of the metrics $\tilde{g}, g$ as well. To emphasize this dependence, sometimes we write the derivative norm as $\| f'(z) \|_{\tilde{g}, g}$.

It is an easy consequence of the chain rule that the derivative norm is multiplicative, that is, if $f_1 : (U_1, g_1) \to (U_2, g_2)$ and $f_2 : (U_2, g_2) \to (U_3, g_3)$ are holomorphic, then

$$(4.21) \qquad \|(f_2 \circ f_1)'(z)\|_{g_1, g_3} = \| f_2'(f_1(z)) \|_{g_2, g_3}\, \| f_1'(z) \|_{g_1, g_2}.$$

Lengths and distances can be estimated in terms of the derivative norm. Suppose, as before, that $f : (\tilde{U}, \tilde{g}) \to (U, g)$ is holomorphic and $\tilde{\rho}, \rho$ are the density functions of $\tilde{g}, g$. For any piecewise C^1 curve $\gamma : [a, b] \to \tilde{U}$, the change of variable formula (see problem 6 in chapter 1) together with (4.20) shows that

$$(4.22) \qquad \text{length}_g(f \circ \gamma) = \int_{f \circ \gamma} \rho(w) \, |dw| = \int_\gamma \rho(f(z)) \, |f'(z)| \, |dz|$$

$$= \int_\gamma \|f'(z)\| \, \tilde{\rho}(z) \, |dz|.$$

LEMMA 4.30. *Suppose $f : (\tilde{U}, \tilde{g}) \to (U, g)$ is holomorphic.*

(i) For every piecewise C^1 curve γ in $\tilde{U}$,

$$\text{length}_g(f \circ \gamma) \leq \sup_{z \in |\gamma|} \|f'(z)\| \ \text{length}_{\tilde{g}}(\gamma).$$

(ii) For every $p, q \in \tilde{U}$,

$$\text{dist}_g(f(p), f(q)) \leq \sup_{z \in \tilde{U}} \|f'(z)\| \ \text{dist}_{\tilde{g}}(p, q).$$

PROOF. Part (i) follows immediately from (4.22). Part (ii) holds trivially if $\sup_{z \in \tilde{U}} \|f'(z)\| = +\infty$. Otherwise, it follows from (i) by taking the infimum over all piecewise C^1 curves γ in $\tilde{U}$ joining p to q. □

THEOREM 4.31 (Characterization of local isometries). *The following conditions on a non-constant holomorphic map $f : (\tilde{U}, \tilde{g}) \to (U, g)$ are equivalent:*

(i) $\|f'(z)\| = 1$ *for all $z \in \tilde{U}$.*
(ii) $f^* g = \tilde{g}$.
(iii) $\rho(f(z)) |f'(z)| = \tilde{\rho}(z)$ *for all $z \in \tilde{U}$.*
(iv) $\text{length}_g(f \circ \gamma) = \text{length}_{\tilde{g}}(\gamma)$ *for every piecewise C^1 curve γ in $\tilde{U}$.*

In the special case $\tilde{U} = U$ and $\tilde{g} = g$, these conditions are expressed by saying that the metric g is **invariant** under f.

PROOF. The equivalence of (i)–(iii) follows at once from the definition while the implication (i) $\implies$ (iv) follows from (4.22). It remains to verify (iv) $\implies$ (i).

Suppose (iv) holds; so by (4.22),

$$\int_\gamma \|f'(z)\| \, \tilde{\rho}(z) \, |dz| = \int_\gamma \tilde{\rho}(z) \, |dz|$$

for every piecewise C^1 curve γ in $\tilde{U}$. If $p \in \tilde{U}$, if $\varepsilon > 0$ is small, and if $\gamma(t) = p + t$ for $0 \leq t \leq \varepsilon$, it follows that

$$\frac{1}{\varepsilon} \int_0^\varepsilon \|f'(p+t)\| \, \tilde{\rho}(p+t) \, dt = \frac{1}{\varepsilon} \int_0^\varepsilon \tilde{\rho}(p+t) \, dt.$$

Letting $\varepsilon \to 0$ and using continuity gives $\|f'(p)\| \tilde{\rho}(p) = \tilde{\rho}(p)$, which implies $\|f'(p)\| = 1$ since $\tilde{\rho}(p) \neq 0$. $\qquad\square$

REMARK 4.32. It can be shown that (i)–(iv) are equivalent to the following condition: Every point in $\tilde{U}$ has a neighborhood $D \subset \tilde{U}$ such that

$$\mathrm{dist}_g(f(p), f(q)) = \mathrm{dist}_{\tilde{g}}(p, q) \qquad \text{for all } p, q \in D.$$

In other words, a local isometry preserves distances of *nearby* points. However, large distances may not be preserved. For example, consider the conformal metric

$$g = -\frac{1}{|w| \log |w|} \, |dw|$$

on the punctured disk $\mathbb{D}^* = \mathbb{D} \smallsetminus \{0\}$. The squaring map $f : \mathbb{D}^* \to \mathbb{D}^*$ defined by $f(z) = z^2$ is a local isometry with respect to g:

$$f^*g = -\frac{2|z|}{|z^2| \log |z^2|} \, |dz| = -\frac{1}{|z| \log |z|} \, |dz| = g.$$

But f does not preserve all distances since it is not injective.

COROLLARY 4.33. *Every local isometry $f : (\tilde{U}, \tilde{g}) \to (U, g)$ which maps $\tilde{U}$ biholomorphically onto U is a global isometry in the sense that*

$$\mathrm{dist}_g(f(p), f(q)) = \mathrm{dist}_{\tilde{g}}(p, q) \qquad \text{for all } p, q \in \tilde{U}.$$

PROOF. The inverse map $f^{-1} : (U, g) \to (\tilde{U}, \tilde{g})$ is a local isometry since $\|(f^{-1})'(z)\| = 1/\|f'(f^{-1}(z))\| = 1$ for all $z \in U$. The claim follows from Lemma 4.30 applied to f and f^{-1}. $\qquad\square$

Conformal metrics can be defined in every domain of the Riemann sphere in a similar fashion, once we take special care of the point ∞. If $U \subset \hat{\mathbb{C}}$ is a domain containing ∞, a conformal metric g in U can be expressed as $g = \rho(z) \, |dz|$, where ρ is continuous and positive in $U \smallsetminus \{\infty\}$. To make sure that g is well defined at ∞, we require that the pull-back of g under $\iota(z) = 1/z$ be well defined in a neighborhood of 0. It is readily seen that this happens precisely when

(4.23) $$\lim_{z \to 0} \frac{1}{|z|^2} \, \rho\!\left(\frac{1}{z}\right)$$

exists and is positive.

EXAMPLE 4.34. The conformal metric

$$\sigma = \frac{2}{1 + |z|^2} \, |dz|$$

is well defined on $\hat{\mathbb{C}}$ since its density $\rho(z) = 2/(1 + |z|^2)$ satisfies

$$(4.24) \qquad\qquad \frac{1}{|z|^2}\, \rho\!\left(\frac{1}{z}\right) = \rho(z) \qquad \text{whenever } z \neq 0,$$

so the limit in (4.23) is 2. We call σ the **spherical metric** on $\hat{\mathbb{C}}$. The relation (4.24) means that the inversion $\iota(z) = 1/z$ is an isometry of $(\hat{\mathbb{C}}, \sigma)$.

It can be shown that under the stereographic projection, σ corresponds to the standard metric of the 2-sphere induced from $\mathbb{R}^3$; see §5.3 for details.

4.5 The hyperbolic metric

This section will present a geometric interpretation of the Schwarz lemma due to G. Pick. The main ingredient will be a preferred conformal metric in the unit disk $\mathbb{D}$ with respect to which all automorphisms are isometries. This will lead to the notion of hyperbolic geometry in $\mathbb{D}$ which resembles the Euclidean geometry at small scales, but has very different global features.

We begin by asking if there is a conformal metric $g = \rho(z)\,|dz|$ in $\mathbb{D}$ which is invariant under all automorphisms. By Theorem 4.31, the invariance of g means

$$(4.25) \qquad\qquad \rho = |\varphi'|\,(\rho \circ \varphi) \qquad \text{for every } \varphi \in \text{Aut}(\mathbb{D}).$$

Suppose such g exists. Then, if $p \in \mathbb{D}$,

$$(4.26) \qquad \rho(p) = |\varphi'(p)|\,\rho(0) \qquad \text{for every } \varphi \in \text{Aut}(\mathbb{D}) \text{ with } \varphi(p) = 0.$$

The right side is independent of the choice of φ since by Theorem 4.16 any $\psi \in \text{Aut}(\mathbb{D})$ with $\psi(p) = 0$ must be of the form $\psi = \alpha\varphi$ with $|\alpha| = 1$, so $|\psi'(p)| = |\varphi'(p)|$. It follows that $\rho(p)$ is completely determined by the positive number $\rho(0)$. Any conformal metric whose density function satisfies (4.26) at every $p \in \mathbb{D}$ satisfies (4.25) and therefore is invariant under the full automorphism group $\text{Aut}(\mathbb{D})$. To see this, take any $\varphi \in \text{Aut}(\mathbb{D})$, and let $p \in \mathbb{D}$, $q = \varphi(p) \in \mathbb{D}$. Using the notation (4.12), the automorphism φ_q sends q to 0, so by (4.26)

$$\rho(q) = |\varphi_q'(q)|\,\rho(0).$$

Similarly, the automorphism $\varphi_q \circ \varphi$ sends p to 0, so

$$\rho(p) = |(\varphi_q \circ \varphi)'(p)|\,\rho(0) = |\varphi_q'(q)|\,|\varphi'(p)|\,\rho(0).$$

Comparing the two relations, we obtain

$$\rho(p) = |\varphi'(p)|\,\rho(\varphi(p)),$$

which is the invariance relation (4.25).

As to the explicit form of the density ρ, take $\varphi(z) = \varphi_p(z) = (z - p)/(1 - \bar{p}z)$ in (4.26) to obtain

$$\rho(p) = \frac{\rho(0)}{1 - |p|^2}.$$

Traditionally, one chooses the normalization $\rho(0) = 2$. The resulting conformal metric

(4.27)
$$g_{\mathbb{D}} = \frac{2}{1 - |z|^2} \, |dz|$$

The choice of the constant 2 is to make the "curvature" of $g_{\mathbb{D}}$ equal to -1; see §11.4.

is called the **hyperbolic** or **Poincaré metric** of the unit disk. We have proved the following

THEOREM 4.35. *There exists a conformal metric $g_{\mathbb{D}}$ in the unit disk which is invariant under the action of* $\mathrm{Aut}(\mathbb{D})$. *This metric, given by (4.27), is C^∞-smooth and unique up to multiplication by a positive constant.*

If we denote the derivative norm of a holomorphic function $f : \mathbb{D} \to \mathbb{D}$ with respect to $g_{\mathbb{D}}$ (in both domain and target) by $\|f'\|$, it follows from (4.27) that

$$\|f'(z)\| = \frac{2/(1 - |f(z)|^2)}{2/(1 - |z|^2)} \, |f'(z)| = \frac{1 - |z|^2}{1 - |f(z)|^2} \, |f'(z)|.$$

Using Theorem 4.31 and Corollary 4.33, we obtain

COROLLARY 4.36. *Every $f \in \mathrm{Aut}(\mathbb{D})$ is a hyperbolic isometry, so it satisfies*

$$\|f'(z)\| = 1, \quad or \quad |f'(z)| = \frac{1 - |f(z)|^2}{1 - |z|^2}$$

for all $z \in \mathbb{D}$.

EXAMPLE 4.37 (Hyperbolic metric of $\mathbb{H}$). Since the unit disk $\mathbb{D}$ and the upper half-plane $\mathbb{H}$ are biholomorphic, it is clear that there is a corresponding hyperbolic metric $g_{\mathbb{H}}$ which is invariant under $\mathrm{Aut}(\mathbb{H})$. Furthermore, $g_{\mathbb{H}} = \varphi^* g_{\mathbb{D}}$ for any biholomorphism $\varphi : \mathbb{H} \to \mathbb{D}$ (why?). For example, taking φ to be the Cayley map $w = \varphi(z) = (i - z)/(i + z)$ of Example 4.5, we see that

$$g_{\mathbb{H}} = \varphi^* \left(\frac{2}{1 - |w|^2} \, |dw| \right) = \frac{2}{1 - |\varphi(z)|^2} \, |\varphi'(z)| \, |dz|$$

$$= \frac{2}{1 - \left| \frac{i-z}{i+z} \right|^2} \, \frac{2}{|i+z|^2} \, |dz| = \frac{4}{|i+z|^2 - |i-z|^2} \, |dz|,$$

which simplifies to

$$g_{\mathbb{H}} = \frac{1}{\mathrm{Im}(z)} \, |dz|.$$

This shows, for example, that a unit (in the hyperbolic sense) tangent vector at $x + iy \in \mathbb{H}$ has Euclidean length y, so its visual size shrinks as $y \to 0$.

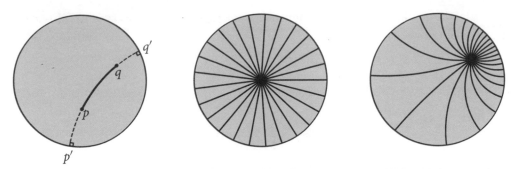

Figure 4.8. Left: The hyperbolic geodesic through $p, q \in \mathbb{D}$. Middle and right: The family of hyperbolic geodesics in $\mathbb{D}$ starting at the origin and at a generic point.

Our next goal is to show the existence and uniqueness of minimal geodesics in the hyperbolic metric of $\mathbb{D}$ and describe them explicitly. To simplify the notation, we denote the distance and length with respect to the hyperbolic metric $g_\mathbb{D}$ by $\text{dist}_\mathbb{D}(\cdot, \cdot)$ and $\text{length}_\mathbb{D}(\cdot)$, respectively.

THEOREM 4.38. *Two distinct points $p, q \in \mathbb{D}$ can be joined by a unique (up to reparametrization) minimal $g_\mathbb{D}$-geodesic. This geodesic is an arc of the Euclidean circle passing through p, q which is orthogonal to the unit circle $\mathbb{T} = \partial\mathbb{D}$. Moreover,*

$$\text{dist}_\mathbb{D}(p, q) = \log\left(\frac{|1 - \bar{p}q| + |p - q|}{|1 - \bar{p}q| - |p - q|}\right).$$

See Fig. 4.8.

PROOF. First consider the case where $p = 0$ and q is a real number $0 < r < 1$. Take any C^1 curve $\gamma : [0, 1] \to \mathbb{D}$ from $\gamma(0) = 0$ to $\gamma(1) = r$. Let $x(t) = \text{Re}(\gamma(t))$, so $(x(t))^2 \leq |\gamma(t)|^2$ and $x'(t) \leq |\gamma'(t)|$ for all t. We have

$$\text{length}_\mathbb{D}(\gamma) = \int_0^1 \|\gamma'(t)\|_{g_\mathbb{D}} \, dt = \int_0^1 \frac{2|\gamma'(t)|}{1 - |\gamma(t)|^2} \, dt \geq \int_0^1 \frac{2x'(t)}{1 - (x(t))^2} \, dt$$

$$= \int_0^r \frac{2 \, dx}{1 - x^2} = \log\left(\frac{1 + r}{1 - r}\right),$$

with equality if and only if $x(t) = |\gamma(t)|$ and $x'(t) = |\gamma'(t)|$, which happens when $\gamma(t) = x(t)$ is positive and increasing from 0 to r. This proves that

$$\text{dist}_\mathbb{D}(0, r) = \log\left(\frac{1 + r}{1 - r}\right) \qquad \text{for } 0 < r < 1$$

and that the minimal geodesic between 0 and r is the straight segment. Since rigid rotations about 0 are hyperbolic isometries, it follows that

(4.28)
$$\text{dist}_\mathbb{D}(0, z) = \log\left(\frac{1 + |z|}{1 - |z|}\right) \qquad z \in \mathbb{D}$$

and the minimal geodesic between 0 and z is the radial segment.

For arbitrary $p, q \in \mathbb{D}$, observe that since the automorphism φ_p of (4.12) is an isometry by Corollary 4.36,

$$\text{dist}_{\mathbb{D}}(p, q) = \text{dist}_{\mathbb{D}}(\varphi_p(p), \varphi_p(q)) = \text{dist}_{\mathbb{D}}\left(0, \frac{q-p}{1-\bar{p}q}\right)$$

$$= \log\left(\frac{1+|\frac{q-p}{1-\bar{p}q}|}{1-|\frac{q-p}{1-\bar{p}q}|}\right) = \log\left(\frac{|1-\bar{p}q|+|p-q|}{|1-\bar{p}q|-|p-q|}\right)$$

as required. The image under $(\varphi_p)^{-1} = \varphi_{-p}$ of the radial segment from 0 to $\varphi(q)$ is clearly the unique minimal geodesic between p and q. The Möbius map φ_{-p} preserves the unit circle $\mathbb{T}$ and sends the line L passing through 0 and $\varphi(q)$ to a circle passing through p and q. This circle meets $\mathbb{T}$ orthogonally since L meets $\mathbb{T}$ orthogonally and since by Theorem 4.4 Möbius maps preserve angles.   □

We can now answer a question raised in Remark 4.17:

COROLLARY 4.39. *Let (p_1, p_2) and (q_1, q_2) be given pairs of distinct points in $\mathbb{D}$. There exists a (unique) element of $\text{Aut}(\mathbb{D})$ which sends p_k to q_k for $k = 1, 2$ if and only if $\text{dist}_{\mathbb{D}}(p_1, p_2) = \text{dist}_{\mathbb{D}}(q_1, q_2)$.*

PROOF. The "only if" part follows from Corollary 4.36 since every disk automorphism is a hyperbolic isometry. For the "if" part, suppose $\text{dist}_{\mathbb{D}}(p_1, p_2) = \text{dist}_{\mathbb{D}}(q_1, q_2)$, map p_1 by an automorphism φ to the origin, and let $p = \varphi(p_2)$. Similarly, map q_1 by an automorphism ψ to the origin and let $q = \psi(q_2)$. As automorphisms are hyperbolic isometries, we have $\text{dist}_{\mathbb{D}}(0, p) = \text{dist}_{\mathbb{D}}(0, q)$. It follows from (4.28) that $|p| = |q|$, so $f : z \mapsto (q/p)z$ is a rigid rotation which maps p to q. Evidently the automorphism $\psi^{-1} \circ f \circ \varphi$ maps the pair (p_1, p_2) to the pair (q_1, q_2).   □

As an application of the above constructions, we give Pick's interpretation of the classical Schwarz lemma (Theorem 4.13) in terms of the hyperbolic metric. It is often called the "invariant" form of the Schwarz lemma because the origin $z = 0$ plays no special role in it:

THEOREM 4.40 (Pick, 1916). *Every holomorphic map $f : \mathbb{D} \to \mathbb{D}$ contracts the hyperbolic metric unless it is an isometry. In other words,*

$$\|f'(z)\| \le 1, \quad \text{or} \quad |f'(z)| \le \frac{1-|f(z)|^2}{1-|z|^2}$$

for all $z \in \mathbb{D}$, with equality at some $z \in \mathbb{D}$ if and only if $f \in \text{Aut}(\mathbb{D})$. In terms of the hyperbolic distance,

$$\text{dist}_{\mathbb{D}}(f(p), f(q)) \le \text{dist}_{\mathbb{D}}(p, q) \qquad \text{for all } p, q \in \mathbb{D},$$

with equality for a pair $p \ne q$ if and only if $f \in \text{Aut}(\mathbb{D})$.

Georg Alexander Pick
(1859–1942)

PROOF. Take any $p \in \mathbb{D}$ and let $a = f(p)$. The holomorphic function $h = \varphi_a \circ f \circ \varphi_{-p} : \mathbb{D} \to \mathbb{D}$ fixes the origin, so by the Schwarz lemma $|h'(0)| \leq 1$. Applying (4.21) to $f = \varphi_{-a} \circ h \circ \varphi_p$ and using the fact that φ_{-a} and φ_p are hyperbolic isometries, we obtain

$$\|f'(p)\| = \|\varphi'_{-a}(0)\| \, \|h'(0)\| \, \|\varphi'_p(p)\| = \|h'(0)\| = |h'(0)| \leq 1.$$

Equality occurs if and only if h satisfies $|h'(0)| = 1$, which by the Schwarz lemma happens precisely when h is a rotation. The latter is equivalent to $f = \varphi_{-a} \circ h \circ \varphi_p \in \mathrm{Aut}(\mathbb{D})$.

The inequality on distances follows from Lemma 4.30 since $\|f'(z)\| \leq 1$ for all z. Given a pair $p \neq q$ in $\mathbb{D}$, let $a = f(p)$ and $h = \varphi_a \circ f \circ \varphi_{-p}$ as above. Then $b = \varphi_p(q)$ is non-zero and

$$\mathrm{dist}_{\mathbb{D}}(0, h(b)) = \mathrm{dist}_{\mathbb{D}}(f(p), f(q)) \leq \mathrm{dist}_{\mathbb{D}}(p, q) = \mathrm{dist}_{\mathbb{D}}(0, b).$$

By (4.28), equality occurs if and only if $|h(b)| = |b|$, which by the Schwarz lemma happens precisely when h is a rotation, or $f \in \mathrm{Aut}(\mathbb{D})$. □

EXAMPLE 4.41. For each integer $n \geq 1$, consider the power map $f : \mathbb{D} \to \mathbb{D}$ defined by $f(z) = z^n$. The inequality $\|f'(z)\| \leq 1$ is equivalent to

$$n|z|^{n-1} \leq \frac{1 - |z|^{2n}}{1 - |z|^2} \qquad \text{for } |z| < 1.$$

For $n = 1$ the map is an automorphism and we have equality everywhere. For $n > 1$ the map is not an automorphism (since it is not injective) and there is strict inequality everywhere. Note however that $\|f'(z)\| \to 1$ as $|z| \to 1$, so the strict inequality becomes an asymptotic equality as $|z| \to 1$. This shows that f does not contract the hyperbolic metric *uniformly* over the disk.

Pick's geometric interpretation of the classical Schwarz lemma was not the final word on the subject. In 1938, Ahlfors realized that what makes the Schwarz-Pick lemma valid is not the particular form of the hyperbolic metric, but negativity of its curvature. This deep insight and some of its applications will be briefly discussed in chapter 11.

Problems

(1)   Let $\Phi : \mathbb{S}^2 \to \hat{\mathbb{C}}$ be the stereographic projection defined by (3.2). Verify the following statements:

(i) Φ maps every Euclidean circle on $\mathbb{S}^2$ (i.e., the intersection of a plane in $\mathbb{R}^3$ with $\mathbb{S}^2$) to a Euclidean circle or line in $\mathbb{C}$.

(ii) Φ is angle-preserving.

(iii)   Under Φ, the reflection $(x_1, x_2, x_3) \mapsto (x_1, x_2, -x_3)$ corresponds to $z \mapsto 1/\bar{z}$ and the antipodal map $(x_1, x_2, x_3) \mapsto (-x_1, -x_2, -x_3)$ corresponds to $z \mapsto -1/\bar{z}$.

For a beautiful visual proof of the statement (i), watch episode 9 of the series *Dimensions* at www.dimensions-math.org

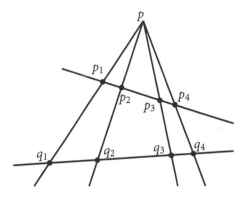

Figure 4.9. Illustration of problem 7.

(Hint: For (i), let Γ be the circle of intersection of the plane $ax_1 + bx_2 + cx_3 + d = 0$ with $\mathbb{S}^2$, where $a^2 + b^2 + c^2 = 1$ and $-1 \leq d \leq 1$. Substitute for x_1, x_2, x_3 in terms of z using (3.3) to obtain an equation for $\Phi(\Gamma)$ which is a circle or a line according as $c + d \neq 0$ or $c + d = 0$.)

(2) (i) Show that $f \in \text{Möb}$ restricts to an element of $\text{Aut}(\mathbb{D})$ if and only if it has the form $f(z) = (az + b)/(\bar{b}z + \bar{a})$ with $|a|^2 - |b|^2 = 1$.

   (ii) Show that $f \in \text{Möb}$ corresponds to a rigid rotation of the sphere $\mathbb{S}^2$ under the stereographic projection if and only if $f(z) = (az + b)/(-\bar{b}z + \bar{a})$ with $|a|^2 + |b|^2 = 1$.

(Hint: For (ii), it would be easier to first find the Möbius maps that correspond to rotations of $\mathbb{S}^2$ about the x_3-axis. Those corresponding to other rotations are then obtained by conjugating with maps of the form $z \mapsto (z - p)/(1 + \bar{p}z)$.)

(3) Find geometric interpretations for the homeomorphisms of $\mathbb{S}^2$ that correspond under the stereographic projection to the Cayley map $\varphi : \mathbb{H} \to \mathbb{D}$ in (4.9) and the similar map $\psi : \{z : \text{Re}(z) > 0\} \to \mathbb{D}$ defined by $\psi(z) = (1 - z)/(1 + z)$.

(4) Prove that every $f \in \text{Möb}$ can be uniquely decomposed as $f = \rho \circ L \circ T$, where ρ corresponds to a rigid rotation of the sphere, $L(z) = \alpha z$ for some $\alpha > 0$, and $T(z) = z + \beta$ for some $\beta \in \mathbb{C}$. (Hint: First treat the case where $f(\infty) = \infty$; the general case can be reduced to this case by postcomposing f with a rigid rotation.)

(5) Suppose $f \in \text{Möb}$ does not fix ∞, so $f(z) = (az + b)/(cz + d)$ with $ad - bc = 1$ and $c \neq 0$.

   (i) Show that the locus $\Gamma_f = \{z \in \mathbb{C} : |f'(z)| = 1\}$ is a Euclidean circle and find its center and radius. This is called the **isometric circle** of f since f preserves the Euclidean length of tangent vectors at every point of Γ_f.

   (ii) Verify that $|f'(z)| > 1$ if z is in the interior of Γ_f, and $|f'(z)| < 1$ if z is in the exterior of Γ_f.

   (iii) Conclude, without any computation, that $f(\Gamma_f) = \Gamma_{f^{-1}}$.

(6) Verify that the cross ratio satisfies the **cocycle relation**

$$[p_1, p_2, p_3, p_4] = [p_1, p_2, z, p_4] \cdot [p_1, z, p_3, p_4]$$

for all $z \in \hat{\mathbb{C}} \setminus \{p_1, p_2, p_3, p_4\}$.

(7)   Referring to Fig. 4.9, prove that

$$[p_1, p_2, p_3, p_4] = [q_1, q_2, q_3, q_4].$$

This means the cross ratio is invariant under projective transformations of the line. (Hint: The cross ratio can be expressed solely in terms of sines of the angles formed at p.)

(8)   Suppose $p, q \in \hat{\mathbb{C}}$, $p \neq q$, and $p \neq -1/\bar{q}$. Use the cross ratio to show that the four points $p, q, -1/\bar{p}, -1/\bar{q}$ lie on a circle. Can you find a simpler explanation for this using the stereographic projection?

(9)   Under the Cayley map $\varphi : \mathbb{H} \to \mathbb{D}$ in (4.9), what element of $\mathrm{Aut}(\mathbb{H})$ would correspond to the rotation $z \mapsto e^{i\theta} z$ in $\mathrm{Aut}(\mathbb{D})$?

(10)   Every element of $\mathrm{Aut}(\mathbb{D})$ extends uniquely to an orientation-preserving homeomorphism of the boundary circle $\mathbb{T}$. Show that under this action of $\mathrm{Aut}(\mathbb{D})$ on $\mathbb{T}$, a triple can be mapped to another triple if and only if they have the same cyclic order. Contrast this with Corollary 4.39.

(11)   Consider the subgroups of $\mathrm{Aut}(\mathbb{C})$ generated by the linear maps $L_\alpha : z \mapsto \alpha z$ and by the translations $T_\beta : z \mapsto z + \beta$, isomorphic to the multiplicative group $\mathbb{C}^*$ and the additive group $\mathbb{C}$, respectively. Show that $\mathrm{Aut}(\mathbb{C})$ is isomorphic to the **semi-direct product** $\mathbb{C}^* \rtimes \mathbb{C}$ defined by the group law

$$(\alpha, \beta) \cdot (\alpha', \beta') = (\alpha\alpha', \alpha\beta' + \beta).$$

(12)   Prove that the group $\mathrm{Aut}(\mathbb{C}^*)$ is generated by the linear maps $L_\alpha : z \mapsto \alpha z$ and the inversion $\iota : z \mapsto 1/z$. More precisely, consider the subgroups of $\mathrm{Aut}(\mathbb{C}^*)$ generated by the L_α and by ι, isomorphic to the multiplicative groups $\mathbb{C}^*$ and $\{\pm 1\}$, respectively. Show that $\mathrm{Aut}(\mathbb{C}^*)$ is isomorphic to the semi-direct product $\mathbb{C}^* \rtimes \{\pm 1\}$ defined by the group law

$$(\alpha, n) \cdot (\alpha', n') = (\alpha(\alpha')^n, nn').$$

(13)   For any $p \neq 0, 1, \infty$, the automorphism group G of the 4-punctured sphere $\hat{\mathbb{C}} \smallsetminus \{0, 1, p, \infty\}$ contains the subgroup

$$H = \left\{ z \mapsto z, \quad z \mapsto \frac{p}{z}, \quad z \mapsto \frac{z-p}{z-1}, \quad z \mapsto \frac{pz-p}{z-p} \right\}.$$

Show that $G = H \cong \mathbb{Z}_2 \times \mathbb{Z}_2$ except for the following five special values of p where G is larger due to extra symmetries:
   • $p = -1$, where $G \cong D_4$ (the dihedral group of order 8) is generated by H and $z \mapsto -z$.
   • $p = 2$, where $G \cong D_4$ is generated by H and $z \mapsto 2 - z$.
   • $p = 1/2$, where $G \cong D_4$ is generated by H and $z \mapsto 1 - z$.
   • $p = e^{\pm i\pi/3}$, where $G \cong D_6$ (the dihedral group of order 12) is generated by H and $z \mapsto 1/(1-z)$.

Show that in the first three cases there is a Möbius map that sends the quadruple $(0, 1, p, \infty)$ to the vertices of a square, while in the last two cases there is a Möbius map that sends the quadruple to the vertices and centroid of an equilateral triangle. (Hint: Elements of H correspond to permutations of $\{0, 1, p, \infty\}$ that preserve the cross ratio. For permutations that change the cross ratio, use the table at the end of §4.1 to find the exceptional values of p.)

(14)   A *finite Blaschke product* is a rational function of the form

$$B(z) = \alpha z^m \prod_{k=1}^{n} \left(\frac{z - p_k}{1 - \overline{p_k} z} \right),$$

where $|\alpha| = 1$, $m \in \mathbb{Z}$, and the p_k are non-zero complex numbers with $|p_k| \neq 1$.

(i)   Show that any such B satisfies $B(1/\overline{z}) = 1/\overline{B(z)}$ for all z. In other words, B commutes with the reflection $z \mapsto 1/\overline{z}$. In particular, z is a zero of B if and only if $1/\overline{z}$ is a pole of B, and $|B(z)| = 1$ whenever $|z| = 1$.

(ii)   Show that if all zeros of B belong to $\mathbb{D}$, then $B(\mathbb{D}) \subset \mathbb{D}$ and $B(\hat{\mathbb{C}} \smallsetminus \overline{\mathbb{D}}) \subset \hat{\mathbb{C}} \smallsetminus \overline{\mathbb{D}}$.

(iii)   Show that z is a critical point of B if and only if $1/\overline{z}$ is a critical point of B.

(15)   Suppose f is a rational function such that $|f(z)| = 1$ whenever $|z| = 1$. Prove that f is a finite Blaschke product. (Hint: $\overline{f(1/\overline{z})}$ is a rational function which agrees with $1/f(z)$ on the unit circle.)

(16)   Suppose that $f : \mathbb{H} \to \mathbb{C}$ is holomorphic and $|f| \leq 1$ everywhere. If $f(i) = 0$, how large or small can $|f(2i)|$ be? Show that your answers are the best possible estimates.

(17)   A generalization of the classical Schwarz lemma: Let $f : \mathbb{D} \to \mathbb{D}$ be holomorphic and $f(p_k) = 0$ for $k = 1, \ldots, n$. Show that for all $z \in \mathbb{D}$,

$$|f(z)| \leq \prod_{k=1}^{n} \left| \frac{z - p_k}{1 - \overline{p_k} z} \right|.$$

What can you say about f if equality holds for some $z \in \mathbb{D} \smallsetminus \{p_1, \ldots, p_n\}$?

(18)   Suppose $f : \mathbb{D}(0, r) \to \mathbb{C}$ is holomorphic and $|f(z)| \leq M$ for all $z \in \mathbb{D}(0, r)$. If f vanishes at $p_1, \ldots, p_n \in \mathbb{D}(0, r) \smallsetminus \{0\}$, prove that

$$\frac{r^n |f(0)|}{|p_1 \cdots p_n|} \leq M.$$

(19)   Suppose $f : \mathbb{D} \to \mathbb{D}$ is holomorphic and $f(0) \neq 0$. Use the Schwarz lemma to prove that f has no zeros in the disk $\mathbb{D}(0, |f(0)|)$.

(20)   Let $f \in \mathcal{O}(\mathbb{D})$ satisfy $f(0) = 0$ and $|f(z) + zf'(z)| < 1$ for all $z \in \mathbb{D}$. Show that $|f(z)| \leq |z|/2$ for all $z \in \mathbb{D}$. (Hint: Consider the function $zf(z)$.)

(21)   Suppose $f : \mathbb{D} \to \mathbb{D}$ is holomorphic and has zeros of even order only. Show that $|f'(0)|^2 \leq 4|f(0)|$. (Hint: f has a holomorphic square root in $\mathbb{D}$.)

(22)   If $f \in \mathcal{O}(\mathbb{D})$, show that there is a sequence $\{z_n\}$ in $\mathbb{D}$, with $|z_n| \to 1$, such that $\{f(z_n)\}$ is a bounded sequence in $\mathbb{C}$. (Hint: Assuming f has finitely many zeros and $|f| \to +\infty$ as $|z| \to 1$, consider B/f for a suitable finite Blaschke product B to get a contradiction.)

(23)   Recall from problem 2 that every $f \in \text{Aut}(\mathbb{D})$ has the form $f(z) = (az + b)/(\overline{b}z + \overline{a})$ with $|a|^2 - |b|^2 = 1$. Express the trace-squared invariant $\tau(f)$ in terms of a, b. Use your formula to find another proof for the fact that $\text{Aut}(\mathbb{D})$ contains no loxodromic element.

(24)   Let f, g be non-identity elements of Möb. If f, g have the same set of fixed points, show that they commute: $f \circ g = g \circ f$. Conversely, if f, g commute, show that either they have the same set of fixed points, or they are involutions (i.e., $f \circ f = g \circ g = \text{id}$) which swap each other's fixed points, such as the pair $f(z) = -z, g(z) = 1/z$.

(25)   Prove the following properties of the hyperbolic metric in $\mathbb{D}$:
  (i)   For every $r > 0$, $\{z \in \mathbb{D} : \mathrm{dist}_{\mathbb{D}}(z, 0) < r\} = \mathbb{D}(0, \tanh(r/2))$. By applying elements of $\mathrm{Aut}(\mathbb{D})$, it follows that every hyperbolic ball in $\mathbb{D}$ is a Euclidean disk, perhaps with a different center.
  (ii)   Closed balls in the metric space $(\mathbb{D}, \mathrm{dist}_{\mathbb{D}})$ are compact. Hence $(\mathbb{D}, \mathrm{dist}_{\mathbb{D}})$ is a complete metric space.

(26)   The hyperbolic geodesic through distinct points $p, q \in \mathbb{D}$ meets the circle $\mathbb{T}$ at p', q' as in Fig. 4.8. Show that

$$\mathrm{dist}_{\mathbb{D}}(p, q) = \log [p', p, q, q'],$$

where, as usual, $[p', p, q, q']$ is the cross ratio of p', p, q, q'.

(27)   Let $f : \mathbb{D} \to \mathbb{D}$ be holomorphic. Show that for every $z \in \mathbb{D}$,

$$\| f'(z) \| = \lim_{w \to z} \frac{\mathrm{dist}_{\mathbb{D}}(f(w), f(z))}{\mathrm{dist}_{\mathbb{D}}(w, z)}.$$

(28)   Find a formula for the hyperbolic distance between two points in the upper half-plane.

(29)   If $f : \mathbb{D} \to \mathbb{D}$ is holomorphic, show that

$$\left| \frac{f(z) - f(w)}{1 - \overline{f(z)} f(w)} \right| \le \left| \frac{z - w}{1 - \bar{z} w} \right| \qquad \text{for } z, w \in \mathbb{D}.$$

(30)   If $f : \mathbb{H} \to \mathbb{H}$ is holomorphic, show that

$$|f'(z)| \le \frac{\mathrm{Im}(f(z))}{\mathrm{Im}(z)} \qquad \text{for } z \in \mathbb{H},$$

with equality at some z if and only if $f \in \mathrm{Aut}(\mathbb{H})$. Verify the inequality

$$\left| \frac{f(z) - f(w)}{\overline{f(z)} - f(w)} \right| \le \left| \frac{z - w}{\bar{z} - w} \right| \qquad \text{for } z, w \in \mathbb{H}.$$

(31)   If $f \in \mathrm{Aut}(\mathbb{H})$, show that

$$|f(z) - f(w)| = |z - w| \, |f'(z) f'(w)|^{1/2} \qquad \text{for } z, w \in \mathbb{H}.$$

(32)   If $f : \mathbb{D} \to \mathbb{D}^*$ is holomorphic, prove that $|f'(0)| \le 2/e$.

5

Convergence and normality

Our main objective in this chapter is to equip the spaces of holomorphic and mero-morphic functions in a domain with a metrizable topology and to study the resulting notions of convergence and compactness. In order to treat both holomorphic and meromorphic functions in a unified fashion, it will be convenient to first build the necessary machinery in the space of all continuous maps from an open subset of the plane into an arbitrary metric space. The opening section of this chapter is devoted to this topological prelude.

5.1 Compact convergence

Throughout this section, $U \subset \mathbb{C}$ will be a non-empty open set and X will be a metric space with the distance function d. In our applications X will be either $\mathbb{C}$ equipped with the Euclidean metric or $\hat{\mathbb{C}}$ equipped with the spherical metric (see §5.3). The space of all continuous maps $U \to X$ will be denoted by $\mathscr{C}(U, X)$.

DEFINITION 5.1. A **basic neighborhood** of $f \in \mathscr{C}(U, X)$ is a set of the form

$$B(f, K, \varepsilon) = \{g \in \mathscr{C}(U, X) : \sup_{p \in K} d(f(p), g(p)) < \varepsilon\},$$

where $K \subset U$ is compact and $\varepsilon > 0$. The topology generated by the collection of all basic neighborhoods (for various choices of f, K, ε) is called the **compact convergence topology** on $\mathscr{C}(U, X)$.

This is a special case of the "compact-open topology" on the space of continuous maps between two topological spaces.

The basic neighborhoods form a basis for this topology. In fact, their union is $\mathscr{C}(U, X)$ since $f \in B(f, K, \varepsilon)$ for every f, K, ε. Moreover, if $g \in B(f_1, K_1, \varepsilon_1) \cap B(f_2, K_2, \varepsilon_2)$ and if we choose ε such that

$$0 < \varepsilon < \min_{j=1,2} \left\{\varepsilon_j - \sup_{p \in K_j} d(f_j(p), g(p))\right\},$$

then the basic neighborhood $B(g, K_1 \cup K_2, \varepsilon)$ is contained in the intersection $B(f_1, K_1, \varepsilon_1) \cap B(f_2, K_2, \varepsilon_2)$. It follows that a subset of $\mathscr{C}(U, X)$ is open if and only if it is a union of basic neighborhoods.

One can verify that the compact convergence topology on $\mathscr{C}(U, X)$ depends only on the topology of X, not on the choice of the metric d (see problem 1).

DEFINITION 5.2. A sequence $\{f_n\}$ in $\mathscr{C}(U, X)$ **converges compactly** to a map $f \in \mathscr{C}(U, X)$ (often abbreviated "$f_n \to f$ compactly in U") if every basic neighborhood of f contains all but finitely many of the f_n. In other words, if for every compact set $K \subset U$ and every $\varepsilon > 0$ there exists an integer $N \geq 1$ (depending on K and ε) such that

$$\sup_{p \in K} d(f_n(p), f(p)) < \varepsilon \qquad \text{for all } n \geq N.$$

The phrase **uniform convergence on compact subsets** is widely used in this context, but we prefer compact convergence for its brevity. One should have no trouble verifying that $f_n \to f$ compactly in U if and only if every $p \in U$ has a neighborhood $V_p \subset U$ such that $f_n \to f$ uniformly in V_p. Because of this, compact convergence is also referred to as **local uniform convergence**.

The compact convergence topology of $\mathscr{C}(U, X)$ is metrizable. In other words, there is a metric $\mathbf{d}$ on $\mathscr{C}(U, X)$ such that the open sets in $\mathscr{C}(U, X)$ are precisely the unions of open balls defined by $\mathbf{d}$, and therefore $f_n \to f$ compactly in U if and only if $\mathbf{d}(f_n, f) \to 0$. The construction of $\mathbf{d}$ (which is far from unique) makes use of an **exhaustion** of U, that is, a sequence $\{K_n\}$ of compact sets such that

The exhaustion defined here has an additional topological property that will be crucial in Runge's theory. See Lemma 9.23.

(i) $U = \bigcup_{n=1}^{\infty} K_n$, and
(ii)   each K_n is contained in the interior of K_{n+1}.

For example, define

$$K_n = \left\{ p \in \mathbb{C} : |p| \leq n \text{ and } \mathbb{D}(p, 1/n) \subset U \right\}.$$

Evidently K_n is a compact subset of U and property (i) holds. Let $p \in K_n$. If $|z - p| < 1/n - 1/(n+1)$, then $\mathbb{D}(z, 1/(n+1)) \subset \mathbb{D}(p, 1/n) \subset U$ and $|z| \leq |p| + 1 \leq n + 1$, which show $z \in K_{n+1}$. This means p is an interior point of K_{n+1}, so (ii) holds as well.

The most useful property of an exhaustion $\{K_n\}$ of U is that every compact set $K \subset U$ is contained in some K_n. In fact, K is covered by the union of the interiors of K_n, and since these interiors are nested, one of them should already cover K.

Given an exhaustion $\{K_n\}$ of U and $f, g \in \mathscr{C}(U, X)$, define

$$\mathbf{d}_n(f, g) = \min \left\{ 1, \sup_{p \in K_n} d(f(p), g(p)) \right\}$$

$$\mathbf{d}(f,g) = \sum_{n=1}^{\infty} 2^{-n} \mathbf{d}_n(f,g).$$

Note that $0 \le \mathbf{d}_n(f,g) \le \mathbf{d}_{n+1}(f,g) \le 1$ for all n, so the series is convergent and $0 \le \mathbf{d}(f,g) \le 1$. It is easily checked that each $\mathbf{d}_n$ is symmetric and satisfies the triangle inequality, so the same must be true of $\mathbf{d}$. Moreover, if $\mathbf{d}(f,g) = 0$, then $\mathbf{d}_n(f,g) = 0$ for all n, so $f = g$ on every K_n, and therefore $f = g$ in U. It follows that $\mathbf{d}$ is a metric on $\mathscr{C}(U,X)$.

THEOREM 5.3. *The topology induced by the metric* $\mathbf{d}$ *is the compact convergence topology on* $\mathscr{C}(U,X)$.

PROOF. Let $D(f,\varepsilon) = \{g \in \mathscr{C}(U,X) : \mathbf{d}(f,g) < \varepsilon\}$, the ε-ball centered at f in the metric $\mathbf{d}$. It suffices to show that every basic neighborhood in $\mathscr{C}(U,X)$ is a union of such balls and vice versa. First suppose $g \in B(f,K,\varepsilon)$. Since basic neighborhoods form a basis, we can find a compact set $H \subset U$ and some $0 < \delta < 1$ such that $B(g,H,\delta) \subset B(f,K,\varepsilon)$. Choose n large enough such that $H \subset K_n$. If $\mathbf{d}(g,h) < 2^{-n}\delta$, then $\mathbf{d}_n(g,h) < \delta$, so

$$\sup_{p \in H} d(g(p),h(p)) \le \sup_{p \in K_n} d(g(p),h(p)) = \mathbf{d}_n(g,h) < \delta,$$

which proves $h \in B(g,H,\delta)$. Thus, $D(g,2^{-n}\delta) \subset B(f,K,\varepsilon)$.

Conversely, suppose $g \in D(f,\varepsilon)$ and take $0 < \delta < 1$ such that $D(g,2\delta) \subset D(f,\varepsilon)$. Choose $n \ge 1$ so that $2^{-n} < \delta$. If $h \in B(g,K_n,\delta)$, then $\mathbf{d}_n(g,h) < \delta$, so

$$\mathbf{d}(g,h) = \left(\sum_{k=1}^{n} + \sum_{k=n+1}^{\infty} \right) 2^{-k} \mathbf{d}_k(g,h) < \delta \sum_{k=1}^{n} 2^{-k} + \sum_{k=n+1}^{\infty} 2^{-k} < 2\delta,$$

which proves $h \in D(g,2\delta)$. It follows that $B(g,K_n,\delta) \subset D(f,\varepsilon)$. $\qquad\square$

Let us call a subset of a topological space "precompact" if it has compact closure. (Caution: Some authors use the term "relatively compact" instead, which is somewhat misleading because unlike openness or closedness, compactness is not a relative property.) Since in metric spaces compactness and sequential compactness are equivalent, it follows that a subset of a metric space is precompact if and only if every sequence in it has a convergent subsequence.

> **DEFINITION 5.4.** A family $\mathscr{F} \subset \mathscr{C}(U,X)$ is **precompact** if it has compact closure. Equivalently, if every sequence in $\mathscr{F}$ has a subsequence which converges compactly in U (the limit map need not be in $\mathscr{F}$).

THEOREM 5.5 (Precompactness is a local property). *A family* $\mathscr{F} \subset \mathscr{C}(U,X)$ *is precompact if and only if every* $p \in U$ *has a neighborhood* $V \subset U$ *such that the restricted family* $\mathscr{F}|_V = \{f|_V : f \in \mathscr{F}\} \subset \mathscr{C}(V,X)$ *is precompact.*

PROOF. Clearly precompactness of $\mathscr{F}$ implies that of $\mathscr{F}|_V$ for every open set $V \subset U$. For the converse, suppose every $p \in U$ has a neighborhood $V_p \subset U$ such that $\mathscr{F}|_{V_p}$ is precompact. Choose an open disk D_p centered at p with $\overline{D_p} \subset V_p$. Let $\{K_n\}$ be an exhaustion of U. Since each K_n is covered by finitely many of the D_p and since $U = \bigcup K_n$, there is a sequence $p_1, p_2, p_3, \ldots$ in U for which the corresponding disks $D_{p_1}, D_{p_2}, D_{p_3}, \ldots$ cover U. To simplify the notation, set $V_k = V_{p_k}$ and $D_k = D_{p_k}$.

Let $\{f_n\}$ be a sequence in $\mathscr{F}$. By the assumption $\{f_n|_{V_1}\}$ is precompact, so there is an increasing sequence S_1 of positive integers such that $\{f_n\}_{n \in S_1}$ converges uniformly on D_1. Similarly, since $\{f_n|_{V_2}\}_{n \in S_1}$ is precompact, there is an increasing sequence $S_2 \subset S_1$ such that $\{f_n\}_{n \in S_2}$ converges uniformly on D_2. Continuing inductively, for every $k \geq 2$, there is an increasing sequence $S_k \subset S_{k-1}$ such that $\{f_n\}_{n \in S_k}$ converges uniformly on D_k. Let S be the sequence formed from the first element of S_1 followed by the second element of S_2 followed by the third element of S_3 and so on (this is an example of Cantor's "diagonal argument"). Evidently, $\{f_n\}_{n \in S}$ converges uniformly on each D_k. Since every compact subset of U is covered by finitely many of the D_k, we conclude that $\{f_n\}_{n \in S}$ converges compactly in U. $\qquad\square$

A classical theorem of Arzelà and Ascoli gives a necessary and sufficient condition for a family in $\mathscr{C}(U, X)$ to be precompact. This condition makes use of the following notion:

> **DEFINITION 5.6.** A family $\mathscr{F} \subset \mathscr{C}(U, X)$ is **equicontinuous** if for every $p \in U$ and $\varepsilon > 0$ there is a $\delta > 0$ such that
> $$|p - q| < \delta \text{ and } f \in \mathscr{F} \quad \text{imply} \quad d(f(p), f(q)) < \varepsilon.$$

The point of the definition is that δ depends only on the given $p \in U$ and $\varepsilon > 0$, not on the choice of the map $f \in \mathscr{F}$.

The notion of equicontinuity as defined above appears to be a pointwise condition, but in fact it is equivalent to *uniform equicontinuity on compact subsets*:

LEMMA 5.7. *A family $\mathscr{F} \subset \mathscr{C}(U, X)$ is equicontinuous if and only if for every compact set $K \subset U$ and $\varepsilon > 0$ there is a $\delta > 0$ such that*

$$p \in K, \ |p - q| < \delta, \text{ and } f \in \mathscr{F} \quad \text{imply} \quad d(f(p), f(q)) < \varepsilon.$$

PROOF. The "if" part is trivial. For the "only if" part, suppose $\mathscr{F}$ is equicontinuous but the above condition fails. Then there is a compact set $K \subset U$, an $\varepsilon > 0$, sequences $\{p_n\}$ in K and $\{q_n\}$ in U, and a sequence $\{f_n\}$ in $\mathscr{F}$ such that

$$|p_n - q_n| < \frac{1}{n} \quad \text{but} \quad d(f_n(p_n), f_n(q_n)) \geq 3\varepsilon \quad \text{for all } n.$$

After passing to a subsequence, we may assume $p_n \to p \in K$. It then follows that $q_n \to p$ as well. Let $\delta > 0$ be an equicontinuity number corresponding to p and ε. Since

$|p_n - p| < \delta$ and $|q_n - p| < \delta$ for all large n, it follows that

$$d(f_n(p_n), f_n(q_n)) \leq d(f_n(p_n), f_n(p)) + d(f_n(p), f_n(q_n)) < 2\varepsilon,$$

which is a contradiction. □

THEOREM 5.8 (Arzelà-Ascoli, 1895). *A family $\mathscr{F} \subset \mathscr{C}(U, X)$ is precompact if and only if the following hold:*

> *(i) for every $p \in U$, the set $\{f(p) : f \in \mathscr{F}\}$ is precompact in X, and*
> *(ii) $\mathscr{F}$ is equicontinuous.*

The notion of equicontinuity was introduced by Ascoli in 1884. A version of Theorem 5.8 for functions $[a, b] \to \mathbb{R}$ was proved by Arzelà in 1895.

PROOF. First assume $\mathscr{F}$ is precompact. Then clearly (i) must hold. If $\mathscr{F}$ is not equicontinuous, we can find a point $p \in U$, an $\varepsilon > 0$, a sequence $\{q_n\}$ in U converging to p and a sequence $\{f_n\}$ in $\mathscr{F}$ such that

$$(5.1) \qquad\qquad d(f_n(p), f_n(q_n)) \geq \varepsilon \qquad \text{for all } n.$$

Since $\mathscr{F}$ is precompact, after passing to a subsequence we can assume that $\{f_n\}$ converges compactly to some $f \in \mathscr{C}(U, X)$. By the triangle inequality,

$$d(f_n(p), f_n(q_n)) \leq d(f_n(p), f(p)) + d(f(p), f(q_n)) + d(f(q_n), f_n(q_n)).$$

As $n \to \infty$, the middle term on the right tends to 0 since f is continuous at p while the remaining two terms tend to 0 since $f_n \to f$ uniformly on any neighborhood of p with compact closure in U (which necessarily contains q_n for all large n). This would imply $\lim_{n \to \infty} d(f_n(p), f_n(q_n)) = 0$, contradicting (5.1). Thus (ii) holds as well.

For the converse, suppose $\mathscr{F}$ satisfies (i) and (ii). Take an arbitrary sequence $\{f_n\}$ in $\mathscr{F}$. Let E be a countable dense subset of U. Since by (i) the sequence $\{f_n(p)\}$ is precompact for every $p \in E$, a diagonal argument (similar to the proof of Theorem 5.5) gives an increasing sequence S of integers for which $\{f_n\}_{n \in S}$ converges at every point of E. We complete the proof by showing that $\{f_n\}_{n \in S}$ converges compactly in U.

To this end, fix a compact set $K \subset U$ and an $\varepsilon > 0$, and let $\delta > 0$ be a uniform equicontinuity number corresponding to K, ε from Lemma 5.7. Cover K by finitely many disks $D_1, \ldots, D_s$ of radius $\delta/2$ and for each $1 \leq j \leq s$ choose a point $p_j \in D_j \cap E$. Since $\{f_n\}_{n \in S}$ converges at each p_j, we can find an integer N such that

$$(5.2) \qquad d(f_n(p_j), f_m(p_j)) < \varepsilon \qquad \text{if } n, m \in S, \ n, m \geq N, \text{ and } 1 \leq j \leq s.$$

Each $p \in K$ belongs to some D_j, so $|p - p_j| < \delta$. Thus, if $n, m \in S$ and $n, m \geq N$,

$$d(f_n(p), f_m(p)) \leq d(f_n(p), f_n(p_j)) + d(f_n(p_j), f_m(p_j)) + d(f_m(p_j), f_m(p))$$

$$< \varepsilon + \varepsilon + \varepsilon = 3\varepsilon.$$

Here the middle term is less than ε because of (5.2) while the remaining two terms are less than ε because of equicontinuity. This shows $\{f_n\}_{n \in S}$ satisfies a uniform Cauchy condition on K, hence converges uniformly on K. □

When the target space X is compact (such as the Riemann sphere), the condition (i) in Theorem 5.8 automatically holds. On the other hand, when X is a metric space in which compactness is equivalent to being bounded and closed (such as the complex plane $\mathbb{C}$), the condition (i) can be described as $\mathscr{F}$ being **pointwise bounded**: For every $p \in U$, the set $\{f(p) : f \in \mathscr{F}\} \subset X$ is bounded.

EXAMPLE 5.9. Fix constants $0 < \alpha < 1, C > 0$. Let $\mathscr{F}$ be the family of all functions $f : \mathbb{D} \to \mathbb{C}$, with $f(0) = 0$, which satisfy the **Hölder condition**

$$|f(z) - f(w)| \leq C|z - w|^\alpha \qquad \text{for all } z, w \in \mathbb{D}.$$

This is weaker than complex differentiability but stronger than mere continuity. Evidently $\mathscr{F}$ is an equicontinuous family in $\mathscr{C}(\mathbb{D}, \mathbb{C})$. Moreover, the inequality

$$|f(z)| \leq C|z|^\alpha \qquad \text{for all } z \in \mathbb{D} \text{ and } f \in \mathscr{F}$$

shows that $\mathscr{F}$ is pointwise (in fact uniformly) bounded. By Theorem 5.8, $\mathscr{F}$ is precompact. Since the defining conditions of $\mathscr{F}$ persist under compact convergence, $\mathscr{F}$ is also closed in $\mathscr{C}(\mathbb{D}, \mathbb{C})$. It follows that $\mathscr{F}$ is a compact family.

REMARK 5.10. The assumption that the maps in this section were defined in an open subset of the plane was not essential. Indeed, we could replace $U \subset \mathbb{C}$ by any Hausdorff space that is **locally compact** (each point has a neighborhood with compact closure), σ-**compact** (a countable union of compact sets), and **separable** (having a countable dense subset). The definition of compact convergence, the construction of the metric **d** via an exhaustion, and the proof of the Arzelà-Ascoli theorem could then be adapted to this general setting with very minor changes.

5.2 Convergence in the space of holomorphic functions

We now return to the familiar setting of holomorphic functions defined in an open set U in the plane. We consider $\mathscr{O}(U)$ as a subspace of $\mathscr{C}(U, \mathbb{C})$ where the target space $\mathbb{C}$ is equipped with the Euclidean metric $d(z, w) = |z - w|$. The induced notion of compact convergence in $\mathscr{O}(U)$ is thus uniform convergence on compact subsets of U with respect to the Euclidean metric.

It is well known that the limit of a compactly convergent sequence of smooth functions in $\mathscr{C}(U, \mathbb{C})$ need not be smooth. For example, the theorem of Stone-Weierstrass shows that for any $f \in \mathscr{C}(U, \mathbb{C})$ (no matter how non-differentiable) there is a sequence of complex-valued polynomials $P_n(x, y)$ in real variables x, y such that $P_n \to f$ compactly in U (see for example [**Ru1**]). Fortunately, the situation is much simpler when dealing with holomorphic functions.

THEOREM 5.11 (Weierstrass, 1894). *Suppose $f_n \in \mathscr{O}(U)$ and $f_n \to f$ compactly in U as $n \to \infty$. Then $f \in \mathscr{O}(U)$ and $f'_n \to f'$ compactly in U as $n \to \infty$.*

In a fancier language, $\mathscr{O}(U)$ is a closed subspace of $\mathscr{C}(U,\mathbb{C})$ and complex differentiation $f \mapsto f'$ is a continuous operator on $\mathscr{O}(U)$.

PROOF. Evidently f is continuous in U. Let $T \subset U$ be any closed triangle. Since the oriented boundary ∂T is a compact subset of U, $f_n \to f$ uniformly on ∂T. Cauchy's theorem applied to each f_n gives

$$\int_{\partial T} f(z)\, dz = \lim_{n \to \infty} \int_{\partial T} f_n(z)\, dz = 0.$$

By Morera's Theorem 1.39, $f \in \mathscr{O}(U)$.

Now fix a compact set $K \subset U$ and an $\varepsilon > 0$. Choose $r > 0$ small enough to guarantee that the closed r-neighborhood $K_r = \{z \in \mathbb{C} : \operatorname{dist}(z, K) \leq r\}$ is contained in U. Find an integer N such that

$$\sup_{z \in K_r} |f_n(z) - f(z)| \leq r\varepsilon \qquad \text{if } n \geq N.$$

The closed disk $\overline{\mathbb{D}}(z, r)$ is contained in K_r if $z \in K$. Thus, Theorem 1.43 applied to $f_n - f$ gives

$$\sup_{z \in K} |f_n'(z) - f'(z)| \leq \frac{r\varepsilon}{r} = \varepsilon \qquad \text{if } n \geq N.$$

This proves $f_n' \to f'$ compactly in U as $n \to \infty$. $\qquad \square$

Karl Theodor Wilhelm Weierstrass (1815–1897)

EXAMPLE 5.12. Let $f(z) = \sum_{k=0}^{\infty} a_k z^k$ be holomorphic in $\mathbb{D}$. Define $f_n(z) = \sum_{k=0}^{n} a_k z^k$. Then $f_n \to f$ compactly in $\mathbb{D}$ (Theorem 1.17). However, the convergence is not necessarily uniform in $\mathbb{D}$. For instance, when $f(z) = 1/(1 - z)$ we have $f_n(z) = (1 - z^{n+1})/(1 - z)$, and

$$\sup_{z \in \mathbb{D}} |f_n(z) - f(z)| = \sup_{z \in \mathbb{D}} \left| \frac{z^{n+1}}{1 - z} \right| = +\infty.$$

The following provides a very useful sufficient condition for compact convergence of infinite series of holomorphic functions:

COROLLARY 5.13 (Weierstrass M-test). *Suppose $f_n \in \mathscr{O}(U)$ and for every compact set $K \subset U$ there are constants $M_n \geq 0$ such that*

$$\sup_{z \in K} |f_n(z)| \leq M_n \quad \text{and} \quad \sum_{n=1}^{\infty} M_n < +\infty.$$

Then the series $\sum_{n=1}^{\infty} f_n$ converges compactly in U to some $f \in \mathscr{O}(U)$. Moreover, the term-by-term differentiated series $\sum_{n=1}^{\infty} f_n'$ converges compactly in U to f'.

PROOF. Take a compact set $K \subset U$ and find the corresponding constants M_n. For a given $\varepsilon > 0$ find an integer $N \geq 1$ such that $\sum_{n=N}^{\infty} M_n < \varepsilon$. If $z \in K$ and $k > m \geq N$, then

$$\left| \sum_{n=m}^{k} f_n(z) \right| \le \sum_{n=m}^{k} |f_n(z)| \le \sum_{n=m}^{k} M_n < \varepsilon.$$

This shows that the sequence $\{s_m = \sum_{n=1}^{m} f_n\}$ of partial sums is uniformly Cauchy on K, so it converges uniformly there as $m \to \infty$. Since K was an arbitrary compact set in U, it follows that the series $\sum_{n=1}^{\infty} f_n$ converges compactly in U to the function $f = \lim_{m \to \infty} s_m$. The assertion $f \in \mathscr{O}(U)$ and the term-by-term differentiation claim both follow from Theorem 5.11 applied to the sequence $\{s_m\}$. $\qquad\square$

EXAMPLE 5.14. The series $\sum_{n=1}^{\infty} z^n/n^2$ converges compactly in $\mathbb{D}$ to a holomorphic function f. This follows from the Weierstrass M-test since $|z^n/n^2| \le 1/n^2$ for $z \in \mathbb{D}$ and the series $\sum 1/n^2$ converges. As a result, f' is obtained by term-by-term differentiation:

$$f'(z) = \sum_{n=1}^{\infty} \frac{z^{n-1}}{n} = -\frac{1}{z} \operatorname{Log}(1-z) \qquad \text{for } |z| < 1,$$

where Log is the principal branch of logarithm defined in Example 2.23. Observe that the series defining f converges uniformly on $\mathbb{D}$, so f has a continuous extension to the unit circle. However, the series defining f' converges uniformly only on compact subsets of $\mathbb{D}$. In fact, if this series did converge uniformly on $\mathbb{D}$, it would have to converge on the unit circle, which is not the case since it clearly diverges at $z = 1$.

EXAMPLE 5.15 (Riemann's Zeta Function). Consider the series

$$\zeta(z) = \sum_{n=1}^{\infty} \frac{1}{n^z},$$

where, as usual, $n^z = \exp(z \log n)$. If $\operatorname{Re}(z) \ge t > 1$, then

$$\left| \frac{1}{n^z} \right| = \frac{1}{n^{\operatorname{Re}(z)}} \le \frac{1}{n^t}.$$

Since $\sum_{n=1}^{\infty} 1/n^t$ converges for every $t > 1$, it follows from the Weierstrass M-test that ζ is holomorphic in the half-plane $\{z \in \mathbb{C} : \operatorname{Re}(z) > 1\}$. Moreover,

$$\zeta'(z) = \sum_{n=1}^{\infty} \left(\frac{1}{n^z} \right)' = -\sum_{n=1}^{\infty} \frac{\log n}{n^z} \qquad \text{if } \operatorname{Re}(z) > 1.$$

Next, we consider the question of whether the mapping properties of a sequence of holomorphic functions would persist under compact convergence. We begin with a basic result which states that the number of times a non-constant holomorphic function takes on a value in a domain does not change under small perturbations of the function. Recall from Notation 3.41 that $\mathrm{N}_f(D, 0)$ is the number of times, counting multiplicities, that a holomorphic function f takes on the value 0 in a domain D.

LEMMA 5.16 (Persistence of degree). *Let $U \subset \mathbb{C}$ be a domain and γ be a Jordan curve in U such that $D = \text{int}(\gamma) \subset U$. Suppose $f \in \mathcal{O}(U)$ and $f(z) \neq 0$ for all $z \in |\gamma|$. Then there exists a $\delta > 0$ such that for every $g \in \mathcal{O}(U)$,*

$$\text{if} \quad \sup_{z \in |\gamma|} |f(z) - g(z)| < \delta, \quad \text{then} \quad N_f(D, 0) = N_g(D, 0).$$

PROOF. Let $\delta = \inf_{z \in |\gamma|} |f(z)| > 0$. If $g \in \mathcal{O}(U)$ satisfies the condition $\sup_{z \in |\gamma|} |f(z) - g(z)| < \delta$, then $|f(z) - g(z)| < \delta \leq |f(z)|$ for all $z \in |\gamma|$. The result now follows from Rouché's Theorem 3.46. $\square$

EXAMPLE 5.17. Consider the family $\{P_t(z) = tz + z^m\}_{t \in \mathbb{C}}$ of polynomials, where $m \geq 2$ is a fixed integer. The polynomial P_0 has a zero of order m at $z = 0$. By Lemma 5.16, for each $r > 0$ there exists a $\delta > 0$ such that P_t has m zeros counting multiplicities in the disk $\mathbb{D}(0, r)$ whenever $|t| < \delta$. If fact, we can verify directly that under small perturbations, the zero of order m at the origin splits into m simple zeros nearby: Given $r > 0$, choose $0 < \delta < r^{m-1}$ and note that if $0 < |t| < \delta$, the polynomial P_t vanishes at $z = 0$ and the $(m-1)$-st roots of $-t$, all contained in $\mathbb{D}(0, r)$.

Suppose a sequence $\{f_n\}$ in $\mathscr{C}(U, \mathbb{C})$ converges compactly to f, and $f \neq 0$ throughout U. An easy exercise shows that for every compact set $K \subset U$ there is an integer N such that $f_n \neq 0$ on K whenever $n \geq N$. Intuitively, the zeros of f_n get pushed out of the way and disappear in the limit. Lemma 5.16 allows us to prove the converse statement when the f_n are holomorphic:

THEOREM 5.18 (Hurwitz, 1889). *Let $U \subset \mathbb{C}$ be a domain, $f_n \in \mathcal{O}(U)$, and $f_n \to f$ compactly in U as $n \to \infty$. Suppose for each compact set $K \subset U$ there is an integer N such that $f_n \neq 0$ on K if $n \geq N$. Then either $f \neq 0$ everywhere in U, or $f = 0$ everywhere in U.*

Adolf Hurwitz
(1859–1919)

The example $f_n(z) = e^z / n$ shows that the second case can actually occur.

PROOF. Assume that f is not identically 0 but $f(p) = 0$ for some $p \in U$. Then p is an isolated zero, so we can choose a disk D centered at p such that $\overline{D} \subset U$ and $f(z) \neq 0$ for all $z \in \partial D$. Since $f_n \to f$ uniformly on ∂D, Lemma 5.16 shows that $N_{f_n}(D, 0) = N_f(D, 0) \geq 1$ for all large n. This contradicts the assumption that $N_{f_n}(D, 0) = 0$ for all large n. $\square$

EXAMPLE 5.19. There is no sequence $\{P_n\}$ of polynomials such that $e^{P_n(z)} \to z$ compactly in $\mathbb{C}$ as $n \to \infty$. This follows from Theorem 5.18 since $\exp(P_n)$ is non-vanishing while the identity function has an isolated zero at the origin. Note, however, that for every disk D that does not contain the origin there are indeed polynomials P_n such that $e^{P_n(z)} \to z$ compactly in D as $n \to \infty$. To see this, let $f \in \mathcal{O}(D)$ be a holomorphic branch of $\log z$ in D, which exists by Corollary 2.22, and let P_n be the n-th partial sum of the power series expansion of f in D.

Here are two corollaries of Hurwitz's theorem:

COROLLARY 5.20. *Let $U \subset \mathbb{C}$ be a domain, $f_n \in \mathcal{O}(U)$, and $f_n \to f$ compactly in U as $n \to \infty$. Suppose there is an open set V such that $f_n(U) \subset V$ for all n. Then either $f(U) \subset V$ or f is a constant function taking its value on the boundary ∂V.*

PROOF. Evidently $f(U) \subset \overline{V} = V \cup \partial V$. Suppose f takes on a value $q \in \partial V$. The functions $f_n - q$ are non-vanishing and converge compactly to $f - q$, which does vanish somewhere in U. By Theorem 5.18, $f = q$ everywhere in U. □

EXAMPLE 5.21. Here is an example of the second possibility in the above corollary. Consider the sequence $f_n \in \text{Aut}(\mathbb{D})$ defined by $f_n(z) = (z - p_n)/(1 - p_n z)$, where $0 < p_n < 1$ and $p_n \to 1$ as $n \to \infty$. Then f_n converges compactly in $\mathbb{D}$ to the constant function -1. In fact,

$$|f_n(z) + 1| = \frac{|(1+z)(1-p_n)|}{|1 - p_n z|} \leq \frac{(1+|z|)(1-p_n)}{1 - p_n|z|} \leq \frac{1+|z|}{1-|z|}(1-p_n).$$

If $0 < r < 1$, it follows that

$$\sup_{|z| \leq r} |f_n(z) + 1| \leq \frac{1+r}{1-r}(1-p_n)$$

and the latter tends to 0 as $n \to \infty$.

COROLLARY 5.22. *Let $U \subset \mathbb{C}$ be a domain, $f_n \in \mathcal{O}(U)$, and $f_n \to f$ compactly in U as $n \to \infty$. If each f_n is injective, then f is either injective or constant.*

The example $f_n(z) = z/n$ shows that the second case can occur.

PROOF. Suppose there are distinct points $p, q \in U$ such that $f(p) = f(q)$. Take a disk D centered at q such that $\overline{D} \subset U \smallsetminus \{p\}$. The functions $f_n - f_n(p)$ are non-vanishing in D since the f_n are injective, and they converge uniformly in D to $f - f(p)$, which vanishes at $q \in D$. By Theorem 5.18, $f = f(p)$ everywhere in D, hence in U by the identity theorem. □

We now turn to the question of precompactness of families of holomorphic functions, a central issue with important applications. For instance, the solution of many extremal problems in complex function theory is obtained as the limit of a sequence of approximate solutions (a historically significant example is the Riemann mapping theorem to be discussed in chapter 6). It would be desirable in such situations to be able to guarantee that the sequence of approximate solutions has a convergent subsequence. The direct application of the Arzelà-Ascoli theorem to prove precompactness runs into the problem of checking equicontinuity, which can be difficult to deal with. As a first reduction of this problem, we can use differentiability of holomorphic functions to replace equicontinuity with a boundedness condition on complex derivatives.

Let us say that a family $\mathcal{F} \subset \mathcal{C}(U, \mathbb{C})$ is ***compactly bounded*** if for every compact set $K \subset U$ there is a constant $M > 0$ such that $|f(z)| \leq M$ whenever $z \in K$ and $f \in \mathcal{F}$.

Precompact families of holomorphic functions are often called "normal." We prefer to reserve normality for a slightly more general property which is well adapted to convergence of meromorphic functions. See §5.3.

THEOREM 5.23. *A family $\mathscr{F} \subset \mathscr{O}(U)$ is precompact if and only if the following hold:*

(i) $\{f(p) : f \in \mathscr{F}\}$ *is a bounded subset of $\mathbb{C}$ for every $p \in U$, and*

(ii) *the family $\mathscr{F}' = \{f' : f \in \mathscr{F}\}$ of derivatives is compactly bounded.*

PROOF. Suppose $\mathscr{F}$ is precompact. Then (i) clearly holds. If (ii) fails, we can find a compact set $K \subset U$, a sequence $\{z_n\}$ in K, and a sequence $\{f_n\}$ in $\mathscr{F}$ such that $|f_n'(z_n)| > n$ for all n. Take a subsequence of $\{f_n\}$ that converges compactly in U. By Theorem 5.11 the corresponding subsequence of $\{f_n'\}$ converges compactly in U, hence uniformly on K, which contradicts $|f_n'(z_n)| \to +\infty$. Thus (ii) holds as well.

Conversely, suppose (i) and (ii) hold. By the Arzelà-Ascoli Theorem 5.8, it suffices to check that $\mathscr{F}$ is equicontinuous. Take $p \in U$ and choose $r > 0$ such that $\overline{\mathbb{D}}(p, r) \subset U$. Use (ii) to find an $M > 0$ such that $|f'(\zeta)| \leq M$ whenever $\zeta \in \mathbb{D}(p, r)$ and $f \in \mathscr{F}$. This gives the estimate

$$|f(z) - f(p)| = \left| \int_{[p,z]} f'(\zeta)\, d\zeta \right| \leq M|z - p|$$

for all $z \in \mathbb{D}(p, r)$ and $f \in \mathscr{F}$, and equicontinuity follows immediately. $\square$

REMARK 5.24. When U is a domain, the pointwise boundedness condition (i) in the above theorem can be weakened to $\{f(p) : f \in \mathscr{F}\} \subset \mathbb{C}$ being bounded for *some* $p \in U$. To see this, let $z \in U$ and use path-connectivity of U to find a C^1 curve γ in U from p to z. Find $M_1, M_2 > 0$ such that $|f(p)| \leq M_1$ and $|f'(\zeta)| \leq M_2$ for every $\zeta \in |\gamma|$ and $f \in \mathscr{F}$. By the ML-inequality,

$$|f(z)| = \left| f(p) + \int_\gamma f'(\zeta)\, d\zeta \right| \leq M_1 + M_2 \cdot \text{length}(\gamma)$$

for all $f \in \mathscr{F}$. This proves that $\mathscr{F}$ is pointwise bounded.

Paul Antoine Aristide Montel (1876–1975)

Locally, a bound on the size of a holomorphic function imposes a bound on the size of its derivative (compare Cauchy's estimates in Theorem 1.42, or Theorem 1.43). This observation enabled Montel to reduce Theorem 5.23 to a simpler characterization of precompact families in $\mathscr{O}(U)$:

THEOREM 5.25 (Montel, 1907). *A family $\mathscr{F} \subset \mathscr{O}(U)$ is precompact if and only if it is compactly bounded.*

PROOF. If $\mathscr{F}$ is not compactly bounded, we can find a compact set $K \subset U$, a sequence $\{z_n\}$ in K, and a sequence $\{f_n\}$ in $\mathscr{F}$ such that $|f_n(z_n)| > n$ for all n. Evidently no subsequence of $\{f_n\}$ can converge uniformly on K. Hence $\mathscr{F}$ is not precompact.

Conversely, suppose $\mathscr{F}$ is compactly bounded. Take a compact set $K \subset U$ and choose $r > 0$ so that the closed r-neighborhood $K_r = \{z \in \mathbb{C} : \text{dist}(z, K) \leq r\}$ is contained in U. Find $M > 0$ such that $|f(z)| \leq M$ for all $z \in K_r$ and all $f \in \mathscr{F}$. Since the

The theorem first appeared in Montel's thesis. Koebe proved it independently in 1908.

closed disk $\overline{\mathbb{D}}(z, r)$ is contained in K_r whenever $z \in K$, Theorem 1.43 gives the estimate $\sup_{z \in K} |f'(z)| \leq M/r$ for all $f \in \mathscr{F}$. This proves $\mathscr{F}'$ is compactly bounded, and we conclude from Theorem 5.23 that $\mathscr{F}$ is precompact. $\qquad\square$

EXAMPLE 5.26. Let $\mathscr{F} \subset \mathscr{O}(\mathbb{D})$ be the family of functions $f(z) = \sum_{n=1}^{\infty} a_n z^n$ for which $|a_n| \leq n$ for all $n \geq 1$. Then, for every $f \in \mathscr{F}$ and $z \in \mathbb{D}$,

$$|f(z)| \leq \sum_{n=1}^{\infty} n|z|^n = \frac{|z|}{(1 - |z|)^2}.$$

This shows that $\mathscr{F}$ is compactly bounded, and it follows from Montel's Theorem 5.25 that $\mathscr{F}$ is precompact.

REMARK 5.27. While the notion of precompactness and the Arzelà-Ascoli theorem are topological, it is important to realize that Montel's result depends heavily on complex analyticity. Consider for example the sequence $f_n : [0, 2\pi] \to \mathbb{R}$ defined by $f_n(x) = \sin(nx)$. This is a uniformly bounded family of C^∞-smooth functions, but it has no uniformly convergent subsequence whatsoever. To see why, simply note that for any open interval $I \subset [0, 2\pi]$,

$$-1 = \inf_{x \in I} f_n(x) < \sup_{x \in I} f_n(x) = 1$$

provided n is large enough.

Even more dramatic is the fact that for any increasing sequence S of positive integers, the set of points $x \in [0, 2\pi]$ for which $\{f_n(x)\}_{n \in S}$ converges has Lebesgue measure zero (see problem 17).

Here is an elementary application of Montel's theorem in which one deduces compact convergence from pointwise convergence:

The theorem in this form was proved by Vitali in 1903. M. B. Porter reproved it independently in 1904.

THEOREM 5.28 (Vitali-Porter, 1903). *Let $U \subset \mathbb{C}$ be a domain and $E \subset U$ be a set which has an accumulation point in U. Suppose $\{f_n\} \subset \mathscr{O}(U)$ is compactly bounded and $\lim_{n \to \infty} f_n(p)$ exists for every $p \in E$. Then $\{f_n\}$ converges compactly in U.*

PROOF. By Theorem 5.25, $\{f_n\}$ is precompact so it has a subsequence which converges compactly to some $f \in \mathscr{O}(U)$. We claim that the full sequence $\{f_n\}$ converges compactly to f. If not, there exists a compact set $K \subset U$, an $\varepsilon > 0$, and an increasing sequence S of positive integers such that

$$\sup_{z \in K} |f_n(z) - f(z)| \geq \varepsilon \qquad \text{for } n \in S.$$

Since the subsequence $\{f_n\}_{n \in S}$ is also precompact, it has a further subsequence which converges compactly to some $g \in \mathscr{O}(U)$. Evidently, $\sup_{z \in K} |g(z) - f(z)| \geq \varepsilon$, which implies $f \neq g$. On the other hand, the full sequence $\{f_n\}$ converges pointwise on E, so $f = g$ on E, hence in U by the identity theorem. This is a contradiction. $\qquad\square$

5.3 Normal families of meromorphic functions

For families of holomorphic functions the precompactness condition is a bit too strong because it means avoiding ∞ as a limit. An example as simple as the family of translations $\mathscr{F} = \{z \mapsto z + \beta : \beta \in \mathbb{C}\} \subset \mathcal{O}(\mathbb{C})$ is not precompact, even though it is quite well behaved: Every sequence in $\mathscr{F}$ has a subsequence which either converges to a translation or to ∞. We can however render $\mathscr{F}$ precompact by viewing it as a family of *meromorphic* functions, that is, as a subspace of $\mathscr{C}(\mathbb{C}, \hat{\mathbb{C}})$, where now the target space $\hat{\mathbb{C}}$ is compact and equipped with the spherical metric. This point of view turns ∞ into a friend, not an outcast to avoid.

Before going into details, we digress to review a few basic facts about the ***spherical metric***

$$\sigma = \frac{2}{1 + |z|^2}\, |dz|$$

on $\hat{\mathbb{C}}$ (see Example 4.34). Recall from §3.2 that the stereographic projection Φ from the 2-sphere $\mathbb{S}^2 = \{(x_1, x_2, x_3) \in \mathbb{R}^3 : x_1^2 + x_2^2 + x_3^2 = 1\}$ to $\hat{\mathbb{C}}$ is defined by

$$\Phi(x_1, x_2, x_3) = \begin{cases} \dfrac{x_1 + ix_2}{1 - x_3} & x_3 \neq 1 \\ \infty & x_3 = 1. \end{cases}$$

Under Φ, the standard metric on $\mathbb{S}^2$ induced from the Euclidean metric of $\mathbb{R}^3$ corresponds to the spherical metric σ on $\hat{\mathbb{C}}$. To verify this, it suffices to show that for any smooth curve $t \mapsto \gamma(t)$ in $\mathbb{S}^2 \setminus \{(0, 0, 1)\}$, the Euclidean norm of the tangent vector $\gamma'(t)$ equals the σ-norm of the tangent vector $\eta'(t)$ of the image curve $\eta = \Phi \circ \gamma$ in $\hat{\mathbb{C}}$. Write $\gamma = (a, b, c)$, where a, b, c are smooth functions of t. Differentiating $a^2 + b^2 + c^2 = 1$ with respect to t gives

$$(5.3) \qquad\qquad aa' + bb' + cc' = 0.$$

Since $\eta = (a + ib)/(1 - c)$, we have

$$\eta' = \frac{(a' + ib')(1 - c) + (a + ib)c'}{(1 - c)^2} = \frac{[a'(1 - c) + ac'] + i[b'(1 - c) + bc']}{(1 - c)^2}.$$

Using (5.3), we compute

$$|\eta'|^2 = \frac{[a'(1 - c) + ac']^2 + [b'(1 - c) + bc']^2}{(1 - c)^4}$$

$$= \frac{(a'^2 + b'^2)(1 - c)^2 + (a^2 + b^2)c'^2 + 2c'(1 - c)(aa' + bb')}{(1 - c)^4}$$

$$= \frac{(a'^2 + b'^2)(1 - c)^2 + (1 - c^2)c'^2 - 2(1 - c)cc'^2}{(1 - c)^4}$$

(5.4)
$$= \frac{(a'^2 + b'^2)(1 - c)^2 + c'^2(1 - c)^2}{(1 - c)^4} = \frac{a'^2 + b'^2 + c'^2}{(1 - c)^2}.$$

Moreover,

(5.5)
$$1 + |\eta|^2 = 1 + \frac{a^2 + b^2}{(1 - c)^2} = \frac{1 - 2c + c^2 + a^2 + b^2}{(1 - c)^2} = \frac{2}{1 - c}.$$

It follows from (5.4) and (5.5) that

$$\|\eta'\|_\sigma = \frac{2|\eta'|}{1 + |\eta|^2} = (a'^2 + b'^2 + c'^2)^{1/2},$$

which is the Euclidean norm of γ'.

The above observation implies that the standard notions of length and distance in $\mathbb{S}^2$ correspond under Φ to those of the metric σ in $\hat{\mathbb{C}}$. Any two points $X, Y \in \mathbb{S}^2$ can be joined by a minimal geodesic Γ which is the shorter arc of a great circle on $\mathbb{S}^2$ passing through X, Y (there are infinitely many such arcs if $X = -Y$; otherwise Γ is unique). Setting $z = \Phi(X), w = \Phi(Y)$, it follows that the spherical distance $\mathrm{dist}_\sigma(z, w)$ is the length of the arc Γ, namely the angle $\angle XOY$ measured in radians (see Fig. 5.1). In particular, $0 \le \mathrm{dist}_\sigma(z, w) \le \pi$. Alternatively, the stereographic image $\gamma = \Phi(\Gamma)$ is a minimal σ-geodesic that joins z, w in $\hat{\mathbb{C}}$, so

$$\mathrm{dist}_\sigma(z, w) = \mathrm{length}_\sigma(\gamma) = \int_\gamma \frac{2|dz|}{1 + |z|^2}.$$

For a description of the σ-geodesic γ and an explicit formula for $\mathrm{dist}_\sigma(z, w)$, see problems 18 and 20.

A closely related notion of distance on the Riemann sphere is the ***chordal distance*** defined by

$$\chi(z, w) = \begin{cases} \dfrac{2|z - w|}{\sqrt{1 + |z|^2}\sqrt{1 + |w|^2}} & z, w \in \mathbb{C} \\[3ex] \dfrac{2}{\sqrt{1 + |z|^2}} & z \in \mathbb{C}, w = \infty, \end{cases}$$

which is often more convenient to work with. The terminology is justified by the fact that under the stereographic projection, the chordal distance corresponds to the Euclidean length of the straight line segment (or "chord") in $\mathbb{R}^3$ that joins a pair of points on the 2-sphere. To verify this, take two points $X = (x_1, x_2, x_3)$ and $Y = (y_1, y_2, y_3)$ on $\mathbb{S}^2$ and set $z = \Phi(X), w = \Phi(Y)$. We may assume without loss of generality that X and Y are not the north pole $(0, 0, 1)$ (otherwise take a limit as x_3 or y_3 tends to 1 in the following computation). Then,

$$z = \frac{x_1 + ix_2}{1 - x_3} \qquad \text{and} \qquad w = \frac{y_1 + iy_2}{1 - y_3}$$

and a computation identical to (5.5) shows that

$$1 + |z|^2 = \frac{2}{1 - x_3} \qquad \text{and} \qquad 1 + |w|^2 = \frac{2}{1 - y_3}.$$

Hence,

$$\chi^2(z, w) = \frac{4|z - w|^2}{(1 + |z|^2)(1 + |w|^2)} = (1 - x_3)(1 - y_3) \left| \frac{x_1 + ix_2}{1 - x_3} - \frac{y_1 + iy_2}{1 - y_3} \right|^2$$

$$= \frac{|(1 - y_3)(x_1 + ix_2) - (1 - x_3)(y_1 + iy_2)|^2}{(1 - x_3)(1 - y_3)}$$

$$= \frac{[(1 - y_3)x_1 - (1 - x_3)y_1]^2 + [(1 - y_3)x_2 - (1 - x_3)y_2]^2}{(1 - x_3)(1 - y_3)}$$

$$= \frac{(1 - y_3)^2(x_1^2 + x_2^2) + (1 - x_3)^2(y_1^2 + y_2^2) - 2(1 - x_3)(1 - y_3)(x_1y_1 + x_2y_2)}{(1 - x_3)(1 - y_3)}$$

$$= \frac{(1 - y_3)^2(1 - x_3^2) + (1 - x_3)^2(1 - y_3^2) - 2(1 - x_3)(1 - y_3)(x_1y_1 + x_2y_2)}{(1 - x_3)(1 - y_3)}$$

$$= (1 - y_3)(1 + x_3) + (1 - x_3)(1 + y_3) - 2(x_1y_1 + x_2y_2)$$

$$= 2 - 2(x_1y_1 + x_2y_2 + x_3y_3) = (x_1 - y_1)^2 + (x_2 - y_2)^2 + (x_3 - y_3)^2,$$

as claimed.

The above geometric interpretations prove the relation

$$\chi(z, w) = 2 \sin\left(\frac{1}{2} \operatorname{dist}_\sigma(z, w)\right)$$

between the spherical and chordal distance (in the isosceles triangle $\triangle XOY$ in Fig. 5.1, the side XY has length $\chi(z, w)$ and the angle $\angle XOY$ is $\operatorname{dist}_\sigma(z, w)$). Since

$$\frac{2}{\pi} t \leq \sin t \leq t \qquad \text{for } 0 \leq t \leq \frac{\pi}{2},$$

we obtain the pair of inequalities

(5.6) $$\frac{2}{\pi} \operatorname{dist}_\sigma(z, w) \leq \chi(z, w) \leq \operatorname{dist}_\sigma(z, w)$$

for all $z, w \in \hat{\mathbb{C}}$.

Thus, *the spherical and chordal distance on $\hat{\mathbb{C}}$ are comparable* in the sense that their ratio is bounded from above and below by positive constants.

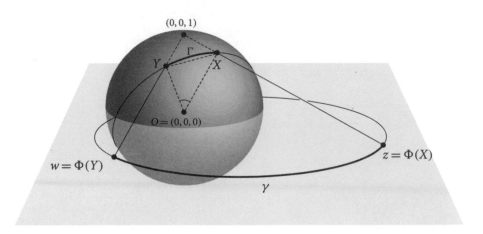

Figure 5.1. Under the stereographic projection $\Phi : \mathbb{S}^2 \to \hat{\mathbb{C}}$, the shorter arc Γ of the great circle passing through X and Y maps to the minimal σ-geodesic γ joining the image points z and w. The spherical distance $\mathrm{dist}_\sigma (z, w)$ is the length of the arc Γ which is equal to the angle $\angle XOY$, while the chordal distance $\chi (z, w)$ is the Euclidean distance between X and Y.

LEMMA 5.29 (Spherical vs. Euclidean distance).

(i) *The inversion $\iota(z) = 1/z$ is a spherical isometry:*

$$\mathrm{dist}_\sigma \left(\frac{1}{z}, \frac{1}{w} \right) = \mathrm{dist}_\sigma (z, w) \qquad \text{for all } z, w \in \hat{\mathbb{C}}.$$

(ii) *On the complex plane, the spherical distance is dominated by the Euclidean distance:*

$$\mathrm{dist}_\sigma (z, w) \le 2|z - w| \qquad \text{for all } z, w \in \mathbb{C}.$$

(iii) *On every bounded subset of the complex plane, the spherical distance dominates the Euclidean distance. More precisely, given any $R > 0$,*

$$\mathrm{dist}_\sigma (z, w) \ge \frac{2}{1 + R^2} |z - w| \qquad \text{for all } z, w \in \mathbb{D}(0, R).$$

It follows in particular that *the spherical and Euclidean distances are comparable on every bounded subset of* $\mathbb{C}$.

PROOF. For (i), simply compute the pull-back of σ under the inversion:

$$\iota^* \sigma = \frac{2|\iota'(z)|}{1 + |\iota(z)|^2} |dz| = \frac{2|z|^{-2}}{1 + |z|^{-2}} |dz| = \sigma.$$

This, by Corollary 4.33, proves that ι is an isometry. (Alternatively, observe that under the stereographic projection, ι corresponds to the 180° rotation $(x_1, x_2, x_3) \mapsto (x_1, -x_2, -x_3)$ of $\mathbb{S}^2$ which is certainly a Euclidean isometry.)

The estimate in (ii) is straightforward since for every $z, w \in \mathbb{C}$,

$$\mathrm{dist}_\sigma (z, w) \le \mathrm{length}_\sigma ([z, w]) = \int_{[z,w]} \frac{2|d\zeta|}{1 + |\zeta|^2} \le 2 \int_{[z,w]} |d\zeta| = 2|z - w|.$$

Finally, (iii) follows from (5.6), since if $|z|, |w| < R$,

$$\text{dist}_\sigma(z, w) \geq \chi(z, w) = \frac{2|z - w|}{\sqrt{1 + |z|^2}\sqrt{1 + |w|^2}} \geq \frac{2|z - w|}{1 + R^2}. \qquad \square$$

Recall from §3.2 that elements of $\mathcal{M}(U)$, the field of meromorphic functions in a domain $U \subset \mathbb{C}$, can be identified with holomorphic maps $U \to \hat{\mathbb{C}}$ that are not identically ∞. In what follows we always consider $\mathcal{M}(U)$ as a subspace of $\mathscr{C}(U, \hat{\mathbb{C}})$ with the compact convergence topology, where the target space $\hat{\mathbb{C}}$ is equipped with the spherical metric $d(z, w) = \text{dist}_\sigma(z, w)$.

THEOREM 5.30. *Let $U \subset \mathbb{C}$ be a domain, $f_n \in \mathcal{M}(U)$, and $f_n \to f$ in $\mathscr{C}(U, \hat{\mathbb{C}})$. Then either $f \in \mathcal{M}(U)$ or $f = \infty$ everywhere.*

This is the analog of the Weierstrass Theorem 5.11 for meromorphic functions. The claim is that $\mathcal{M}(U)$ union the constant function $U \to \{\infty\}$ forms a closed subspace of $\mathscr{C}(U, \hat{\mathbb{C}})$.

PROOF. Assume f is not identically ∞. We show $f \in \mathcal{M}(U)$ by proving that every $p \in U$ has a neighborhood in which f is meromorphic.

First consider the case where $f(p) \neq \infty$. By continuity, there is a disk D centered at p, with $\overline{D} \subset U$, and some $\varepsilon > 0$ such that $\text{dist}_\sigma(f(z), \infty) \geq 2\varepsilon$ for all $z \in D$. Find $N \geq 1$ such that $\text{dist}_\sigma(f_n(z), f(z)) < \varepsilon$ for all $z \in D$ and $n \geq N$. The triangle inequality then guarantees that $\text{dist}_\sigma(f_n(z), \infty) \geq \varepsilon$ for all $z \in D$ and $n \geq N$. In particular, f_n has no poles in D and therefore $f_n \in \mathcal{O}(D)$ if $n \geq N$. Both f and f_n for $n \geq N$ map D into the bounded set $\{w \in \mathbb{C} : \text{dist}_\sigma(w, \infty) \geq \varepsilon\}$ on which the spherical distance dominates the Euclidean distance by Lemma 5.29(iii). Since $f_n \to f$ uniformly on D in the spherical metric, it follows that $f_n \to f$ uniformly on D in the Euclidean metric. Theorem 5.11 now shows that $f \in \mathcal{O}(D)$.

Next, assume $f(p) = \infty$. Then, since $1/f_n \to 1/f$ in $\mathscr{C}(U, \hat{\mathbb{C}})$ by Lemma 5.29(i), we can repeat the above argument to conclude that there is a disk D centered at p, with $\overline{D} \subset U$, such that $1/f_n \in \mathcal{O}(D)$ for all large n and $1/f_n \to 1/f$ uniformly on D in the Euclidean metric, which shows $1/f \in \mathcal{O}(D)$. Since $1/f$ is not identically 0 (otherwise f would be identically ∞), it follows that $f \in \mathcal{M}(D)$ having p as a pole. $\qquad \square$

We record a byproduct of the above proof as a separate result:

COROLLARY 5.31. *Suppose $f_n \to f$ in $\mathcal{M}(U)$. Then, for every compact set $K \subset U \setminus f^{-1}(\infty)$ there is an integer $N \geq 1$ such that f_n has no poles in K if $n \geq N$, and $f_n \to f$ and $f'_n \to f'$ uniformly on K in the Euclidean metric.*

Roughly speaking, this means that away from the poles of f the situation is similar to a compactly convergent sequence of holomorphic functions. In the process, the poles of f_n either accumulate on the poles of f, or simply disappear by accumulating on the boundary of U or by marching off to infinity.

EXAMPLE 5.32. The meromorphic functions $f_n(z) = n^2/((n-z)(nz-1))$ tend to $f(z) = 1/z$ in $\mathcal{M}(\mathbb{C})$ as $n \to \infty$. By (5.6), we only need to verify that for every $R > 0$, the chordal distance $\chi(f_n(z), f(z))$ tends to 0 uniformly on the disk $\mathbb{D}(0, R)$. Direct computation shows

$$\chi(f_n(z), f(z)) = \frac{2|f_n(z) - f(z)|}{\sqrt{1 + |f_n(z)|^2}\sqrt{1 + |f(z)|^2}} = \frac{2|(z^2+1)n - z|}{\sqrt{n^4 + |n-z|^2|nz-1|^2}\sqrt{|z|^2 + 1}},$$

which gives the estimate

$$\sup_{|z| < R} \chi(f_n(z), f(z)) \le \frac{2((R^2+1)n + R)}{n^2},$$

which proves that the supremum on the left tends to 0 as $n \to \infty$. Note that in this example, the pole of f_n at $z = 1/n$ tends to the pole of f at $z = 0$, while the pole of f_n at $z = n$ tends to infinity and disappears in the limit.

The following theorem treats the special case of Theorem 5.30 when the f_n are holomorphic:

THEOREM 5.33. *Let $U \subset \mathbb{C}$ be a domain, $f_n \in \mathcal{O}(U)$, and $f_n \to f$ in $\mathscr{C}(U, \hat{\mathbb{C}})$. Then either $f \in \mathcal{O}(U)$ or $f = \infty$ everywhere. In the first case, $f_n \to f$ compactly in U in the Euclidean metric.*

PROOF. Suppose f is not identically ∞, so $f \in \mathcal{M}(U)$ by Theorem 5.30. If f has a pole at some $p \in U$, then as in the proof of Theorem 5.30, we can find a disk D centered at p such that $1/f_n \in \mathcal{O}(D)$ for all n, and $1/f_n \to 1/f$ uniformly on D in the Euclidean metric. Since $1/f_n \ne 0$ in D and $1/f$ is not identically 0, Hurwitz's Theorem 5.18 implies that $1/f \ne 0$ in D. This contradicts $1/f(p) = 0$. Thus, f has no poles and $f \in \mathcal{O}(U)$. Corollary 5.31 now shows that $f_n \to f$ compactly in U in the Euclidean metric. □

REMARK 5.34. Suppose $f_n \in \mathcal{O}(U)$ and $f_n \to \infty$ in $\mathscr{C}(U, \hat{\mathbb{C}})$. Then, in the Euclidean metric, $\{f_n\}$ *diverges to* ∞ uniformly on every compact set $K \subset U$. In fact, given any $R > 0$ we can find $N \ge 1$ such that $\text{dist}_\sigma(f_n(z), \infty) < 2/\sqrt{1+R^2}$ for all $z \in K$ and $n \ge N$. Then, by (5.6),

$$\frac{2}{\sqrt{1 + |f_n(z)|^2}} = \chi(f_n(z), \infty) \le \frac{2}{\sqrt{1 + R^2}},$$

which shows $|f_n(z)| \ge R$ for all $z \in K$ and $n \ge N$.

DEFINITION 5.35. A family $\mathscr{F} \subset \mathcal{M}(U)$ is ***normal*** if it is precompact as a subset of $\mathscr{C}(U, \hat{\mathbb{C}})$. That is, if every sequence in $\mathscr{F}$ has a subsequence which converges compactly in U using the spherical metric in the target.

EXAMPLE 5.36. The sequence of powers z^n for $n = 1, 2, 3, \ldots$ is a normal family in $\mathbb{D}$ and in $\mathbb{C} \setminus \overline{\mathbb{D}}$ since it converges compactly to 0 in the first case and to ∞ in the second case. It cannot be normal in any neighborhood of a point on the unit circle, having a discontinuous limit there.

Normality in families of meromorphic functions can be characterized in terms of derivatives similar to Theorem 5.23. There our holomorphic functions were viewed as elements of $\mathscr{C}(U, \mathbb{C})$ and we deduced equicontinuity from bounds on $|f'(z)|$, which can be thought of as the derivative norm of the map $f : (U, g_0) \to (\mathbb{C}, g_0)$, g_0 being the Euclidean metric $|dz|$. In the present context, our meromorphic functions are viewed as elements of $\mathscr{C}(U, \hat{\mathbb{C}})$ and we can similarly deduce equicontinuity from bounds on the **spherical derivative norm**

$$f^{\#}(z) = \frac{1}{2} \| f'(z) \|_{g_0, \sigma} = \frac{|f'(z)|}{1 + |f(z)|^2}$$

of the map $f : (U, g_0) \to (\hat{\mathbb{C}}, \sigma)$. The definition of $f^{\#}$ extends to the poles of f by continuity; in fact, a brief computation shows that if p is a pole of f of order m, then $f^{\#}(p) = 1/(\lim_{z \to p}(z - p)f(z))$ for $m = 1$ and $f^{\#}(p) = 0$ for $m \geq 2$. The important point to keep in mind is that $f^{\#}$ is well defined and non-negative everywhere, even at poles.

The spherical derivative norm satisfies the chain rule

$$(f \circ g)^{\#} = (f^{\#} \circ g) \, |g'(z)|$$

provided that g is holomorphic. It also has the useful property

$$(5.7) \qquad\qquad (1/f)^{\#} = f^{\#}.$$

This can be verified by direct computation, or by observing that $\iota(z) = 1/z$ is an isometry $(\hat{\mathbb{C}}, \sigma) \to (\hat{\mathbb{C}}, \sigma)$, so

$$\| (1/f)' \|_{g_0, \sigma} = \| (\iota \circ f)' \|_{g_0, \sigma} = \| \iota' \circ f \|_{\sigma, \sigma} \, \| f' \|_{g_0, \sigma} = \| f' \|_{g_0, \sigma}.$$

In particular, when p is a pole of f, we can conveniently compute $f^{\#}(p)$ as the norm $g^{\#}(p) = |g'(p)|/(1 + |g(p)|^2)$ for the function $g = 1/f$ which has a zero at p.

THEOREM 5.37 (Marty, 1931). *Let $U \subset \mathbb{C}$ be a domain. A family $\mathscr{F} \subset \mathscr{M}(U)$ is normal if and only if the family*

$$\mathscr{F}^{\#} = \left\{ f^{\#} = \frac{|f'|}{1 + |f|^2} : f \in \mathscr{F} \right\}$$

of spherical derivative norms is compactly bounded.

PROOF. Suppose $\mathscr{F}$ is normal but $\mathscr{F}^{\#}$ is not compactly bounded, so there is a compact set $K \subset U$, a sequence $\{z_n\}$ in K, and a sequence $\{f_n\}$ in $\mathscr{F}$ such that

$$(5.8) \qquad \lim_{n \to \infty} f_n^{\#}(z_n) = +\infty.$$

After passing to a subsequence, we may assume that $z_n \to p \in K$ and $f_n \to f$ in $\mathscr{C}(U, \hat{\mathbb{C}})$. By Theorem 5.30, either $f \in \mathscr{M}(U)$, or $f = \infty$ everywhere.

First suppose $f \in \mathscr{M}(U)$. If $f(p) \neq \infty$, take an open disk D centered at p such that $\overline{D} \subset U$ and f has no poles in $\overline{D}$. By Corollary 5.31, f_n has no poles in $\overline{D}$ for large n, and $f_n \to f$ and $f_n' \to f'$ uniformly on K in the Euclidean metric. Hence,

$$f_n^{\#}(z_n) = \frac{|f_n'(z_n)|}{1 + |f_n(z_n)|^2} \to \frac{|f'(p)|}{1 + |f(p)|^2} = f^{\#}(p),$$

Frédéric Ladislas Joseph Marty (1911–1940)

which contradicts (5.8). If $f(p) = \infty$, apply the same argument to $g_n = 1/f_n \to g = 1/f$ and use (5.7) to conclude that $f_n^{\#}(z_n) = g_n^{\#}(z_n) \to g^{\#}(p) = f^{\#}(p)$ and reach a similar contradiction.

Next, suppose $f = \infty$ everywhere so $g_n = 1/f_n \to 0$ in $\mathscr{C}(U, \hat{\mathbb{C}})$. By Corollary 5.31, $g_n \to 0$ and $g_n' \to 0$ uniformly on K in the Euclidean metric. Hence $f_n^{\#}(z_n) = g_n^{\#}(z_n) \to 0$, again contradicting (5.8).

Conversely, suppose $\mathscr{F}^{\#}$ is compactly bounded. We need to show that $\mathscr{F}$ is precompact in $\mathscr{C}(U, \hat{\mathbb{C}})$. By the Arzelà-Ascoli Theorem 5.8, it suffices to check that $\mathscr{F}$ is equicontinuous. The reasoning here is virtually identical to the proof of Theorem 5.23: Take $p \in U$ and choose $r > 0$ such that $\overline{\mathbb{D}}(p, r) \subset U$. Find an $M > 0$ such that $\|f'(\zeta)\|_{g_0, \sigma} = 2f^{\#}(\zeta) \leq M$ whenever $\zeta \in \mathbb{D}(p, r)$ and $f \in \mathscr{F}$. Then, by Lemma 4.30,

$$\text{dist}_\sigma(f(z), f(p)) \leq \sup_{\zeta \in \mathbb{D}(p,r)} \|f'(\zeta)\|_{g_0, \sigma} \cdot |z - p| \leq M|z - p|$$

for all $z \in \mathbb{D}(p, r)$ and $f \in \mathscr{F}$, and equicontinuity follows. $\qquad\square$

EXAMPLE 5.38. Marty's theorem gives an alternative take on Example 5.36: The spherical derivative norm of $f_n(z) = z^n$ is

$$f_n^{\#}(z) = \frac{n|z|^{n-1}}{1 + |z|^{2n}}.$$

If $|z| \leq r < 1$, then $f_n^{\#}(z) \leq nr^{n-1}$, and if $|z| \geq r > 1$, then $f_n^{\#}(z) \leq nr^{-n-1}$. It follows that $f_n^{\#} \to 0$ compactly in $\mathbb{D}$ and in $\mathbb{C} \setminus \overline{\mathbb{D}}$ and therefore $\{f_n\}$ is normal when restricted to either of these domains. However, since $f_n^{\#}(z) = n/2 \to +\infty$ if $|z| = 1$, we see that $\{f_n\}$ is not normal in any neighborhood of a point on the unit circle.

Some applications of the theorems of Montel and Marty will be discussed in chapter 11.

Problems

(1) Verify that the compact convergence topology on $\mathscr{C}(U,X)$ depends only on the topology of X, not on the particular choice of the metric d on X. (Hint: A condition of the form $d(f(p),g(p)) < \varepsilon$ can be replaced by the condition that $(f(p),g(p))$ belongs to a neighborhood of the diagonal in $X \times X$.)

(2) Show that X is a complete metric space if and only if $\mathscr{C}(U,X)$ (with the metric $\mathbf{d}$ defined in 5.1) is a complete metric space.

(3) The compact convergence topology can be defined analogously on the space $\mathscr{C}(Y,X)$ of all continuous maps $Y \to X$ where X is a metric space and Y is a topological space. Show that this topology is metrizable whenever Y is locally compact and σ-compact. (Hint: Start by constructing an exhaustion of Y by compact sets $\{K_n\}$.)

(4) Prove that a sequence $\{f_n\}$ converges in $\mathscr{C}(U,X)$ if and only if for every convergent sequence $\{z_n\}$ in U the sequence $\{f_n(z_n)\}$ converges in X.

(5) Use the Weierstrass M-test to show that the series $\sum_{n=1}^{\infty}(1 - \cos(z/n))$ defines an entire function.

(6) Show that both series

$$f(z) = \sum_{n=1}^{\infty} \frac{nz^n}{1 - z^n} \qquad \text{and} \qquad g(z) = \sum_{n=1}^{\infty} \frac{z^n}{(1 - z^n)^2}$$

converge compactly in $\mathbb{D}$, so $f, g \in \mathscr{O}(\mathbb{D})$. Verify that $f = g$ everywhere in $\mathbb{D}$. (Hint: For the second statement, write the series of f as a double series and reverse the order of summation.)

(7) Suppose $\{f_n\} \subset \mathscr{O}(\mathbb{C})$ converges uniformly on $\mathbb{C}$ as $n \to \infty$. What can you conclude about the f_n for large n?

(8) Let r_n be the absolute value of the root of the polynomial

$$1 + z + \frac{z^2}{2!} + \cdots + \frac{z^n}{n!}$$

that is closest to the origin. Use Hurwitz's Theorem 5.18 to show that $\lim_{n \to \infty} r_n = +\infty$. Can you prove this by finding an explicit lower bound for r_n?

(9) Find a sequence $f_n : \mathbb{D} \to \mathbb{C}$ of non-vanishing C^{∞}-smooth functions which converges compactly to a C^{∞}-smooth function $f : \mathbb{D} \to \mathbb{C}$ with $f^{-1}(0) = 0$. This shows that the analog of Hurwitz's Theorem 5.18 for smooth functions is false.

(10) Let $\{f_n\}$ be a sequence in $\mathscr{O}(\mathbb{D})$ with the power series $f_n(z) = \sum_{k=0}^{\infty} a_{k,n} z^k$.
    (i) If $f_n \to 0$ compactly in $\mathbb{D}$, show that

(5.9) $\lim_{n \to \infty} a_{k,n} = 0$ for each fixed $k \geq 0$.

    (ii) Give an example where (5.9) holds, but $\{f_n\}$ does not converge compactly in $\mathbb{D}$.
    (iii) Show however that if $\{f_n\}$ is compactly bounded, then (5.9) implies $f_n \to 0$ compactly in $\mathbb{D}$.

(11) Let $f \in \mathscr{O}(\mathbb{C})$. If the series $\sum_{n=0}^{\infty} f^{(n)}(z)$ converges at some $p \in \mathbb{C}$, show that it converges compactly in $\mathbb{C}$ to some $F \in \mathscr{O}(\mathbb{C})$ [**PS**]. What simpler relationship is there between F and f? (Hint: Let for simplicity $p = 0$ and $c_n = f^{(n)}(0)$. Use the power series expansion

$f^{(n)}(z) = \sum_{k=0}^{\infty} c_{n+k} z^k / k!$ to estimate the sum $f^{(m)}(z) + f^{(m+1)}(z) + \cdots + f^{(n)}(z)$ for large $n > m > 0$.)

(12)   Let $\mathscr{F} \subset \mathscr{O}(U)$ be a precompact family. Show that for each integer $k \geq 1$ the family of k-th derivatives $\mathscr{F}^{(k)} = \{f^{(k)} : f \in \mathscr{F}\}$ is precompact.

(13)   Suppose $\sum_{n=0}^{\infty} b_n z^n$ has radius of convergence ≥ 1. Let $\mathscr{F}$ be the family of all $f(z) = \sum_{n=0}^{\infty} a_n z^n$ in $\mathscr{O}(\mathbb{D})$ for which $|a_n| \leq |b_n|$ for all n. Show that $\mathscr{F}$ is precompact.

(14)   The family of all holomorphic functions $\mathbb{D} \to \mathbb{D}$ is precompact by Montel's theorem, and every limit point of this family is either a holomorphic function $\mathbb{D} \to \mathbb{D}$ or a constant function taking its value on the unit circle. Discuss possible limits of a sequence of Möbius maps $z \mapsto (z - p_n)(1 - \overline{p_n} z)$ in $\mathrm{Aut}(\mathbb{D})$ when $p_n \to p \in \mathbb{D}$ and when $p_n \to p \in \mathbb{T}$ (compare Example 5.21).

(15)   Let $U \subsetneq \mathbb{C}$ be a domain. Fix $\alpha \geq 1$ and $M > 0$. Let $\mathscr{F}$ be the family of all $f \in \mathscr{O}(U)$ such that $\iint_U |f(z)|^\alpha \, dx \, dy \leq M$. Show that $\mathscr{F}$ is precompact. (Hint: Verify that there exists a constant $C = C(\alpha, M) > 0$ such that $|f(z)| \leq C \, \mathrm{dist}(z, \partial U)^{-2/\alpha}$ for all $f \in \mathscr{F}$ and all $z \in U$. Compare problem 13 in chapter 2.)

(16)   (Osgood) Let $\{f_n\}$ be a sequence in $\mathscr{O}(U)$ such that $f(z) = \lim_{n \to \infty} f_n(z)$ exists for every $z \in U$. Show that there is an open dense set $V \subset U$ such that $f \in \mathscr{O}(V)$. (Hint: Use Baire's category theorem to prove that every disk in U contains a subdisk in which $\{f_n\}$ is uniformly bounded. Theorem 5.25 then shows that f is holomorphic in this subdisk.)

(17)   This exercise relates to Remark 5.27 and requires a modest background in measure theory [**Ru2**]. Take an infinite sequence $n_1 < n_2 < n_3 < \cdots$ of positive integers and let E be the set of points $x \in [0, 2\pi]$ for which $f(x) = \lim_{k \to \infty} \sin(n_k x)$ exists. For any measurable set $A \subset [0, 2\pi]$, the integrals $\int_A \cos(nx) \, dx$ and $\int_A \sin(nx) \, dx$ tend to 0 as $n \to \infty$ since they are the Fourier coefficients of the characteristic function of A.

    (i)   Let E^+ and E^- be the measurable subsets of E on which $f > 0$ and $f < 0$. Show that $\int_{E^\pm} f(x) \, dx = 0$, so $f = 0$ a.e. on E.

    (ii) Use the identity $2 \sin^2 t = 1 - \cos(2t)$ to show that $\int_E f^2(x) \, dx = |E|/2$. Conclude that $|E| = 0$.

(18)   Under the stereographic projection $\Phi : \mathbb{S}^2 \to \hat{\mathbb{C}}$, the antipodal map $X \mapsto -X$ on $\mathbb{S}^2$ corresponds to $z \mapsto -1/\bar{z}$ on $\hat{\mathbb{C}}$, and Φ maps circles in $\mathbb{S}^2$ to circles in $\hat{\mathbb{C}}$ (see problem 1 in chapter 4). Use these facts to show that the minimal σ-geodesic joining a distinct pair $z, w \in \hat{\mathbb{C}}$ with $z \neq -1/\bar{w}$ is an arc of the circle passing through the four points $z, w, -1/\bar{z}, -1/\bar{w}$.

(19)   Describe the infinitely many minimal σ-geodesics joining -1 to 1, and find their common spherical length.

(20)   Show that the spherical distance between $z, w \in \hat{\mathbb{C}}$ is given by the formula

$$\mathrm{dist}_\sigma(z, w) = 2 \arctan \left| \frac{z - w}{1 + z\bar{w}} \right|.$$

(Hint: First treat the case where $w = 0$, so the minimal σ-geodesic joining z to 0 is the radial line. For the general case, take a rotation of the sphere that carries w to 0.)

(21)   Show that $f \in \mathrm{Möb}$ is an isometry of $(\hat{\mathbb{C}}, \sigma)$ (equivalently, $f^* \sigma = \sigma$) if and only if it has the form $f(z) = (az + b)/(-\bar{b}z + \bar{a})$ with $|a|^2 + |b|^2 = 1$. This gives another approach to characterizing Möbius maps that correspond to rigid rotations of $\mathbb{S}^2$ (compare problem 2 of chapter 4).

(22)   The sequence of Möbius maps $f_n(z) = 1/(n^2 z - n)$ tends to 0 at every $z \in \mathbb{C}$. Verify that $\{f_n\}$ is not normal in $\mathbb{C}$, but its restriction to the punctured plane $\mathbb{C}^*$ is normal.

(23)   Prove that the family of all $f \in \mathcal{O}(\mathbb{D})$ which satisfy $\mathrm{Re}(f) > 0$ is normal (i.e., precompact in $\mathscr{C}(\mathbb{D}, \hat{\mathbb{C}})$), but it is not precompact in $\mathscr{C}(\mathbb{D}, \mathbb{C})$. (Hint: The right half-plane can be mapped biholomorphically onto $\mathbb{D}$.)

(24)   Suppose $f \in \mathcal{O}(\mathbb{D}^*)$ has an essential singularity at 0. Show that the sequence $\{f_n\}$ in $\mathcal{O}(\mathbb{D}^*)$ defined by $f_n(z) = f(z/n)$ is not normal.

(25)   Verify that

$$f^{\#}(z) = \frac{1}{2} \lim_{w \to z} \frac{\mathrm{dist}_{\sigma}(f(z), f(w))}{|z - w|}.$$

(Hint: Use problem 20).

(26)   If $f_n \to f$ in $\mathscr{M}(U)$, prove that $f_n' \to f'$ in $\mathscr{M}(U)$ and $f_n^{\#} \to f^{\#}$ compactly in U.

(27)   If $f_n \in \mathscr{M}(U)$ and $f_n \to \infty$ in $\mathscr{C}(U, \hat{\mathbb{C}})$, we cannot say much about what happens to the derivatives f_n', as the examples $f_n(z) = z + n$ on $U = \mathbb{C}$ and $f_n(z) = nz$ on $U = \mathbb{C}^*$ demonstrate. Show however that under these assumptions $f_n^{\#} \to 0$ compactly in U.

(28)   Let P be a complex polynomial of degree ≥ 2. Show that the sequence of iterates $\{P^{\circ n}\}$ cannot be normal in the whole plane. (Hint: On the one hand, $P^{\circ n}(z) \to \infty$ if $|z|$ is large; on the other hand, P has fixed points.)

(29)   Let P be a complex polynomial of degree ≥ 2, with $P(0) = 0$ and $P'(0) = \alpha$. Verify that the sequence of iterates $\{P^{\circ n}\}$ is normal in some neighborhood of the fixed point 0 if $|\alpha| < 1$. By contrast, show that $\{P^{\circ n}\}$ cannot be normal in any neighborhood of 0 if $|\alpha| > 1$ or if α is a root of unity.

6

Conformal maps

Let us begin with a remark on terminology. So far we have used the term "biholomorphism" for a bijective holomorphic map f between two open sets. For every p in the domain of f, we have seen that the derivative $f'(p)$ is non-zero (Corollary 1.63), that the real derivative $Df(p)$ acts as multiplication by the complex number $f'(p)$ so it is a composition of a rotation and a dilation (Corollary 1.10), and that f preserves angles between smooth curves passing through p (Theorem 4.4). Thus, at an infinitesimal level f preserves shapes (but not necessarily scales). To emphasize this geometric feature, biholomorphisms have been traditionally referred to as "conformal maps."

In keeping with this tradition, we will use the term ***conformal map*** as synonymous with biholomorphism. We will say that two domains are ***conformally isomorphic*** if there is a conformal map from one onto the other.

6.1 The Riemann mapping theorem

The following is arguably the deepest result in basic complex analysis:

The theorem was conjectured and partially proved by Riemann in 1851. Several proofs came later, most notably by Schwarz (1870), Osgood (1900), Koebe (1907), and Carathéodory (1912). The modern proof was given by F. Riesz and Fejér in 1923.

THEOREM 6.1 (The Riemann mapping theorem, 1851). *Every simply connected domain in the complex plane* $\mathbb{C}$, *other than* $\mathbb{C}$ *itself, is conformally isomorphic to the unit disk* $\mathbb{D}$. *More precisely, if* $U \subsetneq \mathbb{C}$ *is simply connected and* $p \in U$, *there exists a conformal map* $f : U \to \mathbb{D}$ *with* $f(p) = 0$. *Such* f *is unique up to postcomposition with a rigid rotation* $z \mapsto \alpha z$, $|\alpha| = 1$.

Clearly $\mathbb{C}$ is not conformally isomorphic to $\mathbb{D}$ because every holomorphic map $\mathbb{C} \to \mathbb{D}$ is constant by Liouville's Theorem. The theorem can be expressed equivalently by saying that every simply connected domain in $\hat{\mathbb{C}}$ which misses at least two (hence infinitely many) points is conformally isomorphic to $\mathbb{D}$.

PROOF. The uniqueness issue is easy to settle: if $f, g : U \to \mathbb{D}$ are both conformal maps which send p to 0, then $g \circ f^{-1}$ is an automorphism of the disk which fixes 0, so it must be a rigid rotation $z \mapsto \alpha z$ with $|\alpha| = 1$ by the Schwarz lemma. Hence $g = \alpha f$.

We will find a conformal map $U \to \mathbb{D}$ as the solution to an extremal problem, that is, we prove that there is a map which injects U holomorphically into $\mathbb{D}$ and carries p to 0, whose derivative at p has the largest possible absolute value among all such maps, and that this map is indeed *onto* $\mathbb{D}$. Consider the family

$$\mathscr{F} = \{f : U \to \mathbb{D} : f \text{ is holomorphic and injective, with } f(p) = 0\}.$$

We divide the argument into three steps:

• STEP 1. $\mathscr{F}$ is nonempty. Since $U \neq \mathbb{C}$, we can find some $a \in \mathbb{C} \smallsetminus U$. Since U is simply connected, Corollary 2.22 shows that the non-vanishing function $z \mapsto z - a$ has a holomorphic square root h in U. Thus, $h \in \mathscr{O}(U)$ and $h^2(z) = z - a$ for all $z \in U$. Evidently h is injective, so $h(U)$ is a domain in $\mathbb{C}$ by the open mapping theorem. Moreover, the domains $h(U)$ and $-h(U)$ are disjoint: If $h(z) = -h(w)$ for some $z, w \in U$, then $z - a = h^2(z) = h^2(w) = w - a$, or $z = w$. This implies $h(z) = 0$, hence $z = a$, which is a contradiction since $a \notin U$. Now take a disk $\mathbb{D}(q, r) \subset -h(U)$ so $\mathbb{D}(q, r) \cap h(U) = \emptyset$. The Möbius map $\varphi(z) = r/(z - q)$ maps $\mathbb{D}(q, r)$ onto $\hat{\mathbb{C}} \smallsetminus \overline{\mathbb{D}}$, and therefore $h(U)$ into $\mathbb{D}$. If ψ is an automorphism of $\mathbb{D}$ which sends $\varphi(h(p))$ to 0, it follows that $\psi \circ \varphi \circ h \in \mathscr{F}$.

Georg Friedrich Bernhard Riemann (1826–1866)

• STEP 2. If $f \in \mathscr{F}$ and $f(U) \neq \mathbb{D}$, then there exists a $g \in \mathscr{F}$ such that $|g'(p)| > |f'(p)|$. This can be deduced most easily from the **square root trick** of Carathéodory and Koebe as follows. Recall from Theorem 4.15 that for $a \in \mathbb{D}$, the map $\varphi_a(z) = (z - a)/(1 - \bar{a}z)$ is a disk automorphism which sends a to 0 and has the inverse $(\varphi_a)^{-1} = \varphi_{-a}$. Suppose $f(U) \neq \mathbb{D}$ and take a point $a \in \mathbb{D} \smallsetminus f(U)$. Then $\varphi_a \circ f : U \to \mathbb{D}$ is injective and non-vanishing, so it has a holomorphic square root h in U. Thus, $h \in \mathscr{O}(U)$ and $s \circ h = \varphi_a \circ f$, where $s(z) = z^2$. Evidently h is injective and maps U into $\mathbb{D}$. If $b = h(p)$, it follows that the composition $g = \varphi_b \circ h$ belongs to $\mathscr{F}$, and

$$f = \varphi_{-a} \circ s \circ h = (\varphi_{-a} \circ s \circ \varphi_{-b}) \circ g$$

(see Fig. 6.1). The map $\varphi_{-a} \circ s \circ \varphi_{-b} : \mathbb{D} \to \mathbb{D}$ fixes the origin and is *not* injective because s is not. Hence, the strict inequality $|(\varphi_{-a} \circ s \circ \varphi_{-b})'(0)| < 1$ must hold by the Schwarz lemma. It follows that

$$|f'(p)| = |(\varphi_{-a} \circ s \circ \varphi_{-b})'(0)| \cdot |g'(p)| < |g'(p)|.$$

• STEP 3. Define
$$M = \sup\{|f'(p)| : f \in \mathscr{F}\}.$$

Note that $0 < M < +\infty$. In fact, positivity holds since every $f \in \mathscr{F}$ is injective, so $f'(p) \neq 0$, while finiteness follows from Theorem 1.43 since if $\mathbb{D}(p, r) \subset U$, then $M \leq 1/r$. Take a sequence $\{f_n\}$ in $\mathscr{F}$ such that $|f_n'(p)| \to M$. Since the family $\mathscr{F}$ is uniformly bounded (by the constant 1), Montel's Theorem 5.25 guarantees the existence of a subsequence of $\{f_n\}$ which converges compactly in U to some $f \in \mathscr{O}(U)$. The proof will be complete once we show that $f \in \mathscr{F}$ and $f(U) = \mathbb{D}$.

Clearly $f(p) = 0$ and $f(U) \subset \overline{\mathbb{D}}$. If f takes a value on $\partial \mathbb{D}$, it must be constant by the open mapping theorem, which cannot be the case since $|f'(p)| = M > 0$. Thus,

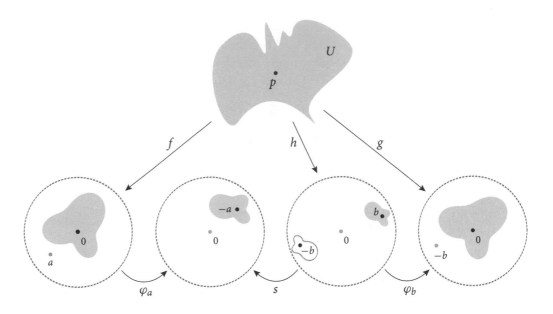

Figure 6.1. Illustration of STEP 2 in the proof of the Riemann mapping theorem.

$f(U) \subset \mathbb{D}$. Moreover, since each f_n is injective in U and f is non-constant, Corollary 5.22 shows that f is injective. This proves $f \in \mathscr{F}$. Finally, STEP 2 and the extremal property of f show that $f(U) = \mathbb{D}$. $\qquad\square$

Let $U \subsetneq \mathbb{C}$ be a simply connected domain. We refer to any conformal map $f : \mathbb{D} \to U$ as a **Riemann map** of U. Such f is unique up to precomposition with an automorphism of the disk, for if $f, g : \mathbb{D} \to U$ are conformal, then $g^{-1} \circ f \in \mathrm{Aut}(\mathbb{D})$. Given a point $p \in U$, there is a unique conformal map $f_p : \mathbb{D} \to U$ which satisfies $f_p(0) = p$ and $f_p'(0) > 0$, which we call the **normalized Riemann map** of U with the marked center p. In fact, let $\varphi : U \to \mathbb{D}$ be any conformal map carrying p to 0, whose existence is guaranteed by the Riemann mapping theorem. Then

$$f_p(z) = \varphi^{-1}(\alpha z), \quad \text{where} \quad \alpha = \frac{\varphi'(p)}{|\varphi'(p)|}.$$

EXAMPLE 6.2 (Riemann map of a semidisk). Let us find an explicit conformal map $\mathbb{D} \to \mathbb{D}^+ = \{z \in \mathbb{D} : \mathrm{Im}(z) > 0\}$ as a composition of simple transformations. First map $\mathbb{D}$ conformally to the upper half-plane $\mathbb{H}$ by the Möbius map $z \mapsto w = i(1-z)/(1+z)$ (this is the inverse of the Cayley map described in Example 4.5). Then map $\mathbb{H}$ conformally to the first quadrant $Q = \{\zeta \in \mathbb{C} : \mathrm{Re}(\zeta) > 0, \mathrm{Im}(\zeta) > 0\}$ by the square-root map $w = re^{i\theta} \mapsto \zeta = \sqrt{w} = \sqrt{r}e^{i\theta/2}$ ($0 < \theta < \pi$). Finally, map Q conformally to the semidisk $\mathbb{D}^+$ by the Möbius map $\zeta \mapsto \xi = (i-\zeta)/(i+\zeta)$. The composition $f : z \mapsto w \mapsto \zeta \mapsto \xi$ is a conformal map $\mathbb{D} \to \mathbb{D}^+$ given by the formula

$$f(z) = \frac{i - \sqrt{i\dfrac{1-z}{1+z}}}{i + \sqrt{i\dfrac{1-z}{1+z}}}.$$

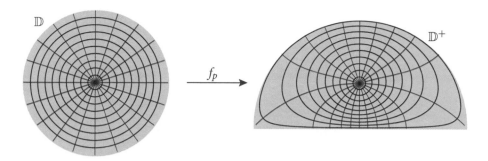

Figure 6.2. Images of the polar coordinate lines under the normalized Riemann map f_p of $\mathbb{D}^+ = \{z \in \mathbb{D} : \mathrm{Im}(z) > 0\}$ with the marked center $p = i(\sqrt{2} - 1)$.

A brief computation shows that $f(0) = i(\sqrt{2} - 1)$ and $f'(0) = 2 - \sqrt{2} > 0$. It follows that f is the normalized Riemann map of $\mathbb{D}^+$ with the marked center $i(\sqrt{2} - 1)$ (see Fig. 6.2).

The question of when the Riemann map of a given domain extends continuously to the closed disk will be addressed in §6.3.

6.2 Schlicht functions

We now study certain universal phenomena in the class of suitably normalized Riemann maps of simply connected domains. These results are generally viewed as cornerstones of geometric function theory.

DEFINITION 6.3. A holomorphic function $f : \mathbb{D} \to \mathbb{C}$ is called **schlicht** if it is injective and normalized by $f(0) = 0$, $f'(0) = 1$. The class of all schlicht functions is denoted by $\mathscr{S}$.

The German word "schlicht" roughly translates to "plain" or "simple."

Thus, every $f \in \mathscr{S}$ has a power series representation of the form

$$f(z) = z + a_2 z^2 + a_3 z^3 + \cdots \qquad \text{for } |z| < 1.$$

Note that by injectivity $f : \mathbb{D} \to f(\mathbb{D})$ is a conformal map (Corollary 1.63). We can think of f as the normalized Riemann map of the simply connected domain $f(\mathbb{D})$ with the marked center 0.

EXAMPLE 6.4. The *Koebe function* $K : \mathbb{D} \to \mathbb{C}$ defined by

(6.1)
$$K(z) = \sum_{n=1}^{\infty} n\, z^n = \frac{z}{(1-z)^2}$$

is schlicht (compare Example 1.21). Injectivity of K is not hard to verify using the above formula, but this approach would give little insight into what the image domain $K(\mathbb{D})$ looks like. A preferred

Figure 6.3. Images of the concentric circles $|z| = r$ under the Koebe function $K(z) = z/(1-z)^2$. These Jordan curves look more or less circular for small r but gradually develop "fjords" near $-1/4$ as r gets close to 1. Their union fills up the slit plane $K(\mathbb{D}) = \mathbb{C} \smallsetminus (-\infty, -1/4]$.

approach is to show that K is the normalized Riemann map of the slit plane $\mathbb{C} \smallsetminus (-\infty, -1/4]$ with the marked center 0 (see Fig. 6.3). To see this, consider the following transformations: First map $\mathbb{D}$ conformally to the right half-plane $U = \{\zeta : \mathrm{Re}(\zeta) > 0\}$ by the Möbius map

$$\zeta = \frac{1}{2} \left(\frac{1+z}{1-z} \right).$$

Then map U conformally to the slit plane $\mathbb{C} \smallsetminus (-\infty, 0]$ by the squaring map $\xi = \zeta^2$. Finally, shift by the map $w = \xi - 1/4$. The composition $z \mapsto \zeta \mapsto \xi \mapsto w$ is a conformal map $\mathbb{D} \to \mathbb{C} \smallsetminus (-\infty, -1/4]$ with the formula

$$w = \xi - \frac{1}{4} = \zeta^2 - \frac{1}{4} = \frac{1}{4}\left[\left(\frac{1+z}{1-z} \right)^2 - 1 \right] = \frac{z}{(1-z)^2},$$

which is the Koebe function K.

The study of schlicht functions is facilitated by the following construction. If $f \in \mathscr{S}$, the image $f(\mathbb{D})$ is a simply connected domain containing 0 which may be unbounded. We can look at the action of f in the coordinate $w = 1/z$ by considering the associated **inverted map** $\hat{f} : w \mapsto 1/f(1/w)$ which maps $\hat{\mathbb{C}} \smallsetminus \overline{\mathbb{D}}$ conformally onto a simply connected domain containing ∞ whose complement is a compact subset of the plane. The simple fact that this complement has finite area will turn out to be a powerful tool.

For future reference, let us find the relation between the power series of $f \in \mathscr{S}$ and the Laurent series of its inverted map $\hat{f}$. If $f(z) = z + \sum_{n=2}^{\infty} a_n z^n$, then

$$\hat{f}(w) = \frac{1}{f\left(\dfrac{1}{w}\right)} = \frac{1}{\dfrac{1}{w} + \dfrac{a_2}{w^2} + \dfrac{a_3}{w^3} + \cdots}$$

$$= \frac{w}{1 + \dfrac{a_2}{w} + \dfrac{a_3}{w^2} + \cdots}$$

$$= w\left(1 - \frac{a_2}{w} + \frac{a_2^2 - a_3}{w^2} + \cdots\right).$$

Thus,

(6.2) $$\hat{f}(w) = w - a_2 + \frac{a_2^2 - a_3}{w} + \cdots \qquad \text{if } |w| > 1.$$

> **DEFINITION 6.5.** We define $\hat{\mathscr{S}}$ to be the collection of all inverted maps of schlicht functions:
> $$\hat{\mathscr{S}} = \{\hat{f} : f \in \mathscr{S}\}.$$
> Thus, $\psi \in \hat{\mathscr{S}}$ if ψ is holomorphic and injective in $\hat{\mathbb{C}} \smallsetminus \overline{\mathbb{D}}$, with a Laurent series of the form
> $$\psi(w) = w + \sum_{n=0}^{\infty} b_n\, w^{-n} \qquad |w| > 1.$$

EXAMPLE 6.6. The degree 2 rational function $Z : w \mapsto w + 1/w$, known as the **Zhukovskii map**, is important in conformal mapping theory as well as applications such as aerodynamics. It is immediate from the definition that $Z(w_1) = Z(w_2)$ if and only if $w_1 = w_2$ or $w_1 = 1/w_2$. This proves injectivity of Z in $\hat{\mathbb{C}} \smallsetminus \overline{\mathbb{D}}$ and shows that $Z \in \hat{\mathscr{S}}$.

The Zhukovskii map Z and the Koebe function K are closely related: Since

$$\hat{K}(w) = \frac{1}{K(1/w)} = w\left(1 - \frac{1}{w}\right)^2 = w + \frac{1}{w} - 2,$$

we have

(6.3) $$Z = \hat{K} + 2.$$

Nikolai Egorovich Zhukovskii (1847–1921)

This relation sheds light on how Z maps the sphere over itself. Since K maps $\mathbb{D}$ conformally onto $\mathbb{C} \smallsetminus (-\infty, -1/4]$, the inverted map $\hat{K}$ maps $\hat{\mathbb{C}} \smallsetminus \overline{\mathbb{D}}$ conformally onto $\hat{\mathbb{C}} \smallsetminus [-4, 0]$. It follows from (6.3) that the restriction $Z : \hat{\mathbb{C}} \smallsetminus \overline{\mathbb{D}} \to \hat{\mathbb{C}} \smallsetminus [-2, 2]$ is a conformal map which fixes ∞. The symmetry relation $Z(1/w) = Z(w)$ then shows that the restriction $Z : \mathbb{D} \to \hat{\mathbb{C}} \smallsetminus [-2, 2]$ is also conformal and sends 0 to ∞.

Note that when $|w| = 1$, $Z(w) = w + \bar{w} = 2\,\mathrm{Re}(w)$, showing that Z projects the unit circle 2-to-1 onto the interval $[-2, 2]$. For $r \neq 1$, Z maps the circle $|w| = r$ homeomorphically onto an ellipse with foci ± 2 (see Fig. 6.4). To see this, use $|w| = r$ to write

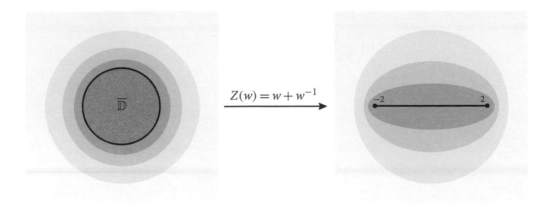

Figure 6.4. The Zhukovskii map restricts to a conformal map from $\hat{\mathbb{C}} \smallsetminus \overline{\mathbb{D}}$ to $\hat{\mathbb{C}} \smallsetminus [-2, 2]$, carrying the circles $|w| = $ const. to the ellipses with foci ± 2.

$$
\begin{aligned}
|Z(w) - 2| + |Z(w) + 2| &= |w + w^{-1} - 2| + |w + w^{-1} + 2| \\
&= r^{-1}|w^2 + 1 - 2w| + r^{-1}|w^2 + 1 + 2w| \\
&= r^{-1}(|w - 1|^2 + |w + 1|^2) \\
&= r^{-1}((w - 1)(\bar{w} - 1) + (w + 1)(\bar{w} + 1)) \\
&= r^{-1}(2|w|^2 + 2) = 2(r + r^{-1}).
\end{aligned}
$$

The area theorem was rediscovered by Bieberbach in 1916.

THEOREM 6.7 (Grönwall's area theorem, 1914). *Suppose $\psi \in \hat{\mathscr{S}}$ has the Laurent series*

$$
\psi(w) = w + \sum_{n=0}^{\infty} b_n w^{-n} \qquad \text{for } |w| > 1.
$$

If U denotes the image $\psi(\hat{\mathbb{C}} \smallsetminus \overline{\mathbb{D}})$, then

$$
\text{area}(\hat{\mathbb{C}} \smallsetminus U) = \pi \left(1 - \sum_{n=1}^{\infty} n |b_n|^2 \right).
$$

PROOF. For each $r > 1$, consider the smooth Jordan curve $\gamma_r(t) = \psi(re^{it}), 0 \le t \le 2\pi$, which bounds the compact set $E_r = \hat{\mathbb{C}} \smallsetminus \psi(\{w : |w| > r\})$. By Green's theorem,

$$
\begin{aligned}
\int_{\gamma_r} \bar{z} \, dz &= \int_{\gamma_r} (x - iy)(dx + i \, dy) \\
&= \int_{\gamma_r} (x \, dx + y \, dy) + i \int_{\gamma_r} (-y \, dx + x \, dy) \\
&= \iint_{E_r} \left(\frac{\partial y}{\partial x} - \frac{\partial x}{\partial y} \right) dx \, dy + i \iint_{E_r} \left(\frac{\partial x}{\partial x} + \frac{\partial y}{\partial y} \right) dx \, dy = 2i \, \text{area}(E_r),
\end{aligned}
$$

or

$$\text{area}(E_r) = \frac{1}{2i} \int_{\gamma_r} \bar{z}\, dz$$

(compare problems 14 and 15 in chapter 2). For simplicity, write $\psi(w) = \sum_{n=-1}^{\infty} b_n w^{-n}$, where $b_{-1} = 1$. Then

$$\gamma_r(t) = \sum_{n=-1}^{\infty} b_n r^{-n} e^{-int} \qquad \text{and} \qquad \gamma_r'(t) = -i \sum_{n=-1}^{\infty} n b_n r^{-n} e^{-int},$$

where both series converge uniformly on the interval $[0, 2\pi]$ by Theorem 3.21 (applied to ψ and ψ'). This allows term-by-term integration to evaluate the above complex integral:

$$\text{area}(E_r) = \frac{1}{2i} \int_0^{2\pi} \overline{\gamma_r(t)}\, \gamma_r'(t)\, dt$$

$$= \frac{1}{2i} \int_0^{2\pi} \left(\sum_{n=-1}^{\infty} \overline{b_n}\, r^{-n} e^{int} \right) \left(\sum_{m=-1}^{\infty} -im b_m\, r^{-m} e^{-imt} \right) dt$$

$$= -\frac{1}{2} \sum_{n,m=-1}^{\infty} m b_m \overline{b_n}\, r^{-(m+n)} \int_0^{2\pi} e^{i(n-m)t}\, dt.$$

Thomas Hakon Grönwall
(1877–1932)

The integral $\int_0^{2\pi} e^{i(n-m)t}\, dt$ is 0 if $n \neq m$ and is 2π if $n = m$. This gives

$$\text{area}(E_r) = -\pi \sum_{n=-1}^{\infty} n |b_n|^2 r^{-2n}.$$

The compact sets $\{E_r\}_{r>1}$ are nested in the sense that $E_r \subset E_s$ whenever $1 < r < s$, and their intersection over all $r > 1$ is precisely $\hat{\mathbb{C}} \smallsetminus U$. Hence,

$$\text{area}(\hat{\mathbb{C}} \smallsetminus U) = \lim_{r \to 1} \text{area}(E_r) = -\pi \sum_{n=-1}^{\infty} n |b_n|^2 = \pi \left(1 - \sum_{n=1}^{\infty} n |b_n|^2 \right). \qquad \square$$

EXAMPLE 6.8. Let $r > 1$. The map $w \mapsto r^{-1} Z(rw) = w + r^{-2} w^{-1} \in \hat{\mathscr{S}}$ maps $\hat{\mathbb{C}} \smallsetminus \overline{\mathbb{D}}$ conformally onto the exterior of the ellipse with the major axis $[-1 - r^{-2}, 1 + r^{-2}]$ and the minor axis $[-1 + r^{-2}, 1 - r^{-2}]$, whose area is therefore $\pi(1 + r^{-2})(1 - r^{-2}) = \pi(1 - r^{-4})$ (compare Example 6.6). This is consistent with the area theorem, since for this map $b_1 = r^{-2}$ and $b_n = 0$ for all $n \neq 1$, hence

$$\pi \left(1 - \sum_{n=1}^{\infty} n |b_n|^2 \right) = \pi(1 - r^{-4}).$$

The area theorem leads to universal estimates on the size of the coefficients $\{b_n\}$:

COROLLARY 6.9. *In the situation of Theorem 6.7,*

(6.4)
$$|b_n| \leq \frac{1}{\sqrt{n}} \qquad \text{for all } n \geq 1.$$

Moreover, the following conditions are equivalent:

(i) $|b_1| = 1$;

(ii) *Some translation of ψ is conjugate to the Zhukovskii map by a rotation:*

$$\psi(w) = b_0 + \alpha \, Z(\alpha^{-1}w),$$

where $|\alpha| = 1$;

(iii) $\hat{\mathbb{C}} \smallsetminus U$ *is a closed line segment of length 4 centered at b_0.*

PROOF. Since area$(\hat{\mathbb{C}} \smallsetminus U) \geq 0$, Theorem 6.7 shows that $\sum_{n=1}^{\infty} n \, |b_n|^2 \leq 1$, which immediately gives (6.4). In particular, $|b_1| \leq 1$.

The equality $|b_1| = 1$ occurs if and only if $b_n = 0$ for all $n \geq 2$, in which case ψ takes the form

$$\psi(w) = w + b_0 + \frac{b_1}{w}.$$

This is just the formula in (ii) if we take α to be a square root of b_1. Thus, the conditions (i) and (ii) are equivalent.

Next, suppose (ii) holds. Since Z maps $\hat{\mathbb{C}} \smallsetminus \overline{\mathbb{D}}$ conformally onto $\hat{\mathbb{C}} \smallsetminus [-2, 2]$ (Example 6.6), ψ must map $\hat{\mathbb{C}} \smallsetminus \overline{\mathbb{D}}$ conformally onto $\hat{\mathbb{C}} \smallsetminus [b_0 - 2\alpha, b_0 + 2\alpha]$, which implies (iii).

Finally, suppose (iii) holds. Then the segment $\hat{\mathbb{C}} \smallsetminus U$ has the form $[b_0 - 2\alpha, b_0 + 2\alpha]$ for some α with $|\alpha| = 1$. It follows that

$$\varphi : w \mapsto \alpha^{-1}(\psi(\alpha w) - b_0)$$

maps $\hat{\mathbb{C}} \smallsetminus \overline{\mathbb{D}}$ conformally onto $\hat{\mathbb{C}} \smallsetminus [-2, 2]$. Since $\varphi(\infty) = Z(\infty) = \infty$ and $\varphi'(\infty) = Z'(\infty) = 1$, the uniqueness part of the Riemann mapping theorem shows that $\varphi = Z$, so (ii) holds. $\qquad \square$

THEOREM 6.10 (Bieberbach's inequality, 1916). *If $f(z) = z + \sum_{n=2}^{\infty} a_n z^n$ is in $\mathscr{S}$, then*

$$|a_2| \leq 2.$$

Moreover, the following conditions are equivalent:

(i) $|a_2| = 2$;

(ii) *f is conjugate to the Koebe function by a rotation:*

$$f(z) = \alpha \, K(\alpha^{-1}z) \qquad \text{where } |\alpha| = 1;$$

(iii) *$f(\mathbb{D})$ is a slit plane of the form $\mathbb{C} \smallsetminus \{-r\alpha : r \geq 1/4\}$, where $|\alpha| = 1$.*

Proof. An obvious plan of proof would be to apply the case $n = 1$ of (6.4) to the inverted map $\hat{f} \in \mathscr{S}$ which by (6.2) has the Laurent series $\hat{f}(w) = w - a_2 + (a_2^2 - a_3)/w + \cdots$, but this would only yield the inequality

$$(6.5) \qquad\qquad |a_2^2 - a_3| \leq 1$$

which is not what we are after.

The trick is to modify the above argument by applying it to a square root of $f(z^2)$ instead of f. Write $f(z) = z f_1(z)$, where $f_1(z) = 1 + \sum_{n=2}^{\infty} a_n z^{n-1}$ is holomorphic in $\mathbb{D}$. Note that f_1 is non-vanishing since $f_1(0) = 1$ and f is non-vanishing in $\mathbb{D}^*$. By Corollary 2.22, f_1 has a holomorphic square root $h \in \mathscr{O}(\mathbb{D})$, with $h(0) = 1$. Define

$$(6.6) \qquad\qquad g(z) = z h(z^2).$$

Ludwig Bieberbach
(1886–1982)

Since $f(z^2) = z^2 f_1(z^2) = z^2 h^2(z^2)$, we have

$$(6.7) \qquad\qquad f(z^2) = g^2(z).$$

Let us verify that $g \in \mathscr{S}$. Clearly $g \in \mathscr{O}(\mathbb{D})$, $g(0) = 0$, and $g'(0) = h(0) = 1$. To show injectivity of g, suppose $g(z) = g(w)$ for some $z, w \in \mathbb{D}$. Then by (6.7), $f(z^2) = f(w^2)$, hence $z = \pm w$ since f is injective. If $z = w$, we are done. If $z = -w$, then by (6.6), $g(z) = -g(w)$. Hence $g(z) = g(w) = 0$ which, by another application of (6.7), gives $z = w = 0$.

Now by (6.6), g has a power series representation $g(z) = z + c z^3 + \cdots$ containing odd powers of z only. Comparing this with the power series of f using (6.7), we see that $c = a_2/2$, hence

$$g(z) = z + \frac{a_2}{2} z^3 + \cdots .$$

By (6.2), the inverted map $\hat{g}(w) = 1/g(1/w)$ has the Laurent series

$$\hat{g}(w) = w - \frac{a_2}{2w} + \cdots .$$

Applying the case $n = 1$ of (6.4) to $\hat{g}$, we obtain $|a_2| \leq 2$.

Let us now discuss the case of equality. Corollary 6.9 shows that $|a_2| = 2$ is equivalent to the condition that $\hat{g}(w) = \lambda Z(\lambda^{-1} w)$ for some λ with $|\lambda| = 1$ (note that there is no constant term b_0 in the Laurant series of $\hat{g}$). Setting $\alpha = -\lambda^{-2}$, this gives

$$\hat{g}(w) = w + \lambda^2 w^{-1} = w - \alpha^{-1} w^{-1},$$

which is equivalent to

$$g(z) = \frac{1}{z^{-1} - \alpha^{-1} z} = \frac{z}{1 - \alpha^{-1} z^2},$$

or

$$f(z^2) = g^2(z) = \frac{z^2}{(1 - \alpha^{-1} z^2)^2},$$

or

$$f(z) = \frac{z}{(1 - \alpha^{-1} z)^2} = \alpha K(\alpha^{-1} z).$$

This shows the equivalence of (i) and (ii). Similarly, Corollary 6.9 shows that $|a_2| = 2$ is equivalent to $\hat{g}(\hat{\mathbb{C}} \smallsetminus \overline{\mathbb{D}}) = \hat{\mathbb{C}} \smallsetminus [-2\lambda, 2\lambda]$ for some λ with $|\lambda| = 1$. This, in turn, is equivalent to $g(\mathbb{D})$ being the doubly-slit plane $\mathbb{C} \smallsetminus \{r\lambda^{-1} : |r| \geq 1/2\}$, or $f(\mathbb{D}) = g^2(\mathbb{D})$ being the slit plane $\mathbb{C} \smallsetminus \{r\lambda^{-2} : r \geq 1/4\}$. This shows the equivalence of (i) and (iii). □

Let us now consider $\mathscr{S}$ as a subspace of $\mathscr{O}(\mathbb{D})$ equipped with the compact convergence topology.

THEOREM 6.11. *The class $\mathscr{S}$ of schlicht functions is compact.*

PROOF. $\mathscr{S}$ is closed in $\mathscr{O}(\mathbb{D})$. To see this, suppose $\{f_k\}$ is a sequence in $\mathscr{S}$ and $f_k \to f$ compactly in $\mathbb{D}$. Clearly $f(0) = 0$ and $f'(0) = 1$; in particular f is not constant. Injectivity of f then follows from Corollary 5.22, so $f \in \mathscr{S}$.

It is therefore enough to prove $\mathscr{S}$ is precompact, that is, every sequence $\{f_k\}$ in $\mathscr{S}$ has a subsequence which converges compactly in $\mathbb{D}$. Let f_k have the power series $z + \sum_{n=2}^{\infty} a_{n,k} z^n$ in $\mathbb{D}$. By (6.2), the inverted map $\hat{f}_k$ has a Laurent series of the form

$$\hat{f}_k(w) = w - a_{2,k} + \sum_{n=1}^{\infty} b_{n,k} w^{-n} \qquad \text{for } |w| > 1,$$

in which $|a_{2,k}| \leq 2$ by Theorem 6.10 and $|b_{n,k}| \leq 1/\sqrt{n}$ for all n by Corollary 6.9. The functions $g_k(z) = \hat{f}_k(1/z) - 1/z = -a_{2,k} + \sum_{n=1}^{\infty} b_{n,k} z^n$ are holomorphic in $\mathbb{D}$ (their singularity at $z = 0$ is removable) and satisfy

$$|g_k(z)| \leq 2 + \left| \sum_{n=1}^{\infty} b_{n,k} z^n \right| \leq 2 + \sum_{n=1}^{\infty} \frac{r^n}{\sqrt{n}} < +\infty$$

whenever $|z| \leq r < 1$. This shows $\{g_k\}$ is compactly bounded in $\mathbb{D}$ and therefore precompact by Montel's Theorem 5.25. Thus, there is a subsequence $\{g_{k_j}\}$ which converges compactly in $\mathbb{D}$ to some $g \in \mathscr{O}(\mathbb{D})$. Since $zg_k(z) + 1 = z\hat{f}_k(1/z) = z/f_k(z) \neq 0$ for all $z \in \mathbb{D}$, Hurwitz's Theorem 5.18 shows that $zg(z) + 1 \neq 0$ for all $z \in \mathbb{D}$. It follows that the subsequence $\{f_{k_j}\}$ converges compactly in $\mathbb{D}$ to the function $f \in \mathscr{O}(\mathbb{D})$ defined by $f(z) = z/(zg(z) + 1)$. □

Several a priori estimates involving schlicht functions can be deduced immediately from compactness of $\mathscr{S}$. The following theorem describes three such estimates:

THEOREM 6.12 (Existence of universal bounds in $\mathscr{S}$).

(i) *For each $0 < r < 1$ there exist constants $B_r, C_r > 1$ such that*

$$\frac{1}{B_r} \leq |f(z)| \leq B_r \quad \text{and} \quad \frac{1}{C_r} \leq |f'(z)| \leq C_r$$

whenever $f \in \mathscr{S}$ and $|z| = r$.

(ii) There exists a sequence $\{A_n\}$ of positive numbers such that

$$|a_n| \leq A_n \qquad (n = 2, 3, \ldots)$$

for every $f(z) = z + \sum_{n=2}^{\infty} a_n z^n$ in $\mathscr{S}$.

(iii) There exists a constant $\beta > 0$ such that

$$f(\mathbb{D}) \supset \mathbb{D}(0, \beta)$$

for every $f \in \mathscr{S}$.

PROOF. For (i), assume by way of contradiction that such B_r does not exist for some $0 < r < 1$. Then we can find sequences $\{f_k\}$ in $\mathscr{S}$ and $\{z_k\}$ with $|z_k| = r$ such that $|f_k(z_k)| \to 0$ or ∞. By compactness, after passing to a subsequence, we can assume that $f_k \to f \in \mathscr{S}$ compactly in $\mathbb{D}$ and $z_k \to p$ with $|p| = r$. This would imply $f(p) = 0$ or ∞, which is a contradiction. This proves the existence of B_r. The case of C_r follows from a similar argument since by Theorem 5.11 $f_k \to f$ implies $f'_k \to f'$, and f' cannot take the value 0 or ∞ in $\mathbb{D}$.

The existence of $\{A_n\}$ in (ii) follows at once from (i): By Cauchy's estimates (Theorem 1.42),

$$|a_n| = \frac{|f^{(n)}(0)|}{n!} \leq 2^n \sup_{|z|=1/2} |f(z)|,$$

so it suffices to take $A_n = 2^n B_{1/2}$.

For (iii), take $\beta = 1/B_{1/2}$. Evidently f maps the disk $\mathbb{D}(0, 1/2)$ conformally onto a domain D containing 0 bounded by the Jordan curve $\partial D = f(\mathbb{T}(0, 1/2))$. By (i), the distance between 0 and ∂D is at least β, so $\mathbb{D}(0, \beta) \subset D \subset f(\mathbb{D})$. $\qquad \square$

The following lemma, which we will use frequently here and in the next section, shows that a conformal map f distorts infinitesimal lengths by the factor $|f'|$ and infinitesimal areas by the factor $|f'|^2$.

LEMMA 6.13 (Length and area formulas). *Let $f : U \to V$ be conformal.*

(i) For any piecewise C^1 curve $\gamma : [a, b] \to U$,

$$\text{length}(f \circ \gamma) = \int_{\gamma} |f'(z)| \, |dz|.$$

(ii) For any measurable set $E \subset U$,

$$\text{area}(f(E)) = \iint_E |f'(z)|^2 \, dx \, dy.$$

PROOF. (i) follows at once from the definition of length:

$$\text{length}(f \circ \gamma) = \int_a^b |(f \circ \gamma)'(t)| \, dt$$

$$= \int_a^b |f'(\gamma(t))| \, |\gamma'(t)| \, dt = \int_\gamma |f'(z)| \, |dz|.$$

For (ii), write $f = u + iv$ and note that the Jacobian of f viewed as a map $U \to \mathbb{R}^2$ is

$$J_f = \det(Df) = \det \begin{bmatrix} u_x & u_y \\ v_x & v_y \end{bmatrix} = u_x v_y - u_y v_x.$$

Since $u_x = v_y$ and $u_y = -v_x$ by the Cauchy-Riemann equations (Theorem 1.13), the Jacobian can be written as $u_x^2 + v_x^2 = |f_x|^2$. Since f is holomorphic, $f_x = f'$ and we obtain

$$J_f = |f'|^2.$$

The standard change of variables formula in calculus now shows that

$$\text{area}(f(E)) = \iint_{f(E)} du \, dv = \iint_E |J_f(z)| \, dx \, dy = \iint_E |f'(z)|^2 \, dx \, dy. \qquad \square$$

It follows from the above lemma and Theorem 6.12(i) that a schlicht function distorts lengths and areas inside the disk $\mathbb{D}(0, r)$ by factors that only depend on r. Let $f \in \mathscr{S}$ and $0 < r < 1$. Since $f' \neq 0$ in $\mathbb{D}$, the maximum principle gives $1/C_r \leq |f'(z)| \leq C_r$ whenever $|z| \leq r$. It follows that if γ is a curve whose image lies in the disk $\mathbb{D}(0, r)$, then

$$\text{length}(f \circ \gamma) = \int_\gamma |f'(z)| \, |dz| \leq C_r \int_\gamma |dz| = C_r \, \text{length}(\gamma)$$

and

$$\text{length}(f \circ \gamma) \geq (1/C_r) \int_\gamma |dz| = (1/C_r) \, \text{length}(\gamma),$$

which prove the length distortion estimate

$$\frac{1}{C_r} \leq \frac{\text{length}(f \circ \gamma)}{\text{length}(\gamma)} \leq C_r.$$

A similar reasoning shows that if E is a measurable subset of $\mathbb{D}(0, r)$, then

$$\frac{1}{C_r^2} \leq \frac{\text{area}(f(E))}{\text{area}(E)} \leq C_r^2.$$

Of course the significance of these bounds is that they are independent of the choice of the schlicht function f.

The sole existence of the universal bounds in Theorem 6.12 is what matters in "soft" applications where the optimal values for the constants involved is not a major concern. However, all three parts of this theorem can be sharpened quantitatively. We begin with part (iii):

THEOREM 6.14 (Koebe's 1/4-theorem, 1916). *If $f \in \mathscr{S}$, then $f(\mathbb{D}) \supset \mathbb{D}(0, 1/4)$.*

The theorem was conjectured by Koebe in 1907, and proved by Bieberbach in 1916.

The Koebe function (Example 6.4) shows that the constant $1/4$ is optimal.

PROOF. Let p be any point outside $f(\mathbb{D})$; we want to show $|p| \geq 1/4$. Consider $\varphi(z) = pz/(p - z)$, the unique Möbius map which fixes the origin, has derivative 1 there, and sends p to ∞. The composition $\varphi \circ f$ is clearly in $\mathscr{S}$. If $f(z) = z + \sum_{n=2}^{\infty} a_n z^n$, then

$$(\varphi \circ f)(z) = \frac{pf(z)}{p - f(z)} = \frac{p(z + a_2 z^2 + \cdots)}{p - (z + a_2 z^2 + \cdots)} = \frac{z + a_2 z^2 + \cdots}{1 - \left(\dfrac{1}{p}z + \dfrac{a_2}{p}z^2 + \cdots\right)}$$

$$= (z + a_2 z^2 + \cdots)\left(1 + \frac{1}{p}z + \cdots\right) = z + \left(a_2 + \frac{1}{p}\right)z^2 + \cdots.$$

By Theorem 6.10 applied to f and $\varphi \circ f$, both inequalities $|a_2| \leq 2$ and $|a_2 + 1/p| \leq 2$ hold. Hence

$$\frac{1}{|p|} = \left|a_2 + \frac{1}{p} - a_2\right| \leq \left|a_2 + \frac{1}{p}\right| + |a_2| \leq 4. \qquad \square$$

The optimal constants in Theorem 6.12(ii) were conjectured by Bieberbach in 1916 and had to wait 68 years for a proof:

THEOREM 6.15 (de Branges, 1984). *If $f(z) = z + \sum_{n=2}^{\infty} a_n z^n$ is in $\mathscr{S}$, then*

$$|a_n| \leq n \qquad (n = 2, 3, \ldots).$$

Equality holds for some n if and only if f is rotationally conjugate to the Koebe function K.

We shall not attempt to prove this deep theorem here, but refer to [**DeB**] for the original proof and [**FK**] for a lively account of the problem and various attempts to solve it.

Louis de Branges de Bourcia (1932–)

Finally, the sharp bounds in Theorem 6.12(i) are given by the following classical result:

THEOREM 6.16 (Koebe's distortion bounds). *If $f \in \mathscr{S}$, then for every $z \in \mathbb{D}$,*

(6.8)
$$\frac{|z|}{(1 + |z|)^2} \leq |f(z)| \leq \frac{|z|}{(1 - |z|)^2}$$

and

(6.9)
$$\frac{1 - |z|}{(1 + |z|)^3} \leq |f'(z)| \leq \frac{1 + |z|}{(1 - |z|)^3}.$$

Again, the example of Koebe function shows that all four inequalities are optimal.

PROOF. We will first prove the inequalities in (6.9) and integrate them to obtain the ones in (6.8). Let $f \in \mathscr{S}$ and $\zeta = re^{it} \in \mathbb{D}$. Consider the disk automorphism $\varphi(z) = (z + \zeta)/(1 + \bar{\zeta}z)$ which sends 0 to ζ. A brief computation shows that

(6.10) $\varphi'(0) = 1 - r^2$ and $\varphi''(0) = -2(1 - r^2)\bar{\zeta}$.

The function

$$g(z) = \frac{f(\varphi(z)) - f(\zeta)}{f'(\zeta)(1 - r^2)}$$

is clearly in $\mathscr{S}$. By Theorem 6.10, $|g''(0)| \leq 4$. Since

$$g''(z) = \frac{f''(\varphi(z))(\varphi'(z))^2 + f'(\varphi(z))\varphi''(z)}{f'(\zeta)(1 - r^2)},$$

we obtain, in view of equation (6.10), the inequality

$$|g''(0)| = \left| \frac{f''(\zeta)(1 - r^2)}{f'(\zeta)} - 2\bar{\zeta} \right| \leq 4.$$

Dividing by $1 - r^2$ and multiplying by $|\zeta| = r$ then gives

$$\left| \frac{\zeta f''(\zeta)}{f'(\zeta)} - \frac{2r^2}{1 - r^2} \right| \leq \frac{4r}{1 - r^2},$$

Paul Koebe (1882–1945)

from which it follows that

(6.11) $$\frac{2r^2 - 4r}{1 - r^2} \leq \mathrm{Re}\left(\frac{\zeta f''(\zeta)}{f'(\zeta)} \right) \leq \frac{2r^2 + 4r}{1 - r^2}.$$

The term in the middle has an alternative description which is more suitable for our purposes. Since f' is non-vanishing in $\mathbb{D}$, it has a holomorphic logarithm there, that is, a function $h = u + iv \in \mathscr{O}(\mathbb{D})$ which satisfies $\exp(h) = f'$. The derivative formula $h' = f''/f'$ together with the Cauchy-Riemann equations for the pair u, v shows that

$$\mathrm{Re}\left(\frac{\zeta f''(\zeta)}{f'(\zeta)} \right) = \mathrm{Re}(\zeta h'(\zeta)) = \mathrm{Re}((x + iy)(u_x + iv_x))$$

$$= x\, u_x - y\, v_x = x\, u_x + y\, u_y$$

$$= r\frac{\partial u}{\partial r} = r\frac{\partial}{\partial r}\log|f'(re^{it})|.$$

Thus, (6.11) can be written as

$$\frac{2r - 4}{1 - r^2} \leq \frac{\partial}{\partial r}\log|f'(re^{it})| \leq \frac{2r + 4}{1 - r^2}   \text{for } re^{it} \in \mathbb{D}.$$

Now given $z = |z|e^{it} \in \mathbb{D}$ integrate the above inequalities from $r = 0$ to $r = |z|$ to obtain

$$\log\left(\frac{1 - |z|}{(1 + |z|)^3} \right) \leq \log|f'(z)| \leq \log\left(\frac{1 + |z|}{(1 - |z|)^3} \right),$$

from which (6.9) follows by exponentiation.

The upper bound in (6.8) follows by integrating the upper bound in (6.9):

$$|f(z)| = \left| \int_{[0,z]} f'(\zeta)\, d\zeta \right| \le \int_{[0,z]} |f'(\zeta)|\, |d\zeta|$$

$$\le \int_0^{|z|} \frac{1+r}{(1-r)^3}\, dr = \frac{|z|}{(1-|z|)^2}.$$

To obtain the lower bound in (6.8), fix $z \in \mathbb{D}^*$ and let p be a point where $|f(\zeta)|$ reaches a minimum on the circle $|\zeta| = |z|$. Let $\gamma : [0,1] \to \mathbb{D}$ be the smooth curve from 0 to p which maps under f to the line segment $[0, f(p)]$. If $r(t)\, e^{i\theta(t)}$ is the polar representation of $\gamma(t)$, then

$$|\gamma'(t)| = |r'(t) + ir(t)\theta'(t)| \ge |r'(t)|$$

for all $t \in (0,1)$. This, together with the lower bound in (6.9), yields

$$|f(p)| = \text{length}(f \circ \gamma) = \int_\gamma |f'(\zeta)|\, |d\zeta|$$

$$= \int_0^1 |f'(\gamma(t))|\, |\gamma'(t)|\, dt \ge \int_0^1 |f'(\gamma(t))|\, |r'(t)|\, dt$$

$$\ge \int_0^1 \frac{1 - r(t)}{(1 + r(t))^3} |r'(t)|\, dt \ge \int_0^1 \frac{1 - r(t)}{(1 + r(t))^3} r'(t)\, dt$$

$$= \int_0^{|z|} \frac{1 - r}{(1 + r)^3}\, dr = \frac{|z|}{(1 + |z|)^2},$$

which implies the lower bound in (6.8). $\qquad\square$

REMARK 6.17. The upper bounds in (6.8) and (6.9) can be easily deduced from de Branges's Theorem 6.15, although this is certainly an overkill: If $f(z) = z + \sum_{n=2}^\infty a_n z^n$ is in $\mathscr{S}$, then

$$|f(z)| \le |z| + \sum_{n=2}^\infty |a_n|\, |z|^n \le \sum_{n=1}^\infty n\, |z|^n = K(|z|) = \frac{|z|}{(1 - |z|)^2}$$

and

$$|f'(z)| \le 1 + \sum_{n=2}^\infty n|a_n|\, |z|^{n-1} \le \sum_{n=1}^\infty n^2\, |z|^{n-1} = K'(|z|) = \frac{1 + |z|}{(1 - |z|)^3}.$$

6.3 Boundary behavior of Riemann maps

The time has come to investigate the question of when the Riemann map of a simply connected domain extends continuously to the closed disk. To put things in perspective, first consider the case of an arbitrary homeomorphism between spherical domains. Let us say that a sequence $\{p_n\}$ in a domain $U \subset \hat{\mathbb{C}}$ is **escaping** if it has no accumulation point in U. An easy exercise shows that the following conditions are equivalent:

(i) $\{p_n\}$ is an escaping sequence in U;

(ii) $\{p_n\}$ eventually leaves every compact subset of U. More precisely, for every compact set $K \subset U$ there is an integer $N \geq 1$ such that $p_n \notin K$ if $n > N$;

(iii)   The spherical distance between p_n and ∂U tends to 0 as $n \to \infty$.

Given a homeomorphism $f : U \to V$ between spherical domains, there is a weak sense in which the boundary of U corresponds to the boundary of V under f:

(†)   *For every escaping sequence $\{p_n\}$ in U, the image $\{f(p_n)\}$ is an escaping sequence in V.*

The proof is immediate.

COROLLARY 6.18. *Suppose $f : U \to V$ is a homeomorphism between domains in $\hat{\mathbb{C}}$. If f has a continuous extension to the closure $\overline{U}$, then $f(\partial U) = \partial V$.*

PROOF. That $f(\partial U) \subset \partial V$ follows immediately from (†). If $q \in \partial V$, take a sequence $\{q_n\}$ in V which tends to q and let $p_n = f^{-1}(q_n)$. Since $\overline{U}$ is compact, after passing to a subsequence we may assume that p_n tends to some $p \in \overline{U}$, where by continuity $f(p) = q$. Applying (†) to f^{-1} shows that $p \in \partial U$. This proves the reverse inclusion $\partial V \subset f(\partial U)$. □

Of course an arbitrary homeomorphism $f : U \to V$ need not extend continuously to any boundary point, even if we assume it is smooth and the domains U, V are as nice as round disks. For example, take any smooth increasing function $\vartheta : [0, 1) \to [0, +\infty)$ such that $\vartheta(0) = 0$ and $\lim_{r \to 1} \vartheta(r) = +\infty$. Then the map $f : \mathbb{D} \to \mathbb{D}$ defined by $f(re^{it}) = re^{i(t+\vartheta(r))}$ is a smooth diffeomorphism that does not extend continuously to any boundary point (see Fig. 6.5).

The situation improves drastically when we consider conformal maps. For the case of the Riemann map of a simply connected domain $U \subset \hat{\mathbb{C}}$, Carathéodory provided a definitive answer to the extension question which turns out to depend solely on the topology of the boundary ∂U. Recall that a set $X \subset \hat{\mathbb{C}}$ is **locally connected** if every point of X has arbitrarily small connected neighborhoods. More precisely, if for every $p \in X$ and every open neighborhood Ω of p there is an open neighborhood $V \subset \Omega$ of p such that $V \cap X$ is connected.

THEOREM 6.19 (Carathéodory, 1913). *A Riemann map $\mathbb{D} \to U$ extends to a continuous map $\overline{\mathbb{D}} \to \overline{U}$ if and only if ∂U is locally connected.*

There is a sharper version of the Carathéodory theorem under a stronger hypothesis:

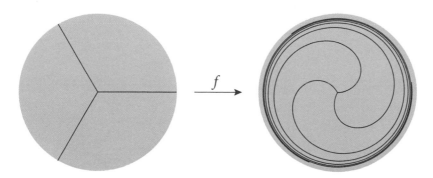

Figure 6.5. The radial lines at angles $0, 2\pi/3, 4\pi/3$ and their images under the diffeomorphism $f : \mathbb{D} \to \mathbb{D}$ defined by $f(re^{it}) = re^{i(t+\vartheta(r))}$, where $\vartheta(r) = \tan(\pi r/2)$. The image of each radial line spirals around infinitely often as it accumulates on the unit circle $\mathbb{T}$. As a result, f does not extend continuously to any point of $\mathbb{T}$.

THEOREM 6.20. *A Riemann map $\mathbb{D} \to U$ extends to a homeomorphism $\overline{\mathbb{D}} \to \overline{U}$ if and only if ∂U is a Jordan curve.*

Theorem 6.20 was proved independently by W. Osgood and E. Taylor in 1913 [**Z1**].

See Fig. 6.6 for an illustration of these theorems.

The proof of the Carathéodory theorem will depend on a few preliminary results. We begin with a topological statement on local connectivity. Let $\mathrm{dist}_\sigma(\cdot, \cdot)$ and $\mathrm{diam}_\sigma(\cdot)$ denote the distance and diameter measured in the spherical metric σ.

LEMMA 6.21 (Characterizations of local connectivity). *The following conditions on a compact set $X \subset \hat{\mathbb{C}}$ are equivalent:*

(i) *X is locally connected.*
(ii) *For every $\varepsilon > 0$ there exists a $\delta > 0$ such that any two points $p, q \in X$ with $\mathrm{dist}_\sigma(p, q) < \delta$ are contained in a connected set $E \subset X$ with $\mathrm{diam}_\sigma(E) \leq \varepsilon$.*
(iii) *For every open set $\Omega \subset \hat{\mathbb{C}}$, the connected components of $\Omega \cap X$ are relatively open in X.*

If $\infty \notin X$ we can replace the spherical distance and diameter in (ii) with the Euclidean distance and diameter since in this case the two metrics are comparable on X (see Lemma 5.29).

PROOF. (i) $\Longrightarrow$ (ii): Let $\varepsilon > 0$. By local connectivity, for every $p \in X$ the open ball $\{z \in \hat{\mathbb{C}} : \mathrm{dist}_\sigma(z, p) < \varepsilon/2\}$ has an open subset V_p containing p such that $V_p \cap X$ is connected. By compactness, we can find finitely many points $p_1, \ldots, p_n \in X$ such that $X \subset V_{p_1} \cup \cdots \cup V_{p_n}$. Let $\delta > 0$ be a Lebesgue number for this covering (see Lemma 2.17). If $p, q \in X$ and $\mathrm{dist}_\sigma(p, q) < \delta$, we can find some $1 \leq j \leq n$ such that $p, q \in V_{p_j}$. The set $E = V_{p_j} \cap X$ is connected, contains p and q, and has diameter $\leq \varepsilon$.

(ii) $\Longrightarrow$ (iii): First we show that for every $p \in X$ and every $\varepsilon > 0$ there is an open neighborhood V of p and a connected set C with $V \cap X \subset C \subset X$ such that $\mathrm{diam}_\sigma(C) \leq \varepsilon$. In fact, let δ be the constant corresponding to $\varepsilon/2$ in the condition (ii)

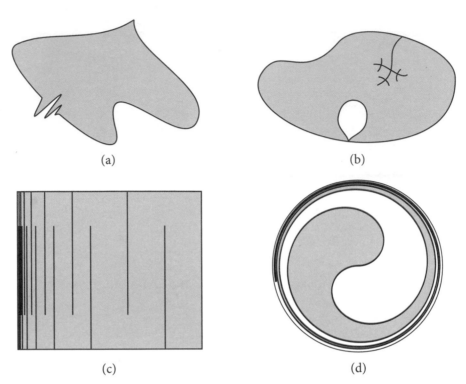

Figure 6.6. Examples of simply connected domains. Any Riemann map in (a) extends homeomorphically to the closed unit disk, while in (b) it extends continuously but not homeomorphically, as there are points on the boundary with two or four preimages (can you identify all such points?). In (c) local connectivity fails at every point of the left vertical segment of the boundary, and in (d) it fails at every point of the circle on which the tail of the Yin Yang domain accumulates. In both cases Riemann maps fail to extend continuously to the closed unit disk.

and $V = \{z \in \hat{\mathbb{C}} : \mathrm{dist}_\sigma(z, p) < \delta\}$. For each $q \in V \cap X$ there is a connected set $E_q \subset X$ containing both p and q such that $\mathrm{diam}_\sigma(E_q) \le \varepsilon/2$. The union $C = \bigcup_{q \in V \cap X} E_q$ will then have the desired property.

Now suppose $\Omega \subset \hat{\mathbb{C}}$ is open and H is a connected component of $\Omega \cap X$. Applying the above observation to each $q \in H$ with $\varepsilon > 0$ sufficiently small (depending on q), we find an open neighborhood V_q of q and a connected set C_q such that $V_q \cap X \subset C_q \subset \Omega \cap X$. Since H is the maximal connected subset of $\Omega \cap X$ containing q, we have the inclusion $C_q \subset H$. Now the open set $V = \bigcup_{q \in H} V_q$ has the property that $V \cap X = H$, proving that H is relatively open in X.

(iii) $\implies$ (i): Let Ω be any open neighborhood of $p \in X$. Let H be the connected component of $\Omega \cap X$ containing p. By the assumption H is relatively open in X, so there is an open set $V \subset \hat{\mathbb{C}}$ such that $V \cap X = H$. The open neighborhood $W = \Omega \cap V$ of p is contained in Ω, and $W \cap X = H$ is connected. □

COROLLARY 6.22. *Let $X \subset \hat{\mathbb{C}}$ be compact and locally connected. If $f : X \to \hat{\mathbb{C}}$ is continuous, then $f(X)$ is locally connected as well.*

PROOF. We use the equivalence (i) $\Longleftrightarrow$ (iii) in Lemma 6.21. Set $Y = f(X)$. Let $\Omega \subset \hat{\mathbb{C}}$ be open and H be a connected component of $\Omega \cap Y$. For each $q \in H$, let C_q be a connected component of the relatively open set $f^{-1}(\Omega \cap Y)$ which meets the set $f^{-1}(q)$. By local connectivity of X, C_q is relatively open in X, so $X \smallsetminus C_q$ is compact. It follows that $f(X \smallsetminus C_q)$ is compact, so $V_q = Y \smallsetminus f(X \smallsetminus C_q)$ is relatively open in Y, with $V_q \subset f(C_q)$. But the image $f(C_q) \subset \Omega \cap Y$ is connected and contains q, so $f(C_q) \subset H$. This shows $V_q \subset H$. It follows that $H = \bigcup_{q \in H} V_q$ is relatively open in Y. $\square$

Next, we consider open-ended curves $\gamma : [a, b) \to \hat{\mathbb{C}}$. The (spherical) length of such a curve is defined as the limit of the lengths of restrictions $\gamma|_{[a,b-\varepsilon]}$ as $\varepsilon \to 0$, which is either finite or $+\infty$. In the case γ is piecewise C^1, the length can still be expressed as the integral

$$\text{length}_\sigma(\gamma) = \lim_{\varepsilon \to 0} \int_a^{b-\varepsilon} \frac{2|\gamma'(t)|}{1 + |\gamma(t)|^2} \, dt = \int_a^b \frac{2|\gamma'(t)|}{1 + |\gamma(t)|^2} \, dt.$$

It will be convenient to say that the open end of γ **lands** at $p \in \hat{\mathbb{C}}$ if $\lim_{t \to b} \gamma(t) = p$.

LEMMA 6.23 (Finite length implies landing). *If $\gamma : [a, b) \to \hat{\mathbb{C}}$ is a curve with finite length, then $\lim_{t \to b} \gamma(t)$ exists.*

PROOF. Let E be the set of all accumulation points of $\gamma(t)$ as $t \to b$:

$$E = \bigcap_{a < t < b} \overline{\gamma((t, b))}.$$

As a nested intersection of non-empty compact sets, E must be non-empty and compact. Suppose E contains two distinct points p, q and choose sequences

$$a < t_1 < s_1 < t_2 < s_2 < \cdots < t_n < s_n < \cdots < b$$

such that $t_n, s_n \to b$, $\gamma(t_n) \to p$ and $\gamma(s_n) \to q$ as $n \to \infty$. Then,

(6.12) $$\text{dist}_\sigma(\gamma(t_n), \gamma(s_n)) \geq \frac{1}{2} \text{dist}_\sigma(p, q) > 0$$

for all large n. But

$$\text{length}_\sigma(\gamma) \geq \text{length}_\sigma(\gamma|_{[a,s_n]}) \geq \sum_{k=1}^n \text{dist}_\sigma(\gamma(t_k), \gamma(s_k))$$

for every $n \geq 1$ and the last sum diverges as $n \to \infty$ because of (6.12). This contradicts our finite length assumption. Thus E consists of a single point p and it follows easily that $\lim_{t \to b} \gamma(t) = p$. $\square$

The length of a curve $\gamma : (a, b) \to \hat{\mathbb{C}}$ can be defined as the sum of lengths of the two open-ended curves $\gamma|_{(a,c]}$ and $\gamma|_{[c,b)}$ in the above sense, where $a < c < b$. It is then clear from Lemma 6.23 that when such γ has finite length, both limits $\lim_{t \to a} \gamma(t)$ and $\lim_{t \to b} \gamma(t)$ exist, so both ends of γ land.

> **DEFINITION 6.24.** Let $U \subsetneq \hat{\mathbb{C}}$ be a simply connected domain. A C^1 curve $\gamma : (a, b) \to U$ is called a ***cross-cut*** in U if γ is injective and both limits $\lim_{t \to a} \gamma(t)$ and $\lim_{t \to b} \gamma(t)$ exist and belong to ∂U.

LEMMA 6.25. *Suppose $U \subsetneq \hat{\mathbb{C}}$ is a simply connected domain and $\gamma : (a, b) \to U$ is a cross-cut. Then $U \smallsetminus |\gamma|$ has exactly two connected components U_1, U_2. Moreover, $\partial U_1 \cap U = \partial U_2 \cap U = |\gamma|$.*

PROOF. By the Riemann mapping theorem, U is homeomorphic to the complex plane. Fix one such homeomorphism $\varphi : U \to \mathbb{C}$. The curve $\eta = \varphi \circ \gamma : (a, b) \to \mathbb{C}$ tends to ∞ in both directions (see (†) before Corollary 6.18), so the extended map $\hat{\eta} : [a, b] \to \hat{\mathbb{C}}$ which sends a, b to ∞ is a Jordan curve. By the Jordan curve theorem, $\hat{\mathbb{C}} \smallsetminus |\hat{\eta}| = \mathbb{C} \smallsetminus |\eta|$ has two connected components Ω_1, Ω_2 such that $\partial \Omega_1 = \partial \Omega_2 = |\hat{\eta}|$. Pulling back under φ, it follows that $U \smallsetminus |\gamma|$ has two connected components $U_1 = \varphi^{-1}(\Omega_1)$, $U_2 = \varphi^{-1}(\Omega_2)$. Moreover, $\partial U_i \cap U = \varphi^{-1}(\partial \Omega_i \cap \mathbb{C}) = \varphi^{-1}(|\eta|) = |\gamma|$, for $i = 1, 2$. □

COROLLARY 6.26. *Suppose $f : U \to V$ is a conformal map between simply connected domains and γ is a cross-cut in U whose image $\eta = f \circ \gamma$ has finite length. Then η is a cross-cut in V. Furthermore, f maps each connected component of $U \smallsetminus |\gamma|$ onto a connected component of $V \smallsetminus |\eta|$.*

PROOF. Clearly η is C^1 and injective, with both ends landing by Lemma 6.23. The landing points both lie on ∂V since f is a homeomorphism. This shows that η is a cross-cut in V.

Now let U_1 be one of the two components of $U \smallsetminus |\gamma|$. Then $f(U_1)$ is a connected subset of $V \smallsetminus |\eta|$, so it must be contained in a connected component V_1 of $V \smallsetminus |\eta|$. Applying the same argument to the inverse map f^{-1}, we see that $f^{-1}(V_1) \subset U_1$. Together, these inclusions imply $f(U_1) = V_1$. □

We now have all the ingredients of the proof of the Carathéodory theorem.

PROOF OF THEOREM 6.19. Take a Riemann map $f : \mathbb{D} \to U$. First suppose f extends continuously to $\overline{\mathbb{D}}$. Then $f(\mathbb{T}) = \partial U$ by Corollary 6.18. Since $\mathbb{T}$ is clearly locally connected, Corollary 6.22 shows that ∂U is locally connected as well.

The converse needs more work. Suppose ∂U is locally connected. After post-composing f with a Möbius map, we can assume $f(0) = \infty$. With this normalization, the image $f(\{z : 1/2 < |z| < 1\})$ misses the neighborhood $f(\mathbb{D}(0, 1/2))$ of ∞, so is a bounded domain in $\mathbb{C}$ on which the spherical and Euclidean metrics are comparable.

Constantin Carathéodory
(1873–1950)

This means we can conveniently carry out our computations near the boundary in the Euclidean metric. Fix $p = e^{i\theta} \in \mathbb{T}$ and for $0 < r < 1$, let

$$\gamma_r(t) = p + re^{it}, \qquad a_r < t < b_r$$

be a parametrization of the open circular arc $\mathbb{T}(p, r) \cap \mathbb{D}$. Here a_r, b_r are uniquely determined by the condition $\theta + \pi/2 < a_r < b_r < \theta + 3\pi/2$. Let $L(r)$ denote the Euclidean length of $f \circ \gamma_r$. By Lemma 6.13(i) and the Cauchy-Schwarz inequality,

$$L^2(r) = \left(\int_{\gamma_r} |f'(z)| \, |dz| \right)^2 \leq \left(\int_{\gamma_r} |dz| \right) \left(\int_{\gamma_r} |f'(z)|^2 \, |dz| \right)$$

$$\leq \pi r \int_{\gamma_r} |f'(z)|^2 \, |dz| = \pi r \int_{a_r}^{b_r} |f'(p + re^{it})|^2 \, r \, dt.$$

Integrating over $0 < r < 1/2$ and using Lemma 6.13(ii), we obtain

$$\int_0^{1/2} \frac{L^2(r)}{r} \, dr \leq \pi \int_0^{1/2} \int_{a_r}^{b_r} |f'(p + re^{it})|^2 \, r \, dt \, dr$$

$$= \pi \iint_{\mathbb{D}(p, 1/2) \cap \mathbb{D}} |f'(z)|^2 \, dx \, dy$$

$$= \pi \, \text{area}(f(\mathbb{D}(p, 1/2) \cap \mathbb{D})) < +\infty,$$

where the area is finite since $f(\mathbb{D}(p, 1/2) \cap \mathbb{D}) \subset f(\{z : 1/2 < |z| < 1\})$ and the latter, as noted above, is a bounded domain in $\mathbb{C}$. Integrability of $L^2(r)/r$ implies the existence of a decreasing sequence $r_n \to 0$ such that $L(r_n) \to 0$, for otherwise we could find constants $c > 0$ and $0 < \delta < 1/2$ such that $L(r) > c$ whenever $0 < r < \delta$, which would give

The technique of deducing finite length from finite area is an example of the "length-area method" in conformal geometry.

$$\int_0^{1/2} \frac{L^2(r)}{r} \, dr \geq \int_0^{\delta} \frac{L^2(r)}{r} \, dr \geq c^2 \int_0^{\delta} \frac{1}{r} \, dr = +\infty.$$

Since $L(r_n)$ is finite, Corollary 6.26 shows that the image $\eta_n = f \circ \gamma_{r_n}$ is a cross-cut in U. Let us denote by $p_n, q_n \in \partial U$ the landing points of the ends of η_n. Note that $|p_n - q_n| \leq \text{diam} |\eta_n| \leq L(r_n)$, so $\text{diam} |\eta_n|$ and $|p_n - q_n|$ both tend to 0 as $n \to \infty$. Since ∂U is assumed locally connected, Lemma 6.21 shows that for each n there is a connected set $E_n \subset \partial U$ containing p_n, q_n such that $\text{diam}(E_n) \to 0$ as $n \to \infty$. By replacing E_n with its closure, which is of course connected and has the same diameter, we may as well assume that E_n is closed.

Let $D_n = \mathbb{D}(p, r_n) \cap \mathbb{D}$ and W_n be the bounded component of $U \smallsetminus |\eta_n|$, i.e., the one not containing ∞ (see Fig. 6.7). Since D_n does not contain 0, the image $f(D_n)$ does not contain ∞, so $f(D_n) = W_n$ by Corollary 6.26. We claim that W_n is contained in a bounded component of $\hat{\mathbb{C}} \smallsetminus (|\eta_n| \cup E_n)$. If not, we could take any point $z \in W_n$, join it to ∞ by an embedded arc within $\hat{\mathbb{C}} \smallsetminus (|\eta_n| \cup E_n)$, and then join ∞ back to z by another embedded arc within U which crosses the smooth arc $|\eta_n|$ only once. The

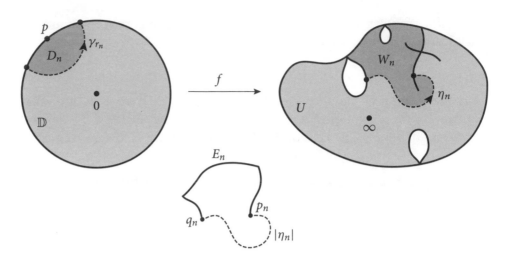

Figure 6.7. Proof of the Carathéodory theorem.

union of these two arcs would be a Jordan curve which avoids E_n and separates p_n from q_n. This would contradict connectedness of E_n and proves our claim. It follows that

$$\text{diam } W_n \leq \text{diam}(|\eta_n| \cup E_n) \to 0 \qquad \text{as } n \to \infty.$$

Since the closures $\overline{W_n}$ are compact and nested, this proves that the intersection $\bigcap_{n=1}^{\infty} \overline{W_n}$ reduces to a single point $q \in \partial U$. This point does not depend on the choice of the sequence $\{r_n\}$ since

$$\bigcap_{n=1}^{\infty} \overline{W_n} = \bigcap_{n=1}^{\infty} \overline{f(D_n)} = \bigcap_{0 < r < 1} \overline{f(\mathbb{D}(p, r) \cap \mathbb{D})}.$$

Thus, we can define $f(p) = q$ unambiguously. This gives an extension of f to a map $f : \overline{\mathbb{D}} \to \overline{U}$.

To see continuity of this extension at a boundary point $p \in \mathbb{T}$, take an arbitrary $\varepsilon > 0$ and choose a neighborhood $D_n = \mathbb{D}(p, r_n) \cap \mathbb{D}$ of p as above, where n is large enough to guarantee $f(D_n) = W_n$ has diameter $< \varepsilon$. If $z \in D_n$, then $f(z) \in W_n$, so $|f(z) - f(p)| \leq \text{diam } \overline{W_n} = \text{diam } W_n < \varepsilon$. On the other hand, if z belongs to the interior of the arc $\overline{D_n} \cap \mathbb{T}$, we can choose a similar neighborhood $D'_m = \mathbb{D}(z, r'_m) \cap \mathbb{D}$ of z with $D'_m \subset D_n$ so that $f(D'_m) = W'_m \subset W_n$. It then follows that $f(z) \in \overline{W'_m} \subset \overline{W_n}$, so again $|f(z) - f(p)| \leq \text{diam } \overline{W_n} < \varepsilon$. $\qquad \square$

To deduce Theorem 6.20 from Theorem 6.19 is an easy task that depends on two simple lemmas. The first is a topological statement about Jordan curves. Recall that for a Jordan curve γ in the plane, $\text{int}(\gamma)$ and $\text{ext}(\gamma)$ denote the bounded and unbounded components of $\mathbb{C} \smallsetminus |\gamma|$, respectively.

LEMMA 6.27. *Suppose γ, η are Jordan curves in the plane that meet at a single point. Then $\text{int}(\gamma)$ and $\text{int}(\eta)$ are either disjoint or nested.*

PROOF. Suppose the claim is false. Then int(γ) meets both int(η) and ext(η), hence it meets $|\eta|$. Similarly, ext(γ) meets both int(η) and ext(η), hence it meets $|\eta|$. If q is the point where the two Jordan curves meet, it follows that $|\eta| \smallsetminus \{q\}$, as the union of the disjoint non-empty (relatively) open sets $|\eta| \cap \text{int}(\gamma)$ and $|\eta| \cap \text{ext}(\gamma)$, is disconnected. This is a contradiction since $|\eta| \smallsetminus \{q\}$ is homeomorphic to an open interval. $\qquad\square$

The second lemma uses an elementary case of the "Schwarz reflection principle" which we will discuss in chapter 10 (see also Theorem 8.36 for a far-reaching generalization of this lemma).

LEMMA 6.28. *Let $I \subset \mathbb{T}$ be an open arc. Suppose $f \in \mathcal{O}(\mathbb{D})$ extends continuously to $\mathbb{D} \cup I$ and $f|_I = 0$. Then $f = 0$ everywhere in $\mathbb{D}$.*

PROOF. To simplify matters, take a Möbius map φ which sends the upper half-plane $\mathbb{H}$ to $\mathbb{D}$, carrying the interval $(-1, 1) \subset \mathbb{R}$ to I. Then $g = f \circ \varphi$ is holomorphic in $\mathbb{H}$, extends continuously to $\mathbb{H} \cup (-1, 1)$, and maps $(-1, 1)$ to 0. Let U be the doubly-slit plane $\mathbb{C} \smallsetminus ((-\infty, -1] \cup [1, +\infty))$. Define $G : U \to \mathbb{C}$ by

$$G(z) = \begin{cases} g(z) & z \in \mathbb{H} \\ 0 & z \in (-1, 1) \\ \overline{g(\bar{z})} & \bar{z} \in \mathbb{H}. \end{cases}$$

Evidently G is continuous in U and holomorphic in $U \smallsetminus \mathbb{R}$. Since lines are removable (Example 1.40), it follows that G is holomorphic in U. The identity theorem then shows that $G = 0$ in U, hence $g = 0$ in $\mathbb{H}$. $\qquad\square$

PROOF OF THEOREM 6.20. Take a Riemann map $f : \mathbb{D} \to U$. If f extends to a homeomorphism $\overline{\mathbb{D}} \to \overline{U}$, the boundary $\partial U = f(\mathbb{T})$ is clearly a Jordan curve. Conversely, suppose ∂U is a Jordan curve. Then ∂U is locally connected by Corollary 6.22, so f extends continuously to a map $\overline{\mathbb{D}} \to \overline{U}$ by Theorem 6.19. To show f is a homeomorphism, it suffices to prove its injectivity, for then by compactness of $\overline{\mathbb{D}}$ the inverse map $f^{-1} : \overline{U} \to \overline{\mathbb{D}}$ will be automatically continuous.

Assume by way of contradiction that there are distinct points $p, p' \in \mathbb{T}$ with the same image $f(p) = f(p') = q$. After postcomposing f with a Möbius map which sends $f(0)$ to 0 and some point outside $\overline{U}$ to ∞, we may assume that $f(0) = 0$ and U is a bounded domain in $\mathbb{C}$. Take a cross-cut γ in $\mathbb{D}$ which avoids 0 and has its ends landing at p, p'. Let D be the component of $\mathbb{D} \smallsetminus |\gamma|$ which does not contain 0. The image $f \circ \gamma$ is a cross-cut in U which avoids 0 and has both ends landing at q (see Fig. 6.8). Let η be the Jordan curve obtained by adding the point q to this cross-cut. Since $|\eta|$ meets U, int(η) should intersect U as well. Since $|\eta| \cap \partial U = \{q\}$, Lemma 6.27 shows that int(η) $\subset U$. In other words, int(η) is one of the two components of $U \smallsetminus |\eta|$. By Corollary 6.26, $f(D) = \text{int}(\eta)$. Now every point of the arc $I = \overline{D} \cap \mathbb{T}$ must map to a point of $|\eta| \cap \partial U$, so f maps I to the single point q. Applying Lemma 6.28 to the function $f - q \in \mathcal{O}(\mathbb{D})$ now gives $f = q$ everywhere in $\mathbb{D}$, which is a contradiction. $\qquad\square$

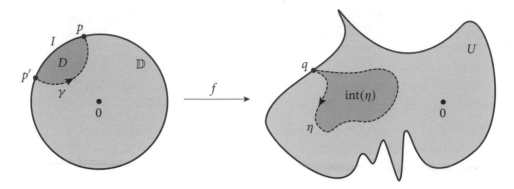

Figure 6.8. Proof of Theorem 6.20.

REMARK 6.29. The inclusion $\text{int}(\eta) \subset U$, a crucial step in the above proof, relies heavily on the assumption of ∂U being a Jordan curve. In general, η may "pinch" ∂U at q, so $\text{int}(\eta)$ may well contain points of ∂U. This is in fact what happens for the Riemann map of a domain such as the one in Fig. 6.6(b).

Let $f : \mathbb{D} \to U$ be a Riemann map. When the boundary ∂U is locally connected, Theorem 6.19 shows that $\lim_{z \to p} f(z)$ exists for every $p \in \mathbb{T}$. In the absence of local connectivity this is no longer true, but one can ask whether a more restricted form of the limit still exists. The affirmative answer, provided by Theorem 6.30 below, turns out to be useful in many applications. Let us say that f has a **radial limit** at $p \in \mathbb{T}$ if

$$f^*(p) = \lim_{r \to 1} f(rp)$$

exists, that is, if the image under f of the radial line at p lands at a well-defined point in $\hat{\mathbb{C}}$.

THEOREM 6.30 (Fatou, 1906). *For every Riemann map $f : \mathbb{D} \to U$ the radial limit f^* exists a.e. on $\mathbb{T}$.*

This theorem holds for all bounded holomorphic functions in $\mathbb{D}$, but the proof is considerably harder (see Theorem 7.44).

PROOF. We use a length-area estimate similar to that in the proof of the Carathéodory theorem. After postcomposing f with a Möbius map, which does not alter the existence of radial limits, we may assume that $f(0) = \infty$, so $f(\{z : 1/2 < |z| < 1\})$ is a bounded domain in $\mathbb{C}$. For each $t \in [0, 2\pi]$, let $L(t)$ denote the length of the curve $r \mapsto f(re^{it})$ restricted to $r \in [1/2, 1]$. By the Cauchy-Schwarz inequality,

$$L^2(t) = \left(\int_{1/2}^1 |f'(re^{it})| \, dr \right)^2 \leq \frac{1}{2} \int_{1/2}^1 |f'(re^{it})|^2 \, dr \leq \int_{1/2}^1 |f'(re^{it})|^2 \, r \, dr.$$

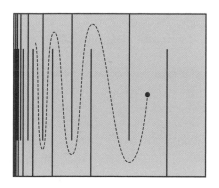 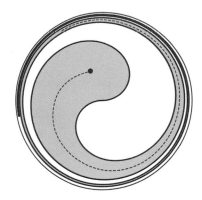

Figure 6.9. For each domain, a single radial limit of the Riemann map fails to exist.

Integrating over t then gives

$$\int_0^{2\pi} L^2(t)\,dt \le \int_0^{2\pi}\int_{1/2}^1 |f'(re^{it})|^2\,r\,dr\,dt$$

$$= \text{area}(f\{z : 1/2 < |z| < 1\}) < +\infty.$$

It follows that $L(t)$ is finite for almost every $t \in [0, 2\pi]$. This proves the theorem since by Lemma 6.23 the radial limit $f^*(e^{it})$ exists whenever $L(t)$ is finite.   □

As an illustration of Theorem 6.30, we have reproduced the domains of Fig. 6.6(c) and (d) in Fig. 6.9. These domains do not have locally connected boundary, so by the Carathéodory theorem their Riemann maps do not extend continuously to the whole unit circle. It can be shown that these Riemann maps do extend continuously everywhere except at a single point $p \in \mathbb{T}$ where the radial limit $f^*(p)$ fails to exist. For the first domain, the image $r \mapsto f(rp)$ of the radial line at p is a topologist's sine curve that accumulates on the left vertical segment. For the second domain, this curve spirals around indefinitely and accumulates on the surrounding circle.

EXAMPLE 6.31 (Modulus of a rectangle). As another application of the length-area method, we show that if f is a conformal map between open rectangles $R = (0, a) \times (0, b)$ and $R' = (0, a') \times (0, b')$ which carries each corner of R to the corresponding corner of R', then $b/a = b'/a'$. Thus, the height-to-width ratio $\text{mod}(R) = b/a$, called the **modulus** of R, is a conformal invariant. By Theorem 6.20, f extends to a homeomorphism between closed rectangles, mapping each side of R to the corresponding side of R'. For each $y \in (0, b)$, let $L(y)$ be the length of the curve $\eta_y : (0, a) \to R'$ defined by $\eta_y(x) = f(x + iy)$. By the Cauchy-Schwarz inequality,

$$L^2(y) = \left(\int_0^a |f'(x + iy)|\,dx\right)^2 \le a\int_0^a |f'(x + iy)|^2\,dx.$$

Integrating over y, we obtain

$$(6.13) \qquad \int_0^b L^2(y)\,dy \le a\int_0^b\int_0^a |f'(x + iy)|^2\,dx\,dy = a\,\text{area}(R') = aa'b'.$$

On the other hand, η_y extends continuously to $[0, a]$, with $\eta_y(0)$ and $\eta_y(a)$ belonging respectively to the left and right vertical sides of R', so $L(y) \geq a'$. Combined with (6.13), this proves $a'^2 b \leq aa'b'$ or $b/a \leq b'/a'$. Applying the same argument to f^{-1} gives the reverse inequality $b'/a' \leq b/a$, which completes the proof.

For an alternative proof of the conformal invariance of modulus based on the Schwarz reflection principle, see Example 10.29. A similar invariant for annuli is discussed in problem 24 as well as Theorem 13.33.

The length-area method was first used systematically by Grötzsch in the 1920s. Decades later, it was incorporated by Beurling and Ahlfors in their study of an important conformal invariant known as the "extremal length" of a family of curves (the modulus of a rectangle being an example). For a concise account of this theory, see [A4].

Problems

(1) Prove that if $U \subset \hat{\mathbb{C}}$ is a simply connected domain, then $\hat{\mathbb{C}} \smallsetminus U$ is connected. (Hint: The result is clear if $\hat{\mathbb{C}} \smallsetminus U$ has at most one point. Otherwise, take a Riemann map $f : \mathbb{D} \to U$ and use the fact that

$$\hat{\mathbb{C}} \smallsetminus U = \bigcap_{0 < r < 1} \left(\hat{\mathbb{C}} \smallsetminus f(\mathbb{D}(0, r)) \right)$$

is a nested intersection of compact connected sets.)

(2) Find explicit Riemann maps for the strip $\{z \in \mathbb{C} : 0 < \mathrm{Im}(z) < 1\}$ as well as the half-strip $\{z \in \mathbb{C} : 0 < \mathrm{Im}(z) < 1 \text{ and } \mathrm{Re}(z) > 0\}$.

(3) Let $f : \mathbb{D} \to U$ be a Riemann map with $f(0) = 0$. If $g : \mathbb{D} \to U$ is any holomorphic function with $g(0) = 0$, show that $g(\mathbb{D}(0, r)) \subset f(\mathbb{D}(0, r))$ for all $0 < r < 1$. This is known as the **subordination principle**.

(4) Let $U \subsetneq \mathbb{C}$ be simply connected, $p \in U$, and $f : U \to \mathbb{D}$ be a conformal map with $f(p) = 0$. If $g : U \to \mathbb{D}$ is any holomorphic function, show that $|g'(p)| \leq |f'(p)|$, with equality if and only if $g = \alpha f$ for some constant α with $|\alpha| = 1$.

(5) Suppose $U \subsetneq \mathbb{C}$ is a simply connected domain which has a $2\pi/N$ rotational symmetry about the origin for some $N \geq 2$, that is, $z \in U$ if and only if $\omega z \in U$, where $\omega = e^{2\pi i/N}$. Let $f : \mathbb{D} \to U$ be any Riemann map with $f(0) = 0$. If $f(z) = \sum_{n=1}^{\infty} a_n z^n$, prove that

$$a_n = 0 \qquad \text{for } n \not\equiv 1 \ (\text{mod } N).$$

As an example, when U has a $180°$ symmetry about the origin, the Riemann map of U which fixes the origin contains only the odd powers of z. (Hint: Show that $f(\omega z) = \omega f(z)$ for all $z \in \mathbb{D}$.)

(6) Let $U \subsetneq \mathbb{C}$ be a simply connected domain and $f : U \to U$ be a holomorphic function.
   (i) Show that f has at most one fixed point in U unless it is the identity map.
   (ii) If $p = f(p)$, show that $|f'(p)| \leq 1$.
   (iii) If $p = f(p)$ and $|f'(p)| = 1$, show that $f \in \mathrm{Aut}(U)$.
   (iv) If $p = f(p)$ and $f'(p) = 1$, show that f is the identity map.

(v) Repeat part (iv) only assuming U is a bounded domain. This is the one-dimensional case of a lemma in several complex variables due to H. Cartan.

(Hint: For (i)–(iv), use the Riemann mapping theorem to transfer the problem to the unit disk. For (v), use the fact that the iterates $\{f^{\circ n}\}$ form a normal family, and look at the power series of $f^{\circ n}$ about the fixed point p.)

(7) Suppose $U \subset \mathbb{C}$ is a convex domain, $f \in \mathcal{O}(U)$, and $\operatorname{Re}(f'(z)) > 0$ for all $z \in U$. Show that $f : U \to f(U)$ is a conformal map. (Hint: For $p, q \in U$, write $f(q) - f(p) = \int_{[p,q]} f'(z)\,dz$.)

(8) Suppose $f \in \mathcal{O}(\mathbb{D})$ is injective and for some $0 < r < 1$ the domain $f(\mathbb{D}(0,r))$ is convex. Show that for every $0 < s < r$, the domain $f(\mathbb{D}(0,s))$ is convex. (Hint: Suppose $f(0) = 0$, $0 \le |a| \le |b| < s < r$, and $0 < t < 1$. Apply the subordination principle of problem 3 to the holomorphic function $g : \mathbb{D}(0,r) \to \mathbb{C}$ defined by $g(z) = tf(az/b) + (1-t)f(z)$.)

(9) Let $P(z) = z + a_2 z^2 + \cdots + a_n z^n$ be a polynomial in $\mathcal{S}$. Prove that $|a_n| \le 1/n$. Show by an example that this is the best upper bound for each $n \ge 2$. (Hint: The derivative of P does not vanish in $\mathbb{D}$.)

(10) Suppose $f \in \mathcal{S}$, and for $0 < r < 1$ let $A(r)$ denote the area of the image domain $f(\mathbb{D}(0,r))$. Show that $A(r) \ge \pi r^2$. (Hint: Substitute the power series of f into the formula

$$A(r) = \iint_{\mathbb{D}(0,r)} |f'(z)|^2 \, dx\, dy = \int_0^r \int_0^{2\pi} f'(\rho e^{it}) \overline{f'(\rho e^{it})} \, \rho \, d\rho \, dt$$

and estimate from below.)

(11) Show that a function in $\mathcal{S}$ is odd (in the sense that it commutes with $z \mapsto -z$) if and only if it is a holomorphic square root of $f(z^2)$ for some $f \in \mathcal{S}$.

(12) Prove that for every $0 < r < 1$ and $n \ge 2$ there is a constant $C_{r,n} > 0$ such that

$$|f^{(n)}(z)| < C_{r,n} \qquad \text{if } f \in \mathcal{S} \text{ and } |z| = r.$$

Show by an example that for each $n \ge 2$ the n-th derivative of a schlicht function can have zeros in $\mathbb{D}$, so no positive lower bound can exist in the above statement.

(13) The function $f : \mathbb{D} \to \mathbb{C}$ defined by $f(z) = (e^{5z} - 1)/5$ is holomorphic, with $f(0) = 0$ and $f'(0) = 1$. However, f does not take the value $-1/5$. Why doesn't this contradict Koebe's $1/4$-theorem?

(14) Let $U \subsetneq \mathbb{C}$ be a simply connected domain and $p \in U$. Define the **conformal radius** of U at p by $\operatorname{rad}(U, p) = |f'(0)|$, where $f : \mathbb{D} \to U$ is any Riemann map which sends 0 to p.
  (i)   Verify that the definition of $\operatorname{rad}(U, p)$ does not depend on the choice of f.
  (ii) If $V \subsetneq \mathbb{C}$ is another simply connected domain and $\varphi : U \to V$ is holomorphic with $\varphi(p) = q$, show that $|\varphi'(p)| \le \operatorname{rad}(V, q)/\operatorname{rad}(U, p)$, with equality if and only if φ is a conformal isomorphism.
  (iii) Determine $\operatorname{rad}(\mathbb{D}, p)$ for $p \in \mathbb{D}$.
  (iv) Show that for every $p \in U$, $1 \le \operatorname{rad}(U, p)/\operatorname{dist}(p, \partial U) \le 4$.

(15) Suppose $f : \mathbb{D} \to U$ is a Riemann map and $0 \notin U$.
  (i) Prove that $|f'(0)| \le 4|f(0)|$.
  (ii) More generally, prove that

$$|f'(z)| \le \frac{4|f(z)|}{1 - |z|^2} \qquad \text{for all } z \in \mathbb{D}.$$

(iii)   Show that for every $z \in \mathbb{D}$,

$$\left(\frac{1-|z|}{1+|z|}\right)^2 \leq \frac{|f(z)|}{|f(0)|} \leq \left(\frac{1+|z|}{1-|z|}\right)^2.$$

(Hint: For (ii), precompose f with a suitable disk automorphism and apply (i). For the upper bound in (iii), estimate the integral of $|f'/f|$ along the radial segment $[0, z]$ using (ii).)

(16)   Suppose $f \in \mathcal{O}(\mathbb{D}), f(0) = 0$, and $f(\mathbb{D}) \cap (-\infty, 1/4] = \emptyset$. Show that

$$|f(z)| \leq \frac{|z|}{(1-|z|)^2} \qquad \text{for all } z \in \mathbb{D}.$$

(Hint: Use the subordination principle of problem 3.)

(17)   Use Koebe's distortion bound (6.8) to show that if $f \in \mathscr{S}$, then

$$\frac{1-|z|}{1+|z|} \leq \left|\frac{zf'(z)}{f(z)}\right| \leq \frac{1+|z|}{1-|z|} \qquad \text{for all } z \in \mathbb{D}^*.$$

(18)   Use Koebe's distortion bound (6.8) and Cauchy's estimates (Theorem 1.42) to show that if $f(z) = z + \sum_{n=2}^{\infty} a_n z^n$ is in $\mathscr{S}$, then

$$|a_n| \leq en^2 \qquad \text{for all } n \geq 2.$$

(19)   (Dieudonné) Prove Theorem 6.15 under the additional hypothesis that the coefficients $\{a_n\}$ are all real, or equivalently, that the domain $f(\mathbb{D})$ is symmetric with respect to the real axis. (Hint: Write $f = u + iv$ and first show that the equality

$$a_n = \frac{1}{\pi r^n} \int_0^{2\pi} v(re^{it}) \sin(nt)\, dt$$

holds for every $0 < r < 1$.)

(20)   For a simply connected domain $U \subsetneq \hat{\mathbb{C}}$, show that the following conditions are equivalent:
   (i)   Any Riemann map $\mathbb{D} \to U$ has a continuous extension to $\overline{\mathbb{D}}$.
   (ii) ∂U is a closed curve.
   (iii) ∂U is locally connected.
   (iv) $\hat{\mathbb{C}} \smallsetminus U$ is locally connected.
   This can be viewed as a slightly sharper formulation of the Carathéodory theorem [**P**].

(21)   Suppose $f(z) = z + \sum_{n=2}^{\infty} a_n z^n$ is holomorphic in $\mathbb{D}$. If $\sum_{n=2}^{\infty} n|a_n| < 1$, show that f extends to a homeomorphism $\overline{\mathbb{D}} \to f(\overline{\mathbb{D}})$. (Hint: Look at the power series expansion of $(f(z) - f(\varsigma))/(z - \varsigma)$ for $z \neq \varsigma$.)

(22)   Let S be the open square $\{x + iy \in \mathbb{C} : 0 < x, y < 1\}$, U be a bounded simply connected domain in $\mathbb{C}$, and $f : S \to U$ be a conformal map. For each $0 < y < 1$, the curve $\eta_y :$ $x \mapsto f(x + iy)$ parametrizes the image of the horizontal segment $(0, 1) \times \{y\} \subset S$ under f. Let $L(y)$ denote the length of η_y. Use the length-area method to verify the following statements:
   (i) $L(y)$ is finite for almost every y.
   (ii)   The majority of the η_y are not too long: The measure of the set of $0 < y < 1$ for which $L(y) \leq \sqrt{2 \operatorname{area}(U)}$ is at least $1/2$.

(23)   Let γ be a Jordan curve in $\mathbb{C}$ and (p_1, p_2, p_3, p_4) be an ordered quadruple of points, chosen counterclockwise on γ. Show that there is a conformal map $\operatorname{int}(\gamma) \to \mathbb{D}$ which carries

(p_1, p_2, p_3, p_4) to the vertices of a rectangle inscribed in $\mathbb{D}$, and that this map is unique up to postcomposition with a rotation. (Hint: It suffices to show that any quadruple on $\mathbb{T}$ can be mapped by an element of $\text{Aut}(\mathbb{D})$ to the vertices of a rectangle which is unique up to a rotation. You may want to use the fact that two quadruples can be mapped to one another by a Möbius map if and only if they have the same cross ratios (Theorem 4.7).)

(24)   (Modulus of a round annulus) Prove that the annuli $A = \{z : a < |z| < b\}$ and $A' = \{z : a' < |z| < b'\}$ are conformally isomorphic if and only if $b/a = b'/a'$. The conformal invariant

$$\text{mod}(A) = \frac{1}{2\pi} \log\left(\frac{b}{a}\right)$$

is called the **modulus** of A. (Hint: Imitate the length-area argument in Example 6.31 as follows: Take a conformal map $f : A \to A'$ and assume without loss of generality that $\lim_{|z| \to a} |f(z)| = a'$ and $\lim_{|z| \to b} |f(z)| = b'$. For $t \in [0, 2\pi]$, let $L(t)$ be the length of the curve $\eta_t : (a, b) \to A'$ defined by $\eta_t(r) = f(re^{it})$ with respect to the conformal metric $g = |dz|/|z|$. Show that

$$\int_0^{2\pi} L^2(t)\, dt \le 2\pi \log\left(\frac{b'}{a'}\right) \log\left(\frac{b}{a}\right).$$

On the other hand, $L(t) \ge \log(b'/a')$ since in the metric g each η_t is at least as long as a radial segment connecting the boundary components of A'.)

(25)   Give an example of a simply connected domain U with locally connected boundary whose Riemann map $\mathbb{D} \to U$ sends a Cantor set in $\mathbb{T}$ to a single point on ∂U.

(26)   Although a typical Riemann map may fail to have radial limits at some points of the unit circle, show that the *inverse* of a Riemann map behaves much nicer in the following sense: If $g : U \to \mathbb{D}$ is conformal and the curve $\xi : [0, 1) \to U$ lands at $q \in \partial U$ (i.e., $\lim_{t \to 1} \xi(t) = q$), then the image curve $g \circ \xi : [0, 1) \to \mathbb{D}$ lands at a well-defined point of $\mathbb{T}$. Moreover two curves in U with distinct landing points on ∂U map to two curves in $\mathbb{D}$ with distinct landing points on $\mathbb{T}$. (Hint: Assuming $p, p' \in \mathbb{T}$ are distinct accumulation points of $g(\xi(t))$ as $t \to 1$, construct cross-cuts γ_n, γ_n' in $\mathbb{D}$ near p, p' with shrinking images $\eta_n = g^{-1} \circ \gamma_n, \eta_n' = g^{-1} \circ \gamma_n'$, as in the proof of the Carathéodory theorem, so η_n, η_n' both lie in an arbitrarily small neighborhood of q for large n. Draw a contradiction from this.)

(27)   The **Schwarzian derivative** of a meromorphic function f is defined by

$$S_f = \left(\frac{f''}{f'}\right)' - \frac{1}{2}\left(\frac{f''}{f'}\right)^2.$$

Verify the following properties:
   (i)   If f and g are meromorphic in $\mathbb{C}$, then

$$S_{g \circ f} = (S_g \circ f) \cdot (f')^2 + S_f.$$

   (ii) If $f \in \text{Möb}$, then $S_f = 0$.
   (iii) Conversely, if f is meromorphic in a domain U, and if $S_f = 0$ in U, then f is the restriction of a Möbius map.
   (Hint: For (iii), solve the separable differential equation $R' - (1/2)R^2 = 0$, where $R = f''/f'$.)

(28)   This exercise gives an alternative description for the Schwarzian derivative. Suppose f is meromorphic in a domain U. If $p \in U$ and $f'(p) \ne 0$, show that there is a unique Möbius

map φ such that

$$\lim_{z \to p} \frac{(\varphi \circ f)(z) - z}{(z - p)^3}$$

is finite, and that this limit is $(1/6)S_f(p)$.

(29)   (Kraus) Suppose $f \in \mathcal{O}(\mathbb{D})$ is injective. Show that

$$|Sf(z)| \le \frac{6}{(1 - |z|^2)^2} \qquad \text{for all } z \in \mathbb{D}.$$

(Hint: After postcomposing with an affine map, which does not affect the Schwarzian derivative, we can assume f is schlicht. The inequality $|Sf(0)| \le 6$ then follows immediately from (6.5). Propagate this special case to other points of $\mathbb{D}$ using disk automorphisms.)

(30)   (Bernstein) Here is an interesting application of the Zhukovskii map [**PS**]. Consider a polynomial $P(z) = a_0 + a_1 z + \cdots + a_n z^n$, with $a_n \ne 0$. Let $M = \sup_{x \in [-1,1]} |P(x)|$.

   (i)   Show that
$$|P(z)| \le M (a + b)^n \qquad \text{for all } z \in \mathbb{C} \smallsetminus [-1, 1],$$

   where a and b are the major and minor semi-axes of the ellipse through z with the foci ± 1.

   (ii)   Show that
$$2^{-n} |a_n| \le M.$$

(Hint: For (i), note that the rational function $w \mapsto Z(w)/2 = (w + w^{-1})/2$ maps the region $|w| > 1$ conformally to $\mathbb{C} \smallsetminus [-1, 1]$, sending the family of circles $|w| = \text{const.}$ to the family of ellipses with the foci ± 1 (compare Example 6.6). The function $f(w) = w^{-n}P(Z(w)/2)$ is holomorphic for $|w| > 1$, remains bounded as $w \to \infty$, and has its modulus bounded by M on the circle $|w| = 1$. Apply a suitable maximum principle to f. For (ii), let $z \to \infty$ in (i).)

(31)   The identity $Z(w^2) = Z(w)^2 - 2$ is easily verified from the definition of the Zhukovskii map. More generally, prove that for every $n \ge 1$ there is a degree n monic polynomial P_n with real coefficients which satisfies the identity

$$Z(w^n) = P_n(Z(w)) \qquad \text{for all } w \in \hat{\mathbb{C}}.$$

Verify that P_n maps $[-2, 2]$ onto itself and satisfies

$$P_n(2 \cos \theta) = 2 \cos(n\theta) \qquad \text{for all } \theta \in \mathbb{R}.$$

The normalized map $T_n(x) = P_n(2x)/2$ is known as the **Chebyshev polynomial of degree** n.

7

Harmonic functions

This chapter introduces a basic theory of harmonic functions in planar domains. Although the subject can be developed without the intervention of complex analysis, the beautiful interplay between harmonic and holomorphic functions in dimension 2 provides a much quicker path to the essential results of the theory. It also illustrates some applications of the ideas developed in previous chapters.

7.1 Elementary properties of harmonic functions

Recall that the **Laplace operator**, or simply the **Laplacian**, acting on C^2-smooth complex-valued functions is defined by

$$\Delta = \frac{\partial^2}{\partial x^2} + \frac{\partial^2}{\partial y^2}.$$

It is easy to express Δ in terms of the complex differential operators $\partial/\partial z$ and $\partial/\partial \bar{z}$ introduced in chapter 1:

$$(7.1) \qquad \Delta = 4\, \frac{\partial}{\partial z}\, \frac{\partial}{\partial \bar{z}} = 4\, \frac{\partial}{\partial \bar{z}}\, \frac{\partial}{\partial z}.$$

Pierre-Simon Laplace (1749–1827)

For example,

$$4\, \frac{\partial}{\partial z}\, \frac{\partial}{\partial \bar{z}} = \left(\frac{\partial}{\partial x} - i \frac{\partial}{\partial y} \right) \left(\frac{\partial}{\partial x} + i \frac{\partial}{\partial y} \right)$$

$$= \frac{\partial^2}{\partial x^2} + i \frac{\partial^2}{\partial x \partial y} - i \frac{\partial^2}{\partial y \partial x} + \frac{\partial^2}{\partial y^2} = \frac{\partial^2}{\partial x^2} + \frac{\partial^2}{\partial y^2}.$$

Here we have used the well-known fact that the mixed partial derivatives of C^2 functions are equal:

$$\frac{\partial^2}{\partial x \partial y} = \frac{\partial^2}{\partial y \partial x}.$$

The theorem on equality of the mixed partial derivatives is attributed to various people, including Clairaut, Euler, and Schwarz.

As usual, U will denote a non-empty open subset of $\mathbb{C}$ unless otherwise stated.

DEFINITION 7.1. A C^2 function $h : U \to \mathbb{C}$ is called ***harmonic*** if it satisfies the Laplace equation $\Delta h = 0$ throughout U.

Notice that we have not restricted harmonicity to real-valued functions. Linearity of Δ shows that the space of all harmonic functions in U is a complex vector space: If h_1 and h_2 are harmonic in U, so is $c_1 h_1 + c_2 h_2$ for every $c_1, c_2 \in \mathbb{C}$.

As in chapter 1, we will use the subscript notation for our differential operators when convenient. Let us call a function $f : U \to \mathbb{C}$ ***anti-holomorphic*** if $f_z = 0$ in U; equivalently, if $\bar{f} \in \mathcal{O}(U)$ (compare problem 5 in chapter 1).

THEOREM 7.2. *The following conditions on a C^2 function $h : U \to \mathbb{C}$ are equivalent:*

(i) *h is harmonic.*
(ii) *$\mathrm{Re}(h)$ and $\mathrm{Im}(h)$ are harmonic.*
(iii) *$\bar{h}$ is harmonic.*
(iv) *$z \mapsto h(\bar{z})$ is harmonic.*
(v) *$h_z = \frac{1}{2}(h_x - ih_y)$ is holomorphic.*
(vi) *$h_{\bar{z}} = \frac{1}{2}(h_x + ih_y)$ is anti-holomorphic.*

PROOF. The equivalence (i) $\Longleftrightarrow$ (ii) follows from

$$\Delta h = \Delta \, \mathrm{Re}(h) + i \Delta \, \mathrm{Im}(h)$$

and the fact that $\Delta \, \mathrm{Re}(h)$ and $\Delta \, \mathrm{Im}(h)$ are real-valued functions. Similarly, (ii) $\Longleftrightarrow$ (iii) follows from

$$\Delta \bar{h} = \Delta \, \mathrm{Re}(h) - i \Delta \, \mathrm{Im}(h).$$

Think of h as a function of two real variables x and y, so $h(\bar{z})$ can be identified with the function $\psi(x, y) = h(x, -y)$. Then,

$$\Delta \psi \, (x, y) = \psi_{xx}(x, y) + \psi_{yy}(x, y) = h_{xx}(x, -y) + h_{yy}(x, -y) = \Delta h \, (x, -y),$$

which proves the equivalence (i) $\Longleftrightarrow$ (iv).

By (7.1), $\Delta h = 0$ if and only if $h_{z\bar{z}} = (h_z)_{\bar{z}} = 0$. By the complex Cauchy-Riemann equation (Theorem 1.13), this happens precisely when h_z is holomorphic. This shows the equivalence (i) $\Longleftrightarrow$ (v). Similarly, $\Delta h = 0$ if and only if $h_{\bar{z}z} = (h_{\bar{z}})_z = 0$, which is the case precisely when $h_{\bar{z}}$ is anti-holomorphic, proving (i) $\Longleftrightarrow$ (vi). $\qquad\square$

EXAMPLE 7.3. Direct computation of Δu shows that $u(x, y) = e^x \cos y$ is harmonic in the plane. Alternatively, we can deduce harmonicity by observing that

$$u_z = \frac{1}{2}(u_x - iu_y) = \frac{1}{2}(e^x \cos y + ie^x \sin y) = \frac{1}{2} e^z$$

is holomorphic.

THEOREM 7.4.

  (i) *A holomorphic or anti-holomorphic function is harmonic.*
  (ii) *If h is harmonic and f is holomorphic or anti-holomorphic, then h ∘ f (wherever it is defined) is harmonic.*

PROOF. If h is holomorphic, so is $h_z = h'$, so h is harmonic by Theorem 7.2. If h is anti-holomorphic, then $\bar{h}$ is holomorphic and therefore harmonic, so h is harmonic, again by Theorem 7.2. This proves (i).

For (ii), first assume f is holomorphic. By the chain rule,

$$(h \circ f)_z = (h_z \circ f)f_z + (h_{\bar{z}} \circ f)\bar{f}_z = (h_z \circ f)f_z,$$

where we have used $\bar{f}_z = 0$ since f is holomorphic. It follows that $(h \circ f)_z$ is holomorphic because f, $f_z = f'$, and h_z are. By Theorem 7.2, $h \circ f$ is harmonic.

When f is anti-holomorphic, $h \circ f$ is the composition of the holomorphic function $\bar{f}$ followed by $z \mapsto h(\bar{z})$, which is harmonic by Theorem 7.2. It follows from the first case above that $h \circ f$ is harmonic.   □

COROLLARY 7.5. *The real and imaginary parts of a holomorphic function are harmonic.*

EXAMPLE 7.6. The function

$$u(z) = \begin{cases} \operatorname{Im}\left(\dfrac{1}{z^2}\right) & z \neq 0 \\ 0 & z = 0 \end{cases}$$

is harmonic in the punctured plane $\mathbb{C}^*$ since it is the imaginary part of the holomorphic function $1/z^2$. The restriction of u to the coordinate axes $x = 0$ and $y = 0$ is identically zero, so the second partial derivatives u_{xx} and u_{yy} both vanish at the origin. It follows that $\Delta u = 0$ *everywhere* in the plane. However, u is not harmonic in the plane because it is not even continuous at the origin.

EXAMPLE 7.7. The function $u : \mathbb{C}^* \to \mathbb{R}$ defined by $u(z) = \log|z|$ is harmonic:

$$\Delta u = 4 \frac{\partial}{\partial z} \frac{\partial}{\partial \bar{z}} \log|z| = 2 \frac{\partial}{\partial z} \frac{\partial}{\partial \bar{z}} \log(z\bar{z}) = 2 \frac{\partial}{\partial z}\left(\frac{1}{\bar{z}}\right) = 0.$$

It follows from Theorem 7.4(ii) that *if $f \in \mathscr{O}(U)$ is non-vanishing, then $\log|f|$ is harmonic in U.*

In view of Corollary 7.5, it is natural to ask whether every real-valued harmonic function in a domain is the real part of a holomorphic function in that domain. The answer is generally negative, as the following example shows.

EXAMPLE 7.8. The harmonic function $u(z) = \log|z|$ in $\mathbb{C}^*$ is not the real part of a holomorphic function in any annulus $U = \{z : a < |z| < b\}$ where $0 \leq a < b \leq +\infty$. Assuming there is an

$f \in \mathcal{O}(U)$ with $u = \mathrm{Re}(f)$, consider the function $g(z) = f(e^z)$ which is holomorphic in the vertical strip $\{z \in \mathbb{C} : \log a < \mathrm{Re}(z) < \log b\}$ and satisfies $\mathrm{Re}(g(z)) = \mathrm{Re}(z)$. The difference $g(z) - z$ is holomorphic in this strip and has vanishing real part, so it must be constant by the open mapping theorem, that is, $f(e^z) = z + c$ for some (purely imaginary) constant c. But this is impossible since, as a function of z, the left side is periodic with period $2\pi i$ and the right side is not. The contradiction shows that no such f exists.

Here is an alternative argument which is well adapted to more general situations: Assuming $u = \mathrm{Re}(f)$ for some $f \in \mathcal{O}(U)$, use $u = (f + \bar{f})/2$ and the Cauchy-Riemann equations to deduce

$$u_z = \frac{1}{2}(f_z + \bar{f}_z) = \frac{1}{2}f'.$$

It follows from Theorem 1.29 that $\int_{\mathbb{T}(0,r)} u_z\, dz = 0$ for every $a < r < b$. This is a contradiction since

$$\int_{\mathbb{T}(0,r)} u_z\, dz = \frac{1}{2}\int_{\mathbb{T}(0,r)} \frac{dz}{z} = \pi i.$$

By contrast, every real-valued harmonic function is *locally* the real part of a holomorphic function. In fact, this holds in every simply connected domain:

THEOREM 7.9. *Suppose $U \subset \mathbb{C}$ is a simply connected domain and $u : U \to \mathbb{R}$ is harmonic. Then there exists an $f \in \mathcal{O}(U)$ such that $u = \mathrm{Re}(f)$ in U. Such f is a primitive of the holomorphic function $2u_z$ and is unique up to addition of a purely imaginary constant.*

PROOF. The function $2u_z$ is holomorphic by Theorem 7.2, so it has a primitive $g \in \mathcal{O}(U)$ by Theorem 2.46. From $(g - 2u)_z = g' - 2u_z = 0$ we see that $g - 2u$ is antiholomorphic. Taking complex conjugates and noting that u is real-valued, it follows that $\bar{g} - 2u \in \mathcal{O}(U)$. Thus, the sum $g + \bar{g} - 2u$ is holomorphic in U. Since this function is real-valued, the open mapping theorem shows that it must be a constant $2c$. It follows that u is the real part of the holomorphic function $f = g - c$. If $\tilde{f} \in \mathcal{O}(U)$ too has its real part equal to u, the holomorphic function $f - \tilde{f}$ has vanishing real part, so it is a purely imaginary constant, again by the open mapping theorem. ☐

The definition of a harmonic function assumed only C^2 smoothness. But since by Theorem 7.9 every real-valued harmonic function is locally the real part of a holomorphic function, it now follows that

COROLLARY 7.10. *Harmonic functions are C^∞-smooth.*

COROLLARY 7.11 (The identity theorem for harmonic functions). *Suppose $U \subset \mathbb{C}$ is a domain and $h, \tilde{h} : U \to \mathbb{C}$ are harmonic. If $h = \tilde{h}$ in a non-empty open subset of U, then $h = \tilde{h}$ everywhere in U.*

PROOF. It suffices to consider the case where h is real-valued and $\tilde{h} = 0$; the general case follows by considering the real and imaginary parts of $h - \tilde{h}$. Let $V \neq \emptyset$ be the maximal open subset of U in which $h = 0$. If $V \neq U$, choose a point $p \in \partial V \cap U$ and a disk $\mathbb{D}(p, r) \subset U$. Since disks are simply connected, Theorem 7.9 gives a function $f \in \mathcal{O}(\mathbb{D}(p, r))$ such that $h = \mathrm{Re}(f)$. Since f takes purely imaginary values in the non-empty open set $\mathbb{D}(p, r) \cap V$, it must be constant in this set by the open mapping theorem, hence constant in the disk $\mathbb{D}(p, r)$ by the identity theorem. This shows that h vanishes in the open set $\mathbb{D}(p, r) \cup V$, contradicting maximality of V. It follows that $V = U$ and h vanishes identically in U. □

REMARK 7.12. Unlike the case of holomorphic functions, the above theorem is no longer true under the weaker assumption that the harmonic functions match along a set with accumulation points. For example, the harmonic function $\mathrm{Re}(z)$ vanishes along the imaginary axis without being identically zero.

A modified form of Theorem 7.9 holds for some non-simply connected domains. Here is a statement for the case of a round annulus; for the generalization to all "finitely connected" domains, see Theorem 9.32.

THEOREM 7.13. *Let U be the annulus $\{z \in \mathbb{C} : a < |z - p| < b\}$, where $0 \leq a < b \leq +\infty$. For any harmonic function $u : U \to \mathbb{R}$ there is a constant $\alpha \in \mathbb{R}$ and a function $f \in \mathcal{O}(U)$ such that*

$$u(z) = \mathrm{Re}(f(z)) + \alpha \log |z - p| \qquad \text{for all } z \in U.$$

Moreover, α is unique but f is determined only up to addition of a purely imaginary constant.

The real constant α is called the **period** of u in U.

PROOF. The argument is a variation of the proof of Theorem 7.9. Fix $a < r < b$ and define

$$\alpha = \frac{1}{\pi i} \int_{\mathbb{T}(p,r)} u_z \, dz.$$

Since u is harmonic, the function u_z is holomorphic in U, so by Cauchy's theorem the value of α depends only on the homology class of $\mathbb{T}(p, r)$ in U. In particular, it is independent of r. Since

$$\alpha = \frac{1}{2\pi i} \int_{\mathbb{T}(p,r)} (u_x - iu_y) \, (dx + idy),$$

we can express $\mathrm{Im}(\alpha)$ as the classical line integral

$$-\frac{1}{2\pi} \int_{\mathbb{T}(p,r)} (u_x \, dx + u_y \, dy) = -\frac{1}{2\pi} \int_{\mathbb{T}(p,r)} du$$

which vanishes since $\mathbb{T}(p, r)$ is a closed curve. Thus, $\alpha \in \mathbb{R}$.

Now consider the function $2u_z - \alpha/(z-p)$ which is holomorphic in U. If γ is any closed curve in U and $n = \mathrm{W}(\gamma, p)$, then the cycle $\gamma - n\mathbb{T}(p,r)$ is null-homologous in U. Hence, by Cauchy's theorem,

$$\int_\gamma \left(2u_z - \frac{\alpha}{z-p}\right) dz = n \int_{\mathbb{T}(p,r)} \left(2u_z - \frac{\alpha}{z-p}\right) dz$$

$$= 2n \int_{\mathbb{T}(p,r)} u_z\, dz - n\alpha \int_{\mathbb{T}(p,r)} \frac{dz}{z-p}$$

$$= 2n \cdot \pi i\alpha - n\alpha \cdot 2\pi i = 0.$$

It follows from Theorem 1.29 that $2u_z - \alpha/(z-p)$ has a primitive $g \in \mathscr{O}(U)$. The computation

$$\frac{\partial}{\partial z}\Big(g - 2u + 2\alpha \log|z-p|\Big) = \frac{\partial}{\partial z}\Big(g - 2u + \alpha \log\big((z-p)(\bar{z}-\bar{p})\big)\Big)$$

$$= g' - 2u_z + \frac{\alpha}{z-p} = 0$$

shows that the function $g - 2u + 2\alpha \log|z-p|$ is anti-holomorphic, so its complex conjugate $\bar{g} - 2u + 2\alpha \log|z-p|$ is holomorphic in U. Thus, the sum

$$g + \bar{g} - 2u + 2\alpha \log|z-p| = 2(\mathrm{Re}(g) - u + \alpha \log|z-p|)$$

is also holomorphic in U and since it is real-valued, it must be a constant $2c$ by the open mapping theorem. The function $f = g - c \in \mathscr{O}(U)$ now satisfies $\mathrm{Re}(f) = u - \alpha \log|z-p|$.

If $\mathrm{Re}(f(z)) + \alpha \log|z-p| = \mathrm{Re}(\tilde{f}(z)) + \tilde{\alpha} \log|z-p|$ for some $f, \tilde{f} \in \mathscr{O}(U)$ and $\alpha, \tilde{\alpha} \in \mathbb{R}$, then $(\alpha - \tilde{\alpha}) \log|z-p| = \mathrm{Re}(\tilde{f}(z) - f(z))$ for all $z \in U$. By Example 7.8, $\alpha - \tilde{\alpha} = 0$ and therefore $\tilde{f} - f$ is a purely imaginary constant by the open mapping theorem. □

Here is an application of the above result:

THEOREM 7.14 (Removable singularity theorem for harmonic functions). *A bounded harmonic function in the punctured disk $\mathbb{D}^*(p,r)$ extends to a harmonic function in $\mathbb{D}(p,r)$.*

PROOF. It suffices to consider real-valued functions. Suppose $u : \mathbb{D}^*(p,r) \to \mathbb{R}$ is bounded and harmonic. By Theorem 7.13, there is an $f \in \mathscr{O}(\mathbb{D}^*(p,r))$ and a real constant α such that

$$\mathrm{Re}(f(z)) = u(z) - \alpha \log|z-p| \qquad \text{if } 0 < |z-p| < r.$$

We show that f has a removable singularity at p. This implies that $\alpha \log|z-p| = u(z) - \mathrm{Re}(f(z))$ stays bounded as $z \to p$, which forces $\alpha = 0$. The harmonic function $\mathrm{Re}(f)$ in $\mathbb{D}(p,r)$ will then provide the extension of u.

Assume by way of contradiction that p is not removable for f, so it is either a pole or an essential singularity. In the first case, the image $f(\mathbb{D}^*(p,r))$ is a punctured neighborhood of ∞ by the open mapping theorem. Hence there are sequences $z_n, w_n \to p$ such that $f(z_n) = n$ and $f(w_n) = -n$ for all large n, so $\exp(f(z_n)) \to \infty$ and $\exp(f(w_n)) \to 0$. In the second case, $f(\mathbb{D}^*(p,r))$ is dense in $\mathbb{C}$ by the Casorati-Weierstrass theorem. Hence there are sequences $z_n, w_n \to p$ such that $f(z_n) \to 0$ and $f(w_n) \to 1$, so $\exp(f(z_n)) \to 1$ and $\exp(f(w_n)) \to e$. In either case, $\lim_{z \to p} \exp(f(z))$ fails to exist and we conclude that p must be an essential singularity of $\exp(f)$. But the equation

$$| \exp(f(z))| = \exp(u(z)) \, |z - p|^{-\alpha}$$

together with the assumption that u is bounded shows that $\exp(f)$ has at worst a pole at p. □

DEFINITION 7.15. A continuous function $h : U \to \mathbb{C}$ is said to have the ***mean value property*** if

$$(7.2) \qquad h(p) = \frac{1}{2\pi} \int_0^{2\pi} h(p + re^{it}) \, dt$$

whenever $\overline{\mathbb{D}}(p, r) \subset U$. We say that h has the ***local mean value property*** if for every $p \in U$ there is a $\delta > 0$ such that (7.2) holds whenever $0 < r < \delta$.

The mean value property is a priori stronger than the local mean value property, but we will see in Theorem 7.32 that both conditions are indeed equivalent to harmonicity. One direction of this statement is easy to prove:

THEOREM 7.16. *Harmonic functions have the mean value property.*

PROOF. It suffices to consider real-valued functions; the general case follows by taking the real and imaginary parts. Suppose $u : U \to \mathbb{R}$ is harmonic and $\overline{\mathbb{D}}(p, r) \subset U$. Choose $s > r$ such that $\mathbb{D}(p, s) \subset U$. By Theorem 7.9 there is an $f \in \mathcal{O}(\mathbb{D}(p, s))$ such that $u = \operatorname{Re}(f)$. Using the parametrization $\zeta(t) = p + re^{it}$ for the oriented circle $\mathbb{T}(p, r)$ and applying Cauchy's integral formula, we see that

$$f(p) = \frac{1}{2\pi i} \int_{\mathbb{T}(p,r)} \frac{f(\zeta)}{\zeta - p} \, d\zeta = \frac{1}{2\pi i} \int_0^{2\pi} \frac{f(p + re^{it})}{re^{it}} \, ire^{it} \, dt$$

$$= \frac{1}{2\pi} \int_0^{2\pi} f(p + re^{it}) \, dt.$$

The mean value property now follows by taking the real part of both sides of this equation. □

THEOREM 7.17 (The maximum principle for harmonic functions). *Suppose $U \subset \mathbb{C}$ is a bounded domain, and h is continuous on $\overline{U}$ and harmonic in U.*

(i) If h is real-valued, then

$$(7.3) \qquad \inf_{\zeta \in \partial U} h(\zeta) \leq h(z) \leq \sup_{\zeta \in \partial U} h(\zeta) \qquad \text{for all } z \in U.$$

(ii) If h is complex-valued, then

$$|h(z)| \leq \sup_{\zeta \in \partial U} |h(\zeta)| \qquad \text{for all } z \in U.$$

In either case, if equality occurs at some $z \in U$, then h is constant.

PROOF. (i) Suppose $h(z_0) \geq \sup_{\zeta \in \partial U} h(\zeta)$ for some $z_0 \in U$. Then the supremum of h on the compact set $\overline{U}$ occurs at some point $p \in U$. Consider the non-empty set $E = \{z \in U : h(z) = h(p)\}$. The local mean value property of h, guaranteed by Theorem 7.16, shows that for any $z \in E$ there is a $\delta > 0$ such that if $0 < r < \delta$,

$$\int_0^{2\pi} \left(h(p) - h(z + re^{it}) \right) dt = 2\pi h(p) - 2\pi h(z) = 0.$$

Since the integrand is continuous and non-negative, it must be identically zero. In other words, $h(z + re^{it}) = h(p)$ for every $0 < r < \delta$ and $0 \leq t \leq 2\pi$, or $\mathbb{D}(z, \delta) \subset E$. This shows that E is open. Since E is clearly closed by continuity of h, and since U is connected, we conclude that $E = U$, so h is constant in U. The left inequality in (7.3) follows by applying the same argument to $-h$.

(ii) Now suppose h is complex-valued and $|h(z_0)| \geq \sup_{\zeta \in \partial U} |h(\zeta)|$ for some $z_0 \in U$. Then the supremum of $|h|$ on $\overline{U}$ occurs at some point $p \in U$. If $h(p) = 0$, then h vanishes identically and there is nothing to prove. If $h(p) \neq 0$, consider the continuous function $\psi(z) = h(z)/h(p)$ which is still harmonic. We have $|\psi| \leq 1$ and $\psi(p) = 1$, hence $\operatorname{Re}(\psi)$ takes its maximum value at p. Since $\operatorname{Re}(\psi)$ is harmonic, by the first case treated above, we must have $\operatorname{Re}(\psi) = 1$ everywhere. The conditions $|\psi| \leq 1$ and $\operatorname{Re}(\psi) = 1$ together imply that $\psi(z) = 1$, or $h(z) = h(p)$, for all $z \in U$. $\qquad \square$

REMARK 7.18. An examination of the above proof shows that the same argument goes through only assuming the local mean value property of h. This observation will play a role in the proof of Theorem 7.32.

There are more general versions of the maximum principle in which the domains are not necessarily bounded and continuity along the boundary is not assumed. Below we will discuss one result of this type (see problem 20 for another). Recall from §6.3 that a sequence $\{z_n\}$ in a domain U is escaping if it has no accumulation point in U. Equivalently, if the spherical distance between z_n and the boundary ∂U (taken in $\hat{\mathbb{C}}$) tends to 0 as $n \to \infty$.

THEOREM 7.19. *Suppose h is harmonic in a domain $U \subset \mathbb{C}$.*

(i) If h is real-valued and

$$\limsup_{n \to \infty} h(z_n) \leq M \quad \text{for every escaping sequence } \{z_n\} \text{ in } U,$$

then $h \leq M$ in U.

(ii) If h is complex-valued and

$$\limsup_{n \to \infty} |h(z_n)| \leq M \quad \text{for every escaping sequence } \{z_n\} \text{ in } U,$$

then $|h| \leq M$ in U.

PROOF. (i) We prove that for every $\varepsilon > 0$ the open set $V_\varepsilon = \{z \in U : h(z) > M + \varepsilon\}$ is empty. Suppose $V_\varepsilon \neq \emptyset$ for some $\varepsilon > 0$. Then the closure $\overline{V_\varepsilon}$ would be a compact subset of U; otherwise we could find a sequence $\{z_n\}$ in V_ε which is escaping in U, and this would imply the absurd inequality

$$M + \varepsilon \leq \limsup_{n \to \infty} h(z_n) \leq M.$$

Thus, $\overline{V_\varepsilon} \subset U$ and continuity of h shows that $h = M + \varepsilon$ on ∂V_ε. Applying Theorem 7.17 to each connected component of V_ε, we see that for every $z \in V_\varepsilon$,

$$M + \varepsilon < h(z) \leq \sup_{\zeta \in \partial V_\varepsilon} h(\zeta) = M + \varepsilon,$$

which is a contradiction.

(ii) We can of course repeat the above argument for $|h|$. Alternatively, observe that if $|\alpha| = 1$, then $\text{Re}(\alpha h)$ is harmonic and for every escaping sequence $\{z_n\}$ in U,

$$\limsup_{n \to \infty} \text{Re}(\alpha h(z_n)) \leq \limsup_{n \to \infty} |\alpha h(z_n)| = \limsup_{n \to \infty} |h(z_n)| \leq M.$$

It follows from part (i) that $\text{Re}(\alpha h) \leq M$ in U. Since this holds for every α on the unit circle, we must have $|h| \leq M$ in U. ∎

EXAMPLE 7.20. Suppose f is holomorphic and non-vanishing in $\mathbb{D}$, with $|f(z)| \to 1$ as $|z| \to 1$. Applying Theorem 7.19 to f and $1/f$ shows that $|f| = 1$ everywhere in $\mathbb{D}$. It follows from the open mapping theorem that f is constant.

EXAMPLE 7.21. The function $h(z) = \text{Im}(z)$ is harmonic in the upper half-plane $\mathbb{H}$ and tends to 0 as z tends to any boundary point on $\mathbb{R}$. However, the conclusion $h \leq 0$ is certainly false since the opposite inequality $h > 0$ holds in $\mathbb{H}$. This does not contradict Theorem 7.19 since there are escaping sequences $\{z_n\}$ in $\mathbb{H}$ which tend to ∞ (such as $z_n = n + i$) for which $\limsup_{n \to \infty} h(z_n) \leq 0$ fails.

7.2 Poisson's formula in a disk

We begin with a fundamental representation theorem for harmonic functions in the unit disk, which can be viewed as a generalization of the mean value property or an analog of Cauchy's integral formula.

THEOREM 7.22 (The Poisson integral formula). *If h is continuous on $\overline{\mathbb{D}}$ and harmonic in $\mathbb{D}$, then*

$$(7.4) \qquad h(z) = \frac{1}{2\pi} \int_0^{2\pi} \frac{1-|z|^2}{|e^{it}-z|^2} h(e^{it})\, dt \qquad \text{for all } z \in \mathbb{D}.$$

The formula reveals the remarkable property that the values of h inside the disk are uniquely determined by the restriction $h|_{\mathbb{T}}$ to the boundary circle. One can interpret (7.4) as saying that for any $z \in \mathbb{D}$, the value $h(z)$ is an average of the values of $h|_{\mathbb{T}}$, weighted by the factor $(1-|z|^2)/|e^{it}-z|^2$ (for $z=0$ this factor is 1 and we recover the mean value property). The importance of this factor will become abundantly clear in this chapter.

Siméon Denis Poisson
(1781–1840)

PROOF. Fix $z \in \mathbb{D}$ and consider the disk automorphism

$$w = \varphi(\zeta) = \frac{\zeta - z}{1 - \bar{z}\zeta} \quad \text{with the inverse} \quad \zeta = \varphi^{-1}(w) = \frac{w+z}{1+\bar{z}w}.$$

The function $\tilde{h} = h \circ \varphi^{-1}$ is continuous on $\overline{\mathbb{D}}$ and harmonic in $\mathbb{D}$ (Theorem 7.4(ii)). The mean value property of $\tilde{h}$ gives

$$\tilde{h}(0) = \frac{1}{2\pi} \int_0^{2\pi} \tilde{h}(re^{it})\, dt$$

for every $0 < r < 1$. Using the fact that $\tilde{h}(re^{it}) \to \tilde{h}(e^{it})$ uniformly on $[0, 2\pi]$ as $r \to 1$, we obtain

$$h(z) = \tilde{h}(0) = \frac{1}{2\pi} \int_0^{2\pi} \tilde{h}(e^{it})\, dt = \frac{1}{2\pi} \int_{\mathbb{T}} \tilde{h}(w)\, |dw|$$

$$= \frac{1}{2\pi} \int_{\mathbb{T}} h(\zeta)\, |\varphi'(\zeta)|\, |d\zeta|.$$

But when $|\zeta| = 1$,

$$|\varphi'(\zeta)| = \frac{1-|z|^2}{|1-\bar{z}\zeta|^2} = \frac{1-|z|^2}{|\zeta(\bar{\zeta}-\bar{z})|^2} = \frac{1-|z|^2}{|\zeta - z|^2}.$$

It follows that

$$h(z) = \frac{1}{2\pi} \int_{\mathbb{T}} h(\zeta) \frac{1-|z|^2}{|\zeta-z|^2}\, |d\zeta| = \frac{1}{2\pi} \int_0^{2\pi} \frac{1-|z|^2}{|e^{it}-z|^2} h(e^{it})\, dt. \qquad \square$$

By Theorem 7.9, every harmonic function in $\mathbb{D}$ is the real part of an essentially unique holomorphic function. Using (7.4) we can find an explicit formula for this holomorphic function, at least when the given real part extends continuously to the boundary circle:

COROLLARY 7.23. *If u is real-valued, continuous on $\overline{\mathbb{D}}$, and harmonic in $\mathbb{D}$, then u is the real part of the holomorphic function f defined by*

(7.5)
$$f(z) = \frac{1}{2\pi} \int_0^{2\pi} \frac{e^{it} + z}{e^{it} - z} u(e^{it}) \, dt \qquad \text{for } z \in \mathbb{D}.$$

This formula is attributed to Schwarz.

PROOF. That f is holomorphic follows from Theorem 1.46. The relation $\mathrm{Re}(f) = u$ follows from the Poisson integral formula for u and the computation

$$\mathrm{Re}\left(\frac{e^{it} + z}{e^{it} - z}\right) = \mathrm{Re}\left(\frac{(e^{it} + z)(e^{-it} - \bar{z})}{|e^{it} - z|^2}\right) = \frac{1 - |z|^2}{|e^{it} - z|^2}. \qquad \square$$

We now study the weight factor that appeared in (7.4):

DEFINITION 7.24. The **Poisson kernel** in the unit disk $\mathbb{D}$ is defined by

(7.6)
$$P(\zeta, z) = \mathrm{Re}\left(\frac{\zeta + z}{\zeta - z}\right) = \frac{1 - |z|^2}{|\zeta - z|^2},$$

where ζ, z are complex numbers with $|\zeta| = 1$ and $|z| < 1$.

We often write $\zeta = e^{it}$, in which case (7.6) takes the form

(7.7)
$$P(e^{it}, z) = \mathrm{Re}\left(\frac{e^{it} + z}{e^{it} - z}\right) = \frac{1 - |z|^2}{|e^{it} - z|^2}.$$

If we use the polar form $z = re^{i\theta}$ where $0 \leq r < 1$, then

(7.8)
$$P(e^{it}, z) = \frac{1 - r^2}{|1 - re^{-i(t-\theta)}|^2} = \frac{1 - r^2}{1 - 2r\cos(t - \theta) + r^2}.$$

We may think of the Poisson kernel either as a family of 2π-periodic functions $t \mapsto P(e^{it}, z)$ parametrized by the point $z \in \mathbb{D}$, or as a family of functions $z \mapsto P(e^{it}, z)$ in the unit disk parametrized by the point $e^{it} \in \mathbb{T}$. Both points of views turn out to be useful in studying harmonic functions in the disk. Observe that because of the relation

$$P(e^{it}, z) = P(1, e^{-it}z)$$

the functions $z \mapsto P(e^{it}, z)$ all differ from $z \mapsto P(1, z)$ by a rotation.

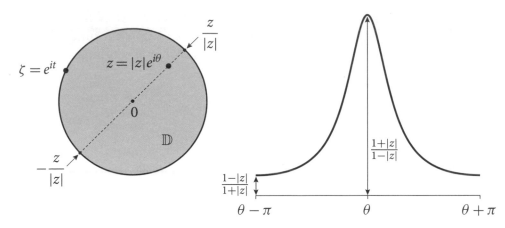

Figure 7.1. Left: The size of the Poisson kernel $P(\zeta, z)$ for fixed $z = |z|e^{i\theta} \in \mathbb{D}$ depends on how close $\zeta = e^{it}$ gets to z as it varies on the unit circle. Right: The graph of $t \mapsto P(e^{it}, z)$ over one period, with a maximum at $t = \theta$ and a minimum at $t = \theta \pm \pi$.

THEOREM 7.25. *Fix $z = |z|e^{i\theta} \in \mathbb{D}$.*

(i) *The function $t \mapsto P(e^{it}, z)$ is positive and 2π-periodic; its graph is symmetric about the line $t = \theta$.*

(ii) $\displaystyle \max_{t \in \mathbb{R}} P(e^{it}, z) = \frac{1 + |z|}{1 - |z|}$ *and* $\displaystyle \min_{t \in \mathbb{R}} P(e^{it}, z) = \frac{1 - |z|}{1 + |z|}.$

(iii) $\displaystyle \frac{1}{2\pi} \int_0^{2\pi} P(e^{it}, z)\, dt = 1.$

PROOF. Positivity of $t \mapsto P(e^{it}, z)$ follows from (7.7) and its symmetry follows from (7.8) since it is an even function of $t - \theta$. This proves (i).

To verify (ii), we may assume that $z \neq 0$ since $P(\zeta, 0) = 1$ for all ζ. By (7.6), $\zeta \mapsto P(\zeta, z)$ takes on its maximum value when $|\zeta - z|$ reaches its minimum value of $1 - |z|$ at $\zeta = z/|z|$. Similarly, $\zeta \mapsto P(\zeta, z)$ takes on its minimum value when $|\zeta - z|$ reaches its maximum value of $1 + |z|$ at $\zeta = -z/|z|$ (compare Fig. 7.1). Substituting these values into the formula for $P(\zeta, z)$ proves (ii).

Part (iii) follows from the Poisson integral formula (7.4) applied to the constant function $h = 1$. □

We now turn to the properties of $P(\zeta, z)$ as a function of z.

THEOREM 7.26. *Fix $\zeta_0 \in \mathbb{T}$.*

(i) *The function $z \mapsto P(\zeta_0, z)$ is harmonic in $\mathbb{D}$.*

(ii) $\displaystyle \lim_{z \to \zeta_0} P(\zeta, z) = 0$ *uniformly on compact subsets of $\mathbb{T} \smallsetminus \{\zeta_0\}$.*

(iii) *For every $c \in [0, +\infty]$ there is a sequence $\{z_n\}$ in $\mathbb{D}$ such that $z_n \to \zeta_0$ and $\displaystyle \lim_{n \to \infty} P(\zeta_0, z_n) = c.$*

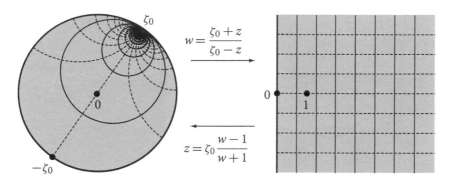

Figure 7.2. Under the Möbius map $w = (\zeta_0 + z)/(\zeta_0 - z)$ the Poisson kernel $P(\zeta_0, z)$ in $\mathbb{D}$ becomes $\mathrm{Re}(w)$ in the right half-plane. Hence $P(\zeta_0, z)$ takes a constant value on every circle in $\mathbb{D}$ that is tangent to $\mathbb{T}$ at ζ_0, with the constant getting larger as the circle shrinks. On the other hand, $P(\zeta_0, z)$ tends $+\infty$ as $z \to \zeta_0$ along a circular arc in $\mathbb{D}$ that is orthogonal to $\mathbb{T}$ at ζ_0.

Since $t \mapsto P(e^{it}, z)$ is positive with average value 1 over the circle by Theorem 7.25, property (ii) above implies that as z tends to $\zeta_0 = e^{it_0}$, the graph of $t \mapsto P(e^{it}, z)$ develops a high spike that is increasingly concentrated near the point t_0.

PROOF. The function $z \mapsto P(\zeta_0, z)$ is the real part of the holomorphic function $z \mapsto (\zeta_0 + z)/(\zeta_0 - z)$, hence is harmonic in $\mathbb{D}$, and (i) holds.

Statement (ii) follows from the fact that as $z \to \zeta_0$, $1 - |z|^2 \to 0$ but $|\zeta - z|^2$ is uniformly bounded away from 0 on compact subsets of $\mathbb{T} \setminus \{\zeta_0\}$. More precisely, let $K \subset \mathbb{T} \setminus \{\zeta_0\}$ be compact and $d > 0$ be the distance between ζ_0 and K. Given $\varepsilon > 0$, let $\delta = \min\{\varepsilon, d/2\}$. Then, if $z \in \mathbb{D}(\zeta_0, \delta) \cap \mathbb{D}$ and $\zeta \in K$,

$$1 - |z|^2 = (1 + |z|)(1 - |z|) < 2\delta \leq 2\varepsilon$$

while

$$|\zeta - z| \geq |\zeta - \zeta_0| - |z - \zeta_0| > d - \delta \geq \frac{d}{2}.$$

Hence,

$$P(\zeta, z) = \frac{1 - |z|^2}{|\zeta - z|^2} < \frac{8\varepsilon}{d^2}.$$

To see (iii), first note that the Möbius map $w = (\zeta_0 + z)/(\zeta_0 - z)$ carries $\mathbb{D}$ conformally to the right half-plane $U = \{w \in \mathbb{C} : \mathrm{Re}(w) > 0\}$, sending ζ_0 to ∞, $-\zeta_0$ to 0, and 0 to 1. The Poisson kernel $P(\zeta_0, z)$ in $\mathbb{D}$ then corresponds to $\mathrm{Re}(w)$ in U. The family of vertical lines in U pulls back under $z = w^{-1} = \zeta_0(w - 1)/(w + 1)$ to the family of circles in $\mathbb{D}$ that are tangent to $\mathbb{T}$ at ζ_0 (see Fig. 7.2). If $z \to \zeta_0$ along the preimage of the vertical line $\mathrm{Re}(w) = c \in (0, +\infty)$, it follows that $P(\zeta_0, z) = c$. On the other hand, the family of horizontal lines in U pulls back to the family of circular arcs in $\mathbb{D}$ that meet $\mathbb{T}$ orthogonally at ζ_0. If $z \to \zeta_0$ along such circular arcs, it follows that $P(\zeta_0, z) \to +\infty$. Finally, any sequence $\{w_n\}$ in the right half-plane with the property $\mathrm{Re}(w_n) \to 0$ and $|\mathrm{Im}(w_n)| \to +\infty$ pulls back to a sequence $\{z_n\}$ in $\mathbb{D}$ which tends to ζ_0 such that $P(\zeta_0, z_n) \to 0$ as $n \to \infty$. $\qquad\square$

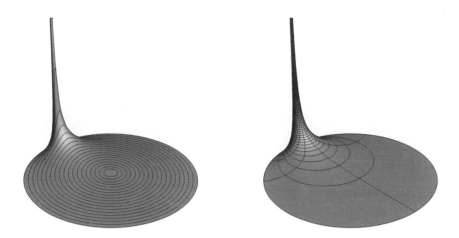

Figure 7.3. Two views of the graph of the Poisson kernel $P(\zeta_0, z)$ over the unit disk $|z| < 1$ for a fixed $\zeta_0 \in \mathbb{T}$. On the left the graph is traced along the circles $|z| = \text{const}$. On the right it is traced along the family of orthogonal circles of Fig. 7.2. Notice that $P(\zeta_0, z) \to 0$ as z tends to any point of the unit circle other than ζ_0 where there is a spike. On the other hand, $P(\zeta_0, z)$ can get arbitrarily close to any number in $[0, +\infty]$ as $z \to \zeta_0$.

Fig. 7.3 further illustrates the behavior of $z \mapsto P(\zeta_0, z)$ on the unit disk.

The Poisson integral formula suggests a way of manufacturing harmonic functions in the disk from integrable functions on the circle. Let us denote by $L^1(\mathbb{T})$ the space of complex-valued Lebesgue integrable functions on the circle, equipped with the norm

$$\|h\|_1 = \frac{1}{2\pi} \int_0^{2\pi} |h(e^{it})| \, dt.$$

DEFINITION 7.27. The *Poisson integral* of $h \in L^1(\mathbb{T})$ is defined by

$$\mathcal{P}[h](z) = \frac{1}{2\pi} \int_{\mathbb{T}} P(\zeta, z) \, h(\zeta) \, |d\zeta|$$

$$= \frac{1}{2\pi} \int_0^{2\pi} \frac{1 - |z|^2}{|e^{it} - z|^2} \, h(e^{it}) \, dt \qquad (z \in \mathbb{D}).$$

THEOREM 7.28. *For every $h \in L^1(\mathbb{T})$, the Poisson integral $\mathcal{P}[h]$ is harmonic in $\mathbb{D}$.*

PROOF. First suppose h is real-valued and consider the function

$$f(z) = \frac{1}{2\pi} \int_0^{2\pi} \frac{e^{it} + z}{e^{it} - z} \, h(e^{it}) \, dt.$$

Then $f \in \mathcal{O}(\mathbb{D})$ by the analog of Theorem 1.46 (see problem 27 in chapter 1). Alternatively, use the expansion

$$\frac{e^{it}+z}{e^{it}-z} = \frac{1+e^{-it}z}{1-e^{-it}z} = (1+e^{-it}z) \sum_{n=0}^{\infty} e^{-int} z^n = 1 + 2 \sum_{n=1}^{\infty} e^{-int} z^n$$

which converges uniformly on $[0, 2\pi]$ for each $z \in \mathbb{D}$, and integrate term-by-term against h to obtain the power series representation

$$f(z) = a_0 + 2 \sum_{n=1}^{\infty} a_n z^n, \qquad \text{where} \quad a_n = \frac{1}{2\pi} \int_0^{2\pi} e^{-int} h(e^{it})\, dt$$

in $\mathbb{D}$, which shows $f \in \mathcal{O}(\mathbb{D})$. Since

$$\text{Re}(f(z)) = \frac{1}{2\pi} \int_0^{2\pi} \text{Re}\left(\frac{e^{it}+z}{e^{it}-z}\right) h(e^{it})\, dt = \mathcal{P}[h](z),$$

Corollary 7.5 shows that $\mathcal{P}[h]$ is harmonic in $\mathbb{D}$. The case of a complex-valued h follows immediately since $\mathcal{P}[h] = \mathcal{P}[\text{Re}(h)] + i\,\mathcal{P}[\text{Im}(h)]$ will be the sum of two harmonic functions. $\qquad\square$

The following result deals with the question of boundary values of Poisson integrals in a simple case. A more general case will be treated in §7.4.

THEOREM 7.29 (Schwarz). *If $h \in L^1(\mathbb{T})$ is continuous at ζ_0, then*

$$\lim_{z \to \zeta_0} \mathcal{P}[h](z) = h(\zeta_0).$$

PROOF. We may assume $h(\zeta_0) = 0$; the general case follows by considering the function $h - h(\zeta_0)$ and noting that

$$\mathcal{P}[h - h(\zeta_0)] = \mathcal{P}[h] - \mathcal{P}[h(\zeta_0)] = \mathcal{P}[h] - h(\zeta_0).$$

Let $\zeta_0 = e^{it_0}$. Given $\varepsilon > 0$, continuity of h at ζ_0 gives an open interval $I \subset \mathbb{R}$ centered at t_0 such that $|h(e^{it})| < \varepsilon$ whenever $t \in I$. Let $J = [t_0 - \pi, t_0 + \pi] \setminus I$, and define

$$h_I(e^{it}) = h(e^{it})\, \mathbb{1}_I(t) \qquad \text{and} \qquad h_J(e^{it}) = h(e^{it})\, \mathbb{1}_J(t),$$

where $\mathbb{1}_I$, $\mathbb{1}_J$ are the indicator functions of I, J. Evidently, $h = h_I + h_J$ so we have the decomposition

$$\mathcal{P}[h] = \mathcal{P}[h_I] + \mathcal{P}[h_J].$$

For every $z \in \mathbb{D}$,

The indicator function of a set E is defined by

$$\mathbb{1}_E(x) = \begin{cases} 1 & x \in E \\ 0 & x \notin E \end{cases}$$

(7.9) $\qquad |\mathcal{P}[h_I](z)| \le \frac{1}{2\pi} \int_I P(e^{it}, z)|h(e^{it})|\, dt \le \frac{\varepsilon}{2\pi} \int_I P(e^{it}, z)\, dt \le \varepsilon,$

where we have used Theorem 7.25(iii) in the last inequality. By Theorem 7.26(ii), there is a $\delta > 0$ such that $P(e^{it}, z) < \varepsilon$ whenever $z \in \mathbb{D}(\zeta_0, \delta) \cap \mathbb{D}$ and $t \in J$. It follows that for such z,

$$(7.10) \qquad |\mathcal{P}[h_J](z)| \leq \frac{1}{2\pi} \int_J P(e^{it}, z) |h(e^{it})| \, dt \leq \frac{\varepsilon}{2\pi} \int_J |h(e^{it})| \, dt \leq \varepsilon \|h\|_1.$$

Putting (7.9) and (7.10) together, we obtain

$$|\mathcal{P}[h](z)| \leq |\mathcal{P}[h_I](z)| + |\mathcal{P}[h_J](z)| \leq (1 + \|h\|_1)\varepsilon$$

whenever $z \in \mathbb{D}(\zeta_0, \delta) \cap \mathbb{D}$. Since $\varepsilon > 0$ was arbitrary, this shows $\lim_{z \to \zeta_0} \mathcal{P}[h](z) = 0$, as required. $\qquad\square$

We now have a complete solution to the **Dirichlet problem** in the disk at our disposal, i.e., the problem of finding a harmonic function in $\mathbb{D}$ with a given continuous boundary value.

COROLLARY 7.30. *Every continuous function* $h : \mathbb{T} \to \mathbb{C}$ *has a unique continuous extension* $H : \overline{\mathbb{D}} \to \mathbb{C}$ *which is harmonic in* $\mathbb{D}$. *It is given by*

$$H(z) = \begin{cases} \mathcal{P}[h](z) & |z| < 1 \\ h(z) & |z| = 1. \end{cases}$$

PROOF. H so defined is harmonic in $\mathbb{D}$ by Theorem 7.28 and continuous on $\overline{\mathbb{D}}$ by Theorem 7.29. If $\tilde{H}$ is another continuous extension of h which is harmonic in $\mathbb{D}$, then Theorem 7.22 shows that $\tilde{H} = \mathcal{P}[\tilde{H}|_{\mathbb{T}}] = \mathcal{P}[h] = H$. $\qquad\square$

7.3 Some applications of Poisson's formula

Everything we have said so far about the Poisson kernel in $\mathbb{D}$ can be transferred to an arbitrary disk $\mathbb{D}(p, r)$ in the plane. This is achieved most conveniently by the affine change of coordinates $z \mapsto (z - p)/r$ which maps $\mathbb{D}(p, r)$ conformally to $\mathbb{D}$. The result will be the formula

$$P\left(e^{it}, \frac{z - p}{r}\right) = \operatorname{Re}\left(\frac{re^{it} + (z - p)}{re^{it} - (z - p)}\right) = \frac{r^2 - |z - p|^2}{|re^{it} - (z - p)|^2}$$

for the Poisson kernel in the disk $\mathbb{D}(p, r)$. The Poisson integral of a function $h \in L^1(\mathbb{T}(p, r))$ then takes the form

$$\mathcal{P}[h](z) = \frac{1}{2\pi} \int_0^{2\pi} \frac{r^2 - |z - p|^2}{|re^{it} - (z - p)|^2} h(p + re^{it}) \, dt \qquad \text{for } z \in \mathbb{D}(p, r).$$

For convenience, we record the following corollary, which is merely a combined restatement of Theorems 7.22, 7.28, and 7.29 for a general disk:

COROLLARY 7.31.

(i) *If h is continuous on $\overline{\mathbb{D}}(p,r)$ and harmonic in $\mathbb{D}(p,r)$, then h is the Poisson integral of its restriction to the boundary circle $\mathbb{T}(p,r)$:*

$$h = \mathcal{P}[h|_{\mathbb{T}(p,r)}].$$

(ii) *If h is continuous on the circle $\mathbb{T}(p,r)$, then $\mathcal{P}[h]$ is harmonic in $\mathbb{D}(p,r)$ and extends h continuously to $\overline{\mathbb{D}}(p,r)$.*

THEOREM 7.32 (Mean value property characterizes harmonicity). *The following conditions on a continuous function $h: U \to \mathbb{C}$ are equivalent:*

(i) *h is harmonic.*
(ii) *h has the mean value property.*
(iii) *h has the local mean value property.*

PROOF. The implication (i) $\Longrightarrow$ (ii) is Theorem 7.16 and (ii) $\Longrightarrow$ (iii) is trivial. To prove (iii) $\Longrightarrow$ (i), take any disk $\mathbb{D}(p,r)$ such that $\overline{\mathbb{D}}(p,r) \subset U$. By Corollary 7.31, the function $\tilde{h} = \mathcal{P}[h|_{\mathbb{T}(p,r)}]$ is harmonic in $\mathbb{D}(p,r)$ and extends $h|_{\mathbb{T}(p,r)}$ continuously to $\overline{\mathbb{D}}(p,r)$. The difference $\psi = h - \tilde{h}$ is continuous on $\overline{\mathbb{D}}(p,r)$ and has the local mean value property in $\mathbb{D}(p,r)$, and $\psi = 0$ on $\mathbb{T}(p,r)$. By the maximum principle (see Remark 7.18), ψ must vanish identically in $\mathbb{D}(p,r)$. Hence $h = \tilde{h}$ is harmonic in this disk. Since U is a union of such disks, we conclude that h is harmonic in U. □

As a useful application of Theorem 7.32, we prove the following extension theorem for harmonic functions, which will be exploited later in the "Schwarz reflection principle" in §10.3:

THEOREM 7.33 (Harmonic extension by reflection). *Let $\mathbb{D}^+ = \{z \in \mathbb{D} : \text{Im}(z) > 0\}$ and $u: \mathbb{D}^+ \to \mathbb{R}$ be a harmonic function such that $u(z) \to 0$ as $z \in \mathbb{D}^+$ tends to a point of $(-1,1) \subset \partial\mathbb{D}^+$. Then u extends uniquely to a harmonic function in $\mathbb{D}$.*

PROOF. Uniqueness of harmonic extensions follows from Corollary 7.11. For existence, set $\mathbb{D}^- = \{z \in \mathbb{D} : \text{Im}(z) < 0\}$. We claim that the function

$$\psi(z) = \begin{cases} u(z) & z \in \mathbb{D}^+ \\ 0 & z \in (-1,1) \\ -u(\bar{z}) & z \in \mathbb{D}^-, \end{cases}$$

which is continuous by our assumption on u, is harmonic in $\mathbb{D}$. The direct verification of this fact is rather tricky, the trouble being harmonicity near the real line. But we can avoid this problem by verifying that ψ has the local mean value property in $\mathbb{D}$. In fact, ψ is harmonic in $\mathbb{D}^+$ by the assumption and in $\mathbb{D}^-$ by Theorem 7.2(iv), hence has the local mean value property in each of these semi-disks. On the other hand, if

$p \in (-1, 1)$ and $\overline{\mathbb{D}}(p, r) \subset \mathbb{D}$, then

$$\int_0^{2\pi} \psi(p + re^{it})\, dt = \left(\int_0^{\pi} + \int_{\pi}^{2\pi} \right) \psi(p + re^{it})\, dt$$

$$= \int_0^{\pi} u(p + re^{it})\, dt - \int_{\pi}^{2\pi} u(p + re^{-it})\, dt$$

$$= \int_0^{\pi} u(p + re^{it})\, dt - \int_0^{\pi} u(p + re^{i(s-2\pi)})\, ds$$

$$= 0 = 2\pi\, \psi(p).$$

This proves that ψ has the local mean value property in $\mathbb{D}$ and therefore is harmonic by Theorem 7.32. □

Let us now consider sequences of harmonic functions in a non-empty open set.

THEOREM 7.34. *Suppose $h_n : U \to \mathbb{C}$ is a sequence of harmonic functions such that $h_n \to h$ compactly in U as $n \to \infty$. Then h is harmonic in U.*

PROOF. Let $\overline{\mathbb{D}}(p, r) \subset U$. By the mean value property of h_n,

$$h_n(p) = \frac{1}{2\pi} \int_0^{2\pi} h_n(p + re^{it})\, dt.$$

Letting $n \to \infty$ and using the fact that $h_n \to h$ uniformly on $\overline{\mathbb{D}}(p, r)$, we obtain

$$h(p) = \frac{1}{2\pi} \int_0^{2\pi} h(p + re^{it})\, dt.$$

This shows that h has the mean value property in U and therefore is harmonic by Theorem 7.32. □

For our next result we need the following quantitative estimates on non-negative harmonic functions:

THEOREM 7.35 (Harnack's inequalities, 1886). *Let u be a non-negative harmonic function in the disk $\mathbb{D}(p, R)$. Then*

$$\frac{R - r}{R + r}\, u(p) \leq u(z) \leq \frac{R + r}{R - r}\, u(p) \qquad \text{if } |z - p| = r < R.$$

PROOF. Fix z with $|z - p| = r$ and choose any s such that $r < s < R$. Since u is continuous on $\overline{\mathbb{D}}(p, s)$, Corollary 7.31 shows that u is the Poisson integral of its restriction to $\mathbb{T}(p, s)$:

$$(7.11) \qquad u(z) = \frac{1}{2\pi} \int_0^{2\pi} \frac{s^2 - r^2}{|se^{it} - (z - p)|^2}\, u(p + se^{it})\, dt.$$

It is easy to see that

$$\frac{s-r}{s+r} \le \frac{s^2 - r^2}{|se^{it} - (z-p)|^2} \le \frac{s+r}{s-r}$$

for all $0 \le t \le 2\pi$ (this is the analog of Theorem 7.25(ii)). It follows from (7.11), the assumption $u \ge 0$, and the mean value property that

$$u(z) \le \frac{s+r}{s-r} \cdot \frac{1}{2\pi} \int_0^{2\pi} u(p + se^{it}) \, dt = \frac{s+r}{s-r} \, u(p)$$

and

$$u(z) \ge \frac{s-r}{s+r} \cdot \frac{1}{2\pi} \int_0^{2\pi} u(p + se^{it}) \, dt = \frac{s-r}{s+r} \, u(p).$$

Letting $s \to R$, we obtain the desired inequalities.   □

Carl Gustav Axel
Harnack (1851–1888)

EXAMPLE 7.36. (Liouville's theorem for harmonic functions) Every harmonic function $u : \mathbb{C} \to \mathbb{R}$ which is bounded from above or below must be constant. To see this, first assume u is bounded from above, say $u \le M$ on $\mathbb{C}$, so $\psi = M - u$ is a non-negative harmonic function in the plane. By Harnack's inequalities, if $|z| = r < R$,

$$\frac{R-r}{R+r} \, \psi(0) \le \psi(z) \le \frac{R+r}{R-r} \, \psi(0).$$

Letting $R \to +\infty$, we obtain $\psi(z) = \psi(0)$. This proves that ψ, hence u, is constant in $\mathbb{C}$. The case where u is bounded from below follows from this by considering the harmonic function $-u$.

Similarly, a bounded complex-valued harmonic function in the plane must be constant, as can be seen by applying the above result to its real and imaginary parts.

THEOREM 7.37. *Suppose $U \subset \mathbb{C}$ is a domain and $u_n : U \to \mathbb{R}$ is a sequence of harmonic functions in U such that $u_1 \le u_2 \le u_3 \le \cdots$. Set $u = \lim_{n \to \infty} u_n$. Then either $u = +\infty$ everywhere in U, or u is harmonic in U. In the latter case, $u_n \to u$ compactly in U as $n \to \infty$.*

PROOF. It suffices to consider the case where the u_n are non-negative; the general case follows by considering the sequence $u_n - u_1$. Let $\mathbb{D}(p, R) \subset U$. By Harnack's inequalities,

$$\frac{R-r}{R+r} \, u_n(p) \le u_n(z) \le \frac{R+r}{R-r} \, u_n(p) \qquad \text{if } |z - p| = r < R.$$

Taking the limit as $n \to \infty$ shows that the same inequalities must hold for u in place of u_n. It follows that if $u(p) = +\infty$, then $u = +\infty$ everywhere in $\mathbb{D}(p, R)$, and if $u(p) < +\infty$, then $u < +\infty$ everywhere in $\mathbb{D}(p, R)$. This proves that the set $\{z \in U : u(z) = +\infty\}$ is both open and closed. Since U is connected, this set is either U, in which case $u = +\infty$ everywhere, or it is empty, in which case $u < +\infty$ everywhere.

Suppose we are in the second situation. Take an $\varepsilon > 0$ and a disk $\mathbb{D}(p, R) \subset U$. Since the sequence $\{u_n(p)\}$ converges to the finite limit $u(p)$, we can find a large enough integer N such that $0 \leq u_n(p) - u_m(p) < \varepsilon$ whenever $n > m > N$. Applying Harnack's inequalities to the harmonic functions $u_n - u_m$ then gives

$$0 \leq u_n(z) - u_m(z) \leq 3\varepsilon \qquad \text{if } |z - p| < \frac{R}{2} \text{ and } n > m > N.$$

This shows that $\{u_n\}$ satisfies a uniform Cauchy condition in $\mathbb{D}(p, R/2)$, hence converges uniformly to u in this disk. Since every compact subset of U is covered by finitely many such disks, we conclude that $u_n \to u$ compactly in U as $n \to \infty$. It now follows from Theorem 7.34 that u is harmonic. $\qquad \square$

7.4 Boundary behavior of harmonic functions

Our goal in this section is to generalize some of the previous results on Poisson integrals to the case where the boundary function is no longer continuous. Recall that the radial limit of a function $f : \mathbb{D} \to \mathbb{C}$ at a point $\zeta \in \mathbb{T}$ is defined by $f^*(\zeta) = \lim_{r \to 1} f(r\zeta)$ whenever this limit exists.

EXAMPLE 7.38. Consider the function

$$u(e^{it}) = \begin{cases} 1 & -\pi < t < 0 \\ 0 & 0 < t < \pi \end{cases}$$

with jump discontinuities at ± 1. By Theorem 7.29,

$$\lim_{z \to \zeta} \mathcal{P}[u](z) = u(\zeta) \qquad \text{for every } \zeta \in \mathbb{T} \smallsetminus \{\pm 1\}.$$

On the other hand, an elementary exercise in integration shows that for $-1 < x < 1$,

$$\mathcal{P}[u](x) = \frac{1}{2\pi} \int_{-\pi}^{0} \frac{1 - x^2}{1 - 2x \cos t + x^2} \, dt = \frac{1}{2}.$$

Thus, both radial limits $\mathcal{P}[u]^*(1)$ and $\mathcal{P}[u]^*(-1)$ exist and equal $1/2$, i.e., the average of the left and right limits of u at ± 1. It is not hard to show that $\mathcal{P}[u]$ does not extend continuously to these boundary points. In fact, by letting $z \to \pm 1$ through carefully chosen paths, we can obtain all numbers in $[0, 1]$ as the limit of $\mathcal{P}[u](z)$ (see problem 26).

It turns out that the averaging property of the radial limit seen in the above example is typical:

THEOREM 7.39. *Suppose $h \in L^1(\mathbb{T})$ has one-sided limits*

$$A^+ = \lim_{t \to t_0^+} h(e^{it}) \qquad \text{and} \qquad A^- = \lim_{t \to t_0^-} h(e^{it})$$

at $\zeta_0 = e^{it_0}$. Then,

$$\mathcal{P}[h]^*(\zeta_0) = \frac{1}{2}(A^+ + A^-).$$

PROOF. We only need a minor modification of the proof of Theorem 7.29. It suffices to consider the case where $A^+ = -A^- = A$ and show that $\mathcal{P}[h]^*(\zeta_0) = 0$; the general case follows by considering the function $h - (A^+ + A^-)/2$. Given $\varepsilon > 0$, there is a open interval $I \subset \mathbb{R}$ centered at t_0 such that if I^+ and I^- are the connected components of $I \setminus \{t_0\}$ to the right and left of t_0, then

(7.12) $$|h(e^{it}) \mp A| < \varepsilon \quad \text{whenever } t \in I^{\pm}.$$

Let $J = [t_0 - \pi, t_0 + \pi] \setminus I$ and define

$$h^{\pm}(e^{it}) = h(e^{it})\, \mathbb{1}_{I^{\pm}}(t) \qquad \text{and} \qquad h_J(e^{it}) = h(e^{it})\, \mathbb{1}_J(t),$$

where, as before, the $\mathbb{1}$s stand for indicator functions. The relation $h = h^+ + h^- + h_J$ on $\mathbb{T} \setminus \{\zeta_0\}$ then gives

$$\mathcal{P}[h] = \mathcal{P}[h^+] + \mathcal{P}[h^-] + \mathcal{P}[h_J].$$

Estimating $\mathcal{P}[h_J]$ is identical to the proof of Theorem 7.29: By Theorem 7.26(ii), there is a $\delta > 0$ such that $P(e^{it}, z) < \varepsilon$ whenever $z \in \mathbb{D}(\zeta_0, \delta) \cap \mathbb{D}$ and $t \in J$. Hence, for all such z,

(7.13) $$|\mathcal{P}[h_J](z)| \leq \frac{1}{2\pi} \int_J P(e^{it}, z)|h(e^{it})|\, dt \leq \varepsilon \|h\|_1.$$

To estimate $\mathcal{P}[h^+]$ and $\mathcal{P}[h^-]$, we use the assumption that z is approaching ζ_0 radially. If $z \in \mathbb{D}$ is on the radial segment $[0, \zeta_0]$, then by the symmetry of the Poisson kernel (Theorem 7.25(i)),

$$\int_{I^+} P(e^{it}, z)\, dt = \int_{I^-} P(e^{it}, z)\, dt,$$

so

$$\mathcal{P}[h^+](z) + \mathcal{P}[h^-](z) = \frac{1}{2\pi} \int_{I^+} P(e^{it}, z)\, (h(e^{it}) - A)\, dt$$

$$+ \frac{1}{2\pi} \int_{I^-} P(e^{it}, z)\, (h(e^{it}) + A)\, dt.$$

By (7.12),

$$\left| \frac{1}{2\pi} \int_{I^{\pm}} P(e^{it}, z)(h(e^{it}) \mp A)\, dt \right| \leq \frac{\varepsilon}{2\pi} \int_{I^{\pm}} P(e^{it}, z)\, dt \leq \varepsilon.$$

It follows that

(7.14) $$|\mathcal{P}[h^+](z) + \mathcal{P}[h^-](z)| \leq 2\varepsilon$$

whenever $z \in \mathbb{D} \cap [0, \zeta_0]$. Putting (7.13) and (7.14) together, we conclude that

$$|\mathcal{P}[h](z)| \leq |\mathcal{P}[h^+](z) + \mathcal{P}[h^-](z)| + |\mathcal{P}[h_J](z)| \leq (2 + \|h\|_1)\varepsilon$$

whenever $z = r\zeta_0$ with $1 - \delta < r < 1$. $\qquad\qquad\qquad\qquad\qquad\qquad$ $\square$

Along the same lines, we now show that the continuity assumption in Theorem 7.29 can be substantially weakened as long as we restrict ourselves to radial limits. Here is the key notion that takes the role of continuity in this context:

> **DEFINITION 7.40.** A point $e^{it_0} \in \mathbb{T}$ is called a **Lebesgue point** of $h \in L^1(\mathbb{T})$ if
>
> $$\lim_{\delta \to 0} \frac{1}{2\delta} \int_{t_0-\delta}^{t_0+\delta} |h(e^{it}) - h(e^{it_0})|\, dt = 0.$$

Roughly speaking, this means the average oscillation of h over intervals centered at a Lebesgue point t_0 tends to 0 as the intervals shrink, a condition that is clearly weaker than continuity at t_0. According to **Lebesgue's differentiation theorem**, for a given integrable function almost every point of the circle is a Lebesgue point [**Ru2**].

THEOREM 7.41 (Radial limits of Poisson integrals). *If $h \in L^1(\mathbb{T})$, then*

$$\mathcal{P}[h]^*(\zeta) = h(\zeta)$$

at every Lebesgue point ζ of h, hence a.e. on $\mathbb{T}$.

PROOF. Again, we closely follow the idea of the proof of Theorem 7.29. Fix a Lebesgue point $\zeta_0 = e^{it_0}$ of h. We may assume $h(\zeta_0) = 0$; the general case follows by considering the function $h - h(\zeta_0)$. Given $\varepsilon > 0$ there is an interval $I \subset \mathbb{R}$ centered at t_0 such that

$$(7.15) \qquad\qquad \int_{I'} |h(e^{it})|\, dt < \varepsilon \operatorname{length}(I')$$

for every interval $I' \subset I$ centered at t_0. Let $J = [t_0 - \pi, t_0 + \pi] \smallsetminus I$ and define

$$h_I(e^{it}) = h(e^{it})\, \mathbb{1}_I(t) \qquad \text{and} \qquad h_J(e^{it}) = h(e^{it})\, \mathbb{1}_J(t),$$

so the relation $\mathcal{P}[h] = \mathcal{P}[h_I] + \mathcal{P}[h_J]$ holds. As in the proofs of Theorem 7.29 and Theorem 7.39, there is a $\delta > 0$ such that

$$(7.16) \qquad\qquad |\mathcal{P}[h_J](z)| \leq \varepsilon \|h\|_1 \qquad \text{if } z \in \mathbb{D}(\zeta_0, \delta) \cap \mathbb{D}.$$

Estimating $\mathcal{P}[h_I]$ requires a little more work since we can only control $|h|$ over I in an average sense. Let $z \in \mathbb{D}$ be on the radial segment $[0, \zeta_0]$, so $t \mapsto P(e^{it}, z)$ is symmetric

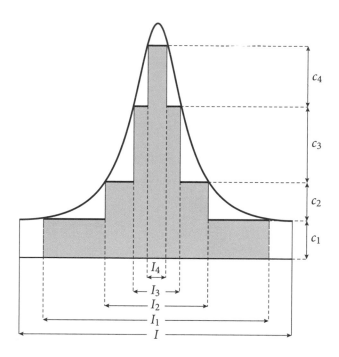

Figure 7.4. Approximating the Poisson kernel from below by symmetric step functions, used in the proof of Theorem 7.41.

about t_0. Let $I \supset I_1 \supset I_2 \supset \cdots \supset I_n$ be any nested collection of intervals centered at t_0. Choose positive numbers $c_1, \ldots, c_n$ such that the step function

$$\psi(e^{it}) = \sum_{k=1}^{n} c_k \mathbb{1}_{I_k}(t)$$

satisfies $\psi(e^{it}) \leq P(e^{it}, z)$ on I (see Fig. 7.4 for the case $n = 4$). We have, by (7.15),

$$\int_I \psi(e^{it}) |h(e^{it})| \, dt = \sum_{k=1}^{n} c_k \int_I |h(e^{it})| \mathbb{1}_{I_k}(t) \, dt = \sum_{k=1}^{n} c_k \int_{I_k} |h(e^{it})| \, dt$$

$$\leq \varepsilon \sum_{k=1}^{n} c_k \, \mathrm{length}(I_k) = \varepsilon \int_I \psi(e^{it}) \, dt$$

$$\leq \varepsilon \int_I P(e^{it}, z) \, dt \leq 2\pi\varepsilon.$$

Since $t \mapsto P(e^{it}, z)$ can be uniformly approximated by such ψ on I, it follows that

$$\int_I P(e^{it}, z) |h(e^{it})| \, dt \leq 2\pi\varepsilon.$$

Thus,

(7.17)
$$|\mathcal{P}[h_I](z)| \le \frac{1}{2\pi} \int_I P(e^{it}, z)|h(e^{it})|\, dt \le \varepsilon$$

whenever $z \in \mathbb{D}$ is on the radial segment $[0, \zeta_0]$. Putting (7.16) and (7.17) together, we conclude that

$$|\mathcal{P}[h](z)| \le |\mathcal{P}[h_I](z)| + |\mathcal{P}[h_J](z)| \le (1 + \|h\|_1)\varepsilon$$

whenever $z = r\zeta_0$ with $1 - \delta < r < 1$. $\square$

The preceding results lead to two natural questions on a given harmonic function h in $\mathbb{D}$:

(1)   Does the radial limit h^* exist a.e. on $\mathbb{T}$?
(2)   Assuming h^* exists and belongs to $L^1(\mathbb{T})$, is it true that h is the Poisson integral of h^*?

The answer to both questions is negative, as the following two examples show.

EXAMPLE 7.42. The idea of this example comes from a problem in [**Ru2**]. Let $n_k = 2^k k!$ and consider the power series
$$f(z) = \sum_{k=1}^{\infty} 5^k z^{n_k}$$

which is easily seen to have radius of convergence 1, so $f \in \mathcal{O}(\mathbb{D})$. Let us estimate $|f(z)|$ on the circle $|z| = 1 - 1/n_j$ for large j. Let z be any point on this circle and write

$$f(z) = \left(\sum_{k<j} + \sum_{k=j} + \sum_{k>j} \right) 5^k z^{n_k} = S_1 + S_2 + S_3.$$

We have

(7.18)
$$|S_1| \le \sum_{k<j} 5^k < \frac{1}{4} 5^j.$$

On the other hand, since $(1 - 1/n_j)^{n_j} \to 1/e$ as $j \to \infty$, we have, for large j,

(7.19)
$$|S_2| = 5^j \left(1 - \frac{1}{n_j}\right)^{n_j} > \frac{1}{3} 5^j.$$

To estimate S_3, note that for $k > j$,

$$\frac{n_k}{n_j} = 2^{k-j} \frac{k!}{j!} \ge k\, 2^{k-j}.$$

Using the inequality $1 - x \le e^{-x}$, we see that

$$|S_3| \le \sum_{k>j} 5^k \left(1 - \frac{1}{n_j}\right)^{n_k} \le \sum_{k>j} 5^k \exp(-n_k/n_j)$$

$$\le \sum_{k>j} 5^k \exp(-k\,2^{k-j}) = 5^j \sum_{k>0} 5^k \exp(-(j+k)2^k)$$

$$\le 5^j \sum_{k>0} \exp(-j\,2^k) \qquad (\text{since } 5^k \exp(-k\,2^k) < 1)$$

$$\le 5^j \sum_{k>0} \exp(-jk) = 5^j \cdot \frac{e^{-j}}{1 - e^{-j}}.$$

Thus, for large j,

(7.20) $$|S_3| < \frac{1}{24}\,5^j.$$

Putting (7.18), (7.19), and (7.20) together, it follows that

$$|f(z)| \ge |S_2| - |S_1| - |S_3| \ge \left(\frac{1}{3} - \frac{1}{4} - \frac{1}{24}\right)5^j = \frac{1}{24}\,5^j$$

whenever $|z| = 1 - 1/n_j$ and j is large. This proves that f does not have a radial limit anywhere on $\mathbb{T}$.

EXAMPLE 7.43. Even if a harmonic function in $\mathbb{D}$ has radial limits a.e. on $\mathbb{T}$, it may not be the Poisson integral of its radial limit function. As a simple example, the Poisson kernel $u(z) = P(1, z)$ is positive and harmonic in $\mathbb{D}$, and has radial limit $u^*(\zeta) = 0$ for all $\zeta \in \mathbb{T} \smallsetminus \{1\}$ (Theorem 7.26(ii)). Hence $\mathcal{P}[u^*]$ is identically zero.

Problem 27 gives a similar example in which $u^* = 0$ *everywhere* on $\mathbb{T}$.

Despite the negative results of the above examples, we can guarantee the existence of radial limits and restore the Poisson representation formula once we make a boundedness assumption. This is the content of the following theorem, a special case of which we have already encountered in Theorem 6.30:

THEOREM 7.44 (Fatou, 1906). *For every bounded harmonic function $h : \mathbb{D} \to \mathbb{C}$ the radial limit h^* exists a.e. on $\mathbb{T}$. Moreover, $h = \mathcal{P}[h^*]$.*

Recall that $L^\infty(\mathbb{T})$ is the space of all measurable functions $h : \mathbb{T} \to \mathbb{C}$ that are essentially bounded in the sense that $|h| \le M$ a.e. for some $M > 0$. The smallest such M is called the L^∞-norm of h and is denoted by $\|h\|_\infty$.

COROLLARY 7.45 (Solution to the $\boldsymbol{L^\infty}$ Dirichlet problem). *If $h \in L^\infty(\mathbb{T})$, the Poisson integral $\mathcal{P}[h]$ is the unique bounded harmonic function in $\mathbb{D}$ with radial limit h a.e. on $\mathbb{T}$. Moreover, $|\mathcal{P}[h](z)| \le \|h\|_\infty$ for every $z \in \mathbb{D}$.*

We note that in the above situation there may be *unbounded* harmonic functions in $\mathbb{D}$ with the same radial limit a.e. as $\mathcal{P}[h]$. For example, the Poisson kernel $P(1, z)$ and the constant function zero both have radial limit 0 a.e. on $\mathbb{T}$.

PROOF. $\mathcal{P}[h]^* = h$ a.e. by Theorem 7.41. Moreover, for every $z \in \mathbb{D}$

$$|\mathcal{P}[h](z)| \leq \frac{1}{2\pi} \int_0^{2\pi} P(e^{it}, z)|h(e^{it})| \, dt$$

$$\leq \|h\|_\infty \cdot \frac{1}{2\pi} \int_0^{2\pi} P(e^{it}, z) \, dt = \|h\|_\infty.$$

For uniqueness, simply observe that if H is any bounded harmonic function with $H^* = h$, then by Fatou's Theorem 7.44, $H = \mathcal{P}[H^*] = \mathcal{P}[h]$. $\square$

For the proof of Fatou's theorem, we first need the following basic version of **Parseval's formula** in Fourier analysis:

LEMMA 7.46. *Suppose the series* $\varphi(t) = \sum_{n=0}^\infty a_n e^{int}$ *converges uniformly on* $[0, 2\pi]$. *Then,*

$$\frac{1}{2\pi} \int_0^{2\pi} |\varphi(t)|^2 \, dt = \sum_{n=0}^\infty |a_n|^2.$$

PROOF. This is similar to the proof of Theorem 6.7. Consider the partial sums $\varphi_k(t) = \sum_{n=0}^k a_n e^{int}$. Write

$$\frac{1}{2\pi} \int_0^{2\pi} |\varphi_k(t)|^2 \, dt = \frac{1}{2\pi} \int_0^{2\pi} \varphi_k(t)\overline{\varphi_k(t)} \, dt$$

$$= \frac{1}{2\pi} \int_0^{2\pi} \left(\sum_{n=0}^k a_n e^{int} \right) \left(\sum_{m=0}^k \overline{a_m} e^{-imt} \right) dt$$

$$= \frac{1}{2\pi} \sum_{n,m=0}^k a_n \overline{a_m} \int_0^{2\pi} e^{i(n-m)t} \, dt.$$

Since $\int_0^{2\pi} e^{i(n-m)t} \, dt$ is 0 if $n \neq m$ and is 2π if $n = m$, we obtain

$$\frac{1}{2\pi} \int_0^{2\pi} |\varphi_k(t)|^2 \, dt = \sum_{n=0}^k |a_n|^2.$$

Since $\varphi_k \to \varphi$ uniformly on $[0, 2\pi]$, the result follows by letting $k \to \infty$. $\square$

PROOF OF THEOREM 7.44. It suffices to prove the theorem for real-valued functions. Suppose $u : \mathbb{D} \to \mathbb{R}$ is harmonic and $|u| \leq M$ in $\mathbb{D}$. By Theorem 7.9 there is an $f \in \mathcal{O}(\mathbb{D})$ such that $\text{Re}(f) = u$. The function $g = \exp(f)$ is holomorphic and satisfies $e^{-M} \leq |g| = \exp(u) \leq e^M$ in $\mathbb{D}$.

For each $0 < r < 1$, the function $g_r(z) = g(rz)$ is holomorphic in some neighborhood of $\overline{\mathbb{D}}$ and in particular is continuous on the unit circle. We claim that as $r \to 1$, the restrictions of $\{g_r\}$ on the unit circle satisfy a Cauchy condition in the L^2-norm

$$\|h\|_2 = \left(\frac{1}{2\pi} \int_0^{2\pi} |h(e^{it})|^2 \, dt \right)^{1/2}.$$

To see this, consider the power series $g(z) = \sum_{n=0}^{\infty} a_n z^n$ in $\mathbb{D}$. By Lemma 7.46,

$$\sum_{n=0}^{\infty} r^{2n} |a_n|^2 = \frac{1}{2\pi} \int_0^{2\pi} |g_r(e^{it})|^2 \, dt \le e^{2M}.$$

Letting $r \to 1$ then shows that the series $\sum_{n=0}^{\infty} |a_n|^2$ is convergent. Take an arbitrary $\varepsilon > 0$, choose $N > 1$ such that $\sum_{n=N+1}^{\infty} |a_n|^2 < \varepsilon$, and find $0 < \delta < 1$ such that $\sum_{n=0}^{N} |r^n - s^n|^2 |a_n|^2 < \varepsilon$ whenever $1 - \delta < r < s < 1$. Then,

$$\|g_r - g_s\|_2^2 = \frac{1}{2\pi} \int_0^{2\pi} |(g_r - g_s)(e^{it})|^2 \, dt$$

$$= \frac{1}{2\pi} \int_0^{2\pi} |g(re^{it}) - g(se^{it})|^2 \, dt$$

$$= \sum_{n=0}^{\infty} |r^n - s^n|^2 |a_n|^2 \qquad \text{(by Lemma 7.46)}$$

$$\le \sum_{n=0}^{N} |r^n - s^n|^2 |a_n|^2 + \sum_{n=N+1}^{\infty} |a_n|^2 < 2\varepsilon$$

Pierre Joseph Louis Fatou
(1878–1929)

whenever $1 - \delta < r < s < 1$, which proves our claim. Thus, we can select an increasing sequence $\{r_n\}$ of radii tending to 1 such that

$$\|g_{r_{k+1}} - g_{r_k}\|_2 \le 2^{-k} \qquad \text{for all } k \ge 1.$$

Define

$$S = |g_{r_1}| + \sum_{k=1}^{\infty} |g_{r_{k+1}} - g_{r_k}|.$$

By the triangle inequality,

$$\|S\|_2 \le \|g_{r_1}\|_2 + \sum_{k=1}^{\infty} \|g_{r_{k+1}} - g_{r_k}\|_2 \le \|g_{r_1}\|_2 + \sum_{k=1}^{\infty} 2^{-k} < +\infty,$$

which shows in particular that $S < +\infty$ a.e. on $\mathbb{T}$. At any point where $S < +\infty$, the telescoping series $\sum_{k=1}^{\infty} (g_{r_{k+1}} - g_{r_k})$ is absolutely convergent, hence convergent. It follows that

$$G = \lim_{n \to \infty} g_{r_n} = g_{r_1} + \sum_{k=1}^{\infty} (g_{r_{k+1}} - g_{r_k})$$

exists a.e. on $\mathbb{T}$. Note that $e^{-M} \le |G| \le e^M$, so $\log |G| \in L^\infty(\mathbb{T})$.

Now each function $u_{r_n}(z) = u(r_n z) = \log |g_{r_n}(z)|$ is harmonic in some neighborhood of $\overline{\mathbb{D}}$. Hence, by Corollary 7.31(i),

$$\lim_{n \to \infty} \mathcal{P}[u_{r_n}] = \lim_{n \to \infty} u_{r_n} = u \qquad \text{in } \mathbb{D}.$$

On the other hand, if $z \in \mathbb{D}$, then

$$\lim_{n \to \infty} P(e^{it}, z)\, u_{r_n}(e^{it}) = P(e^{it}, z) \log |G(e^{it})|$$

a.e. on $\mathbb{T}$, and

$$|P(e^{it}, z)\, u_{r_n}(e^{it})| \le \frac{1 + |z|}{1 - |z|}\, M$$

everywhere on $\mathbb{T}$. By Lebesgue's dominated convergence theorem,

$$\lim_{n \to \infty} \mathcal{P}[u_{r_n}](z) = \lim_{n \to \infty} \frac{1}{2\pi} \int_0^{2\pi} P(e^{it}, z) u_{r_n}(e^{it})\, dt$$

$$= \frac{1}{2\pi} \int_0^{2\pi} P(e^{it}, z) \log |G(e^{it})|\, dt = \mathcal{P}[\log |G|](z).$$

This proves $u = \mathcal{P}[\log |G|]$ in $\mathbb{D}$ and we conclude from Theorem 7.41 that u^* exists and equals $\log |G|$ a.e. on $\mathbb{T}$. $\qquad \square$

7.5 Harmonic measure on the circle

This section introduces a conformally invariant analog of Lebesgue measure on the unit circle $\mathbb{T}$ (for a more general case, see problem 30). We will denote the normalized Lebesgue measure on $\mathbb{T}$ by ℓ, that is, for a measurable set $E \subset \mathbb{T}$,

$$\ell(E) = \frac{1}{2\pi} \int_E |dz|.$$

The constant 2π is chosen so that $\ell(\mathbb{T}) = 1$.

Every element of $\mathrm{Aut}(\mathbb{D})$ induces an orientation-preserving homeomorphism $\mathbb{T} \to \mathbb{T}$, so it preserves measurable subsets of the circle.

DEFINITION 7.47. The *harmonic measure* of a measurable set $E \subset \mathbb{T}$ as seen from the point $z \in \mathbb{D}$ is defined by

$$\omega(z, E) = \ell(\varphi(E)),$$

where φ is any element of $\mathrm{Aut}(\mathbb{D})$ which sends z to 0.

This definition does not depend on the choice of φ, for if $\psi \in \mathrm{Aut}(\mathbb{D})$ also sends z to 0, then by the Schwarz lemma $\varphi \circ \psi^{-1}$ is a rotation about 0, which clearly preserves ℓ, so $\ell(\varphi(E)) = \ell(\psi(E))$. It follows from Definition 7.47 that *the harmonic measure is conformally invariant*:

$$\omega(z, E) = \omega(\varphi(z), \varphi(E)) \qquad \text{for all } \varphi \in \mathrm{Aut}(\mathbb{D}).$$

For each $z \in \mathbb{D}$ the assignment $E \mapsto \omega(z, E)$ defines a probability measure which we often denote by ω_z. Thus, $0 \leq \omega_z \leq 1$ and $\omega_z(\mathbb{T}) = 1$. At the origin, this measure reduces to $\omega_0 = \ell$.

The following gives an explicit formula for $\omega(z, E)$ and justifies the use of the word "harmonic" in its terminology. Recall that $\mathbb{1}_E$ denotes the indicator function of E.

THEOREM 7.48 (Harmonic measure as a Poisson integral). *For every measurable set $E \subset \mathbb{T}$ and every $z \in \mathbb{D}$,*

$$(7.21) \qquad \omega(z, E) = \mathcal{P}[\mathbb{1}_E](z) = \frac{1}{2\pi} \int_E \frac{1 - |z|^2}{|\zeta - z|^2} \, |d\zeta|.$$

In particular, the function $z \mapsto \omega(z, E)$ is harmonic in $\mathbb{D}$.

PROOF. This follows from the same computation as in the proof of Theorem 7.22. The disk automorphism $w = \varphi(\zeta) = (\zeta - z)/(1 - \bar{z}\zeta)$ sends z to 0 and satifies $|\varphi'(\zeta)| = (1 - |z|^2)/|\zeta - z|^2$ when $|\zeta| = 1$. Hence,

$$\omega(z, E) = \ell(\varphi(E)) = \frac{1}{2\pi} \int_{\varphi(E)} |dw|$$

$$= \frac{1}{2\pi} \int_E |\varphi'(\zeta)| \, |d\zeta| = \frac{1}{2\pi} \int_E \frac{1 - |z|^2}{|\zeta - z|^2} \, |d\zeta|.$$

The harmonicity claim follows from Theorem 7.28. $\qquad \square$

REMARK 7.49. Readers familiar with the notion of the "Radon-Nikodym derivative" in measure theory may notice that (7.21) can be written as

$$\frac{d\omega_z}{d\ell} = \frac{1 - |z|^2}{|\zeta - z|^2} = P(\zeta, z),$$

which allows us to write the Poisson integral formula (7.4) as

$$h(z) = \int_{\mathbb{T}} h \, d\omega_z.$$

Thus, $h(z)$ is the average value of h over $\mathbb{T}$ with respect to the measure ω_z.

The following is an immediate corollary of Theorems 7.48 and 7.41:

COROLLARY 7.50. *The function $z \mapsto \omega(z, E)$ has radial limit 1 a.e. on E and 0 a.e. on $\mathbb{T} \smallsetminus E$.*

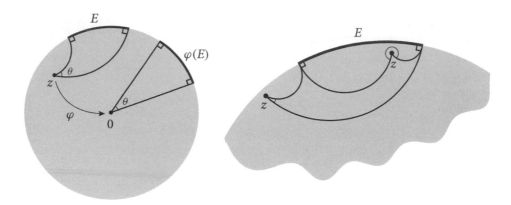

Figure 7.5. Left: The harmonic measure $\omega(z, E)$ is $\theta/(2\pi)$. Right: $\omega(z, E)$ tends to 1 or 0 according as z tends to the interior of E or $\mathbb{T} \smallsetminus E$. However, if z approaches the boundary points of the arc E along suitable paths, $\omega(z, E)$ can tend to any limit in $[0, 1]$.

Here is a useful geometric interpretation of harmonic measure, based on elementary hyperbolic geometry in the disk, as discussed in §4.5. For simplicity, let $E \subset \mathbb{T}$ be an arc. Take a disk automorphism φ which sends $z \in \mathbb{D}$ to 0. Then $\omega(z, E) = \ell(\varphi(E))$ is $\theta/2\pi$, where θ is the angle at which the arc $\varphi(E)$ is seen from the origin (Fig. 7.5 left). The radial lines from 0 to the endpoints of $\varphi(E)$ pull back under φ to hyperbolic geodesics from z to the endpoints of E. By conformality of φ, the angle between these geodesics is θ also. It follows that *the harmonic measure of an arc $E \subset \mathbb{T}$ as seen from $z \in \mathbb{D}$ is the normalized hyperbolic angle at which E is seen from z.* This provides an alternative view on Corollary 7.50 as well (Fig. 7.5 right): As z tends to the interior of E (resp. $\mathbb{T} \smallsetminus E$), the angle between the hyperbolic geodesics from z to the endpoints of E tends to 2π (resp. 0).

> This geometric interpretation extends from arcs to all measurable subsets of the circle; compare the end of the proof of Theorem 7.52.

The normalized Lebesgue measure ℓ is invariant under all rotations of $\mathbb{T}$. By conformal invariance, for each $p \in \mathbb{D}$ the harmonic measure ω_p is invariant under all elements of $\mathrm{Aut}(\mathbb{D})$ which fix p (these elements are elliptic in the sense of §4.3 and form a subgroup of $\mathrm{Aut}(\mathbb{D})$ isomorphic to the multiplicative group $\mathbb{T}$). What is not clear, and perhaps surprising, is that ω_p is invariant under all maps $\mathbb{T} \to \mathbb{T}$ which admit a holomorphic extension $\mathbb{D} \to \mathbb{D}$ fixing p. These maps turn out to be restrictions of special rational functions, so it seems appropriate to briefly introduce them here (compare problem 14 in chapter 4).

As before, let φ_p denote the Möbius map $z \mapsto (z - p)/(1 - \bar{p}z)$, where $|p| \neq 1$. By a ***finite Blaschke product*** is meant a rational function of the form

$$B = \alpha \prod_{k=1}^{n} \varphi_{p_k},$$

where $|\alpha| = 1$ and $|p_k| \neq 1$ for all $1 \leq k \leq n$. Evidently, $B^{-1}(0) = \{p_1, \ldots, p_n\}$ and $B^{-1}(\infty) = \{1/\overline{p_1}, \ldots, 1/\overline{p_n}\}$. Moreover, B acts symmetrically with respect to the unit circle since it commutes with the reflection $z \mapsto 1/\bar{z}$:

$$B\left(\frac{1}{\bar{z}}\right) = \frac{1}{\overline{B(z)}} \qquad \text{for all } z \in \hat{\mathbb{C}}.$$

Wilhelm Johann Eugen
Blaschke (1885–1962)

In particular, $|B(z)| = 1$ when $|z| = 1$.

For the rest of the discussion we assume that the zeros of B belong to $\mathbb{D}$, so $|p_k| < 1$ for all $1 \leq k \leq n$. In this case, the poles $1/\overline{p_k}$ are all in $\hat{\mathbb{C}} \smallsetminus \overline{\mathbb{D}}$ and B is holomorphic in some neighborhood of $\overline{\mathbb{D}}$. Since $|B(z)| = 1$ when $|z| = 1$, the maximum principle implies that $|B(z)| < 1$ if $|z| < 1$. It then follows from symmetry that $|B(z)| > 1$ if $|z| > 1$. Thus, $B(\mathbb{D}) = \mathbb{D}$, $B(\hat{\mathbb{C}} \smallsetminus \overline{\mathbb{D}}) = \hat{\mathbb{C}} \smallsetminus \overline{\mathbb{D}}$, and $B(\mathbb{T}) = \mathbb{T}$.

The map B can have at most one fixed point in $\mathbb{D}$ unless it is the identity map: If B had two distinct fixed points, Pick's version of the Schwarz lemma (Theorem 4.40) would imply that $B \in \text{Aut}(\mathbb{D})$. But an element of $\text{Aut}(\mathbb{D})$ with more than one fixed point in $\mathbb{D}$ must be the identity.

The restriction of B to the unit circle acts as an n-to-1 "covering map" much like $z \mapsto z^n$. To see this, first note that B has n zeros and no poles in $\mathbb{D}$, so by the argument principle $\text{W}(\gamma, 0) = n$, where $\gamma : [0, 2\pi] \to \mathbb{T}$ is defined by $\gamma(t) = B(e^{it})$. If $\tilde{\gamma} : [0, 2\pi] \to \mathbb{R}$ is any lift of γ which satisfies $\gamma(t) = \exp(i\tilde{\gamma}(t))$, it follows that $\tilde{\gamma}(2\pi) = \tilde{\gamma}(0) + 2\pi n$. Moreover, $\tilde{\gamma}$ must be strictly increasing. In fact, if $|z| = 1$,

$$\frac{zB'(z)}{B(z)} = \sum_{k=1}^{n} \frac{z\varphi'_{p_k}(z)}{\varphi_{p_k}(z)} = \sum_{k=1}^{n} \frac{z(1 - |p_k|^2)}{(z - p_k)(1 - \overline{p_k}z)} = \sum_{k=1}^{n} \frac{1 - |p_k|^2}{|z - p_k|^2} > 0,$$

so

$$\tilde{\gamma}'(t) = \frac{\gamma'(t)}{i\gamma(t)} = \frac{e^{it}B'(e^{it})}{B(e^{it})} > 0.$$

This shows in particular that under B every $q \in \mathbb{T}$ has n *distinct* preimages on $\mathbb{T}$. These preimages subdivide the circle into n open arcs $I_1, \ldots, I_n$ and B maps each I_k diffeomorphically onto $\mathbb{T} \smallsetminus \{q\}$, preserving the orientation.

That $B' \neq 0$ on $\mathbb{T}$ also follows from a simple topological argument based on $B^{-1}(\mathbb{T}) = \mathbb{T}$ and the local normal form near critical points. Can you see how?

The restriction $B : \mathbb{T} \to \mathbb{T}$ preserves sets of Lebesgue measure zero:

(7.22)   If $\ell(N) = 0$ for a measurable set $N \subset \mathbb{T}$, then $\ell(B^{-1}(N)) = 0$.

To see this, choose any $q \in \mathbb{T} \smallsetminus N$ and consider the open arcs $I_1, \ldots, I_n$ as above. Then, for each $1 \leq k \leq n$,

$$\frac{1}{2\pi} \int_{B^{-1}(N) \cap I_k} |B'(z)| \, |dz| = \ell(N) = 0$$

and since $|B'| > 0$ on $\mathbb{T}$, we must have $\ell(B^{-1}(N) \cap I_k) = 0$. Thus, $\ell(B^{-1}(N)) = \sum_{k=1}^{n} \ell(B^{-1}(N) \cap I_k) = 0$.

LEMMA 7.51. *Let $f : \mathbb{D} \to \mathbb{D}$ be a non-constant holomorphic function with the property $|f(z)| \to 1$ as $|z| \to 1$. Then f is the restriction of a finite Blaschke product.*

PROOF. First note that f has finitely many zeros; otherwise, since f is non-constant, any infinite sequence of its zeros would accumulate on the boundary circle, contradicting the assumption $\lim_{|z|\to 1}|f(z)|=1$. Let $\{p_1,\ldots,p_n\}$ be the zeros of f, repeated according to their orders, and consider the Blaschke product $B=\prod_{k=1}^{n}\varphi_{p_k}$ which has the same zeros of the same orders as f. The ratio $g=f/B$ has removable singularities at the p_k, so it extends to a non-vanishing holomorphic function in $\mathbb{D}$, which also has the property $\lim_{|z|\to 1}|g(z)|=1$. The maximum principle then implies that g is a constant α with $|\alpha|=1$ (compare Example 7.20). Thus, $f=\alpha B$ is a finite Blaschke product.   □

For another proof of this lemma, see Example 12.31.

It follows from the above lemma that *any map $\mathbb{T}\to\mathbb{T}$ with a holomorphic extension $\mathbb{D}\to\mathbb{D}$ must be the restriction of a finite Blaschke product.* This explains why in Theorem 7.52 below we do not lose generality by considering finite Blaschke products.

THEOREM 7.52. *Let $f:\mathbb{D}\to\mathbb{D}$ be the restriction of a finite Blaschke product with a fixed point $p\in\mathbb{D}$. Then the harmonic measure ω_p is invariant under $f:\mathbb{T}\to\mathbb{T}$.*

To say that ω_p is invariant under f means $\omega_p(f^{-1}(E))=\omega_p(E)$ for every measurable set $E\subset\mathbb{T}$.

PROOF. To simplify the notation we drop the dependence of ω on p. First suppose $A\subset\mathbb{T}$ is an open arc with endpoints a,b. By Theorem 7.29, $\lim_{z\to\zeta}\mathcal{P}[\mathbb{1}_A](z)=\mathbb{1}_A(\zeta)$ for every $\zeta\in\mathbb{T}\smallsetminus\{a,b\}$. Continuity of f then gives

$$\lim_{z\to\zeta}\mathcal{P}[\mathbb{1}_A](f(z))=\mathbb{1}_A(f(\zeta))=\mathbb{1}_{f^{-1}(A)}(\zeta)$$

for every $\zeta\in\mathbb{T}\smallsetminus f^{-1}(\{a,b\})$. In particular,

$$(\mathcal{P}[\mathbb{1}_A]\circ f)^{*}=\mathbb{1}_{f^{-1}(A)}\qquad\text{a.e. on }\mathbb{T}.$$

Since $\mathcal{P}[\mathbb{1}_A]\circ f$ is harmonic and bounded (by 1), it follows from the uniqueness part of Corollary 7.45 that $\mathcal{P}[\mathbb{1}_A]\circ f=\mathcal{P}[\mathbb{1}_{f^{-1}(A)}]$. Hence, by Theorem 7.48,

$$\omega(f^{-1}(A))=\mathcal{P}[\mathbb{1}_{f^{-1}(A)}](p)=(\mathcal{P}[\mathbb{1}_A]\circ f)(p)=\mathcal{P}[\mathbb{1}_A](p)=\omega(A).$$

The passage from open arcs to all measurable sets is a standard argument in measure theory, which we outline here in four steps:

(a)   Since every open subset of $\mathbb{T}$ is a disjoint union of open arcs, countable additivity of ω shows that $\omega(f^{-1}(U))=\omega(U)$ holds when $U\subset\mathbb{T}$ is open.

(b)   If $G=\bigcap_{n\geq 1}U_n$ where $U_1\supset U_2\supset U_3\supset\cdots$ are open sets in $\mathbb{T}$, then

$$\omega(f^{-1}(G))=\omega\left(\bigcap_{n\geq 1}f^{-1}(U_n)\right)$$

$$=\lim_{n\to\infty}\omega(f^{-1}(U_n))=\lim_{n\to\infty}\omega(U_n)=\omega(G).$$

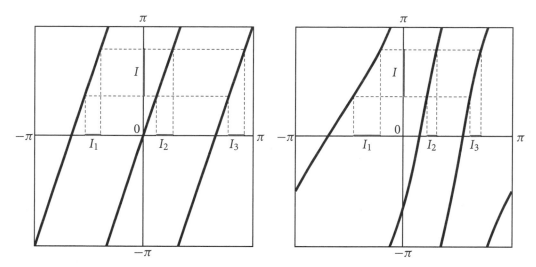

Figure 7.6. Illustration of Example 7.53. Left: The cubic polynomial $z \mapsto z^3$. Right: The cubic Blaschke product $z \mapsto z\,\varphi_p(z)\,\varphi_q(z)$, where $p = 0.5 + 0.3i$ and $q = -0.1 + 0.6i$. The graphs show the action of these maps on the unit circle $\mathbb{T}$, identified with the quotient $\mathbb{R}/(2\pi\mathbb{Z})$. In both cases, the preimage of any interval I consists of three intervals I_1, I_2, I_3 with $\ell(I_1) + \ell(I_2) + \ell(I_3) = \ell(I)$.

(c) If $N \subset \mathbb{T}$ is measurable and $\ell(N) = 0$, then $\omega(N) = 0$ by (7.21). Also, $\ell(f^{-1}(N)) = 0$ by (7.22), so $\omega(f^{-1}(N)) = 0$.

(d) Finally, any measurable set $E \subset \mathbb{T}$ can be written as $G \smallsetminus N$ with G as in (b) and $\ell(N) = 0$. Then, by (b) and (c),

$$\omega(f^{-1}(E)) = \omega(f^{-1}(G) \smallsetminus f^{-1}(N)) = \omega(f^{-1}(G))$$

$$= \omega(G) = \omega(E). \qquad \square$$

EXAMPLE 7.53. By Theorem 7.52 the normalized Lebesgue measure $\omega_0 = \ell$ is invariant under the restriction to $\mathbb{T}$ of any finite Blaschke product $\mathbb{D} \to \mathbb{D}$ which fixes the origin. Let us consider two such examples in degree 3. Under the identification $e^{it} \longleftrightarrow t$ between $\mathbb{T}$ and $\mathbb{R}/(2\pi\mathbb{Z})$, the cubic polynomial $z \mapsto z^3$ acts as the linear map $t \mapsto 3t \pmod{2\pi}$. The preimage of every interval of length c consists of three intervals of length $c/3$ and the invariance of ℓ is quite obvious (Fig. 7.6 left). On the other hand, consider the degree 3 Blaschke product $z \mapsto z\,\varphi_p(z)\,\varphi_q(z)$, with $p = 0.5 + 0.3i$ and $q = -0.1 + 0.6i$, whose action on $\mathbb{T}$ corresponds to a nonlinear map of $\mathbb{R}/(2\pi\mathbb{Z})$ (Fig. 7.6 right). Now the preimage of every interval of length c consists of three intervals of different sizes whose lengths still add up to c.

Problems

(1) Use the chain rule to verify the formula

$$\Delta(h \circ f) = (\Delta h \circ f)\,|f'|^2$$

when h is C^2 and f is holomorphic. This gives an alternative proof of the statement that $h \circ f$ is harmonic whenever h is harmonic and f is holomorphic.

(2)   Suppose f and g are holomorphic in a domain U. If the product $f\bar{g}$ is harmonic, prove that either f or g must be constant in U.

(3)   Prove the expression

$$P(e^{it}, re^{i\theta}) = \sum_{n=-\infty}^{\infty} r^{|n|} e^{in(t-\theta)}$$

for the Poisson kernel in $\mathbb{D}$, where for fixed $r \in [0,1)$ the series converges uniformly in $t, \theta \in \mathbb{R}$.

(4)   Give two alternative proofs for $1/(2\pi) \int_0^{2\pi} P(e^{it}, z)\, dt = 1$ by
   (i)   computing the integral

$$\frac{1}{2\pi} \int_0^{2\pi} \frac{e^{it}+z}{e^{it}-z}\, dt = \frac{1}{2\pi i} \int_{\mathbb{T}} \frac{\zeta+z}{\zeta-z} \frac{d\zeta}{\zeta}$$

   using the residue theorem and taking the real parts.
   (ii)   term-by-term integration of the series in the previous problem.

(5)   Suppose f is holomorphic in a neighborhood of the closed unit disk $\overline{\mathbb{D}}$ and $u = \mathrm{Re}(f)$. Show that

$$f(z) = \frac{1}{\pi i} \int_{\mathbb{T}} \frac{u(\zeta)}{\zeta-z}\, d\zeta - \overline{f(0)} \qquad \text{for all } z \in \mathbb{D}.$$

(6)   Derive the Poisson integral formula for a holomorphic function $f: \mathbb{D} \to \mathbb{C}$ with continuous extension to $\overline{\mathbb{D}}$ by applying the residue theorem to the integral

$$\mathcal{P}[f](z) = \frac{1}{2\pi i} \int_{\mathbb{T}} \frac{1-|z|^2}{(\zeta-z)(1-\zeta\bar{z})} f(\zeta)\, d\zeta.$$

(7)   Verify that the Laplace operator in polar coordinates (r, θ) is

$$\Delta = \frac{\partial^2}{\partial r^2} + \frac{1}{r}\frac{\partial}{\partial r} + \frac{1}{r^2}\frac{\partial^2}{\partial \theta^2}.$$

Use this to show that every harmonic function h in an annulus centered at the origin which depends only on r must have the form $h(r) = \alpha \log r + \beta$ for some constants $\alpha, \beta \in \mathbb{C}$.

(8)   Let $h: \{z \in \mathbb{C} : a < |z| < b\} \to \mathbb{C}$ be harmonic and define

$$I(r) = \frac{1}{2\pi} \int_0^{2\pi} h(re^{it})\, dt \qquad \text{for } a < r < b.$$

Show that $I(r) = \alpha \log r + \beta$ for some constants $\alpha, \beta \in \mathbb{C}$. Use this result to give another proof for Theorem 7.14.

(9)   Give an alternative proof of Liouville's theorem for harmonic functions (Example 7.36) by reducing to the case where $u \le 0$, finding an entire function f with $\mathrm{Re}(f) = u$, and applying Liouville's theorem for entire functions.

(10)   What can you say about a continuous function $u: \mathbb{C} \to \mathbb{R}$ for which $\exp(u)$ is harmonic?

(11)   (i)   Verify that the unique harmonic function $h: \mathbb{D} \to \mathbb{C}$ whose boundary value is a trigonometric polynomial

$$h(e^{it}) = \sum_{n=0}^{N} \left(a_n \cos(nt) + b_n \sin(nt) \right)$$

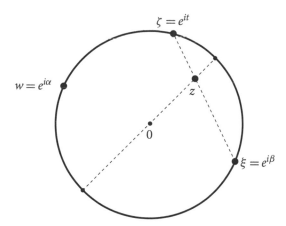

Figure 7.7. Schwarz's geometric interpretation of the Poisson kernel: $P(\zeta, z) = d\beta/dt$. See problem 14.

is given by the formula

$$h(re^{it}) = \sum_{n=0}^{N} r^n \big(a_n \cos(nt) + b_n \sin(nt)\big).$$

    (ii)   Use (i) to find the solution of the Laplace equation $\Delta h = 0$ in $\mathbb{D}$ with the boundary value $h(e^{it}) = \cos^2 t$. Compare this solution with the one given by the Poisson integral formula to find the value of the integral

$$\int_0^{2\pi} \frac{\cos^2 t}{5 - 3\cos t}\, dt.$$

(12)   Show that $h : U \to \mathbb{C}$ is harmonic if and only if

$$h(p) = \frac{1}{\pi r^2} \iint_{\mathbb{D}(p,r)} h(x,y)\, dx\, dy$$

whenever $\overline{\mathbb{D}}(p, r) \subset U$.

(13)   If h is continuous in $\mathbb{D}$ and harmonic in $\mathbb{D} \smallsetminus (-1, 1)$, show that h is harmonic in $\mathbb{D}$.

(14)   This exercise gives a geometric interpretation for the Poisson kernel due to Schwarz. Fix $z \in \mathbb{D}$.
    (i)   Consider the disk automorphism $\zeta \mapsto w = (\zeta - z)/(1 - \bar{z}\zeta)$. Let $\zeta = e^{it} \in \mathbb{T}$ and $w = e^{i\alpha}$ (where the argument α is well defined up to addition of an integer multiple of 2π). Show that $P(\zeta, z) = d\alpha/dt$.
    (ii) Let $\zeta = e^{it} \in \mathbb{T}$ and $\xi = e^{i\beta}$ be the second point where the line through ζ and z intersects $\mathbb{T}$ (see Fig. 7.7). Show that $P(\zeta, z) = d\beta/dt$. Derive the following alternative form of the Poisson formula:

$$h(z) = \frac{1}{2\pi} \int_0^{2\pi} h(e^{i\beta})\, dt.$$

How do you reconcile the formulas for $P(\zeta, z)$ in (i) and (ii)? (Hint: (i) is direct computation. For (ii), use the geometry of Fig. 7.7 to prove $-(\zeta - z)(\bar{\xi} - \bar{z}) = 1 - |z|^2$.)

(15)   Using the Riemann mapping and Carathéodory theorems in chapter 6, formulate and solve the Dirichlet problem on a simply connected domain bounded by a Jordan curve.

(16)   (i)   Show that a harmonic function $h : \mathbb{H} \to \mathbb{C}$ which extends continuously to $\overline{\mathbb{H}}$ has the representation

$$h(x + iy) = \frac{1}{\pi} \int_{-\infty}^{\infty} P(x - t, y)\, h(t)\, dt \qquad \text{for } x + iy \in \mathbb{H},$$

where

$$P(x, y) = \frac{y}{x^2 + y^2}$$

is the Poisson kernel in the upper half-plane.

(ii)   Deduce the following version of the mean value property:

$$\int_{-\infty}^{\infty} h(x + iy)\, dx = \int_{-\infty}^{\infty} h(x)\, dx \qquad \text{for } y > 0.$$

(17)   Let u be a real-valued harmonic function in a domain $U \subset \mathbb{C}$. A **harmonic conjugate** of u is a harmonic function $v : U \to \mathbb{R}$ such that $u + iv$ is holomorphic in U. Such v, if it exists, is unique up to an additive constant.

(i)   If u is continuous on $\overline{\mathbb{D}}$ and harmonic in $\mathbb{D}$, show that

$$v(z) = \frac{1}{2\pi} \int_0^{2\pi} Q(e^{it}, z) u(e^{it})\, dt$$

is a harmonic conjugate of u, where for $z = re^{i\theta} \in \mathbb{D}$,

$$Q(e^{it}, z) = \frac{1}{i} \frac{e^{-it}z - e^{it}\bar{z}}{|e^{it} - z|^2} = \frac{2r \sin(\theta - t)}{1 - 2r\cos(\theta - t) + r^2}$$

is the **conjugate Poisson kernel** in $\mathbb{D}$ (see Fig. 7.8). Verify that for $t \neq 0$,

$$\lim_{r \to 1} Q(e^{it}, r) = -\cot(t/2).$$

(ii)   Similarly, show that if u is continuous on $\overline{\mathbb{H}}$ and harmonic in $\mathbb{H}$, then

$$v(x + iy) = \frac{1}{\pi} \int_{-\infty}^{\infty} Q(x - t, y) u(t)\, dt$$

is a harmonic conjugate of u, where

$$Q(x, y) = \frac{x}{x^2 + y^2}$$

is the conjugate Poisson kernel in $\mathbb{H}$.
(Hint: For (i), take the imaginary part of each side of (7.5).)

(18)   Suppose $\{u_n\}$ is a sequence of non-negative harmonic functions in a domain U. Show that either $u_n \to +\infty$ in U as $n \to \infty$, or there exists a subsequence $\{u_{n_k}\}$ which converges compactly in U as $k \to \infty$.

(19)   Let $U \subset \mathbb{C}$ be a bounded domain. Suppose $h : \overline{U} \to \mathbb{C}$ is continuous and h is harmonic in U. Does the minimum principle

$$|h(z)| \geq \inf_{\zeta \in \partial U} |h(\zeta)| \qquad \text{for all } z \in U$$

hold? What if we assume h is non-vanishing?

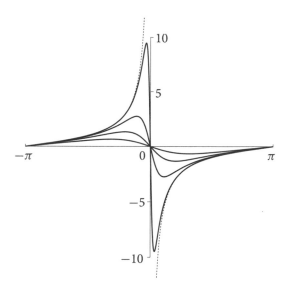

Figure 7.8. Graphs of the conjugate Poisson kernel $t \mapsto Q(e^{it}, r) = -2r \sin t / (1 - 2r \cos t + r^2)$ for $r = 0.3, 0.5, 0.7, 0.9$. As $r \to 1$, these graphs tend to the graph of $t \mapsto -\cot(t/2)$ (the dotted curve). See problem 17.

(20)   (Lindelöf's maximum principle) Let $U \subset \mathbb{C}$ be a domain and E be a finite proper subset of ∂U. Suppose $h : U \to \mathbb{R}$ is harmonic and bounded above. Show that if

$$\limsup_{n \to \infty} h(z_n) \leq M \quad \text{whenever } z_n \to \partial U \smallsetminus E,$$

then $h \leq M$ in U. (Hint: First assume U is bounded, take $\varepsilon > 0$, set $d = \operatorname{diam} U > 0$, and apply Theorem 7.19 to the function

$$h(z) - \varepsilon \sum_{\zeta \in E} \log\left(\frac{d}{|z - \zeta|}\right).$$

Then let $\varepsilon \to 0$ to obtain the result. If U is unbounded but $\mathbb{C} \smallsetminus U$ contains a disk, reduce to the bounded case by applying a Möbius map. For a general unbounded U, remove a small disk from U and reduce to the previous case.)

(21)   Let $r > 0$ and $U = \{z \in \mathbb{C} : \operatorname{Re}(z) > 0 \text{ and } |2z - r| > r\}$. By mapping U conformally onto a suitable domain, solve the Dirichlet problem $\Delta u = 0$ in U with the boundary condition

$$u(z) = \begin{cases} 0 & \operatorname{Re}(z) = 0 \\ 1 & |2z - r| = r, \ z \neq 0. \end{cases}$$

(22)   Let $U \subset \mathbb{C}$ be a domain and $K \subset U$ be compact. Show that there is a constant $C > 0$ depending only on U, K such that for every positive harmonic function u in U and every $z, w \in K$, $u(z)/u(w) \leq C$.

(23)   Let $U \subset \mathbb{C}$ be a domain. For $z, w \in U$, define the **Harnack distance** $\lambda_U(z, w)$ to be the infimum over all $C \geq 1$ for which the inequality $C^{-1} \leq u(z)/u(w) \leq C$ holds for every positive harmonic function u in U.

    (i)   Show that $1 \leq \lambda_U < +\infty$.

    (ii)   Verify that $\log \lambda_U$ satisfies all the properties of a metric in U, except that $\log \lambda_U(z,w)$ may vanish even though $z \neq w$ (one calls such a function a "pseudo-metric").

    (iii) Give an example of a domain U for which $\log \lambda_U$ vanishes identically.

    (iv) Find the explicit formula for $\lambda_{\mathbb{D}}$.

    (v)   Show that if $f : U \to V$ is holomorphic, then

$$\lambda_V(f(z), f(w)) \leq \lambda_U(z, w) \qquad \text{for all } z, w \in U.$$

Conclude that λ_U is a conformal invariant.

(24)   Let $U \subset \mathbb{C}$ be a bounded domain whose oriented boundary ∂U is a finite union of piecewise C^1 Jordan curves. Assume all functions in the following discussion are real-valued and C^2-smooth in some neighborhood of $\overline{U}$. The **normal derivative** of a function u along ∂U is defined by $\partial u / \partial n = \nabla u \cdot n$, where $\nabla u = (u_x, u_y)$ is the gradient of u and n is the outward unit normal vector.

    (i)   Use Green's theorem to prove **Green's first identity**

$$\int_{\partial U} v \frac{\partial u}{\partial n} \, ds = \iint_U (v \Delta u + \nabla u \cdot \nabla v) \, dx \, dy.$$

    Conclude that

$$\int_\gamma \frac{\partial u}{\partial n} \, ds = 0$$

    if u is harmonic in U.

    (ii)   Deduce **Green's second identity**,

$$\int_{\partial U} \left(u \frac{\partial v}{\partial n} - v \frac{\partial u}{\partial n} \right) ds = \iint_U (u \Delta v - v \Delta u) \, dx \, dy.$$

Minimizing the Dirichlet integral is one of the oldest methods of constructing harmonic extensions. The idea played a pivotal rule in the development of functional analysis in the early twentieth century.

    (iii) Use Green's first identity to prove that if u is harmonic in U, and if $v = u$ on ∂U, then

$$\iint_U \|\nabla v\|^2 \, dx \, dy = \iint_U \left(\|\nabla u\|^2 + \|\nabla(u-v)\|^2 \right) dx \, dy.$$

    Conclude that for a given continuous function $\partial U \to \mathbb{R}$, a harmonic extension u has the least **Dirichlet integral** $\iint_U \|\nabla u\|^2 \, dx \, dy$ among all possible extensions.

(25)   Suppose $u : \mathbb{D} \to \mathbb{R}$ is bounded and harmonic. Show that there is a constant $C > 0$ such that

$$\|\nabla u(z)\| \leq \frac{C}{1 - |z|} \qquad \text{for } z \in \mathbb{D}.$$

(Hint: If $u = \operatorname{Re}(f)$ for some $f \in \mathscr{O}(\mathbb{D})$, then $\|\nabla u\| = |f'|$. Consider $g = \exp(f)$ and estimate $|g'|$.)

(26)   This exercise relates to Example 7.38.

    (i)   Show that the function $\tilde{u}(x + iy) = (1/\pi) \arctan(y/x)$ is harmonic in the upper half-plane $\mathbb{H}$, and extends continuously to 0 on the positive real axis and to 1 on the negative real axis.

    (ii)   Let $\varphi(z) = (i - z)/(i + z) : \mathbb{H} \to \mathbb{D}$ be the Cayley map. Show that $u = \tilde{u} \circ \varphi^{-1}$ is harmonic in $\mathbb{D}$ and extends continuously to 0 on the semicircle $\{e^{it} : 0 < t < \pi\}$ and to 1 on the semicircle $\{e^{it} : -\pi < t < 0\}$.

(iii)   Show that u is the Poisson integral of its boundary value. Use the relation between u and $\tilde{u}$ to verify that the level sets of u are circular arcs in $\mathbb{D}$ passing through ± 1, and that there is one such circle corresponding to each value between 0 and 1.

(iv)   Conclude that u does not extend continuously to ± 1.

(27)   Consider the function

$$f(z) = \left(\frac{1-z}{1+z}\right)^2$$

which is the composition of the Möbius map $z \mapsto (1-z)/(1+z)$ followed by the squaring map $z \mapsto z^2$.

(i)   Find the image of the unit disk $\mathbb{D}$ under f. In particular, describe (at least qualitatively) the images under f of the radial lines in $\mathbb{D}$.

(ii)   Consider the harmonic function $u = \mathrm{Im}(f)$ in $\mathbb{D}$. Show that for every $\zeta \neq -1$ on the unit circle, $\lim_{z \to \zeta} u(z) = 0$. Show that the same is true provided that $z \to -1$ radially. Thus $u^* = 0$ everywhere on the circle, even though u is not identically zero. Contrast this with Theorem 7.44.

(28)   A function $f : \mathbb{D} \to \mathbb{C}$ is said to have **non-tangential limit** L at $\zeta \in \mathbb{T}$ if $f(z) \to L$ whenever $z \to \zeta$ through a triangle in $\mathbb{D}$ with one vertex at ζ (equivalently, if $|\zeta - z|/(1 - |z|)$ remains bounded as $z \to \zeta$). The bounded harmonic function u in problem 26 has radial limit $1/2$ at $\zeta = 1$ but it does not have a non-tangential limit there. Show, however, that if f is bounded and *holomorphic* in $\mathbb{D}$, and if $f^*(\zeta) = \lim_{r \to 1} f(r\zeta) = L$, then f has non-tangential limit L at ζ. (Hint: Map $\mathbb{D}$ conformally to the strip $U = \{z : -1 < \mathrm{Im}(z) < 1\}$, the point ζ to ∞, and the radial segment landing at ζ to the real axis. The triangles at ζ then correspond to proper sub-strips $\{z : -1 + \varepsilon < \mathrm{Im}(z) < 1 - \varepsilon\}$. If $g \in \mathcal{O}(U)$ is bounded and $\lim_{x \to \infty} g(x) = L$, apply Montel's theorem to the sequence $g_n(z) = g(z + n)$ to show that $g_n \to L$ compactly in U.)

(29)   Let f be a finite Blaschke product with $f(\mathbb{D}) = \mathbb{D}$ and $f(0) = 0$. Show that $|f'(z)| > 1$ if $z \in \mathbb{T}$, so $\ell(f(I)) > \ell(I)$ for every arc $I \subset \mathbb{T}$. Contrast this with the invariance of ℓ under f claimed by Theorem 7.52.

(30)   Let $f : \mathbb{D} \to U$ be a Riemann map of a simply connected domain $U \subset \hat{\mathbb{C}}$ with locally connected boundary. According to Carathéodory's Theorem 6.19, f extends continuously to a map between the closures. For every measurable set $E \subset \partial U$, define the harmonic measure of E as seen from $p = f(0)$ by

$$\omega(p, E, U) = \ell(f^{-1}(E)).$$

(i)   Show that this definition does not depend on the choice of f, and that $\omega(p, E, \mathbb{D})$ coincides with $\omega(p, E)$ of Definition 7.47.

(ii)   Verify the relation

$$\omega(z, E, U) = \omega(f^{-1}(z), f^{-1}(E)) \qquad \text{for } z \in U,$$

and conclude that the function $z \mapsto \omega(z, E, U)$ is harmonic.

(iii)   For $U = \hat{\mathbb{C}} \smallsetminus [-1, 1]$, show that

$$\omega(\infty, [a, b], U) = \frac{1}{\pi}\big(\arccos(a) - \arccos(b)\big) \qquad \text{if } -1 \leq a < b \leq 1.$$

8

Zeros of holomorphic functions

The problem of constructing holomorphic functions with prescribed zeros can be tackled by considering products of "elementary" factors, each vanishing at one of the designated points. This chapter examines this idea, first in the plane and then in a general domain, and provides conditions under which infinite products of such factors converge and define holomorphic functions with the anticipated zeros. It also illustrates some of the intricate relations between the growth rate of an entire function and how its zeros may proliferate in the plane.

8.1 Infinite products

Let $\{a_n\}_{n=1}^{\infty}$ be a sequence of complex numbers. It would seem natural to define the infinite product $\prod_{n=1}^{\infty} a_n$ as the limit of the partial products $\prod_{n=1}^{k} a_n$ as $k \to \infty$ whenever this limit exists. But there are two problems with this definition. First, if $a_n = 0$ for some n, the infinite product would exist regardless of how wildly the rest of the sequence may behave. Second, the infinite product could be zero even if none of the factors is zero (take $a_n = 1/n$). The latter is not an inherently flawed feature except that for the applications we have in mind, we would like infinite products to behave as much like finite products as possible. For example, we want a definition according to which an infinite product $\prod_{n=1}^{\infty} f_n$ of functions vanishes at some point if and only if some factor f_n vanishes at that point.

Both of these issues are addressed in the following slightly modified definition:

DEFINITION 8.1. Let $\{a_n\}_{n=1}^{\infty}$ be a sequence of complex numbers. We say that the infinite product $\prod_{n=1}^{\infty} a_n$ **converges** if there is an integer n_0 for which the limit

$$p = \lim_{k \to \infty} \prod_{n=n_0}^{k} a_n$$

exists and is *non-zero*. In this case, we set

$$\prod_{n=1}^{\infty} a_n = \lim_{k \to \infty} \prod_{n=1}^{k} a_n = a_1 \cdots a_{n_0-1}\, p.$$

It follows from the definition that $a_n \neq 0$ for all $n \geq n_0$. Note that the value of $\prod_{n=1}^{\infty} a_n$ as defined above is independent of the choice of the integer n_0: If $q = \lim_{k \to \infty} \prod_{n=m_0}^{k} a_n$ is also non-zero, then $a_1 \cdots a_{n_0-1}\, p = a_1 \cdots a_{m_0-1}\, q$.

THEOREM 8.2. *Suppose* $\prod_{n=1}^{\infty} a_n$ *converges. Then*

(i) $\lim_{n \to \infty} a_n = 1$.

(ii) $\prod_{n=1}^{\infty} a_n = 0$ *if and only if* $a_n = 0$ *for some n.*

(iii) $\prod_{n=m}^{\infty} a_n$ *converges for every* $m \geq 1$ *and* $\lim_{m \to \infty} \prod_{n=m}^{\infty} a_n = 1$.

(iv) *If* $\prod_{n=1}^{\infty} b_n$ *converges also, so does* $\prod_{n=1}^{\infty} (a_n b_n)$, *and*

$$\prod_{n=1}^{\infty} (a_n b_n) = \prod_{n=1}^{\infty} a_n \cdot \prod_{n=1}^{\infty} b_n.$$

PROOF. (i) and (ii) are easy to verify. For (iii), choose n_0 and p as in Definition 8.1. For $k > m > n_0$,

$$\prod_{n=m}^{k} a_n = \frac{\prod_{n=n_0}^{k} a_n}{\prod_{n=n_0}^{m-1} a_n}.$$

Letting $k \to \infty$ gives

$$(8.1) \qquad \prod_{n=m}^{\infty} a_n = \frac{p}{\prod_{n=n_0}^{m-1} a_n} \neq 0,$$

which proves convergence of $\prod_{n=m}^{\infty} a_n$ for $m > n_0$, hence for all $m \geq 1$. If we now let $m \to \infty$ in (8.1), we obtain

$$\lim_{m \to \infty} \prod_{n=m}^{\infty} a_n = \frac{p}{p} = 1.$$

To see (iv), choose n_0 such that the limits

$$p = \lim_{k \to \infty} \prod_{n=n_0}^{k} a_n \qquad \text{and} \qquad q = \lim_{k \to \infty} \prod_{n=n_0}^{k} b_n$$

are non-zero. Then

$$\lim_{k \to \infty} \prod_{n=n_0}^{k} (a_n b_n) = pq \neq 0,$$

so $\prod_{n=1}^{\infty}(a_n b_n)$ converges. Moreover,

$$\prod_{n=1}^{\infty}(a_n b_n) = (a_1 b_1) \cdots (a_{n_0-1} b_{n_0-1}) \, pq$$

$$= (a_1 \cdots a_{n_0-1} \, p) \cdot (b_1 \cdots b_{n_0-1} \, q)$$

$$= \prod_{n=1}^{\infty} a_n \cdot \prod_{n=1}^{\infty} b_n. \qquad \square$$

REMARK 8.3. Closely related to Theorem 8.2(iii) is the statement that for every $\varepsilon > 0$ there is an integer N such that $|\prod_{n=m}^{k} a_n - 1| < \varepsilon$ whenever $k \geq m \geq N$ (compare problem 1). To see this, use continuity of the function $(z, w) \mapsto z/w$ at the point $(1,1)$ to find a $\delta > 0$ such that $|z - 1| < \delta$ and $|w - 1| < \delta$ imply $|z/w - 1| < \varepsilon$. Take an integer N such that $|\prod_{n=m}^{\infty} a_n - 1| < \delta$ whenever $m \geq N$. Then, if $k \geq m \geq N$,

$$\left| \prod_{n=m}^{k} a_n - 1 \right| = \left| \frac{\prod_{n=m}^{\infty} a_n}{\prod_{n=k+1}^{\infty} a_n} - 1 \right| < \varepsilon.$$

EXAMPLE 8.4. The infinite product $\prod_{n=2}^{\infty}(1 - 1/n^2)$ converges. In fact, a simple induction shows that

$$\left(1 - \frac{1}{2^2}\right)\left(1 - \frac{1}{3^2}\right) \cdots \left(1 - \frac{1}{k^2}\right) = \frac{k+1}{2k},$$

so

$$\prod_{n=2}^{\infty}\left(1 - \frac{1}{n^2}\right) = \frac{1}{2}.$$

EXAMPLE 8.5. The infinite product $\prod_{n=1}^{\infty}(1 + (-1)^{n+1}/n)$ converges. One easily checks that

$$\left(1 + \frac{1}{1}\right)\left(1 - \frac{1}{2}\right)\left(1 + \frac{1}{3}\right) \cdots \left(1 + \frac{(-1)^{k+1}}{k}\right) = \begin{cases} 1 + \dfrac{1}{k} & \text{if } k \text{ is odd} \\ 1 & \text{if } k \text{ is even,} \end{cases}$$

so

$$\prod_{n=1}^{\infty}\left(1 + \frac{(-1)^{n+1}}{n}\right) = 1.$$

EXAMPLE 8.6. The infinite product $\prod_{n=1}^{\infty}(1 + 1/n)$ does not converge. This follows from the obvious inequality

$$\prod_{n=1}^{k}\left(1 + \frac{1}{n}\right) > \sum_{n=1}^{k} \frac{1}{n}$$

and the fact that the right-hand sum tends to infinity as $k \to \infty$. Similarly, the infinite product $\prod_{n=2}^{\infty}(1 - 1/n)$ does not converge since

$$\prod_{n=2}^{k}\left(1-\frac{1}{n}\right)=\prod_{n=2}^{k}\frac{n-1}{n}=\frac{1}{k}$$

and the right hand side tends to zero as $k\to\infty$ (compare problem 2).

The convergence of infinite products can be formulated in terms of infinite series using logarithms. Let "Log" denote the principal branch of the logarithm function which is holomorphic in $\mathbb{C}\smallsetminus(-\infty,0]$ and maps 1 to 0, as described in Example 2.23. Recall the power series representation

$$\mathrm{Log}(1+z)=\sum_{n=1}^{\infty}\frac{(-1)^{n-1}}{n}z^n=z-\frac{z^2}{2}+\frac{z^3}{3}-\cdots\qquad\text{for }|z|<1.$$

The following properties are easily verified:

(L1) If $|z-1|<1$ and $|w-1|<1$, then $\mathrm{Log}(zw)=\mathrm{Log}\,z+\mathrm{Log}\,w$.

(L2) Since $\mathrm{Log}\,1=0$, for every $\varepsilon>0$ there is a $0<\delta<1$ such that $|z-1|<\delta$ implies $|\mathrm{Log}\,z|<\varepsilon$.

(L3) Since $\lim_{z\to0}\mathrm{Log}(1+z)/z=1$, there is an $0<r_0<1$ such that

$$\frac{1}{2}|z|\leq|\mathrm{Log}(1+z)|\leq\frac{3}{2}|z|\qquad\text{if }|z|<r_0.$$

LEMMA 8.7. $\prod_{n=1}^{\infty}a_n$ *converges if and only if* $\sum_{n=n_0}^{\infty}\mathrm{Log}\,a_n$ *converges for some integer* $n_0\geq1$.

PROOF. First suppose $\prod_{n=1}^{\infty}a_n$ converges. Since $\lim_{n\to\infty}a_n=1$, there is an $n_0\geq1$ such that $|a_n-1|<1$ for all $n\geq n_0$. We show that $\sum_{n=n_0}^{\infty}\mathrm{Log}\,a_n$ converges. Let $\varepsilon>0$ and find the corresponding $0<\delta<1$ from (L2). Find $N>n_0$ such that $|\prod_{n=m}^{k}a_n-1|<\delta$ whenever $k\geq m\geq N$ (see Remark 8.3). It follows from (L1) and (L2) that

$$\left|\sum_{n=m}^{k}\mathrm{Log}\,a_n\right|=\left|\mathrm{Log}\prod_{n=m}^{k}a_n\right|<\varepsilon\qquad\text{whenever }k\geq m\geq N.$$

This proves $\{\sum_{n=n_0}^{k}\mathrm{Log}\,a_n\}$ is a Cauchy sequence, hence it converges as $k\to\infty$.

Conversely, suppose $\sum_{n=n_0}^{\infty}\mathrm{Log}\,a_n$ converges for some n_0. Then,

$$\lim_{k\to\infty}\prod_{n=n_0}^{k}a_n=\lim_{k\to\infty}\exp\left(\sum_{n=n_0}^{k}\mathrm{Log}\,a_n\right)=\exp\left(\sum_{n=n_0}^{\infty}\mathrm{Log}\,a_n\right)\neq0,$$

which shows $\prod_{n=1}^{\infty}a_n$ converges. □

Our next goal is to define a notion of "absolute convergence" for an infinite product as a sufficient condition for convergence. Imitating the case of infinite series, we might be tempted to declare $\prod_{n=1}^{\infty} a_n$ absolutely convergent if $\prod_{n=1}^{\infty} |a_n|$ converges. But this definition would lead to a *weaker* notion in which convergence would always imply absolute convergence, whereas a product like $\prod_{n=1}^{\infty} (-1)^n$ would be absolutely convergent without being convergent. It turns out that the right definition should effectively take into account how close the numbers a_n are to 1. For this reason, it will be convenient to write $a_n = 1 + b_n$ and work with the sequence $\{b_n\}$ instead.

> **DEFINITION 8.8.** The infinite product $\prod_{n=1}^{\infty}(1 + b_n)$ *converges absolutely* if $\prod_{n=1}^{\infty}(1 + |b_n|)$ converges.

Since $1 + |b_n| \neq 0$ for all n, the convergence of $\prod_{n=1}^{\infty}(1 + |b_n|)$ simply means that $\lim_{k \to \infty} \prod_{n=1}^{k}(1 + |b_n|)$ exists. This happens precisely when the monotonic sequence $\{\prod_{n=1}^{k}(1 + |b_n|)\}_{k=1}^{\infty}$ of partial products is bounded above.

THEOREM 8.9. $\prod_{n=1}^{\infty}(1 + b_n)$ *converges absolutely if and only if* $\sum_{n=1}^{\infty} b_n$ *converges absolutely.*

PROOF. Since by Lemma 8.7 the product $\prod_{n=1}^{\infty}(1 + |b_n|)$ converges if and only if the sum $\sum_{n=1}^{\infty} \mathrm{Log}(1 + |b_n|)$ converges, it suffices to check that the latter is equivalent to the convergence of $\sum_{n=1}^{\infty} |b_n|$. But this is clear since by (L3) the two series have eventually comparable terms. □

THEOREM 8.10. *An absolutely convergent infinite product is convergent.*

Combined with Theorem 8.9, this gives a convenient *sufficient* condition for convergence of infinite products: $\prod_{n=1}^{\infty}(1 + b_n)$ converges whenever $\sum_{n=1}^{\infty} |b_n|$ converges.

PROOF. By Lemma 8.7 and Theorem 8.9, it suffices to check that the convergence of $\sum_{n=1}^{\infty} |b_n|$ implies that of $\sum_{n=n_0}^{\infty} \mathrm{Log}(1 + b_n)$, where n_0 is any integer for which $|b_n| < 1$ for all $n \geq n_0$. Take an arbitrary $0 < \varepsilon < r_0$, where r_0 is the constant in (L3). Find $N > n_0$ such that $\sum_{n=m}^{k} |b_n| < \varepsilon$ whenever $k \geq m \geq N$. Then, by (L3),

$$\left| \sum_{n=m}^{k} \mathrm{Log}(1 + b_n) \right| \leq \sum_{n=m}^{k} |\mathrm{Log}(1 + b_n)| \leq \frac{3}{2} \sum_{n=m}^{k} |b_n| < \frac{3}{2} \varepsilon$$

whenever $k \geq m \geq N$. This proves convergence of $\sum_{n=n_0}^{\infty} \mathrm{Log}(1 + b_n)$. □

EXAMPLE 8.11. The convergence of the infinite product $\prod_{n=2}^{\infty}(1 - 1/n^2)$ of Example 8.4 also follows from its absolute convergence, which is guaranteed by Theorem 8.9 since the series $\sum 1/n^2$ converges.

EXAMPLE 8.12. The infinite product $\prod_{n=1}^{\infty}(1+(-1)^{n+1}/n)$ of Example 8.5 is convergent but not absolutely convergent since the series $\sum_{n=1}^{\infty}1/n$ diverges.

We are now ready to pass to infinite products of functions.

> **DEFINITION 8.13.** Let $\{f_n\}_{n=1}^{\infty}$ be a sequence of continuous complex-valued functions defined in an open set $U \subset \mathbb{C}$. We say that the infinite product $f = \prod_{n=1}^{\infty}f_n$ **converges compactly** in U if for every compact set $K \subset U$ there exists an integer n_0 such that the sequence $\{\prod_{n=n_0}^{k}f_n\}$ converges uniformly on K to a non-vanishing function g as $k \to \infty$, so $f = f_1 \cdots f_{n_0-1}g$ on K.

It is readily seen that a compactly convergent infinite product of continuous functions defines a continuous function.

The following theorem gives a useful sufficient condition for compact convergence of infinite products of continuous functions. It essentially follows from what we have already seen for numerical infinite products once we note that all the estimates involved hold uniformly on compact sets.

THEOREM 8.14. *Suppose $\{f_n\}_{n=1}^{\infty}$ is a sequence of continuous complex-valued functions in U such that $\sum_{n=1}^{\infty}|f_n-1|$ converges compactly in U. Then*

(i) *The infinite product $f = \prod_{n=1}^{\infty}f_n$ converges compactly in U.*
(ii) *$f(p) = 0$ if and only if $f_n(p) = 0$ for some n.*
(iii) *$\prod_{n=m}^{\infty}f_n \to 1$ compactly in U as $m \to \infty$.*

Proof. (i) Fix a compact set $K \subset U$. Since $f_n \to 1$ uniformly on K, there is an $n_0 \geq 1$ such that $|f_n-1| < 1$ on K for all $n \geq n_0$. For any $0 < \varepsilon < r_0$ find an integer $N > n_0$ such that $\sum_{n=m}^{k}|f_n-1| < \varepsilon$ on K whenever $k \geq m \geq N$. Then, as in the proof of Theorem 8.10, we have the uniform estimate $|\sum_{n=m}^{k}\mathrm{Log}\,f_n| \leq 3\varepsilon/2$ on K if $k \geq m \geq N$. This shows that $\sum_{n=n_0}^{\infty}\mathrm{Log}\,f_n$ converges uniformly on K to a continuous limit h. It easily follows that $\prod_{n=n_0}^{\infty}f_n$ converges uniformly on K to the non-vanishing function $\exp(h)$.

(ii) follows from Theorem 8.2(ii).

(iii) Take a compact set $K \subset U$ and an arbitrary $\varepsilon > 0$. Choose n_0 as in (i). Find $0 < \delta < r_0$ such that $|z| < 3\delta/2$ implies $|\exp(z)-1| < \varepsilon$. Finally, find an integer $N > n_0$ such that $\sum_{n=m}^{k}|f_n-1| < \delta$ on K whenever $k \geq m \geq N$. Then, as in (i), we have the estimate $|\sum_{n=m}^{k}\mathrm{Log}\,f_n| \leq 3\delta/2$ and therefore

$$\left|\prod_{n=m}^{k}f_n - 1\right| = \left|\exp\left(\sum_{n=m}^{k}\mathrm{Log}\,f_n\right) - 1\right| < \varepsilon$$

on K whenever $k \geq m \geq N$. Letting $k \to \infty$, we obtain the bound $|\prod_{n=m}^{\infty} f_n - 1| \leq \varepsilon$ on K for all $m \geq N$. This shows that the sequence $\{\prod_{n=m}^{\infty} f_n\}$ tends to 1 uniformly on K as $m \to \infty$. $\qquad\square$

As a corollary of the above theorem, we arrive at the following result which we will use repeatedly in this chapter:

THEOREM 8.15. *Suppose $f_n \in \mathcal{O}(U)$ and $\sum_{n=1}^{\infty} |f_n - 1|$ converges compactly in U. Then $f = \prod_{n=1}^{\infty} f_n \in \mathcal{O}(U)$. Moreover, if no f_n is identically zero in any connected component of U, then*

$$\operatorname{ord}(f, p) = \sum_{n=1}^{\infty} \operatorname{ord}(f_n, p) \qquad \text{for all } p \in U.$$

Note that for each p all but finitely many terms $\operatorname{ord}(f_n, p)$ vanish by Theorem 8.14(ii), so the right side is in fact a finite sum.

PROOF. $f = \prod_{n=1}^{\infty} f_n$ converges compactly in U by Theorem 8.14, so $f \in \mathcal{O}(U)$ by Theorem 5.11. Fix $p \in U$ and a disk D centered at p such that $\overline{D} \subset U$. Since $f_n \to 1$ uniformly in D as $n \to \infty$, there is an integer n_0 such that $f_n \neq 0$ in D for all $n > n_0$. The product $g = \prod_{n=n_0+1}^{\infty} f_n$ converges compactly in U, so $g \in \mathcal{O}(U)$. Moreover, $g \neq 0$ in D by the choice of n_0, so $\operatorname{ord}(g, p) = 0$. Since $f = f_1 \cdots f_{n_0} g$ and $\operatorname{ord}(f_n, p) = 0$ for $n > n_0$, we obtain

$$\operatorname{ord}(f, p) = \sum_{n=1}^{n_0} \operatorname{ord}(f_n, p) + \operatorname{ord}(g, p) = \sum_{n=1}^{n_0} \operatorname{ord}(f_n, p) = \sum_{n=1}^{\infty} \operatorname{ord}(f_n, p). \qquad\square$$

EXAMPLE 8.16. The infinite product $f(z) = \prod_{n=1}^{\infty} \cos(z/n)$ defines an entire function. To see this, use the power series representation

$$\cos w = 1 - \frac{1}{2!} w^2 + \frac{1}{4!} w^4 - \frac{1}{6!} w^6 + \cdots$$

to obtain the estimate

$$|\cos w - 1| \leq C|w|^2 \qquad \text{for } |w| \leq 1,$$

where

$$C = \frac{1}{2!} + \frac{1}{4!} + \frac{1}{6!} + \cdots = \cosh(1) - 1 > 0.$$

Given any radius $r > 0$, this yields the estimate

$$\left| \cos\left(\frac{z}{n}\right) - 1 \right| \leq \frac{Cr^2}{n^2} \qquad \text{for } |z| \leq r \text{ and } n \geq r.$$

Since the series $\sum 1/n^2$ converges, it follows from the Weierstrass M-test (Corollary 5.13) that $\sum_{n=1}^{\infty} |\cos(z/n) - 1|$ converges compactly in $\mathbb{C}$. Theorem 8.15 now shows that $f \in \mathcal{O}(\mathbb{C})$.

EXAMPLE 8.17. Let $\{z_n\}$ be a sequence of non-zero complex numbers such that $\sum_{n=1}^{\infty} 1/|z_n|$ converges. By the Weierstrass M-test, the series $\sum_{n=1}^{\infty} |z/z_n|$ converges compactly in $\mathbb{C}$.

It follows from Theorem 8.15 that the infinite product $\prod_{n=1}^{\infty}(1-z/z_n)$ defines an entire function with zeros precisely along the sequence $\{z_n\}$.

For generalizations of this construction, see Theorem 8.21.

The following statement is often useful when dealing with derivatives of infinite products:

THEOREM 8.18 (Logarithmic differentiation). *Suppose $\{f_n\}$ is a sequence of holomorphic functions in a domain U, none being identically zero. If the infinite product $\prod_{n=1}^{\infty} f_n$ converges compactly in U to f, then*

$$\frac{f'}{f} = \sum_{n=1}^{\infty} \frac{f_n'}{f_n} \qquad \text{in } U.$$

More precisely, for every compact set $K \subset U$ the meromorphic function $f'/f - \sum_{n=1}^{k} f_n'/f_n$ has its poles outside K for all large k, and tends to zero uniformly on K as $k \to \infty$.

PROOF. Fix a compact set $K \subset U$ and any $0 < \varepsilon < 1$. Choose $0 < r < 1/2$ small enough to guarantee that the closed r-neighborhood $K_r = \{z \in \mathbb{C} : \text{dist}(z, K) \leq r\}$ is contained in U. Since $g_k = \prod_{n=k+1}^{\infty} f_n \to 1$ uniformly on the compact set K_r by Theorem 8.14(iii), we can find an integer N so that

$$(8.2) \qquad \sup_{z \in K_r} |g_k(z) - 1| < r\varepsilon \qquad \text{for all } k \geq N$$

(in particular, $g_k \neq 0$ on K_r for all $k \geq N$). Since $\overline{\mathbb{D}}(z, r) \subset K_r$ for every $z \in K$, Theorem 1.43 applied to $g_k - 1$ gives

$$(8.3) \qquad \sup_{z \in K} |g_k'(z)| \leq \frac{r\varepsilon}{r} = \varepsilon \qquad \text{for all } k \geq N.$$

Using the relation

$$\frac{f'}{f} = \sum_{n=1}^{k} \frac{f_n'}{f_n} + \frac{g_k'}{g_k},$$

obtained by differentiating the finite product $f = f_1 \cdots f_k\, g_k$, along with estimates (8.2) and (8.3), we obtain

$$\sup_{z \in K} \left| \frac{f'(z)}{f(z)} - \sum_{n=1}^{k} \frac{f_n'(z)}{f_n(z)} \right| = \sup_{z \in K} \left| \frac{g_k'(z)}{g_k(z)} \right| \leq \frac{\varepsilon}{1 - r\varepsilon} < 2\varepsilon$$

for all $k \geq N$, which proves the result. $\qquad \square$

8.2 Weierstrass's theory of elementary factors

The fundamental theorem of algebra (see Examples 1.45 and 1.67) asserts that every non-constant complex polynomial P has a *unique* factorization of the form

$$(8.4) \qquad P(z) = C\,z^m \prod_{n=1}^{k} \left(1 - \frac{z}{z_n} \right),$$

where C is a non-zero constant, $m \geq 0$ is an integer, and $z_1, \ldots, z_k$ are the non-zero roots of P, repeated according to their multiplicity. Thus, two polynomials with the same roots of the same multiplicities agree up to a multiplicative constant.

Naturally, one is interested in knowing whether such a factorization is possible for entire functions. Two problems arise immediately: First, one has to deal with the possibility of having infinitely many zeros leading to infinite products. Second, two entire functions with the same zeros of the same orders coincide up to multiplication by a non-vanishing entire function; the uniqueness of such factorizations can no longer be guaranteed.

The most naive attempt to construct an entire function with zeros along a given sequence $\{z_n\}$ would be to consider $\prod_{n=1}^{\infty}(1 - z/z_n)$. But this infinite product need not converge in the plane unless some condition on the growth of $\{z_n\}$ is imposed (compare Example 8.17). To guarantee convergence when such conditions fail, the linear factors $(1 - z/z_n)$ must be replaced by new types of factors, an idea that led Weierstrass to the notion of elementary factors.

> **DEFINITION 8.19.** The **Weierstrass elementary factors** E_d are the entire functions defined by
>
> $$E_0(z) = 1 - z$$
>
> $$E_d(z) = (1 - z)\,\exp\left(\sum_{n=1}^{d} \frac{z^n}{n} \right) \qquad d = 1, 2, 3, \ldots$$

The elementary factors vanish only at the simple zero $z = 1$. Observe that the power series $\sum_{n=1}^{\infty} z^n/n$ represents the principal branch of the logarithm of $1/(1 - z)$ in the unit disk $\mathbb{D}$ which maps 0 to 0 (see Example 2.23). Hence, when $|z| < 1$ and $d \geq 1$,

$$E_d(z) = (1 - z)\,\exp\left(\sum_{n=1}^{\infty} \frac{z^n}{n} \right) \exp\left(-\sum_{n=d+1}^{\infty} \frac{z^n}{n} \right)$$

$$= \exp\left(-\sum_{n=d+1}^{\infty} \frac{z^n}{n} \right) = 1 - \frac{z^{d+1}}{d+1} + \cdots.$$

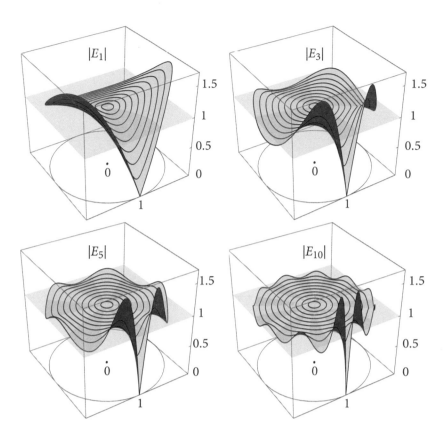

Figure 8.1. Graph of the absolute value $|E_d|$ of the Weierstrass elementary factor over the unit disk for $d = 1, 3, 5, 10$.

Thus, on any compact subset of $\mathbb{D}$ we expect E_d to be close to the constant function 1 when d gets large (compare Fig. 8.1). The following lemma is a quantitative version of this observation:

LEMMA 8.20. *For every $d \geq 0$,*

$$|E_d(z) - 1| \leq |z|^{d+1} \qquad if \ |z| \leq 1.$$

PROOF. We assume $d \geq 1$ since the case $d = 0$ is trivial. The computation

$$E_d(z) - 1 = E_d(z) - E_d(0) = z \int_0^1 E_d'(tz) \, dt$$

suggests that we estimate $|E_d'|$. Differentiation of the formula for E_d gives

$$E_d'(z) = -z^d \exp\left(\sum_{n=1}^{d} \frac{z^n}{n}\right).$$

Using the inequality $|\exp(w)| \leq \exp|w|$, we see that when $|z| \leq 1$ and $0 \leq t \leq 1$,

$$|E_d'(tz)| \leq |z|^d\, t^d \, \exp\left(\sum_{n=1}^d \frac{t^n |z|^n}{n}\right) \leq |z|^d\, t^d \, \exp\left(\sum_{n=1}^d \frac{t^n}{n}\right) = -|z|^d\, E_d'(t).$$

It follows that

$$|E_d(z) - 1| \leq |z| \int_0^1 |E_d'(tz)|\, dt \leq -|z|^{d+1} \int_0^1 E_d'(t)\, dt$$

$$= -|z|^{d+1}(E_d(1) - E_d(0)) = |z|^{d+1}. \qquad \square$$

The following theorem shows that the zeros of an entire function can be pre-scribed arbitrarily.

THEOREM 8.21 (Weierstrass product theorem, 1876). *Suppose $\{z_n\}$ is a sequence in $\mathbb{C}^*$ such that $r_n = |z_n| \to +\infty$ as $n \to \infty$. Let $\{d_n\}$ be any sequence of non-negative integers for which the series*

$$\sum_{n=1}^\infty \left(\frac{r}{r_n}\right)^{d_n+1}$$

converges for every $r > 0$ (for example, $d_n = n$ will always do). Then the infinite product

(8.5) $$f(z) = \prod_{n=1}^\infty E_{d_n}\left(\frac{z}{z_n}\right)$$

defines an entire function which vanishes precisely along $\{z_n\}$. Moreover, if some point appears m times in the sequence $\{z_n\}$, then f has a zero of order m at that point.

To create a zero of order m at $z = 0$, simply multiply the product by z^m.

PROOF. Since $E_{d_n}(z/z_n)$ has a simple zero at z_n, by Theorem 8.15 it suffices to show that the infinite product (8.5) converges compactly in $\mathbb{C}$. For any $r > 0$ there is an integer N such that $r_n \geq r$ for all $n \geq N$. By Lemma 8.20,

$$\left| E_{d_n}\left(\frac{z}{z_n}\right) - 1 \right| \leq \left| \frac{z}{z_n} \right|^{d_n+1} \leq \left(\frac{r}{r_n}\right)^{d_n+1}$$

whenever $|z| \leq r$ and $n \geq N$. Since by the assumption $\sum (r/r_n)^{d_n+1} < +\infty$, the Weierstrass M-test shows that $\sum |E_{d_n}(z/z_n) - 1|$ converges compactly in $\mathbb{C}$. The result now follows from Theorem 8.15. $\qquad \square$

COROLLARY 8.22. *Under the assumptions of the above theorem, if $\sum r_n^{-1} < +\infty$, we can choose $d_n = 0$ for all n, and f takes the simple form*

$$f(z) = \prod_{n=1}^\infty \left(1 - \frac{z}{z_n}\right).$$

More generally, if $\sum r_n^{-(d+1)} < +\infty$ but $\sum r_n^{-d} = +\infty$ for some positive integer d, we can choose $d_n = d$ for all n, and f takes the form

$$f(z) = \prod_{n=1}^{\infty} \left(1 - \frac{z}{z_n}\right) \exp\left[\left(\frac{z}{z_n}\right) + \frac{1}{2}\left(\frac{z}{z_n}\right)^2 + \cdots + \frac{1}{d}\left(\frac{z}{z_n}\right)^d\right].$$

These expressions are traditionally known as **canonical products**.

EXAMPLE 8.23. Let us construct an entire function with simple zeros at the integer points along the real axis and zeros of order 2 at the points $\pm i\sqrt{n}$ for $n \geq 1$ along the imaginary axis, and with no other zeros. Since $\sum 1/n^2 < +\infty$, Theorem 8.21 shows that the infinite product

$$g(z) = z \prod_{n=1}^{\infty} E_1\left(\frac{z}{n}\right) \prod_{n=1}^{\infty} E_1\left(-\frac{z}{n}\right)$$

is an entire function with a simple zero at every integer. Similarly, since $\sum (1/\sqrt{n})^3 < +\infty$, the infinite product

$$h(z) = \prod_{n=1}^{\infty} \left[E_2\left(\frac{z}{i\sqrt{n}}\right)\right]^2 \prod_{n=1}^{\infty} \left[E_2\left(-\frac{z}{i\sqrt{n}}\right)\right]^2$$

is an entire function with zeros of order 2 at the points $\pm i\sqrt{n}$. The product $f = gh$ will then be the required function. Substituting the expressions for E_1 and E_2 in these formulas and using Theorem 8.2(iv), we can combine terms to obtain

$$f(z) = z \prod_{n=1}^{\infty} \left(1 - \frac{z^2}{n^2}\right)\left(1 + \frac{z^2}{n}\right)^2 \exp\left(-\frac{2z^2}{n}\right).$$

We are now ready to prove an analogue of the factorization (8.4) for entire functions.

THEOREM 8.24 (Weierstrass factorization theorem, 1876). *Suppose $f \in \mathcal{O}(\mathbb{C})$ is not identically zero. Arrange the zeros of f in $\mathbb{C}^*$ in a sequence $\{z_n\}$ where each zero is repeated as many times as its order. Then f has a factorization of the form*

$$(8.6) \qquad\qquad f(z) = e^{g(z)} z^m \prod_{n=1}^{\infty} E_n\left(\frac{z}{z_n}\right),$$

where $g \in \mathcal{O}(\mathbb{C})$ and $m = \mathrm{ord}(f, 0)$.

It is understood that the infinite product in this statement will reduce to a polynomial if f has finitely many zeros, and will be missing altogether if f has no zeros in $\mathbb{C}^*$. For a sharper result for certain classes of entire functions, see Theorem 8.52.

PROOF. Evidently $|z_n| \to +\infty$ since $\{z_n\}$ has no accumulation point in $\mathbb{C}$. By Theorem 8.21, the function $h(z) = z^m \prod_{n=1}^{\infty} E_n(z/z_n)$ is entire with the same zeros of the same orders as f. The ratio f/h has removable singularities at the z_n as well as the origin, so it extends to a non-vanishing entire function. Corollary 2.22 then shows that $f/h = \exp(g)$ for some $g \in \mathcal{O}(\mathbb{C})$.   □

It is not hard to generalize the Weierstrass product theorem to arbitrary open sets.

Versions of this result seem to have first appeared in an 1884 paper of Mittag-Leffler and then in an 1894 paper of Picard.

THEOREM 8.25. *Let $U \subset \mathbb{C}$ be open and $\{z_n\} \subset U$ be a sequence with no accumulation point in U. Then there exists an $f \in \mathcal{O}(U)$ which vanishes precisely along $\{z_n\}$. Moreover, if some point appears m times in the sequence $\{z_n\}$, then f has a zero of order m at that point.*

For a more general interpolation result, see Theorem 9.5.

PROOF. The case $U = \mathbb{C}$ has already been dealt with in Theorem 8.21, so let us assume $U \neq \mathbb{C}$. It will be convenient to first apply a Möbius change of coordinates to turn $\{z_n\}$ into a bounded sequence in the plane. To this end, take any $q \in U \setminus \{z_n\}$, let $\varphi(z) = 1/(z - q)$, and consider the images $V = \varphi(U) \subset \hat{\mathbb{C}}$ and $w_n = \varphi(z_n)$. Note that $\infty \in V$, so ∂V is a compact subset of the plane. Moreover, $\{w_n\}$ does not accumulate on $\varphi(q) = \infty$, which means it is a bounded sequence in $\mathbb{C}$. We will construct a holomorphic function $g : V \to \mathbb{C}$, with $g(\infty) = 1$, which vanishes precisely along $\{w_n\}$. The composition $f = g \circ \varphi : U \to \mathbb{C}$ will have the required property.

Let ζ_n be a closest point of ∂V to w_n. Since $\{w_n\}$ is bounded and has no accumulation point in V, we must have $w_n - \zeta_n \to 0$ as $n \to \infty$. Define

$$(8.7) \qquad g(w) = \prod_{n=1}^{\infty} E_n \left(\frac{w_n - \zeta_n}{w - \zeta_n} \right) \qquad \text{for } w \in V.$$

The n-th factor $E_n((w_n - \zeta_n)/(w - \zeta_n))$ is holomorphic in V (it takes the value $E_n(0) = 1$ at $w = \infty$) and has a simple zero at w_n. It suffices to verify that the infinite product converges compactly in V.

Take any compact set $K \subset V$. Since K has a positive distance to ∂V while $w_n - \zeta_n \to 0$ as $n \to \infty$, we can find an integer N such that

$$|w - \zeta_n| \geq 2|w_n - \zeta_n| \qquad \text{if } w \in K \text{ and } n \geq N.$$

It follows from Lemma 8.20 that for all $w \in K$ and $n \geq N$,

$$\left| E_n \left(\frac{w_n - \zeta_n}{w - \zeta_n} \right) - 1 \right| \leq \left| \frac{w_n - \zeta_n}{w - \zeta_n} \right|^{n+1} \leq \frac{1}{2^{n+1}}.$$

Since $\sum 1/2^n$ converges, the series $\sum |E_n((w_n - \zeta_n)/(w - \zeta_n)) - 1|$ converges compactly in V by the Weierstrass M-test. Theorem 8.15 now shows that the infinite product in (8.7) converges compactly in V.   □

We conclude with an important corollary of the above theorem.

COROLLARY 8.26. *Every meromorphic function in an open set $U \subset \mathbb{C}$ is the ratio of two holomorphic functions in U.*

PROOF. Let $f \in \mathcal{M}(U)$ and consider the set $E = f^{-1}(\infty)$ of poles of f. By Theorem 8.25, there exists an $h \in \mathcal{O}(U)$ which has a zero of order $\operatorname{ord}(f, p) \geq 1$ at every $p \in E$ and does not vanish anywhere else. Set $g = f h$. The singularities of g at the points of E are all removable, so $g \in \mathcal{O}(U)$. Evidently $f = g/h$ in $U \smallsetminus E$.   □

8.3 Jensen's formula and its applications

We now establish a fundamental link between the average size of a holomorphic function on a circle and the distribution of its zeros inside. To motivate this result, suppose f is holomorphic and non-vanishing in some disk $\mathbb{D}(0, R)$. Then $\log|f|$ is harmonic (Example 7.7), so by the mean value property (Theorem 7.16), for every $r < R$,

$$\frac{1}{2\pi} \int_0^{2\pi} \log|f(re^{it})| \, dt = \log|f(0)|.$$

In other words, the average value of $\log|f|$ over the circle $|z| = r$ is independent of r as long as there are no zeros of f inside. The following theorem treats the case where zeros are allowed:

THEOREM 8.27 (Jensen's formula, 1898). *Suppose f is holomorphic in the disk $\mathbb{D}(0, R)$, $f(0) \neq 0$, and f has no zeros on the circle $|z| = r < R$. Let $z_1, z_2, \ldots, z_k$ be the zeros of f in $\mathbb{D}(0, r)$, where each zero is repeated as many times as its order. Then*

$$(8.8) \qquad \frac{1}{2\pi} \int_0^{2\pi} \log|f(re^{it})| \, dt = \log|f(0)| + \sum_{n=1}^{k} \log\left(\frac{r}{|z_n|}\right).$$

Johan Ludwig William Valdemar Jensen (1859–1925)

To obtain the corresponding formula when $f(0) = 0$, apply the theorem to $f(z)/z^m$, where $m = \operatorname{ord}(f, 0)$.

Jensen's formula remains valid even if f has zeros on the circle $|z| = r$ (see problem 15). In this case the function $t \mapsto \log|f(re^{it})|$ has logarithmic singularities (similar to $t \mapsto \log|t|$ at $t = 0$) but is nonetheless integrable. We will not need this more general case in what follows.

Before proving Jensen's formula, let us interpret it in a way that is more geometric and easier to remember (compare [**Mi**]): *The average value of $\log|f|$ over the circle $|z| = r$ is a piecewise affine function of $\log r$.* It remains an affine function of $\log r$ as long as the circle $|z| = r$ does not pass through any zero of f, and the slope of this affine function is the number of zeros of f inside the disk $\mathbb{D}(0, r)$. When the circle passes through one or more zeros, the slope jumps by the number of those zeros, counting multiplicities (see Fig. 8.2). The transition is continuous since the zeros on a circle $|z| = r_0$ do not show up in the sum in (8.8) as $r \to r_0^-$ and their contribution to this sum vanishes as $r \to r_0^+$.

PROOF. It suffices to treat the case $r = 1$; the general case will follow by considering the function $f(rz)$ in the unit disk with zeros at z_n/r. The finite Blaschke product

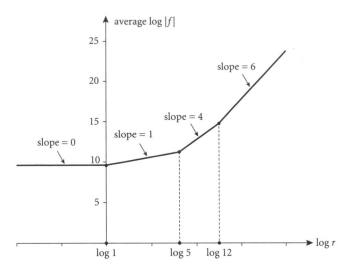

Figure 8.2. The average value of $\log|f(z)|$ on the circle $|z|=r$ as a function of $\log r$, where $f(z)=(z-1)(z+5i)^3(z-12)^2$. Jensen's formula implies that this function is piecewise affine, with a slope equal to the total number of zeros inside $\mathbb{D}(0,r)$.

$$B(z)=\prod_{n=1}^{k}\left(\frac{z-z_n}{1-\bar{z}_n z}\right)$$

has the same zeros of the same orders as f, so the singularities of the ratio $g=f/B$ are removable. It follows that g extends to a non-vanishing holomorphic function in some neighborhood of $\overline{\mathbb{D}}$, which implies $\log|g|$ is harmonic in this neighborhood. By the mean value property,

$$\frac{1}{2\pi}\int_0^{2\pi}\log|g(e^{it})|\,dt=\log|g(0)|.$$

Substituting $g=f/B$ and using the fact that $|B|=1$ on the unit circle, we obtain

$$\frac{1}{2\pi}\int_0^{2\pi}\log|f(e^{it})|\,dt=\log|f(0)|-\log|B(0)|$$

$$=\log|f(0)|+\sum_{n=1}^{k}\log\left(\frac{1}{|z_n|}\right),$$

which is Jensen's formula (8.8) for $r=1$.   □

EXAMPLE 8.28. Theorem 8.27 allows a quick proof of the fundamental theorem of algebra. Suppose $P:\mathbb{C}\to\mathbb{C}$ is a monic polynomial of degree $d\geq 1$. Then, for large r, the average value of $\log|P(z)|$ over the circle $|z|=r$ is asymptotic to $d\log r$. By Jensen's formula, there are d zeros of P in the disk $\mathbb{D}(0,r)$, counting multiplicities.

> **DEFINITION 8.29.** Let f be holomorphic in the disk $\mathbb{D}(0, R)$. For $0 < r < R$, we set
>
> $$\mathrm{M}(r) = \mathrm{M}_f(r) = \sup_{|z|=r} |f(z)|,$$
>
> $$\mathrm{N}(r) = \mathrm{N}_f(r) = \sum_{p \in f^{-1}(0) \cap \mathbb{D}(0,r)} \mathrm{ord}(f, p).$$
>
> Thus, $\mathrm{N}(r)$ is the number of zeros of f in $\mathbb{D}(0, r)$, counting multiplicities.

The function $r \mapsto \mathrm{M}(r)$ is continuous and strictly increasing by the maximum principle unless f is constant. A more clever application of the maximum principle reveals more:

THEOREM 8.30 (Hadamard's 3-circles theorem, 1896). *If f is not identically zero, then $\log \mathrm{M}(r)$ is a convex increasing function of $\log r$.*

Compare this with Jensen's formula which implies the same property when $\log \mathrm{M}(r)$ is replaced with the average value of $\log |f|$ on the circle $|z| = r$.

PROOF. $\log \mathrm{M}(r)$ is an increasing function of $\log r$ since $\mathrm{M}(r)$ is an increasing function of r. To see convexity, take a, b with $0 < a < b < R$. Let n, m be arbitrary integers with $m > 0$. The maximum principle applied to the holomorphic function $z^n f(z)^m$ gives

$$r^n \mathrm{M}(r)^m \le \max\left\{ a^n \mathrm{M}(a)^m, b^n \mathrm{M}(b)^m \right\} \qquad \text{if } a \le r \le b,$$

which can be written as

$$\log \mathrm{M}(r) \le \max\left\{ \frac{n}{m} \log\left(\frac{a}{r}\right) + \log \mathrm{M}(a), \frac{n}{m} \log\left(\frac{b}{r}\right) + \log \mathrm{M}(b) \right\}.$$

Since this holds for every rational number n/m, we can take the limit of the right side of this inequality as $n/m \to \log(\mathrm{M}(a)/\mathrm{M}(b))/\log(b/a)$. In that case, the two quantities on the right tend to a common value and we obtain

$$\log \mathrm{M}(r) \le \frac{\log \mathrm{M}(a) - \log \mathrm{M}(b)}{\log\left(\frac{b}{a}\right)} \log\left(\frac{a}{r}\right) + \log \mathrm{M}(a)$$

$$= \frac{\log\left(\frac{b}{r}\right)}{\log\left(\frac{b}{a}\right)} \log \mathrm{M}(a) + \frac{\log\left(\frac{r}{a}\right)}{\log\left(\frac{b}{a}\right)} \log \mathrm{M}(b)$$

whenever $a \le r \le b$. This proves the asserted convexity. $\qquad \square$

Here is a useful corollary of Jensen's formula:

COROLLARY 8.31. *Suppose f is holomorphic in a neighborhood of the closed disk $\overline{\mathbb{D}}(0, R)$ and $f(0) \neq 0$. Let $z_1, z_2, \ldots, z_k$ be the zeros of f in $\mathbb{D}(0, R)$, arranged so that $0 < |z_1| \leq |z_2| \leq \cdots \leq |z_k|$. Then, for all $0 < r < R$,*

$$\text{(8.9)} \qquad \prod_{n=1}^{N(r)} \frac{R}{|z_n|} \leq \frac{M(R)}{|f(0)|}.$$

As a result,

$$\text{(8.10)} \qquad N(r) \leq \frac{\log M(R) - \log |f(0)|}{\log R - \log r}.$$

PROOF. Choose a radius s such that $r < s < R$ and $f(z) \neq 0$ if $|z| = s$. By (8.8),

$$M(s) \geq \exp\left(\frac{1}{2\pi} \int_0^{2\pi} \log |f(se^{it})| \, dt \right) = |f(0)| \prod_{n=1}^{N(s)} \frac{s}{|z_n|} \geq |f(0)| \prod_{n=1}^{N(r)} \frac{s}{|z_n|}.$$

Letting $s \to R$ (through radii for which $|f| \neq 0$ on the circle $|z| = s$), we obtain (8.9). Since each $|z_n|$ in (8.9) is less than r, it follows that

$$\left(\frac{R}{r} \right)^{N(r)} \leq \frac{M(R)}{|f(0)|},$$

which, after taking logarithms, gives (8.10). □

REMARK 8.32. The comparison between $N(r)$ and $M(r)$ goes in one direction only. By (8.10), any bound on the growth of $M(r)$ will force a bound on $N(r)$, but the information on the growth of $N(r)$ gives no clue as to how big $M(r)$ might be. This point is illustrated by an iterated exponential $\exp \circ \cdots \circ \exp$ for which $M(r)$ can be made to grow faster and faster (by choosing higher and higher iterates) but $N(r) = 0$ for all r. Similarly, multiplying any entire function by a non-vanishing entire function does not alter $N(r)$ but can make $M(r)$ grow much faster.

EXAMPLE 8.33. Suppose $f : \mathbb{D} \to \mathbb{D}$ is holomorphic with $f(0) \neq 0$, so the inequality (8.10) holds whenever $0 < r < R < 1$. It follows by letting $R \to 1$ that

$$N(r) \leq \frac{\log |f(0)|}{\log r} \qquad \text{if } 0 < r < 1.$$

In particular, $N(r) = 0$ whenever $r < |f(0)|$. In other words, f has no zeros in the disk $\mathbb{D}(0, |f(0)|)$, a fact that can also be proved using the Schwarz lemma (see problem 19 in chapter 4).

Jensen's formula and its Corollary 8.31 will be used in the next section where we study entire functions of finite order. For now, let us show two classical applications in which the domain of the function is a disk of finite radius.

THEOREM 8.34 (Zeros of bounded holomorphic functions in $\mathbb{D}$). *Let $\{z_n\}$ be an infinite sequence in $\mathbb{D}$ arranged so that $0 < |z_1| \leq |z_2| \leq |z_3| \leq \cdots < 1$. There is a bounded holomorphic function in $\mathbb{D}$ which vanishes precisely along $\{z_n\}$ if and only if $\sum_{n=1}^{\infty}(1 - |z_n|) < +\infty$.*

PROOF. First suppose the series converges. Consider the function

$$B_n(z) = -\frac{|z_n|}{z_n}\left(\frac{z - z_n}{1 - \bar{z}_n z}\right),$$

i.e., the unique element of $\mathrm{Aut}(\mathbb{D})$ such that $B_n(z_n) = 0$ and $B_n(0) > 0$. We claim that the infinite Blaschke product defined by $B = \prod_{n=1}^{\infty} B_n$ has the desired properties. We have

$$|B_n(z) - 1| = \left|\frac{|z_n|}{z_n}\left(\frac{z - z_n}{1 - \bar{z}_n z}\right) + 1\right| = \left|\frac{|z_n|z + z_n}{z_n(1 - \bar{z}_n z)}\right|(1 - |z_n|),$$

which, in view of $|z_n| < 1$, gives the estimate

$$|B_n(z) - 1| \leq \frac{1 + |z|}{1 - |z|}(1 - |z_n|) \qquad \text{if } z \in \mathbb{D}.$$

Since by the assumption $\sum_{n=1}^{\infty}(1 - |z_n|) < +\infty$, the Weierstrass M-test shows that $\sum_{n=1}^{\infty}|B_n - 1|$ converges compactly in $\mathbb{D}$. Hence, by Theorem 8.15, $B = \prod_{n=1}^{\infty} B_n$ is a holomorphic function in $\mathbb{D}$ with zeros precisely along $\{z_n\}$. Moreover, $|B| < 1$ in $\mathbb{D}$ because $|B_n| < 1$ in $\mathbb{D}$ for all n.

Conversely, suppose there is an $f \in \mathcal{O}(\mathbb{D})$ which vanishes precisely along $\{z_n\}$ such that $M = \sup_{z \in \mathbb{D}}|f(z)|$ is finite. For $0 < r < R < 1$, the inequality (8.9) gives

$$\sum_{n=1}^{\mathrm{N}(r)} \log\left(\frac{R}{|z_n|}\right) \leq \log\left(\frac{M}{|f(0)|}\right).$$

Letting $R \to 1$, and using the elementary inequality $1 - x < \log(1/x)$ for $0 < x < 1$, we obtain the estimate

$$\sum_{n=1}^{\mathrm{N}(r)}(1 - |z_n|) < \sum_{n=1}^{\mathrm{N}(r)} \log\left(\frac{1}{|z_n|}\right) \leq \log\left(\frac{M}{|f(0)|}\right).$$

Letting $r \to 1$, we conclude that $\sum_{n=1}^{\infty}(1 - |z_n|) \leq \log(M/|f(0)|) < +\infty$. $\square$

EXAMPLE 8.35. Here is a restatement of part of Theorem 8.34:

If $f \in \mathcal{O}(\mathbb{D})$ is bounded, if $f(z_n) = 0$ for all n, and if $\sum(1 - |z_n|) = +\infty$, then $f = 0$ everywhere.
For if f is not identically zero, the sum $\sum(1 - |w|)$ over *all* zeros $w \in \mathbb{D}$ of f would be finite by Theorem 8.34, which would imply the convergence of the dominated sum $\sum(1 - |z_n|)$.

Let us illustrate the above statement: By Theorem 8.25 there is a holomorphic function in $\mathbb{D}$ which vanishes precisely along the sequence $\{1 - 1/n\}$. Any such function is necessarily unbounded, for if $f \in \mathcal{O}(\mathbb{D})$ is bounded and $f(1 - 1/n) = 0$ for all n, then $f = 0$ everywhere.

Our second application of Jensen's formula is the following generalization of Lemma 6.28 which was treated using the idea of extending by reflection:

THEOREM 8.36 (F. Riesz and M. Riesz, 1916). *Suppose $f \in \mathcal{O}(\mathbb{D})$ is bounded and the radial limit $f^*(e^{it}) = \lim_{r \to 1} f(re^{it})$ is zero on a set $E \subset [0, 2\pi]$ of positive Lebesgue measure. Then f is identically zero.*

PROOF. Suppose f is not identically zero. Replacing $f(z)$ with $f(z)/z^m$ where $m = \mathrm{ord}(f, 0)$, we may assume without loss of generality that $f(0) \neq 0$. Take any $0 < \varepsilon < 1$. Define, for $n \geq 1$, the sets

$$E_n(\varepsilon) = \left\{ t \in [0, 2\pi] : |f(re^{it})| < \varepsilon \text{ for all } 1 - \frac{1}{n} < r < 1 \right\}.$$

Evidently,

(8.11) $$E_n(\varepsilon) \subset E_{n+1}(\varepsilon) \quad \text{and} \quad E \subset \bigcup_{n=1}^{\infty} E_n(\varepsilon).$$

For each $n \geq 1$ take a radius r so that $1 - 1/n < r < 1$ and use (8.8) to estimate

$$\log |f(0)| \leq \frac{1}{2\pi} \left(\int_{E_n(\varepsilon)} + \int_{[0,2\pi] \setminus E_n(\varepsilon)} \right) \log |f(re^{it})| \, dt$$

$$\leq \frac{1}{2\pi} \ell(E_n(\varepsilon)) \, \log \varepsilon + \log M,$$

where ℓ is Lebesgue measure and $M = \sup_{z \in \mathbb{D}} |f(z)|$. Setting $C = \log(M/|f(0)|)$, which is strictly positive since f is non-constant, we can write the above inequality as

$$\ell(E_n(\varepsilon)) \leq \frac{2\pi C}{\log(1/\varepsilon)} \quad \text{for all } n \geq 1.$$

This, in view of (8.11), implies $\ell(E) \leq 2\pi C / \log(1/\varepsilon)$. Letting $\varepsilon \to 0$, we obtain $\ell(E) = 0$, which contradicts the assumption. □

8.4 Entire functions of finite order

The growth function $M_f(r)$ associated with a non-constant entire function (Definition 8.29) is strictly increasing and tends to $+\infty$ as $r \to +\infty$. The first statement follows from the maximum principle and the second from Liouville's Theorem. The question of how fast $M_f(r)$ tends to $+\infty$ and its implications for the distribution of zeros of f, or the possibilities for representing f as an infinite product, is the main subject of this section.

Throughout this section the symbol "const." will be used for a positive constant whose specific value is not important to us. Different occurrences of const. in an argument will be considered independent. For example, we will lump several const.'s together and write the result as another const. and so on. We often simplify the notation $M_f(r)$ to $M(r)$ if the function f is clear from the context.

We begin with the following elementary observation:

THEOREM 8.37. *Suppose $f \in \mathcal{O}(\mathbb{C})$ and there is a constant $d > 0$ such that*

$$M(r) \leq \text{const. } r^d \qquad \text{for all large } r > 0.$$

Then f is a polynomial of degree $\leq d$.

PROOF. By Cauchy's estimates (Theorem 1.42), the inequality

$$|f^{(n)}(0)| \leq \frac{n!}{r^n} M(r) \leq \text{const. } n! \, r^{d-n}$$

holds for all $n \geq 0$ and all large $r > 0$. If $n > d$, let $r \to +\infty$ to obtain $f^{(n)}(0) = 0$. It follows that $f(z) = \sum_{0 \leq n \leq d} (f^{(n)}(0)/n!) \, z^n$ is a polynomial of degree $\leq d$. $\square$

Thus we can distinguish a polynomial from a **transcendental** (that is, non-polynomial) entire function by checking if $M(r)$ has polynomial or non-polynomial growth as $r \to +\infty$. Of course not all transcendental functions behave the same way near infinity. For example, $M(r)$ is asymptotic to e^r when $f(z) = e^z$ or $\cos z$ or $\sin z$, to $e^{\sqrt{r}}$ when $f(z) = \cos(\sqrt{z})$, and to e^{e^r} when $f(z) = e^{e^z}$ (see Example 8.39 below). It would be desirable to have a measure of growth near infinity that takes into account not only the difference between transcendental and polynomial functions, but the difference between various transcendental functions as well.

DEFINITION 8.38. An entire function f is said to be of **finite order** if there exists an $a \in (0, +\infty)$ such that

$$M(r) \leq \exp(r^a) \qquad \text{for all large } r > 0.$$

The greatest lower bound of all such a is called the **(growth) order** of f and is denoted by $\rho = \rho(f)$. If there is no such a, we say that f is of **infinite order** and set $\rho = +\infty$. Thus, in all cases, $\rho \in [0, +\infty]$.

EXAMPLE 8.39. Here are a few elementary functions and their orders:

- $\rho = 0$ for every polynomial. This is because $M(r)$ is asymptotic to r^d, where d is the degree of the polynomial.
- $\rho = +\infty$ for $f(z) = e^{e^z}$. This simply follows from $M(r) \geq f(r) = e^{e^r}$.
- $\rho = 1$ for $f(z) = e^z$. We have $|f(z)| = |e^z| \leq e^{|z|}$ and $f(r) = e^r$ from which it follows that $M(r) = e^r$.
- $\rho = 1$ for $f(z) = \cos z$. In fact, the estimates

$$|f(z)| = \frac{1}{2}\left|e^{iz} + e^{-iz}\right| \leq e^{|z|} \qquad \text{and} \qquad f(ir) = \frac{1}{2}(e^r + e^{-r}) > \frac{1}{2}e^r$$

show that $e^r/2 \leq M(r) \leq e^r$. We can verify in a similar fashion that the complex sine function has order 1.

- $\rho = 1/2$ for $f(z) = (\sin\sqrt{z})/\sqrt{z}$. Note that f is an entire function even though $\sqrt{z}$ is not holomorphic in $\mathbb{C}$. This is because

$$\frac{\sin z}{z} = 1 - \frac{z^2}{3!} + \frac{z^4}{5!} - \frac{z^6}{7!} + \cdots$$

is an even entire function (the singularity at 0 is removable), so with either of the two choices for $\sqrt{z}$ one obtains a well-defined convergent power series

$$f(z) = \frac{\sin\sqrt{z}}{\sqrt{z}} = 1 - \frac{z}{3!} + \frac{z^2}{5!} - \frac{z^3}{7!} + \cdots.$$

We have

$$|\sin\sqrt{z}| = \frac{1}{2}\left|e^{i\sqrt{z}} - e^{-i\sqrt{z}}\right| \leq e^{\sqrt{|z|}},$$

which shows $M(r) \leq e^{\sqrt{r}}$ and $\rho \leq 1/2$. On the other hand, for large enough $r > 0$,

$$|\sin(\pm i\sqrt{r})| = \frac{1}{2}(e^{\sqrt{r}} - e^{-\sqrt{r}}) > \frac{1}{3}e^{\sqrt{r}},$$

hence

$$|f(-r)| = \frac{|\sin(\pm i\sqrt{r})|}{\sqrt{r}} \geq \frac{e^{\sqrt{r}}}{3\sqrt{r}}.$$

Thus, for any small $\varepsilon > 0$,

$$M(r) \geq \frac{e^{\sqrt{r}}}{3\sqrt{r}} > e^{r^{1/2-\varepsilon}}$$

if r is large enough. This gives $\rho \geq 1/2 - \varepsilon$. Letting $\varepsilon \to 0$, we obtain $\rho \geq 1/2$.

THEOREM 8.40 (Alternative characterizations of order). *The following conditions on $f \in \mathcal{O}(\mathbb{C})$ are equivalent:*

(i) f *is of order* ρ;

(ii) $\rho = \inf\left\{a \in (0, +\infty) : \limsup\limits_{r\to+\infty} \dfrac{\log M(r)}{r^a} = 0\right\}$;

(iii) $\rho = \limsup\limits_{r\to+\infty} \dfrac{\log\log M(r)}{\log r}.$

PROOF. The proof is straightforward. Let μ denote the infimum in (ii). Suppose $\rho = \rho(f)$ is finite and take a, b such that $\rho < a < b$. By the definition of ρ, $M(r) \leq \exp(r^a)$ for all large r, which implies $\limsup_{r\to+\infty} \log M(r)/r^a \leq 1$. This shows $\limsup_{r\to+\infty} \log M(r)/r^b = 0$, so $\mu \leq b$. Letting $b \to \rho$ then gives $\mu \leq \rho$. Now suppose μ is finite and take any $a > \mu$. Then $\limsup_{r\to+\infty} \log M(r)/r^a = 0$, so $M(r) \leq \exp(r^a)$ for all large r, so $\rho \leq a$. Letting $a \to \mu$ then shows $\rho \leq \mu$. Thus $\rho = \mu$ whether they are finite or $+\infty$.

Similarly, let λ denote the limsup in (iii). Suppose ρ is finite and take any $a > \rho$. Then $M(r) \leq \exp(r^a)$ or $\log \log M(r) \leq a \log r$ for all large r. Dividing by $\log r$ and taking the limsup as $r \to +\infty$, we obtain $\lambda \leq a$. Letting $a \to \rho$ then gives $\lambda \leq \rho$. Now suppose λ is finite and take any $a > \lambda$. Then $\log \log M(r) \leq a \log r$ or $M(r) \leq \exp(r^a)$ for all large r, so $\rho \leq a$. Letting $a \to \lambda$ then shows $\rho \leq \lambda$. Thus $\rho = \lambda$ whether they are finite or $+\infty$. $\square$

EXAMPLE 8.41. (Order of an elementary factor) Let us show that the order of the Weierstrass elementary factor E_d in Definition 8.19 is d. As a polynomial, $E_0(z) = 1 - z$ has order zero, so we may assume $d \geq 1$. In this case the formula

$$E_d(z) = (1 - z) \, \exp\left(\sum_{n=1}^{d} \frac{z^n}{n}\right)$$

together with the inequality $|\exp(w)| \leq \exp |w|$ shows that

$$|E_d(z)| \leq (|z| + 1) \, \exp\left(\sum_{n=1}^{d} \frac{|z|^n}{n}\right) \leq (|z| + 1) \, \exp(|z|^d)$$

provided that $|z|$ is sufficiently large. On the other hand, when $r > 1$,

$$|E_d(r)| \geq (r - 1) \, \exp\left(\frac{r^d}{d}\right).$$

Combining the two inequalities, we obtain

$$(r - 1) \, \exp\left(\frac{r^d}{d}\right) \leq M(r) \leq (r + 1) \, \exp(r^d)$$

for all large r. It follows from Theorem 8.40 that

$$\rho = \lim_{r \to +\infty} \frac{\log \log M(r)}{\log r} = d.$$

For the more general case of a canonical product $f(z) = \prod_{n=1}^{\infty} E_d(z/z_n)$ of the type encountered in Corollary 8.22, see Theorem 8.49.

EXAMPLE 8.42. Suppose P is a complex polynomial of degree $d \geq 1$, $f \in \mathcal{O}(\mathbb{C})$, and $g = Pf$. Then $\rho(g) = \rho(f)$. To see this, note that there are constants $C_1, C_2 > 0$ such that $C_1 |z|^d \leq |P(z)| \leq C_2 |z|^d$ for large $|z|$, which shows

$$C_1 r^d M_f(r) \leq M_g(r) \leq C_2 r^d M_f(r) \qquad \text{for all large } r.$$

Thus, $\log M_g(r)$ stays within a bounded distance of $\log M_f(r) + d \log r$ as $r \to +\infty$. This, in view of Theorem 8.40, shows that $\rho(g) = \rho(f)$.

For more general results of this type, see problem 18.

The next theorem gives us a recipe for computing the order of an entire function from the knowledge of its power series.

THEOREM 8.43 (Order of a power series). *Suppose* $f(z) = \sum_{n=0}^{\infty} a_n z^n$ *represents an entire function. Then the order of f is given by*

$$\rho = \limsup_{n \to \infty} \frac{n \log n}{-\log |a_n|}.$$

The fraction is understood to be 0 when $a_n = 0$. The formula shows that the faster the coefficients a_n tend to zero, the smaller the order will be.

PROOF. Let λ denote the right-hand side of the above formula. First assume $\lambda < +\infty$. Take any $a > \lambda$ and find an integer $N \geq 1$ such that $(n \log n)/(-\log |a_n|) < a$ if $n > N$. This gives the estimate

$$(8.12) \qquad\qquad |a_n| \leq n^{-n/a} \qquad \text{if } n > N.$$

Choose a large radius r and let k be the integer part of $(2r)^a$. By the triangle inequality,

$$M(r) \leq \left(\sum_{n=0}^{N} + \sum_{n=N+1}^{k} + \sum_{n=k+1}^{\infty} \right) |a_n| r^n.$$

Let us denote the sum on the right by $S_1 + S_2 + S_3$. Then S_1 is a polynomial in r of degree $\leq N$, so if r is large enough,

$$(8.13) \qquad\qquad S_1 \leq \text{const. } r^N.$$

In S_2, the index n satisfies $n \leq k \leq (2r)^a$, which implies $r^n \leq r^{(2r)^a}$. This, in view of (8.12), gives

$$(8.14) \qquad\qquad S_2 \leq r^{(2r)^a} \sum_{n=N+1}^{k} n^{-n/a} \leq \text{const. } r^{(2r)^a}$$

since $\sum_{n=1}^{\infty} n^{-n/a}$ is convergent (the const. of course depends on a). In S_3, the index n satisfies $n \geq k+1 > (2r)^a$, which implies $2r < n^{1/a}$ or $r^n \leq n^{n/a} 2^{-n}$. Hence $|a_n| r^n \leq n^{-n/a} n^{n/a} 2^{-n} = 2^{-n}$ by (8.12). This shows

$$(8.15) \qquad\qquad S_3 \leq \sum_{n=k+1}^{\infty} 2^{-n} \leq 1.$$

Adding up (8.13), (8.14), and (8.15), we obtain

$$M(r) \leq \text{const. } r^N + \text{const. } r^{(2r)^a} + 1 \leq \text{const. } r^{(2r)^a}$$

if r is large enough. It follows that

$$\rho = \limsup_{r \to +\infty} \frac{\log \log M(r)}{\log r} \leq a.$$

Letting $a \to \lambda$ then gives $\rho \leq \lambda$.

Next, assume $\rho < +\infty$. Take any $a > \rho$, so $M(r) \leq \exp(r^a)$ if r is large enough. By Cauchy's estimates (Theorem 1.42), the inequality

$$|a_n| \leq \frac{M(r)}{r^n} \leq \frac{\exp(r^a)}{r^n}$$

holds for all $n \geq 0$ and all large $r > 0$. Choosing $r = (n/a)^{1/a}$, it follows that

$$|a_n| \leq \left(\frac{e\,a}{n}\right)^{n/a}$$

for all large n. Taking logarithms of both sides and manipulating the inequality gives

$$\frac{n \log n + \text{const.}\, n}{-\log |a_n|} \leq a,$$

which shows

$$\lambda = \limsup_{n\to\infty} \frac{n \log n}{-\log |a_n|} \leq a.$$

Letting $a \to \rho$, we obtain $\lambda \leq \rho$.

We have proved that if λ is finite, then $\rho \leq \lambda$, and if ρ is finite, then $\lambda \leq \rho$. This shows $\lambda = \rho$, whether they are finite or $+\infty$. $\square$

EXAMPLE 8.44. Theorem 8.43 provides a simple method for constructing entire functions of prescribed order. Let us show, for example, that for $\delta > 0$ the power series

$$f(z) = \sum_{n=0}^{\infty} \frac{z^n}{(n!)^\delta}$$

represents an entire function of order $\rho = 1/\delta$. By **Stirling's formula**, one has the asymptotic relation

$$n! \sim \sqrt{2\pi n}\, n^n\, e^{-n} \qquad \text{for large } n,$$

which means

$$\lim_{n\to\infty} \frac{n!}{n^{n+1/2}\, e^{-n}} = \sqrt{2\pi}.$$

Taking logarithms of both sides gives

$$\lim_{n\to\infty} \left[\log n! - \left(n + \frac{1}{2}\right) \log n + n\right] = \frac{1}{2}\log(2\pi),$$

which implies

(8.16)
$$\lim_{n\to\infty} \frac{\log n!}{n \log n} = 1.$$

If R denotes the radius of convergence of the power series defining f, then

$$\log R = \lim_{n\to\infty} \log[(n!)^{\delta/n}] = \lim_{n\to\infty} \frac{\delta}{n}\log n! = +\infty,$$

so f is entire. Moreover, by (8.16) and Theorem 8.43,

$$\rho = \lim_{n \to \infty} \frac{n \log n}{\delta \log n!} = \frac{1}{\delta}.$$

For a different construction of entire functions of prescribed order based on canonical products, see Theorem 8.49.

EXAMPLE 8.45. Theorem 8.43 also shows that *there are transcendental entire functions of order* 0. For example, if

$$f(z) = \sum_{n=1}^{\infty} \frac{z^n}{n^{n^2}},$$

then the coefficients $a_n = n^{-n^2}$ satisfy

$$\lim_{n \to \infty} \sqrt[n]{a_n} = 0 \qquad \text{and} \qquad \lim_{n \to \infty} \frac{n \log n}{-\log a_n} = 0,$$

which imply f is an entire function of order 0.

Next, we consider the distribution of zeros of entire functions of finite order. We continue using the notations $M(r)$ and $N(r)$ in Definition 8.29.

LEMMA 8.46. *Let f be an entire function of finite order ρ and $\{z_n\}$ be the sequence of zeros of f in $\mathbb{C}^*$, where each zero is repeated as many times as its order. If $a > \rho$, then*

(i) $N(r) \le \text{const. } r^a$ *for all large r;*

(ii) $\displaystyle\sum_{n=1}^{\infty} \frac{1}{|z_n|^a} < +\infty.$

PROOF. Assume without loss of generality that $f(0) \ne 0$. If $f(0) = 0$ with $m = \text{ord}(f, 0) \ge 1$, work with the entire function $f(z)/z^m$ instead, which has the same zeros in $\mathbb{C}^*$ as f, the same order as f (see Example 8.42), and $N(r) - m$ zeros in the disk $\mathbb{D}(0, r)$. We may also assume that the zeros are arranged so that $0 < |z_1| \le |z_2| \le |z_3| \le \cdots$.

Since $f(0) \ne 0$, we can apply Corollary 8.31 with $R = 2r$ to obtain

$$N(r) \le \frac{\log M(2r) - \log |f(0)|}{\log 2} \le \text{const. } \log M(2r)$$

for large $r > 0$. Since by the definition of order, $\log M(2r) \le (2r)^a$ for large r, (i) easily follows.

For (ii), let $r_n = |z_n|$ and note that by the definition of $N(r)$,

$$n \le N(r_n + 1)$$

for all $n \geq 1$. Choose $\varepsilon > 0$ so small that $(1 + \varepsilon)\rho < a$. By (i), $N(r) \leq$ const. $r^{a/(1+\varepsilon)}$ for all large r. It follows that

$$n \leq \text{const. } (r_n + 1)^{a/(1+\varepsilon)} \qquad \text{or} \qquad n^{1+\varepsilon} \leq \text{const. } r_n^a$$

for all large n. Since the series $\sum 1/n^{1+\varepsilon}$ converges, we conclude that $\sum 1/r_n^a$ converges as well. $\qquad \square$

Part (ii) of the above theorem suggests the following

> **DEFINITION 8.47.** Let $\{z_n\}$ be a (finite or infinite) sequence in $\mathbb{C}^*$. The greatest lower bound of the set of $a > 0$ for which
>
> $$\sum_n \frac{1}{|z_n|^a} < +\infty$$
>
> is called the ***exponent of convergence*** of $\{z_n\}$ and is denoted by ν. If there is no such a, we set $\nu = +\infty$. When $\{z_n\}$ is the sequence of zeros of an entire function f, we call ν the exponent of convergence of f and denote it by $\nu(f)$.

Observe that if f has at most finitely many zeros, then $\nu(f) = 0$.
The following is an immediate consequence of Lemma 8.46:

COROLLARY 8.48 (Order dominates exponent of convergence). *For every entire function f, $\nu(f) \leq \rho(f)$.*

Of course ν can be strictly smaller than ρ, as the example e^z with $\rho = 1$ and $\nu = 0$ shows. Perhaps this is not a totally satisfactory example since the exponential function has no zeros at all. For a non-trivial case of $\nu < \rho$, see Example 8.50.

THEOREM 8.49 (Order of a canonical product). *Let $\{z_n\}$ be an infinite sequence in $\mathbb{C}^*$, with $|z_n| \to +\infty$, whose exponent of convergence ν is finite. Let d be any integer $> \nu - 1$. Then the canonical product*

$$f(z) = \prod_{n=1}^{\infty} E_d\left(\frac{z}{z_n}\right)$$

defines an entire function of finite order $\rho \leq \max\{d, \nu\}$. In particular, if we choose $d = \lfloor \nu \rfloor$ (the integer part of ν), then $\rho = \nu$.

PROOF. The assumption $d + 1 > \nu$ implies $\sum_{n=1}^{\infty} 1/|z_n|^{d+1} < +\infty$. By Theorem 8.21, the canonical product f defines an entire function.
To prove the upper bound on the order $\rho = \rho(f)$, we start by estimating the size of the elementary factors E_d. Pick any constant a such that $\max\{d, \nu\} < a < d + 1$. The

definition of E_d shows that

$$\log |E_d(z)| \leq \begin{cases} \log(|z|+1) & d=0 \\ \log(|z|+1) + \displaystyle\sum_{n=1}^{d} \frac{|z|^n}{n} & d \geq 1. \end{cases}$$

If $|z| > 1$, it follows that

$$\log |E_d(z)| \leq \begin{cases} \text{const.} + \log |z| & d=0 \\ \text{const. } |z|^d & d \geq 1, \end{cases}$$

which, in view of $a > d \geq 0$ shows

$$(8.17) \qquad\qquad \log |E_d(z)| \leq \text{const. } |z|^a \qquad \text{if } |z| > 1.$$

On the other hand, by Lemma 8.20, if $|z| \leq 1$, then

$$\log |E_d(z)| \leq \log(|z|^{d+1} + 1) \leq |z|^{d+1}.$$

This, in view of $a < d+1$, gives the estimate

$$(8.18) \qquad\qquad \log |E_d(z)| \leq |z|^a \qquad \text{if } |z| \leq 1.$$

Now let $r_n = |z_n|$. For $|z| = r > 0$, we use the estimates (8.17) and (8.18) to obtain

$$\log |f(z)| = \left(\sum_{r_n < r} + \sum_{r_n \geq r} \right) \log \left| E_d \left(\frac{z}{z_n} \right) \right| \leq \text{const.} \sum_{r_n < r} \left(\frac{r}{r_n} \right)^a + \sum_{r_n \geq r} \left(\frac{r}{r_n} \right)^a$$

$$= \left(\text{const.} \sum_{r_n < r} \frac{1}{r_n^a} + \sum_{r_n \geq r} \frac{1}{r_n^a} \right) r^a \leq \text{const. } r^a,$$

where in the last inequality we have used the fact that $a > \nu$, so $\sum 1/r_n^a < +\infty$. This gives the estimate $\log M(r) \leq \text{const. } r^a$, which yields $\rho \leq a$. Letting $a \to \max\{d, \nu\}$ now proves $\rho \leq \max\{d, \nu\}$, as required.

If $d = \lfloor \nu \rfloor$, the above estimate shows $\rho \leq \nu$. The reverse inequality $\nu \leq \rho$ always holds by Corollary 8.48, so $\rho = \nu$. $\qquad\square$

EXAMPLE 8.50. Let us construct an entire function with $\rho = 1$ and $\nu = 1/2$. The sequence $\{z_n = -n^2\}$ has exponent of convergence $1/2$. By Theorem 8.49, the canonical product

$$g(z) = \prod_{n=1}^{\infty} E_0 \left(\frac{z}{z_n} \right) = \prod_{n=1}^{\infty} \left(1 + \frac{z}{n^2} \right)$$

is an entire function of order $\rho(g) = \nu(g) = 1/2$. The entire function $f(z) = e^z g(z)$ has the same exponent of convergence $\nu(f) = 1/2$ but its order is $\rho(f) = 1$. To see the latter, note that by the triangle inequality, $|g(z)| \leq g(|z|)$, which implies $M_g(r) = g(r)$. Similarly, $|f(z)| \leq e^{|z|} g(|z|) = f(|z|)$, which implies $M_f(r) = f(r) = e^r g(r)$. Since $\rho(g) = 1/2$, we have $1 \leq g(r) \leq e^r$ for large r, which gives $e^r \leq f(r) \leq e^{2r}$ for large r, which shows $\rho(f) = 1$.

Our final goal is to prove a theorem of Hadamard which asserts that the Weierstrass factorization of an entire function given by Theorem 8.24 can be replaced by a simpler product if the function is known to be of finite order. The proof will make use of the following classical lemma which is of independent interest:

LEMMA 8.51 (Borel-Carathéodory inequality, 1897). *Let f be holomorphic in a neighborhood of the closed disk $\overline{\mathbb{D}}(0, R)$. Set*

$$M(r) = \sup_{|z|=r} |f(z)| \qquad and \qquad A(r) = \sup_{|z|=r} \mathrm{Re}(f(z)).$$

Then,

(8.19) $$M(r) \leq \frac{2r}{R-r} A(R) + \frac{R+r}{R-r} |f(0)| \qquad if\ 0 \leq r < R.$$

With the constant 2 replaced by 4, this inequality was first proved by Hadamard in 1892.

PROOF. We may assume f is non-constant since otherwise (8.19) is trivial. First suppose $f(0) = 0$. The maximum principle applied to the harmonic function $\mathrm{Re}(f)$ implies $A(R) > \mathrm{Re}(f(0)) = 0$. The linear map $L : \zeta \mapsto R\zeta$ carries the unit disk $\mathbb{D}$ onto the disk $\mathbb{D}(0, R)$, f maps $\mathbb{D}(0, R)$ into the left half-plane $\{w : \mathrm{Re}(w) < A(R)\}$, and the Möbius map $\varphi : w \mapsto w/(2A(R) - w)$ sends this half-plane conformally back to $\mathbb{D}$. It follows that the composition

$$\varphi \circ f \circ L : \zeta \mapsto \frac{f(R\zeta)}{2A(R) - f(R\zeta)}$$

maps $\mathbb{D}$ to $\mathbb{D}$, fixing the origin. By the Schwarz lemma,

$$|(\varphi \circ f \circ L)(\zeta)| \leq \frac{r}{R} \qquad \text{whenever } |\zeta| = \frac{r}{R} < 1.$$

It follows that when $|z| = r$,

$$R|f(z)| \leq r|2A(R) - f(z)| \leq 2rA(R) + r|f(z)|,$$

so

$$|f(z)| \leq \frac{2r}{R-r} A(R),$$

which proves (8.19) when $f(0) = 0$.

If $f(0) \neq 0$, apply the inequality we just proved to the function $f(z) - f(0)$. It follows that when $|z| = r$,

$$|f(z)| \leq |f(z) - f(0)| + |f(0)| \leq \frac{2r}{R-r} (A(R) - \mathrm{Re}(f)(0)) + |f(0)|$$

$$\leq \frac{2r}{R-r} (A(R) + |f(0)|) + |f(0)|$$

$$= \frac{2r}{R-r} \, A(R) + \frac{R+r}{R-r} \, |f(0)|. \qquad \square$$

Two alternative proofs for the Borel-Carathéodory inequality, based on power series and the Poisson integral formula, are outlined in problems 26 and 27.

Jacques Salomon Hadamard
(1865–1963)

THEOREM 8.52 (Hadamard's factorization theorem,1893). *Suppose $f \in \mathcal{O}(\mathbb{C})$ is not identically zero and has finite order ρ and therefore finite exponent of convergence v. Arrange the zeros of f in $\mathbb{C}^*$ in a sequence $\{z_n\}$ where each zero is repeated as many times as its order. Then f has a factorization of the form*

$$(8.20) \qquad f(z) = e^{g(z)} \, z^m \prod_{n=1}^{\infty} E_d\left(\frac{z}{z_n}\right),$$

where $m = \mathrm{ord}(f,0) \geq 0$, $d = \lfloor v \rfloor$, and g is a polynomial of degree at most $\lfloor \rho \rfloor$.

As in Theorem 8.24, it is understood that the canonical product will reduce to a polynomial if f has finitely many zeros, and will be missing if f has no zeros in $\mathbb{C}^*$. Note also that once the choice of $d = \lfloor v \rfloor$ is made, the polynomial g is uniquely determined up to an additive constant.

The proof presented here is due to E. Landau [T].

PROOF. Set $k = \lfloor \rho \rfloor$. By Corollary 8.48, $d \leq v \leq \rho < k+1$. By Corollary 8.22, the canonical product $h(z) = z^m \prod_{n=1}^{\infty} E_d(z/z_n)$ is an entire function with the same zeros of the same orders as f. Hence, f/h extends to a non-vanishing entire function and therefore it is of the form e^g for some entire function g. We show that g is a polynomial of degree $\leq k$ by proving that its $(k+1)$-st derivative vanishes identically. Since the factor z^m does not affect the order of f or the location of the z_n, we may assume without loss of generality that $f(0) \neq 0$ so $m = 0$.

We begin by taking the logarithmic derivative of $f(z) = e^{g(z)} \prod_{n=1}^{\infty} E_d(z/z_n)$, which is justified by Theorem 8.18:

$$(8.21) \qquad \frac{f'(z)}{f(z)} = g'(z) + \sum_{n=1}^{\infty} \frac{1}{z_n} \frac{E_d'(z/z_n)}{E_d(z/z_n)}.$$

Note that by the definition of the elementary factor E_d,

$$\frac{E_d'(\zeta)}{E_d(\zeta)} = \begin{cases} \dfrac{-1}{1-\zeta} & d=0 \\[2ex] \dfrac{-1}{1-\zeta} + \displaystyle\sum_{n=0}^{d-1} \zeta^n & d \geq 1, \end{cases}$$

which has the k-th derivative

$$\left(\frac{E_d'(\zeta)}{E_d(\zeta)}\right)^{(k)} = -\frac{k!}{(1-\zeta)^{k+1}}.$$

Thus, the k-th derivative of (8.21) takes the simple form

(8.22)
$$\left(\frac{f'(z)}{f(z)}\right)^{(k)} = g^{(k+1)}(z) - k! \sum_{n=1}^{\infty} \frac{1}{(z_n - z)^{k+1}}.$$

Here a term-by-term differentiation was legitimate because the series in (8.21) converges compactly in $\mathbb{C}$ in the sense of Theorem 8.18.

Now let $r_n = |z_n|$, take a large $R > 0$, and define an auxiliary function

(8.23)
$$\varphi(z) = \varphi_R(z) = \frac{f(z)}{f(0) \displaystyle\prod_{r_n \leq R} \left(1 - \frac{z}{z_n}\right)},$$

which is entire, with $\varphi(0) = 1$, and has no zeros in $\overline{\mathbb{D}}(0, R)$. By Corollary 2.22, there is a holomorphic logarithm $\psi = \psi_R$ of φ defined in a neighborhood of $\overline{\mathbb{D}}(0, R)$, with $\psi(0) = 0$. Substituting e^{ψ} for φ in (8.23), taking the logarithmic derivative, and then differentiating k times, gives

(8.24)
$$\psi^{(k+1)}(z) = \left(\frac{f'(z)}{f(z)}\right)^{(k)} + k! \sum_{r_n \leq R} \frac{1}{(z_n - z)^{k+1}} \qquad \text{if } |z| \leq R.$$

Comparing (8.22) and (8.24), we obtain

(8.25)
$$g^{(k+1)}(z) = \psi^{(k+1)}(z) + k! \sum_{r_n > R} \frac{1}{(z_n - z)^{k+1}} \qquad \text{if } |z| \leq R.$$

If $|z| \leq R/2 < R < r_n$, then $|z_n - z| > r_n/2$. Hence,

$$|g^{(k+1)}(z)| \leq |\psi^{(k+1)}(z)| + k! \, 2^{k+1} \sum_{r_n > R} \frac{1}{r_n^{k+1}} \qquad \text{if } |z| \leq R/2.$$

Since $k + 1 > \nu$, the series $\sum_{n=1}^{\infty} 1/r_n^{k+1}$ converges and the above sum tends to zero as $R \to +\infty$. Therefore, to prove $g^{(k+1)} = 0$, it suffices to show that the maximum of $|\psi^{(k+1)}|$ on $\mathbb{D}(0, R/2)$ tends to zero as $R \to +\infty$.

To this end, note that if $|z| = 2R$, each factor $(1 - z/z_n)$ in the definition of φ satisfies

$$\left|1 - \frac{z}{z_n}\right| \geq \frac{|z|}{r_n} - 1 \geq \frac{2R}{R} - 1 = 1.$$

Fixing a constant a such that $\rho < a < k + 1$, it follows that

$$|\varphi(z)| \leq \text{const.} \, |f(z)| \leq \text{const.} \, \exp((2R)^a)$$

provided that $|z| = 2R$ and R is large enough. By the maximum principle,

$$|\varphi(z)| \leq \text{const. } \exp((2R)^a) \qquad \text{if } |z| \leq 2R.$$

This gives the estimate

$$\text{Re}(\psi(z)) = \log|\varphi(z)| \leq \text{const. } R^a \qquad \text{if } |z| \leq R.$$

By the Borel-Carathéodory inequality (8.19) applied to ψ with $r = 3R/4$,

$$|\psi(z)| \leq \text{const. } \frac{2(3R/4)}{R - 3R/4} R^a = \text{const. } R^a \qquad \text{if } |z| \leq 3R/4.$$

Now if $|z| \leq R/2$, the disk $\mathbb{D}(z, R/4)$ is contained in the disk $\mathbb{D}(0, 3R/4)$. Hence, by Cauchy's estimates (Theorem 1.42),

$$|\psi^{(k+1)}(z)| \leq \text{const. } \frac{(k+1)!}{(R/4)^{k+1}} R^a = \text{const. } R^{a-k-1} \qquad \text{if } |z| \leq R/2.$$

Since $a < k+1$, we have $R^{a-k-1} \to 0$ as $R \to +\infty$. $\qquad\square$

REMARK 8.53. The same proof shows that f has a factorization of the form (8.20) with the (a priori larger) choice $d = \lfloor \rho \rfloor$ and g a polynomial of degree $\leq d$.

We conclude with a few applications of Hadamard's factorization theorem.

COROLLARY 8.54. *Suppose the order of $f \in \mathcal{O}(\mathbb{C})$ is finite but non-integer. Then f has infinitely many zeros.*

By considering $f - c$ for $c \in \mathbb{C}$, which of course has the same order as f, it follows that f takes every complex value infinitely often.

PROOF. Otherwise f has at most finitely many zeros. Hence, by Theorem 8.52, $f(z) = e^{g(z)} z^m P(z)$, where P is a polynomial having the same non-zero roots as f, $m = \text{ord}(f, 0) \geq 0$, and g is a polynomial. But then the order of f would be the same as that of e^g (Example 8.42), which is evidently the integer $\deg(g)$. This is a contradiction. $\qquad\square$

COROLLARY 8.55. *Suppose $f \in \mathcal{O}(\mathbb{C})$ has order $0 < \rho < 1$. Then f is an infinite product of linear factors*

$$f(z) = C z^m \prod_{n=1}^{\infty} \left(1 - \frac{z}{z_n}\right),$$

where C is a constant, $m = \text{ord}(f, 0) \geq 0$, and $\{z_n\}$ is the sequence of zeros of f in $\mathbb{C}^*$.

PROOF. By Corollary 8.54 f has infinitely many zeros $\{z_n\}$. The result follows immediately from Theorem 8.52 since $d = \lfloor v \rfloor = \lfloor \rho \rfloor = 0$, $E_0(z/z_n) = 1 - z/z_n$, and g, being a polynomial of degree zero, is constant. $\qquad\square$

EXAMPLE 8.56. The entire function $f(z) = (\sin\sqrt{z})/\sqrt{z}$ is of order $\rho = 1/2$ (see Example 8.39) and has a simple zero at $n^2\pi^2$ for each $n \geq 1$. It follows from Corollary 8.55 that

$$\frac{\sin\sqrt{z}}{\sqrt{z}} = C\prod_{n=1}^{\infty}\left(1 - \frac{z}{n^2\pi^2}\right).$$

Comparing the two sides at $z = 0$ gives $C = 1$. Thus,

$$\sin z = z\prod_{n=1}^{\infty}\left(1 - \frac{z^2}{n^2\pi^2}\right) \qquad \text{for all } z \in \mathbb{C}.$$

This is known as **Euler's product formula** (1734). A longer but more elementary derivation is sketched in problem 7.

The image of a non-constant entire function can omit one point in $\mathbb{C}$, as the example $f(z) = e^z$ with $f(\mathbb{C}) = \mathbb{C}^*$ shows. A deep and far-reaching generalization of Liouville's theorem, due to Picard, asserts that this is optimal: An entire function whose image omits two or more points must be constant. We will discuss several approaches to Picard's theorem later in chapters 11 and 13, but here we can give a quick proof under the superfluous assumption of finiteness of order.

COROLLARY 8.57. *Suppose* $f \in \mathcal{O}(\mathbb{C})$ *has finite order. If* $\mathbb{C} \smallsetminus f(\mathbb{C})$ *contains more than one point, then* f *is constant.*

PROOF. Suppose $\mathbb{C} \smallsetminus f(\mathbb{C})$ contains distinct points a, b. The entire function $h = (f-a)/(b-a)$ has the same finite order as f, and does not take the values $0, 1$. By Theorem 8.52, $h = e^g$ for some polynomial g. Since g does not take the value 0, the fundamental theorem of algebra shows that g is constant. Hence h and f must be constant functions as well. $\square$

Problems

(1) Prove the following form of Cauchy's criterion for convergence of infinite products: $\prod_{n=1}^{\infty} a_n$ converges if and only if for every $\varepsilon > 0$ there is an integer N such that $|\prod_{n=m}^{k} a_n - 1| < \varepsilon$ whenever $k \geq m \geq N$.

(2) Recall from Theorem 8.9 that if $b_n \geq 0$, then $\prod_{n=1}^{\infty}(1 + b_n)$ converges if and only if $\sum_{n=1}^{\infty} b_n$ converges.
    (i) Show that the same holds if $b_n \leq 0$.
    (ii) Verify that if $b_n = (-1)^{n+1}/\sqrt{n}$, then $\prod_{n=1}^{\infty}(1 + b_n)$ diverges even though $\sum_{n=1}^{\infty} b_n$ converges.
    (iii) Verify that if $b_n = \exp(i(-1)^{n+1}/\sqrt{n}) - 1$, then $\prod_{n=1}^{\infty}(1 + b_n)$ converges even though $\sum_{n=1}^{\infty} b_n$ diverges.

(3) (Rearrangements of infinite products) Let $\beta : \mathbb{N} \to \mathbb{N}$ be a bijection. Prove the following statements:

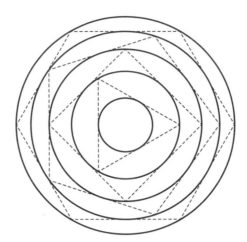

Figure 8.3. Illustration of problem 4: Do the radii stay bounded or tend to $+\infty$?

    (i)   If $\prod_{n=1}^{\infty} a_n$ converges absolutely, so does $\prod_{n=1}^{\infty} a_{\beta(n)}$, and $\prod_{n=1}^{\infty} a_{\beta(n)} = \prod_{n=1}^{\infty} a_n$.

    (ii) If $f_n : U \to \mathbb{C}$ are continuous and $\sum_{n=1}^{\infty} |f_n - 1|$ converges compactly in U, then the infinite product $f = \prod_{n=1}^{\infty} f_{\beta(n)}$ converges compactly in U, where the continuous function f does not depend on β.

(Hint: For (i), use Theorem 8.9 and the fact that absolute convergence of infinite series persists under rearrangements.)

(4)   Consider a sequence $C_1, C_2, C_3, \ldots$ of concentric circles of increasing radii such that C_n is inscribed in a regular $(n+2)$-gon and circumscribes a regular $(n+1)$-gon for every $n \geq 2$ (see Fig. 8.3). Determine whether the radius of C_n stays bounded or tends to infinity as $n \to \infty$ [**Bo**].

(5)   Let $U \subset \mathbb{C}$ be a simply connected domain. Suppose $f \in \mathcal{O}(U)$ is not identically zero. Given a positive integer n, show that $f = g^n$ for some $g \in \mathcal{O}(U)$ if and only if $\mathrm{ord}(f, p)$ is divisible by n for every $p \in U$.

(6)   Let $f(z) = \prod_{n=0}^{\infty}(1 + z^{2^n})$.

    (i)   Show that the infinite product converges compactly in $\mathbb{D}$, so $f \in \mathcal{O}(\mathbb{D})$.

    (ii)   Let $p_k(z) = \prod_{n=0}^{k}(1 + z^{2^n})$. Show that $p_k(z) = (1 + z)p_{k-1}(z^2)$, and justify the functional equation $f(z) = (1 + z)f(z^2)$.

    (iii) Conclude that $f(z) = 1/(1 - z)$.

    (iv) What does the resulting identity

$$(1 + z)(1 + z^2)(1 + z^4)(1 + z^8) \cdots = 1 + z + z^2 + z^3 + \cdots$$

for $|z| < 1$ tell you about binary expansion of integers?

(7)   This exercise will give an elementary proof of Euler's product formula

$$\sin z = z \prod_{n=1}^{\infty} \left(1 - \frac{z^2}{n^2 \pi^2}\right) \qquad \text{for } z \in \mathbb{C}$$

(see Example 8.56). The argument is deliberately divided into small steps for more clarity.

    (i)   Show that the infinite product converges compactly in $\mathbb{C}$ to an entire function f with a simple zero at $n\pi$ for every $n \in \mathbb{Z}$, and with no other zeros.

(ii)   Show that the ratio $\sin z / f(z)$ has removable singularities at every $n\pi$, so it is a non-vanishing entire function. Conclude that for some entire function g with $g(0) = 0$,

$$\sin z = e^{g(z)}\, z \prod_{n=1}^{\infty} \left(1 - \frac{z^2}{n^2\pi^2}\right).$$

(iii)   Use logarithmic differentiation to show that

$$\cot z = g'(z) + \frac{1}{z} + 2z \sum_{n=1}^{\infty} \frac{1}{z^2 - n^2\pi^2},$$

hence

$$g''(z) = -\frac{1}{\sin^2 z} + \sum_{n=-\infty}^{\infty} \frac{1}{(z - n\pi)^2}.$$

Show that the right-hand side is invariant under the translation $z \mapsto z + \pi$, which means $g''(z + \pi) = g''(z)$. Prove the estimate

$$|g''(z)| \leq \frac{1}{\sinh^2 y} + 2 \sum_{n=0}^{\infty} \frac{1}{(n^2\pi^2 + y^2)} \qquad \text{for } z = x + iy$$

provided that $0 \leq x \leq \pi$. Use Liouville's theorem to conclude that $g'' = 0$ and therefore g' is constant.

(iv)   The first identity in (iii) shows that g' is an odd function. Thus $g' = g = 0$ (for another proof of $g = 0$, see the partial fraction expansion of $\cot z$ in problem 2 of chapter 9.)

(8)   Prove **Wallis's formula** (1656):

$$\frac{\pi}{2} = \frac{2}{1} \cdot \frac{2}{3} \cdot \frac{4}{3} \cdot \frac{4}{5} \cdot \frac{6}{5} \cdot \frac{6}{7} \cdot \frac{8}{7} \cdot \frac{8}{9} \cdots.$$

(9)   Use problem 7 to calculate $\prod_{n=1}^{\infty}(1 + 1/n^2)$ and $\sum_{n=1}^{\infty} 1/n^2$.

(10)   Use the identity $\sin(2x) = 2 \sin x \, \cos x$ to show that

$$\prod_{n=2}^{\infty} \cos\left(\frac{\pi}{2^n}\right) = \frac{2}{\pi}.$$

Combine with the identity $\cos(2x) = 2\cos^2 x - 1$ to deduce **Veita's formula** (1579):

$$\frac{2}{\pi} = \frac{\sqrt{2}}{2} \cdot \frac{\sqrt{2 + \sqrt{2}}}{2} \cdot \frac{\sqrt{2 + \sqrt{2 + \sqrt{2}}}}{2} \cdots.$$

(11)   Suppose $f \in \mathscr{O}(\mathbb{D}(0, r))$ has finitely many zeros at $z_1, \ldots, z_k$, where $0 < |z_1| \leq \cdots \leq |z_k| < r$ and each zero is repeated as many times as its order. Show that

$$\sum_{n=1}^{k} \log\left(\frac{r}{|z_n|}\right) = \int_0^r \frac{N(t)}{t}\, dt.$$

This gives an interpretation for the main term on the right side of Jensen's formula (8.8) as the logarithmic density of the number of zeros.

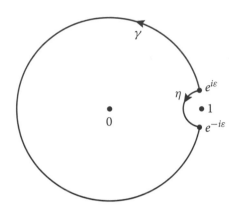

Figure 8.4. Paths of integration in problem 14.

(12)   Prove the following generalization of Jensen's formula for meromorphic functions: Suppose $f \in \mathscr{M}(\mathbb{D}(0, R))$ has no zeros or poles at $z = 0$ or on the circle $|z| = r < R$. Let $z_1, \ldots, z_k$ and $p_1, \ldots, p_m$ denote the zeros and poles of f in $\mathbb{D}(0, r)$, each repeated as many times as its order. Then

$$\frac{1}{2\pi} \int_0^{2\pi} \log |f(re^{it})| \, dt = \log |f(0)| + \sum_{n=1}^{k} \log \left(\frac{r}{|z_n|} \right) - \sum_{n=1}^{m} \log \left(\frac{r}{|p_n|} \right).$$

(13)   Suppose $f \in \mathscr{O}(\mathbb{D}(0, R))$ has no zeros at $z = 0$ or on the circle $|z| = r < R$. Let $z_1, \ldots, z_k$ be the zeros of f in $\mathbb{D}(0, r)$, each repeated as many times as its order. Prove the **Poisson-Jensen formula**

$$\log |f(z)| = \frac{1}{2\pi} \int_0^{2\pi} \frac{r^2 - |z|^2}{|re^{it} - z|^2} \log |f(re^{it})| \, dt + \sum_{n=1}^{k} \log \left| \frac{r(z - z_n)}{r^2 - \bar{z}_n z} \right|$$

for $|z| < r$. (Hint: Form a suitable ratio $g = f/B$, as in the proof of Theorem 8.27, and apply the Poisson integral formula to the harmonic function $\log |g|$.)

(14)   Show that

$$\int_0^{2\pi} \log |1 - e^{it}| \, dt = 0$$

by completing the following outline [**Ru2**]. Consider $g(z) = \mathrm{Log}(1 - z)$, which is well defined and holomorphic in the half-plane $\{z : \mathrm{Re}(z) < 1\}$ and satisfies $g(0) = 0$. For small $\varepsilon > 0$, let γ, η be the circular arcs shown in Fig. 8.4. Then

(i) $\displaystyle \int_\varepsilon^{2\pi - \varepsilon} \log |1 - e^{it}| \, dt = \mathrm{Re} \left(\frac{1}{i} \int_\gamma g(z) \, \frac{dz}{z} \right) = \mathrm{Re} \left(\frac{1}{i} \int_\eta g(z) \, \frac{dz}{z} \right).$

(ii) $|g| \le \mathrm{const.} \, \log(1/\varepsilon)$ on $|\eta|$ and $\mathrm{length}(\eta) \le \mathrm{const.} \, \varepsilon$, so

$$\left| \int_\eta g(z) \, \frac{dz}{z} \right| < \mathrm{const.} \, \varepsilon \log(1/\varepsilon) \to 0 \qquad \text{as } \varepsilon \to 0.$$

(15)   Use the result of the previous problem to prove that Jensen's formula (8.8) holds even if the circle $|z| = r$ contains zeros of f. (Hint: Divide f by the factor $1 - z/a$ for every zero a on the circle $|z| = r$.)

(16) Suppose $f, g \in \mathcal{O}(\mathbb{D})$ are bounded and $f(e^{-1/n}) = g(e^{-1/n})$ for all positive integers n. Show that $f = g$ in $\mathbb{D}$.

(17) Find the order of the following entire functions:

$$f(z) = z^2 e^{2z} - e^{3z}, \qquad g(z) = \cos \sqrt{z}, \qquad h(z) = \sum_{n=0}^{\infty} \frac{z^n}{(7n)!}.$$

(18) Suppose $f, g \in \mathcal{O}(\mathbb{C})$ have finite order. Prove the following statements:
  (i) $\rho(Pf + Q) = \rho(f)$ if P, Q are polynomials and $P \neq 0$.
  (ii) $\rho(f + g) \leq \max\{\rho(f), \rho(g)\}$. Equality holds if $\rho(f) \neq \rho(g)$.
  (iii) $\rho(fg) \leq \max\{\rho(f), \rho(g)\}$ (again, equality holds if $\rho(f) \neq \rho(g)$, but the proof is harder).
  (iv) $\rho(f) = \rho(f')$.

(19) Entire functions can grow arbitrarily fast: Let $\varphi : [0, +\infty) \to [0, +\infty)$ be any increasing function with $\lim_{r \to +\infty} \varphi(r) = +\infty$. Construct an entire function f such that $M_f(r) > \varphi(r)$ for all r.

(20) Suppose $f \in \mathcal{O}(\mathbb{C})$ and $d > 0$. If $\operatorname{Re}(f(z)) \leq$ const. $|z|^d$ for all large $|z|$, show that f is a polynomial of degree at most d.

(21) Suppose $\varphi : [0, 1] \to \mathbb{C}$ is continuous. Show that

$$f(z) = \int_0^1 \varphi(t) e^{zt} \, dt$$

is an entire function of finite order. What can you say about $\rho(f)$?

(22) Prove the following alternative description for the exponent of convergence of a sequence: If $\{z_n\}$ is a sequence in $\mathbb{C}$ such that $0 < |z_1| \leq |z_2| \leq |z_3| \leq \cdots$, then

$$v = \limsup_{n \to \infty} \frac{\log n}{\log |z_n|}.$$

(Hint: First show that if $\{a_n\}$ is a decreasing sequence of positive numbers for which $\sum a_n < +\infty$, then $\lim_{n \to \infty} n a_n = 0$.)

(23) In the factorization (8.20) of Theorem 8.52, the integer $p = \max\{d, \deg(g)\}$ is called the **genus** of f. Show that $p \leq \rho < p + 1$. (Hint: Use Theorem 8.49 and problem 18(iii) above.)

(24) Let $f \in \mathcal{O}(\mathbb{C})$. Show that if $\rho(f)$ is finite but non-integer, then $\rho(f) = v(f)$. (Hint: Use Theorems 8.49 and 8.52.)

(25) Suppose $f(z) = \sum_{n=0}^{\infty} a_n z^n$ is holomorphic in a neighborhood of the closed disk $\overline{\mathbb{D}}(0, R)$. Consider the quantities

$$M(r) = \sup_{|z|=r} |f(z)|, \quad M_1(r) = \sum_{n=0}^{\infty} |a_n| r^n, \quad M_2(r) = \left(\sum_{n=0}^{\infty} |a_n|^2 r^{2n} \right)^{1/2}.$$

Verify the following inequalities for $0 < r < R$:
  (i) $M_2(r)^2 = \dfrac{1}{2\pi} \int_0^{2\pi} |f(re^{it})|^2 \, dt$ (compare Parseval's formula in Lemma 7.46).
  (ii) $M(r) \leq M_1(r) \leq \dfrac{R}{R - r} M(R)$.

(iii) $\dfrac{\sqrt{R^2-r^2}}{R}\,M_1(r)\le M_2(R)\le M(R).$

Conclude that when f is an entire function of finite order, $M(r)$, $M_1(r)$, and $M_2(r)$ have the same growth rate as $r\to+\infty$.

(26)   Give an alternative proof for the Borel-Carathéodory inequality by completing the following outline. Let $u=\operatorname{Re}(f)$ and assume without loss of generality that $f(0)=0$, so f has a power series representation of the form $\sum_{n=1}^{\infty} a_n z^n$.

(i)   Use Cauchy's integral formula to show that
$$a_n\,r^n=\frac{1}{\pi}\int_0^{2\pi} u(re^{it})e^{-int}\,dt=\frac{1}{\pi}\int_0^{2\pi}\left(u(re^{it})-A(r)\right)e^{-int}\,dt$$
if $0\le r\le R$ and $n\ge 1$.

(ii)   Show that
$$|a_n|\,r^n\le 2\,A(r)\qquad \text{if } 0\le r\le R \text{ and } n\ge 1.$$

(iii)   Conclude that
$$M(r)\le\frac{2r}{R-r}\,A(R)\qquad \text{if } 0\le r<R.$$

(27)   Here is an outline of yet another proof for the Borel-Carathéodory inequality. Let, as above, $u=\operatorname{Re}(f)$ and $f(0)=0$.

(i)   Use Corollary 7.23 and the mean value property of u to show that
$$f(z)=\frac{1}{2\pi}\int_0^{2\pi}\frac{2z}{Re^{it}-z}\,u(Re^{it})\,dt\qquad \text{if } |z|<R.$$

(ii)   Show that
$$f(z)=\frac{1}{2\pi}\int_0^{2\pi}\frac{2z}{Re^{it}-z}\left(u(Re^{it})-A(R)\right)dt\qquad \text{if } |z|<R.$$

(iii)   Conclude that
$$|f(z)|\le\frac{1}{2\pi}\int_0^{2\pi}\frac{2r}{R-r}\left(A(R)-u(Re^{it})\right)dt\qquad \text{if } |z|=r<R,$$
from which the desired inequality follows.

9

Interpolation and approximation theorems

9.1 Mittag-Leffler's theorem

This section is devoted to a theorem of Mittag-Leffler on the existence of meromorphic functions in a given domain with prescribed principal parts. Although there are modern approaches to this classical result which generalize to several complex variables and the setting of Riemann surfaces, we prefer an elementary proof due to Weierstrass whose idea parallels that of Theorem 8.25. The problem is easy to solve if the set of prescribed poles is finite because in that case we can simply add all the principal parts to obtain the desired meromorphic function. We may therefore assume that the set of poles is infinite.

We first address the problem in the plane in the following "additive" version of the Weierstrass product theorem:

THEOREM 9.1 (Mittag-Leffler, 1877). *Let $\{z_n\}$ be a sequence of distinct points in $\mathbb{C}$ such that $|z_n| \to +\infty$ as $n \to \infty$. Suppose for each n we are given a polynomial P_n of degree ≥ 1 such that $P_n(0) = 0$. Then there exists a meromorphic function in $\mathbb{C}$ with the principal part $P_n((z - z_n)^{-1})$ at z_n for every n, and with no other poles.*

PROOF. It suffices to prove the theorem when $z_n \neq 0$ for all n. If, say, $z_1 = 0$, simply add $P_1(z^{-1})$ to the function with the prescribed principal parts along the poles $\{z_n\}_{n=2}^{\infty}$.

Let $r_n = |z_n|$. The rational function $R_n(z) = P_n((z - z_n)^{-1})$ is holomorphic in the disk $\mathbb{D}(0, r_n)$, so it has a power series representation

$$R_n(z) = \sum_{k=0}^{\infty} a_{n,k}\, z^k \qquad \text{for } z \in \mathbb{D}(0, r_n).$$

Choose an integer d_n such that the Taylor polynomial $Q_n(z) = \sum_{k=0}^{d_n} a_{n,k}\, z^k$ satisfies

(9.1) $$\sup_{z \in \mathbb{D}(0, r_n/2)} |R_n(z) - Q_n(z)| \leq 2^{-n}.$$

Magnus Gösta
Mittag-Leffler
(1846–1927)

We claim that the series

$$f = \sum_{n=1}^{\infty} (R_n - Q_n)$$

defines the desired meromorphic function. To see this, take an arbitrary $r > 0$ and find an integer N such that $r_n \geq 2r$ whenever $n > N$. The rational function $R_n - Q_n$ is holomorphic in $\mathbb{D}(0, r)$ for all $n > N$, and by (9.1)

$$\sup_{z \in \mathbb{D}(0,r)} |R_n(z) - Q_n(z)| \leq 2^{-n} \quad \text{if } n > N.$$

Since $\sum_{n=N+1}^{\infty} 2^{-n} < +\infty$, the series $g = \sum_{n=N+1}^{\infty} (R_n - Q_n)$ converges uniformly in $\mathbb{D}(0, r)$ by the Weierstrass M-test, hence $g \in \mathscr{O}(\mathbb{D}(0, r))$. It follows that $f = g + \sum_{n=1}^{N} (R_n - Q_n)$ is meromorphic in $\mathbb{D}(0, r)$, has the principal part $R_n(z)$ at z_n if $r_n < r$, and has no other poles in $\mathbb{D}(0, r)$. Since r is arbitrary, we conclude that $f \in \mathscr{M}(\mathbb{C})$ and it has the required property. $\quad\square$

EXAMPLE 9.2. Let us construct an $f \in \mathscr{M}(\mathbb{C})$ with the principal part $1/(z - n)$ at every $n \in \mathbb{Z}$, and with no other poles. It turns out that taking each Taylor polynomial Q_n in the above proof to be of degree $d_n = 0$ already guarantees the convergence of the series that defines the meromorphic function f. In other words,

$$f(z) = \frac{1}{z} + \sum_{n \in \mathbb{Z} \setminus \{0\}} \left(\frac{1}{z - n} + \frac{1}{n} \right)$$

is the desired meromorphic function. To see this, take any $r > 0$ and note that

$$\left| \frac{1}{z - n} + \frac{1}{n} \right| = \frac{|z|}{|n| \, |z - n|} \leq \frac{2r}{n^2} \quad \text{if } |z| \leq r \text{ and } |n| \geq 2r.$$

Since $\sum 1/n^2 < +\infty$, it follows that the series

$$g(z) = \sum_{|n| \geq 2r} \left(\frac{1}{z - n} + \frac{1}{n} \right)$$

converges uniformly in the disk $\mathbb{D}(0, r)$, hence $g \in \mathscr{O}(\mathbb{D}(0, r))$. This shows

$$f(z) = g(z) + \frac{1}{z} + \sum_{0 < |n| < 2r} \left(\frac{1}{z - n} + \frac{1}{n} \right)$$

is meromorphic in $\mathbb{D}(0, r)$ with the principal part $1/(z - n)$ at every $n \in \mathbb{Z}$ for which $|n| < r$. Since r is arbitrary, we conclude that $f \in \mathscr{M}(\mathbb{C})$ and it has the desired property. One can verify that in fact $f(z) = \pi \cot(\pi z)$ (see problem 2 and compare Fig. 9.1).

EXAMPLE 9.3 (The Weierstrass $\wp$-function). Consider the lattice $\Gamma = \{m + in : n, m \in \mathbb{Z}\}$ of Gaussian integers in the plane. According to Theorem 9.1, there is a function in $\mathscr{M}(\mathbb{C})$ with the principal part $(z - \omega)^{-2}$ at every $\omega \in \Gamma$ and with no other poles. As in the previous example, taking

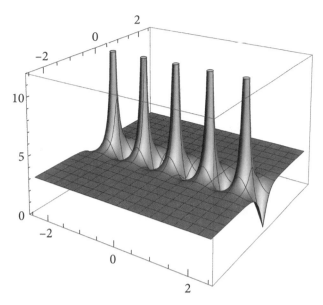

Figure 9.1. Graph of the absolute value of the meromorphic function $f(z) = \pi \cot(\pi z)$ of Example 9.2.

each polynomial Q_n to be of degree $d_n = 0$ turns out to produce a convergent series which defines the desired meromorphic function called the **Weierstrass $\wp$-function** of the lattice Γ:

$$\wp(z) = \frac{1}{z^2} + \sum_{\omega \in \Gamma^*} \left(\frac{1}{(z - \omega)^2} - \frac{1}{\omega^2} \right),$$

where $\Gamma^* = \Gamma \smallsetminus \{0\}$. To show convergence, let us first verify that

$$(9.2) \qquad\qquad \sum_{\omega \in \Gamma^*} \frac{1}{|\omega|^3} < +\infty.$$

Write Γ^* as the disjoint union of the "layers"

$$\Gamma_k = \{m + in : \max\{|m|, |n|\} = k\} \qquad k = 1, 2, 3, \ldots$$

(see Fig. 9.2). It is easy to see that Γ_k has $8k$ points which are at distance $\geq k$ from the origin. Hence,

$$\sum_{\omega \in \Gamma^*} \frac{1}{|\omega|^3} = \sum_{k=1}^{\infty} \sum_{\omega \in \Gamma_k} \frac{1}{|\omega|^3} \leq \sum_{k=1}^{\infty} \frac{1}{k^3} \cdot 8k = 8 \sum_{k=1}^{\infty} \frac{1}{k^2} < +\infty,$$

which proves (9.2). Now take any $r > 0$ and note that if $|z| \leq r$ and if $\omega \in \Gamma$ satisfies $|\omega| \geq 2r$, then

$$\left| \frac{1}{(z - \omega)^2} - \frac{1}{\omega^2} \right| = \frac{|z| \, |2\omega - z|}{|\omega|^2 \, |z - \omega|^2} \leq \frac{r \cdot 5|\omega|/2}{|\omega|^2 \cdot |\omega|^2/4} = \frac{10r}{|\omega|^3}.$$

It follows from (9.2) and the Weierstrass M-test that the series

$$g(z) = \sum_{|\omega| \geq 2r} \left(\frac{1}{(z - \omega)^2} - \frac{1}{\omega^2} \right)$$

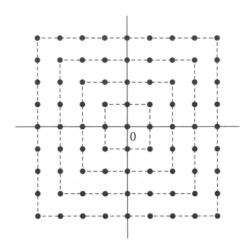

Figure 9.2. The set of non-zero Gaussian integers is partitioned into "layers" $\Gamma_1, \Gamma_2, \Gamma_3, \ldots$ where $\Gamma_k = \{m + in : \max\{|m|, |n|\} = k\}$ contains $8k$ points. The figure shows Γ_k for $1 \leq k \leq 4$.

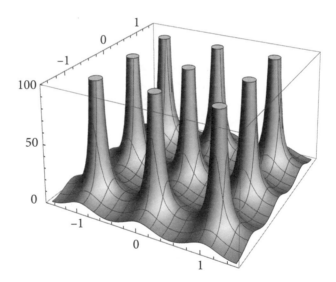

Figure 9.3. Graph of the absolute value of the Weierstrass $\wp$-function associated with the lattice of Gaussian integers.

converges uniformly in the disk $\mathbb{D}(0, r)$, hence $g \in \mathcal{O}(\mathbb{D}(0, r))$. Thus,

$$\wp(z) = g(z) + \frac{1}{z^2} + \sum_{0 < |\omega| < 2r} \left(\frac{1}{(z - \omega)^2} - \frac{1}{\omega^2} \right)$$

is meromorphic in $\mathbb{D}(0, r)$ with the principal part $1/(z - \omega)^2$ at every $\omega \in \Gamma$ for which $|\omega| < r$. Since r is arbitrary, we conclude that $\wp \in \mathcal{M}(\mathbb{C})$ and it has the desired property (see Fig. 9.3).

For generalizations and further properties, see the discussion of elliptic functions in the next section.

We now generalize Theorem 9.1 to arbitrary open sets:

THEOREM 9.4. *Let $U \subset \mathbb{C}$ be open and $\{z_n\} \subset U$ be a sequence of distinct points with no accumulation point in U. Suppose for each n we are given a polynomial P_n of degree ≥ 1 such that $P_n(0) = 0$. Then there exists a meromorphic function in U with the principal part $P_n((z - z_n)^{-1})$ at each z_n, and with no other poles.*

PROOF. The argument resembles the proof of the Weierstrass theorem for open sets (Theorem 8.25). The case $U = \mathbb{C}$ has already been dealt with in Theorem 9.1, so let us assume $U \neq \mathbb{C}$. It will be convenient to apply a preliminary change of coordinates to turn $\{z_n\}$ into a bounded sequence in the plane. To this end, take any $q \in U \smallsetminus \{z_n\}$, let $w = \varphi(z) = 1/(z - q)$, and consider the images $V = \varphi(U) \subset \hat{\mathbb{C}}$ and $w_n = \varphi(z_n)$. Note that $\infty \in V$, so ∂V is a compact subset of the plane. Moreover, $\{w_n\}$ does not accumulate on $\varphi(q) = \infty$, which means it is a bounded sequence in $\mathbb{C}$. Let $R_n(z) = P_n((z - z_n)^{-1})$ be the given principal part at z_n. We will construct a meromorphic function g in V, such that $g(\infty) = 0$, with the principal part

$$S_n(w) = R_n(\varphi^{-1}(w)) - R_n(q)$$

at each w_n and with no other poles. The composition $f = g \circ \varphi$ will then be the desired function in $\mathscr{M}(U)$.

For each n, let ζ_n be a closest point of ∂V to w_n. Then $\delta_n = |w_n - \zeta_n| \to 0$ since $\{w_n\}$ is bounded and does not accumulate in V. The rational function S_n has a single pole at w_n and $\lim_{w \to \infty} S_n(w) = 0$, so it is bounded and holomorphic in the annulus $\{w \in \mathbb{C} : |w - \zeta_n| > \delta_n\}$. As such, it has a Laurent series of the form

$$S_n(w) = \sum_{k=-\infty}^{-1} a_{n,k} (w - \zeta_n)^k$$

in this annulus, with all non-negative powers missing, which converges uniformly on $\{w \in \mathbb{C} : |w - \zeta_n| \geq 2\delta_n\}$ (compare problem 8 in chapter 3). Choose a negative integer d_n such that the rational function $Q_n(w) = \sum_{k=d_n}^{-1} a_{n,k} (w - \zeta_n)^k$ satisfies

(9.3) $|S_n(w) - Q_n(w)| \leq 2^{-n}$ if $|w - \zeta_n| \geq 2\delta_n$.

Let us show that

$$g = \sum_{n=1}^{\infty} (S_n - Q_n)$$

has the property we are looking for. Take any small $r > 0$ and let $V_r = \{w \in V : \text{dist}(w, \partial V) > r\}$. Choose an integer N such that $\delta_n < r/2$ if $n > N$. Then

$$|w - \zeta_n| > 2\delta_n \qquad \text{if } w \in V_r \text{ and } n > N,$$

so by (9.3),

$$\sup_{w \in V_r} |S_n(w) - Q_n(w)| \leq 2^{-n} \qquad \text{if } n > N.$$

Since $\sum_{n=N+1}^{\infty} 2^{-n} < +\infty$, the series $h = \sum_{n=N+1}^{\infty} (S_n - Q_n)$ converges uniformly in V_r by the Weierstrass M-test, hence $h \in \mathcal{O}(V_r)$. Now $g = h + \sum_{n=1}^{N} (S_n - Q_n)$ is meromorphic in V_r, has the principal part S_n at w_n if $w_n \in V_r$, and has no other poles in V_r. Since V is the union of the V_r for $r > 0$, it follows that g is meromorphic in V and it has the desired property. □

As an application of the above result, we prove a general interpolation theorem for holomorphic functions:

THEOREM 9.5. *Let $U \subset \mathbb{C}$ be open and $\{z_n\} \subset U$ be a sequence of distinct points with no accumulation point in U. Suppose for each n we are given a polynomial P_n of degree $d_n \geq 0$. Then there exists an $f \in \mathcal{O}(U)$ whose power series centered at z_n begins with P_n, so*

$$P_n(z) = \sum_{k=0}^{d_n} \frac{f^{(k)}(z_n)}{k!} (z - z_n)^k \qquad \text{for every } n.$$

In other words, one can prescribe the values of a holomorphic function and its finitely many derivatives along the sequence $\{z_n\}$.

PROOF. By Theorem 8.25, there is a $g \in \mathcal{O}(U)$ with a zero of order $d_n + 1$ at z_n for every n. We use Theorem 9.4 to find an $h \in \mathcal{M}(U)$ with suitable principal parts along $\{z_n\}$ so that $f = gh$ has the desired property. If R_n is the principal part of h at z_n, we simply need to make sure that the power series of $g(z)R_n(z)$ centered at z_n is given by $P_n(z)$ plus terms of order $> d_n$. Fix n and write

$$P_n(z) = \sum_{k=0}^{d_n} a_k(z - z_n)^k \quad \text{and} \quad g(z) = \sum_{k=0}^{\infty} b_k(z - z_n)^{k+d_n+1},$$

where $b_0 \neq 0$. We must then find a rational function of the form

$$R_n(z) = \sum_{k=0}^{d_n} c_k(z - z_n)^{k-d_n-1}$$

such that the coefficient of $(z - z_n)^k$ in $g(z)R_n(z)$ is a_k for every $0 \leq k \leq d_n$. This leads to the system of linear equations

$$\sum_{j=0}^{k} b_{k-j} c_j = a_k \qquad \text{for } 0 \leq k \leq d_n.$$

Since $b_0 \neq 0$, these equations can be solved successively for $c_0, c_1, \ldots, c_{d_n}$, and this completes the proof. □

9.2 Elliptic functions

The construction of the Weierstrass $\wp$-function for the lattice of Gaussian integers in Example 9.3 provides a suitable occasion to digress to a short introduction to "elliptic functions," whose theory has played a significant role in the early history of complex analysis. The terminology comes from the connection of these functions to elliptic integrals $\int dx/\sqrt{P(x)}$, where P is a polynomial of degree 3 or 4 with simple roots. Such integrals arise in numerous mathematical contexts such as measuring the arclength along ellipses and lemniscates, as well as physical problems such as the period of an ideal pendulum. They were studied extensively by Euler in the late 18th and Legendre in the early 19th century, but complex function theory did not enter the game until the remarkable discovery of Gauss and Abel that the *inverse* of an elliptic integral defines a doubly periodic meromorphic function in the plane, much the same way as the inverse of the integral $\int dx/\sqrt{1-x^2}$ defines the periodic trigonometric function $\sin z$. The subject has since been a fascinating scene of interaction between analysis, geometry, and arithmetic. For an engaging introduction, we recommend [**MM**].

Our quick excursion focuses on the construction of doubly periodic meromorphic functions. Throughout this section $\Lambda \subset \mathbb{C}$ will denote the lattice

$$\Lambda = \langle 1, \tau \rangle = \{m + n\tau : m, n \in \mathbb{Z}\}$$

for some fixed $\tau \in \mathbb{C}$ with $\operatorname{Im}(\tau) > 0$ (any lattice in $\mathbb{C}$ can be reduced to this form by a linear map $z \mapsto \alpha z$). Algebraically, Λ is an additive subgroup of $\mathbb{C}$ isomorphic to $\mathbb{Z} \oplus \mathbb{Z}$. For $z, w \in \mathbb{C}$, it is customary to write

$$z \equiv w \pmod{\Lambda} \quad \text{whenever} \quad z - w \in \Lambda.$$

A ***fundamental parallelogram*** for Λ is a subset of the form

$$\Pi = \Pi_p = \{p + t + s\tau : 0 \le t < 1, 0 \le s < 1\}$$

(see Fig. 9.4). The basic property of Π is that for every $z \in \mathbb{C}$ there is a *unique* $w \in \Pi$ such that $z \equiv w \pmod{\Lambda}$. A geometric way of thinking about this property is that various images of Π under repeated applications of $z \mapsto z + 1$ and $z \mapsto z + \tau$ cover the whole plane without overlaps. It follows that the quotient group $\mathbb{C}/\Lambda$ is homeomorphic to Π with the opposite sides identified, so it is topologically a torus.

DEFINITION 9.6. A meromorphic function f in the plane is called ***elliptic*** with respect to Λ if it is Λ-periodic in the sense that

$$f(z + \omega) = f(z) \quad \text{for every } z \in \mathbb{C} \text{ and } \omega \in \Lambda.$$

The set of all such functions forms a subfield of $\mathscr{M}(\mathbb{C})$ which we denote by $\mathscr{M}(\mathbb{C}, \Lambda)$.

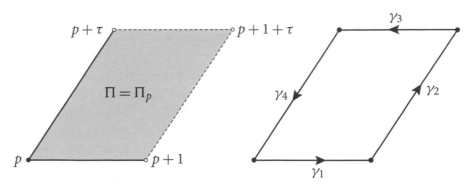

Figure 9.4. Left: A fundamental parallelogram Π for the lattice $\langle 1, \tau \rangle$. Right: The four segments of the oriented boundary $\partial \Pi$.

Observe that Λ-periodicity of f is equivalent to the condition

$$f(z+1) = f(z+\tau) = f(z) \qquad \text{for every } z \in \mathbb{C}.$$

For this reason such f is often called ***doubly periodic***. Evidently f induces a well-defined continuous map $\hat{f} : \mathbb{C}/\Lambda \to \hat{\mathbb{C}}$. Since the quotient $\mathbb{C}/\Lambda$ is compact, the image $f(\mathbb{C}) = \hat{f}(\mathbb{C}/\Lambda)$ must be a compact subset of $\hat{\mathbb{C}}$. Since $f(\mathbb{C})$ is also open whenever f is non-constant, we arrive at the following

THEOREM 9.7. *If $f \in \mathcal{M}(\mathbb{C}, \Lambda)$ is non-constant, then $f(\mathbb{C}) = \hat{\mathbb{C}}$. In particular, every doubly periodic entire function must be constant.*

Suppose $\Pi = \Pi_p$ is a fundamental parallelogram for Λ. The oriented boundary $\partial \Pi$ is homologous to the sum of the four segments

$$(9.4) \qquad \begin{aligned} \gamma_1 &= [p, p+1] & \gamma_2 &= [p+1, p+1+\tau] \\ \gamma_3 &= [p+1+\tau, p+\tau] & \gamma_4 &= [p+\tau, p], \end{aligned}$$

as shown in Fig. 9.4. Suppose $f \in \mathcal{M}(\mathbb{C}, \Lambda)$ has no poles on $\partial \Pi$. By periodicity,

$$\int_{\gamma_1} f(z)\, dz = -\int_{\gamma_3} f(z)\, dz \quad \text{and} \quad \int_{\gamma_2} f(z)\, dz = -\int_{\gamma_4} f(z)\, dz.$$

Thus, the integrals of f along opposite sides of Π cancel out and we obtain

$$(9.5) \qquad \int_{\partial \Pi} f(z)\, dz = 0 \qquad \text{if } f \in \mathcal{M}(\mathbb{C}, \Lambda) \text{ has no poles on } \partial \Pi.$$

THEOREM 9.8. *Suppose $f \in \mathcal{M}(\mathbb{C}, \Lambda)$ is non-constant and Π is a fundamental parallelogram for Λ.*

(i) We have

$$\sum \operatorname{res}(f, p) = 0,$$

where the sum is taken over all poles of f in Π. In particular, f has at least two poles in Π counting multiplicities.

(ii) *There is an integer $d \geq 2$ such that for every $q \in \hat{\mathbb{C}}$ the equation $f(z) = q$ has d solutions in Π counting multiplicities.*

(iii) *If $z_1, \ldots, z_d$ and $p_1, \ldots, p_d$ are the zeros and poles of f in Π, repeated according to multiplicity, then*

(9.6)
$$\sum_{k=1}^{d} z_k \equiv \sum_{k=1}^{d} p_k \quad (\mathrm{mod}\ \Lambda).$$

Abel showed that (9.6) is also sufficient for the existence of an elliptic function with zeros z_k and poles p_k.

The integer d in (ii) is called the **degree** of f and is denoted by $\deg(f)$. Note that the lower bound $d \geq 2$ is an immediate consequence of (i).

PROOF. (i) After a small translation of Π if necessary, we may assume that all poles of f in Π are in the interior of Π. It then follows from (9.5) and the residue theorem that

$$\sum \mathrm{res}(f, p) = 2\pi i \int_{\partial \Pi} f(z)\, dz = 0.$$

Notice that f must have at least one pole in Π since $f(\Pi) = f(\mathbb{C}) = \hat{\mathbb{C}}$. If f had only one simple pole in Π, the residue at that pole would necessarily be non-zero, which would contradict what we just proved. Hence, counting multiplicities, f must have ≥ 2 poles in Π.

(ii) By periodicity, the number of solutions of $f(z) = q$ in Π does not depend on the choice of Π, so we may pick Π such that f has no zeros or poles on $\partial \Pi$. Applying (9.5) to the function $f'/f \in \mathscr{M}(\mathbb{C}, \Lambda)$, we obtain

$$\frac{1}{2\pi i} \int_{\partial \Pi} \frac{f'(z)}{f(z)}\, dz = 0.$$

It follows from the argument principle that f has as many zeros as poles in Π. The same argument applied to the function $f - q \in \mathscr{M}(\mathbb{C}, \Lambda)$ for $q \in \mathbb{C}$ (which has the same poles as f) will then prove the result.

(iii) As in (ii), we may assume that no z_k or p_k is on $\partial \Pi$. By the generalized argument principle (problem 27 in chapter 3),

$$\frac{1}{2\pi i} \int_{\partial \Pi} z \frac{f'(z)}{f(z)}\, dz = \sum_{k=1}^{d} z_k - \sum_{k=1}^{d} p_k.$$

Niels Henrik Abel (1802–1829)

Thus, to prove (9.6), we need to show that the above integral belongs to Λ.

To this end, consider the oriented segments $\gamma_1, \gamma_2, \gamma_3, \gamma_4$ of $\partial \Pi$ defined in (9.4). Since γ_3 is obtained from the the reverse curve γ_1^- by the translation $z \mapsto z + \tau$, periodicity of f'/f shows that

$$\int_{\gamma_3} z \frac{f'(z)}{f(z)}\, dz = \int_{\gamma_1^-} (z + \tau) \frac{f'(z)}{f(z)}\, dz = -\int_{\gamma_1} z \frac{f'(z)}{f(z)}\, dz - \tau \int_{\gamma_1} \frac{f'(z)}{f(z)}\, dz.$$

We claim that $\int_{\gamma_1} f'(z)/f(z)\,dz$ is an integer multiple of $2\pi i$. In fact, if g is a holomorphic logarithm of f in a simply connected neighborhood of $|\gamma_1|$, then

$$\int_{\gamma_1} \frac{f'(z)}{f(z)}\,dz = \int_{\gamma_1} g'(z)\,dz = g(p+1) - g(p).$$

But $e^{g(p+1)-g(p)} = f(p+1)/f(p) = 1$ implies $g(p+1) - g(p) = 2\pi i n$ for some integer n, as claimed. Similarly,

$$\int_{\gamma_2} z\frac{f'(z)}{f(z)}\,dz = \int_{\gamma_4^-} (z+1)\frac{f'(z)}{f(z)}\,dz = -\int_{\gamma_4} z\frac{f'(z)}{f(z)}\,dz - \int_{\gamma_4} \frac{f'(z)}{f(z)}\,dz,$$

where $\int_{\gamma_4} f'(z)/f(z)\,dz = 2\pi i m$ for some integer m. It follows that

$$\frac{1}{2\pi i}\int_{\partial \Pi} z\frac{f'(z)}{f(z)}\,dz = \frac{1}{2\pi i}\sum_{k=1}^{4} \int_{\gamma_k} z\frac{f'(z)}{f(z)}\,dz = -m - n\tau \in \Lambda. \qquad \square$$

EXAMPLE 9.9. Suppose $f \in \mathscr{M}(\mathbb{C}, \Lambda)$ has $\deg(f) = 2$ and a pole of order 2 at 0 (the Weierstrass $\wp$-function of Example 9.3 is such a function for $\Lambda = \langle 1, i \rangle$). Then the two zeros z_1, z_2 of f in every fundamental parallelogram satisfy $z_1 + z_2 \in \Lambda$.

We now address the question of existence of elliptic functions for an arbitrary lattice. Luckily, the Mittag-Leffler construction of Example 9.3 for the lattice of Gaussian integers works in general and gives rise to an elliptic function of degree 2. Although at first glance this example seems very special, it turns out that all elliptic functions can be expressed in terms of this particular one (see problem 10).

> **DEFINITION 9.10.** The **Weierstrass $\wp$-function** associated with a lattice Λ is defined by
>
> $$(9.7) \qquad \wp(z) = \wp_\Lambda(z) = \frac{1}{z^2} + \sum_{\omega \in \Lambda^*} \left(\frac{1}{(z-\omega)^2} - \frac{1}{\omega^2} \right),$$
>
> where $\Lambda^* = \Lambda \smallsetminus \{0\}$.

The convergence of (9.7) is proved in the same manner as Example 9.3. We only need to check the analog of (9.2). This can be done directly or by simply observing that there is an $\mathbb{R}$-linear isomorphism $L : \mathbb{C} \to \mathbb{C}$ that carries Λ onto the lattice Γ of Gaussian integers. Then $|L(z)| \le M|z|$ for all z, where $M = \sup_{|z|=1} |L(z)| > 0$, and (9.2) shows that

$$(9.8) \qquad \sum_{\omega \in \Lambda^*} \frac{1}{|\omega|^3} \le M^3 \sum_{\omega \in \Gamma^*} \frac{1}{|\omega|^3} < +\infty.$$

Now the argument in Example 9.3 shows that $\wp \in \mathcal{M}(\mathbb{C})$, with the principal part $1/(z-\omega)^2$ at every $\omega \in \Lambda$.

THEOREM 9.11 (Basic properties of $\wp$). *Let $\Lambda = \langle 1, \tau \rangle$ and $\wp = \wp_\Lambda$.*

   (i) $\wp$ is meromorphic in $\mathbb{C}$ with poles of order 2 at the points of Λ.
   (ii) $\wp$ is an even function: $\wp(z) = \wp(-z)$ for all $z \in \mathbb{C}$.
   (iii) $\wp \in \mathcal{M}(\mathbb{C}, \Lambda)$ and $\deg(\wp) = 2$.
   (iv) z is a critical point of $\wp$ if and only if $2z \equiv 0 \pmod{\Lambda}$. Every critical point z is simple in the sense that the local degree $\deg(\wp, z)$ is 2.
   (v) The critical points of $\wp$ in Π_0 are $0, c_1 = 1/2, c_2 = (1+\tau)/2, c_3 = \tau/2$. The corresponding critical values $\infty = \wp(0), e_1 = \wp(c_1), e_2 = \wp(c_2), e_3 = \wp(c_3)$ are distinct.
   (vi) $\wp(z) = \wp(w)$ if and only if $z \equiv \pm w \pmod{\Lambda}$.

PROOF. (i) is immediate from what we have just shown, while (ii) follows from (9.7) since $-\Lambda = \Lambda$.

(iii) The derivative

$$(9.9) \qquad \wp'(z) = \frac{-2}{z^3} + \sum_{\omega \in \Lambda^*} \frac{-2}{(z-\omega)^3} = -2 \sum_{\omega \in \Lambda} \frac{1}{(z-\omega)^3}$$

is clearly Λ-periodic. In particular, for $\omega \in \{1, \tau\}$ the function $\wp(z+\omega) - \wp(z)$ must be constant. Calling this constant C and setting $z = -\omega/2$, we see that $\wp(\omega/2) - \wp(-\omega/2) = C$. Since $\wp$ is even and $\pm\omega/2$ is not a pole of $\wp$, we obtain $C = 0$. This proves $\wp(z+1) = \wp(z+\tau) = \wp(z)$ for all z, so $\wp \in \mathcal{M}(\mathbb{C}, \Lambda)$. The claim $\deg(\wp) = 2$ follows from the fact that in every fundamental parallelogram $\wp$ has a unique point of Λ, hence a unique pole of order 2.

(iv) First note that all points of Λ are critical since they are poles at which the local degree is 2. All other critical points of $\wp$ must be zeros of $\wp'$. By (9.9), $\wp'$ has poles of order 3 along Λ, so $\deg(\wp') = 3$. It follows that there are three roots of the equation $\wp'(z) = 0$ in every fundamental parallelogram Π. Together with the unique pole at $\Lambda \cap \Pi$, we must have a total of four critical points of $\wp$ in Π.

To locate the critical points, we use the identity $\wp'(\omega - z) = -\wp'(z)$ which holds for all $z \in \mathbb{C}$ and $\omega \in \Lambda$ because $\wp'$ is Λ-periodic and odd. Setting $z = \omega/2$ implies $\wp'(\omega/2) = 0$ or ∞; either way, it follows that $\omega/2$ is a critical point. Thus, $\wp$ has a critical point at z if $2z \equiv 0 \pmod{\Lambda}$. This equation has four solutions in the fundamental parallelogram Π_0, namely $0, 1/2, (1+\tau)/2, \tau/2$. By the preceding paragraph, these must account for all critical points in Π_0. If z is any critical point of $\wp$, the unique point $w \in \Pi_0$ with $z \equiv w \pmod{\Lambda}$ is also critical, so $2z \equiv 2w \equiv 0 \pmod{\Lambda}$.

Finally, the critical points of $\wp$ are all simple because a critical point of local degree > 2 would violate $\deg(\wp) = 2$.

(v) By what we have seen above, the four points $0, 1/2, (1+\tau)/2, \tau/2$ are critical. The critical values are distinct since if two of the above points map to the same critical

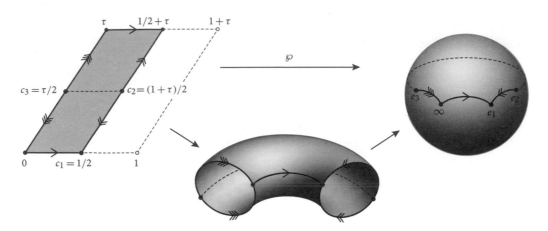

Figure 9.5. A cut-and-paste realization of $\wp$ as the quotient of $\mathbb{C}$ under the action of the affine group generated by $z \mapsto z+1$, $z \mapsto z+\tau$, and $z \mapsto -z$. This amounts to gluing the parallelogram $\{t + s\tau : 0 \le t \le 1/2, 0 \le s \le 1\}$ shown on the left along the indicated pairs of boundary segments to obtain a topological sphere. The quotient map is locally homeomorphic except near the critical points $0, 1/2, (1+\tau)/2, \tau/2 \pmod{\Lambda}$ where it is locally 2-to-1.

value v, the equation $\wp(z) = v$ would have four or more solutions in Π_0 counting multiplicities, contradicting $\deg(\wp) = 2$.

(vi) If $z \equiv -w \pmod{\Lambda}$, then $\wp(z) = \wp(w)$ since $\wp$ is even. Conversely, given $z \in \mathbb{C}$ let ζ be the unique point in Π_z such that $\zeta \equiv -z \pmod{\Lambda}$. If $\zeta \ne z$, then $\wp^{-1}(\wp(z)) \cap \Pi_z = \{z, \zeta\}$ since $\deg(\wp) = 2$. If $\zeta = z$, then $2z \equiv 0 \pmod{\Lambda}$, so z is a critical point by (iv) and $\wp^{-1}(\wp(z)) \cap \Pi_z = \{z\}$ again since $\deg(\wp) = 2$. It follows that every $w \in \wp^{-1}(\wp(z))$ satisfies $w \equiv \pm z \pmod{\Lambda}$. $\qquad\square$

REMARK 9.12. A good way of thinking about the action of $\wp$ is as follows: $\wp$ induces a map $\hat{\wp} : \mathbb{C}/\Lambda \to \hat{\mathbb{C}}$ which is 2-to-1 since $\deg(\wp) = 2$. The preimage $\hat{\wp}^{-1}(q)$ of each $q \in \hat{\mathbb{C}}$ is a pair $\{z, -z\} \pmod{\Lambda}$ on the torus $\mathbb{C}/\Lambda$ since $\wp$ is even. If $z \not\equiv -z \pmod{\Lambda}$, these points are distinct in $\mathbb{C}/\Lambda$ and therefore have to be non-critical. On the other hand, if $z \equiv -z$ or $2z \equiv 0 \pmod{\Lambda}$, then these points collide in $\mathbb{C}/\Lambda$ and the common point is critical.

Thus, at least on a topological level, the action of $\hat{\wp}$ can be identified with taking the quotient of the torus $\mathbb{C}/\Lambda$ under the map $z \mapsto -z \pmod{\Lambda}$ which yields a sphere (see Fig. 9.5).

THEOREM 9.13. *If e_1, e_2, e_3 denote the finite critical values of $\wp$, then*

$$(9.10) \qquad\qquad (\wp')^2 = 4(\wp - e_1)(\wp - e_2)(\wp - e_3).$$

PROOF. The two sides of (9.10) belong to $\mathcal{M}(\mathbb{C}, \Lambda)$, and they have the same zeros (all of order 2) and poles (all of order 6). This implies that their ratio has removable singularities and therefore extends to an entire function in $\mathcal{M}(\mathbb{C}, \Lambda)$. By Theorem 9.7 this ratio must be a constant function. The constant is easily seen to be 1 since near the origin both sides of (9.10) are asymptotic to $4/z^6$. □

The Laurent series of $\wp$ near 0 is easy to compute and reveals other remarkable properties. Let $\delta = \min_{\omega \in \Lambda^*} |\omega| > 0$. If $|z| < \delta$, then for every $\omega \in \Lambda^*$ we can write

$$\frac{1}{(z-\omega)^2} - \frac{1}{\omega^2} = \frac{1}{\omega^2}\left[\frac{1}{(1-\frac{z}{\omega})^2} - 1\right]$$

$$= \frac{1}{\omega^2}\sum_{n=2}^{\infty} n\left(\frac{z}{\omega}\right)^{n-1} = \sum_{n=1}^{\infty}\frac{(n+1)z^n}{\omega^{n+2}}.$$

This gives the expansion

(9.11)
$$\wp(z) = \frac{1}{z^2} + \sum_{\omega \in \Lambda^*}\sum_{n=1}^{\infty}\frac{(n+1)z^n}{\omega^{n+2}}$$

for $0 < |z| < \delta$. The double series in this formula is absolutely convergent. In fact, if we set $C = \sum_{\omega \in \Lambda^*} 1/|\omega|^3$, which is finite by (9.8), then

$$\sum_{\omega \in \Lambda^*}\frac{1}{|\omega|^{n+2}} \le \frac{C}{\delta^{n-1}} \qquad \text{for all } n \ge 1.$$

It follows that

$$\sum_{n=1}^{\infty}\sum_{\omega \in \Lambda^*}\left|\frac{(n+1)z^n}{\omega^{n+2}}\right| \le C\delta \sum_{n=1}^{\infty}(n+1)\left(\frac{|z|}{\delta}\right)^n < +\infty$$

provided $|z| < \delta$. Thus, it is legitimate to switch the order of summations in (9.11) to obtain

$$\wp(z) = \frac{1}{z^2} + \sum_{n=1}^{\infty}\left(\sum_{\omega \in \Lambda^*}\frac{1}{\omega^{n+2}}\right)(n+1)z^n$$

for $0 < |z| < \delta$. Set

(9.12)
$$G_n = \sum_{\omega \in \Lambda^*}\frac{1}{\omega^n} \qquad \text{for } n \ge 3.$$

The G_{2n}, viewed as functions of τ, are examples of "Eisenstein series" in number theory.

By (9.8), every G_n is a well-defined complex number, with $G_n = 0$ for all odd n because of the symmetry $\Lambda = -\Lambda$. We arrive at the Laurent series

(9.13)
$$\wp(z) = \frac{1}{z^2} + \sum_{n=1}^{\infty}(2n+1)G_{2n+2}z^{2n} \qquad \text{for } 0 < |z| < \delta.$$

THEOREM 9.14 (Basic differential equation of $\wp$). *The Weierstrass $\wp$-function satisfies*

(9.14) $$(\wp')^2 = 4\wp^3 - g_2\wp - g_3,$$

where

$$g_2 = 60\, G_4 = 60 \sum_{\omega \in \Lambda^*} \frac{1}{\omega^4},$$

$$g_3 = 140\, G_6 = 140 \sum_{\omega \in \Lambda^*} \frac{1}{\omega^6}.$$

The constants g_2, g_3 are called the **invariants** of the lattice Λ. All higher coefficients G_{2n+2} in (9.13) can be expressed polynomially in terms of g_2, g_3 (see problem 7).

PROOF. Let us denote by $O(z^k)$ any holomorphic function for which 0 is a zero of order $\geq k$. By (9.13), near the origin,

$$\wp(z) = \frac{1}{z^2} + 3G_4\, z^2 + 5G_6\, z^4 + O(z^6),$$

so

$$\wp^3(z) = \frac{1}{z^6} + \frac{9G_4}{z^2} + 15G_6 + O(z^2)$$

and

$$(\wp'(z))^2 = \left(-\frac{2}{z^3} + 6G_4\, z + 20G_6\, z^3 + O(z^5) \right)^2$$

$$= \frac{4}{z^6} - \frac{24G_4}{z^2} - 80G_6 + O(z^2).$$

Thus, in a punctured neighborhood of 0,

$$(\wp'(z))^2 - 4\wp^3(z) + 60G_4\wp(z) = -140G_6 + O(z^2).$$

If f denotes the left-hand side of this equation, we conclude that f has a removable singularity at 0. Since $f \in \mathscr{M}(\mathbb{C}, \Lambda)$, the singularities of f, which are along Λ, must all be removable, so f extends to an entire function. By Theorem 9.7, f is the constant function $-140G_6$. □

COROLLARY 9.15. *The finite critical values e_1, e_2, e_3 of $\wp$ satisfy the following relations:*

- $e_1 + e_2 + e_3 = 0$
- $e_1 e_2 + e_2 e_3 + e_3 e_1 = -\frac{1}{4} g_2$
- $e_1 e_2 e_3 = \frac{1}{4} g_3.$

Moreover,

(9.15) $$g_2^3 - 27g_3^2 \neq 0.$$

PROOF. By Theorems 9.13 and 9.14,

$$4(\wp - e_1)\,(\wp - e_2)\,(\wp - e_3) = 4\wp^3 - g_2\wp - g_3,$$

and the relations follow by comparing the coefficients of similar powers of $\wp$.

If $g_2^3 = 27g_3^2$ and we set $\zeta^2 = g_2/12$, then $\zeta^2(4\zeta^2 - g_2)^2 = g_3^2$, so one of the two choices for ζ would be a common root of the cubic $4w^3 - g_2w - g_3$ and its derivative $12w^2 - g_2$. This would contradict the fact that $4w^3 - g_2w - g_3$ has simple roots. □

REMARK 9.16. The differential equation (9.14) can be interpreted as saying that the map

$$\Phi(z) = (\wp(z), \wp'(z))$$

parametrizes the complex cubic curve $C = \{(X, Y) \in \mathbb{C}^2 : Y^2 = 4X^3 - g_2X - g_3\}$. In fact, it is not hard to check that Φ is a homeomorphism (and in fact a biholomorphism, in a suitable sense) between the torus $\mathbb{C}/\Lambda$ and the closure $\overline{C}$ in the complex projective plane $\mathbb{CP}^2$, with $0 \pmod{\Lambda}$ mapping to the point at infinity on $\overline{C}$ (that is, the intersection of $\overline{C}$ with the line at infinity in $\mathbb{CP}^2$). A classical theorem of Abel asserts that every non-singular cubic curve in $\mathbb{CP}^2$ arises from this construction. This is equivalent to the statement that every $g_2, g_3 \in \mathbb{C}$ subject to the condition (9.15) are the invariants of the lattice $\langle 1, \tau \rangle$ for some $\tau \in \mathbb{H}$; see [**MM**] for details. The correspondence also allows us to equip $\overline{C}$ with the structure of an abelian group, with the point at infinity playing the role of the identity element; see problem 12.

Finally, let us comment on the relation between elliptic functions and elliptic integrals. Let $\Lambda = \langle 1, \tau \rangle$ and consider the open parallelogram

$$P = \left\{ t + s\tau : 0 < t < 1/2, 0 < s < 1/2 \right\}$$

with corners at the critical points $0, c_1 = 1/2, c_2 = (1 + \tau)/2, c_3 = \tau/2$. It is easy to see that $\wp$ maps ∂P homeomorphically onto a Jordan curve in $\hat{\mathbb{C}}$ containing the critical values ∞, e_1, e_2, e_3 (compare Fig. 9.5). Hence $w = \wp(z)$ is a biholomorphism $P \to \Omega$, where Ω is one of the two complementary components of the Jordan curve $\wp(\partial P)$. Since the cubic $4w^3 - g_2w - g_3 = 4(w - e_1)(w - e_2)(w - e_3)$ is non-vanishing in Ω, it has precisely two holomorphic square roots in there (here we use the fact that each complementary component of a Jordan curve is simply connected; see the discussion after Theorem 9.27 below). The differential equation (9.14) shows that one of these square roots, which we simply denote by $\sqrt{4w^3 - g_2w - g_3}$, satisfies $dw/dz = \sqrt{4w^3 - g_2w - g_3}$ in P. It follows that the inverse map $z = \wp^{-1}(w)$, from Ω to P, satisfies $dz/dw = 1/\sqrt{4w^3 - g_2w - g_3}$. This proves

$$(9.16) \qquad z = \wp^{-1}(w) = z_0 + \int_{w_0}^{w} \frac{d\zeta}{\sqrt{4\zeta^3 - g_2\zeta - g_3}},$$

where the integral is taken along any curve in Ω from $w_0 = \wp(z_0)$ to w.

9.3 Rational approximation

A holomorphic function can be uniformly approximated on every closed subdisk of its domain by a sequence of polynomials, namely the partial sums of its power series representation. A natural question is whether uniform approximation by polynomials is possible on more general compact sets.

EXAMPLE 9.17. The function $f(z) = 1/z$, which is holomorphic in the punctured plane, cannot be uniformly approximated by polynomials on the unit circle $\mathbb{T}$. In fact, if there were polynomials P_n such that $P_n \to f$ uniformly on $\mathbb{T}$ as $n \to \infty$, then by Cauchy's theorem we would have

$$2\pi i = \int_{\mathbb{T}} f(z)\, dz = \lim_{n \to \infty} \int_{\mathbb{T}} P_n(z)\, dz = 0.$$

Here is another reasoning without direct reference to integration: If f could be uniformly approximated on $\mathbb{T}$ by polynomials, we could find a polynomial P such that $|1/z - P(z)| \leq 1/2$ or $|1 - zP(z)| \leq 1/2$ for all $z \in \mathbb{T}$. The maximum principle would then imply $|1 - zP(z)| \leq 1/2$ for all $z \in \mathbb{D}$, which is clearly false at $z = 0$.

The above example shows that in general a given function $f \in \mathcal{O}(U)$ may not be uniformly approximated on a compact set $K \subset U$ by polynomials. But, somewhat surprisingly, there is a purely topological condition on K that guarantees uniform approximation by polynomials (Corollary 9.19 below). Even if K fails to satisfy this condition, we can still uniformly approximate every $f \in \mathcal{O}(U)$ on K by rational functions with poles outside K. Moreover, the construction is flexible in that it allows us to prescribe one pole in each component of $\hat{\mathbb{C}} \smallsetminus K$. This is stated more precisely in the following

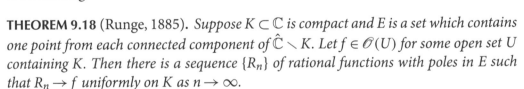

THEOREM 9.18 (Runge, 1885). *Suppose $K \subset \mathbb{C}$ is compact and E is a set which contains one point from each connected component of $\hat{\mathbb{C}} \smallsetminus K$. Let $f \in \mathcal{O}(U)$ for some open set U containing K. Then there is a sequence $\{R_n\}$ of rational functions with poles in E such that $R_n \to f$ uniformly on K as $n \to \infty$.*

Observe that K is *not* assumed to be connected (and in fact this is a great advantage of the theorem). Note also that E is necessarily countable since the open set $\hat{\mathbb{C}} \smallsetminus K$ has countably many components.

Before discussing the proof of Runge's theorem, let us consider an important special case. A compact set $K \subset \mathbb{C}$ is said to be **full** if $\mathbb{C} \smallsetminus K$ is connected, or equivalently, if $\mathbb{C} \smallsetminus K$ has no bounded component in the plane. Intuitively, a full compact set has no "holes" in it. When K is full, we can choose $E = \{\infty\}$ in the statement of Runge's Theorem 9.18 to obtain the following

COROLLARY 9.19 (Uniform approximation by polynomials). *Suppose $K \subset \mathbb{C}$ is a full compact set and $f \in \mathcal{O}(U)$ for some open set U containing K. Then there is a sequence $\{P_n\}$ of polynomials such that $P_n \to f$ uniformly on K as $n \to \infty$.*

Carle David Tolme Runge
(1856–1927)

REMARK 9.20. The assumption of K being full in the above corollary is necessary, as can be seen by an argument similar to Example 9.17. Suppose K is not full. Take a bounded component V of $\mathbb{C} \smallsetminus K$ and some $p \in V$. We claim that the function $f(z) = 1/(z - p)$, which is holomorphic in a neighborhood of K, cannot be uniformly approximated on K by polynomials. Otherwise, we could find a polynomial P which satisfies

$$\sup_{z \in K} |f(z) - P(z)| \leq \frac{1}{2M},$$

where $M = \sup_{z \in K} |z - p|$. Then,

$$|1 - (z - p)P(z)| \leq \frac{1}{2M} |z - p| \leq \frac{1}{2}$$

for all $z \in K$. Since $\partial V \subset K$, the maximum principle would imply

$$|1 - (z - p)P(z)| \leq \frac{1}{2}$$

for all $z \in V$. This is a contradiction since $1 - (z - p)P(z) = 1$ when $z = p$.

The idea of the proof of Theorem 9.18 is simple: We first find a cycle γ in $U \smallsetminus K$ for which Cauchy's integral formula holds:

$$f(z) = \frac{1}{2\pi i} \int_{\gamma} \frac{f(\zeta)}{\zeta - z} \, d\zeta \qquad \text{for } z \in K.$$

By breaking γ into finitely many small pieces γ_k, the integral can be approximated uniformly on K by a Riemann sum

$$\frac{1}{2\pi i} \sum_k \frac{f(q_k)}{q_k - z} \Delta_k,$$

where the q_k are sample points on $|\gamma_k|$ and $\Delta_k = \gamma_k(1) - \gamma_k(0)$. Evidently such Riemann sums are rational functions of z with poles on $|\gamma|$. A classical trick, often known as "pole shifting," then shows how to approximate these rational functions by ones with poles along E.

A more modern proof, which uses basic ideas of functional analysis, is sketched in problem 24.

LEMMA 9.21. *For every compact subset K of an open set $U \subset\subset \mathbb{C}$ there is a null-homologous cycle γ in U that "represents" K in the following sense:*

$$|\gamma| \cap K = \emptyset \quad \text{and} \quad \mathrm{W}(\gamma, p) = 1 \quad \text{for every } p \in K.$$

PROOF. By compactness of K there is an $r > 0$ such that $\mathbb{D}(p, 2r) \subset U$ whenever $p \in K$. Consider the closed squares of side length r whose corners belong to the lattice $r(\mathbb{Z} \oplus \mathbb{Z}) = \{r(m + in) : m, n \in \mathbb{Z}\}$. Let $\{S_1, \ldots, S_n\}$ be the collection of all such squares which intersect K. By the choice of r, each S_k is contained in U. If p is the lower left corner of S_k, consider the oriented segments

$$\beta_k^1 = [p, p+r]$$

$$\beta_k^2 = [p+r, p+(1+i)r]$$

$$\beta_k^3 = [p+(1+i)r, p+ir]$$

$$\beta_k^4 = [p+ir, p].$$

Evidently the cycle $\partial S_k = \sum_{j=1}^4 \beta_k^j$ is homologous to the oriented boundary of S_k. Let $\mathcal{C}$ be the collection of all β_k^j for $1 \le k \le n$ and $1 \le j \le 4$. The chain

$$\eta = \sum_{\xi \in \mathcal{C}} \xi = \sum_{k=1}^n \partial S_k$$

is clearly a cycle. Let $\mathcal{D}$ be the collection obtained from $\mathcal{C}$ by removing every oriented segment ξ whose reverse ξ^- also appears in $\mathcal{C}$. It is immediate that

$$\gamma = \sum_{\xi \in \mathcal{D}} \xi$$

is also a cycle (see Fig. 9.6). If $|\xi| \cap K \ne \emptyset$ for some $\xi \in \mathcal{C}$, then $|\xi|$ must be the common edge of two adjacent squares in $\{S_1, \ldots, S_n\}$. Hence ξ and ξ^- both appear in $\mathcal{C}$, so $\xi \notin \mathcal{D}$. It follows that $|\gamma| \cap K = \emptyset$.

To show that γ has the desired winding number property, first note that for every $p \in \mathbb{C} \smallsetminus |\eta|$,

$$W(\gamma, p) = \sum_{\xi \in \mathcal{D}} W(\xi, p) = \sum_{\xi \in \mathcal{C}} W(\xi, p) = W(\eta, p)$$

since the difference between the two sums consists of terms of the form $W(\xi, p) + W(\xi^-, p)$ which are always zero. Hence

$$W(\gamma, p) = \sum_{k=1}^n W(\partial S_k, p) = \begin{cases} 1 & \text{if } p \text{ is in the interior of some } S_k \\ 0 & \text{if } p \in \mathbb{C} \smallsetminus \bigcup_{k=1}^n S_k. \end{cases}$$

Here we have used the fact that $W(\partial S_k, p)$ is 1 if p is in the interior of S_k and is 0 if p is outside S_k (Theorem 2.28). Since $\bigcup_{k=1}^n S_k \subset U$, this shows that $W(\gamma, p) = 0$ if $p \in \mathbb{C} \smallsetminus U$, so γ is null-homologous. On the other hand, every $p \in K$ is either in the interior of some S_k or is accumulated by such interior points. Since $W(\gamma, \cdot)$ is constant in some neighborhood of p (Theorem 2.33(iii)), it follows that $W(\gamma, p) = 1$. $\qquad \square$

LEMMA 9.22 (Pole shifting). *Suppose $K \subset\subset \mathbb{C}$ is compact, V is a connected component of $\hat{\mathbb{C}} \smallsetminus K$, and $p \in V$. Then, for every $q \in V \smallsetminus \{\infty\}$, the function $1/(z-q)$ can be uniformly approximated on K by rational functions with a single pole at p.*

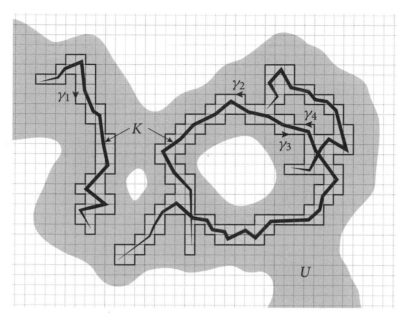

Figure 9.6. Proof of Lemma 9.21. There is a null-homologous cycle γ in U with the property that $W(\gamma, \cdot) = 1$ on the compact set K. Here γ is homologous to the sum of the oriented boundaries of all blue squares, which in turn is homologous to the sum $\sum_{k=1}^{4} \gamma_k$ of the polygonal closed curves shown.

Such rational functions are non-constant polynomials in $1/(z-p)$ when $p \neq \infty$ and in z when $p = \infty$.

PROOF. First assume $p \neq \infty$. Let W be the set of all $q \in V \setminus \{\infty\}$ for which the function $1/(z-q)$ can be uniformly approximated on K by rational functions with a single pole at p. Evidently W is non-empty since $p \in W$. Moreover, W is closed in $V \setminus \{\infty\}$ since if $\{q_n\}$ is a sequence in W which tends to some $q \in V \setminus \{\infty\}$, then $1/(z-q_n) \to 1/(z-q)$ uniformly on K, hence $q \in W$. It suffices to verify that W is also open in $V \setminus \{\infty\}$, for then connectedness of $V \setminus \{\infty\}$ implies that $W = V \setminus \{\infty\}$.

Take $q \in W$ and a disk $\mathbb{D}(q, r) \subset V \setminus \{\infty\}$. If $w \in \mathbb{D}(q, r)$ and $z \in K$, then

$$\frac{1}{z-w} = \frac{1}{(z-q)\left[1 - \left(\dfrac{w-q}{z-q}\right)\right]} = \sum_{n=0}^{\infty} \frac{(w-q)^n}{(z-q)^{n+1}}.$$

Here the geometric series converges uniformly in z because

$$\sup_{z \in K} \left|\frac{w-q}{z-q}\right| \leq \frac{|w-q|}{r} < 1.$$

The partial sums of this series are polynomials in $1/(z-q)$ which uniformly approximate $1/(z-w)$ on K. By the assumption, $1/(z-q)$ and therefore these partial sums can be uniformly approximated on K by rational functions with a single pole at p.

Hence the same must be true of $1/(z-w)$. This proves $\mathbb{D}(q,r) \subset W$. Since $q \in W$ was arbitrary, it follows that W is open.

Now suppose $p = \infty$. Choose $r > 0$ large enough so that $\{z \in \mathbb{C} : |z| \geq r\} \subset V$. If $z \in K$, then

$$\frac{1}{z-r} = \frac{-1}{r(1-z/r)} = -\sum_{n=0}^{\infty} \frac{z^n}{r^{n+1}},$$

where the geometric series converges uniformly in z because $\sup_{z \in K} |z|/r < 1$. The partial sums of this series are polynomials in z which uniformly approximate $1/(z-r)$ on K. By the first case treated above, for every $q \in V \setminus \{\infty\}$ the function $1/(z-q)$ can be uniformly approximated on K by polynomials in $1/(z-r)$. It follows that $1/(z-q)$ can be uniformly approximated on K by polynomials in z.   □

PROOF OF THEOREM 9.18. By Lemma 9.21 there is a null-homologous cycle γ in U that represents K. By Theorem 2.41, Cauchy's integral formula

$$(9.17) \qquad f(z) = \frac{1}{2\pi i} \int_\gamma \frac{f(\zeta)}{\zeta - z} \, d\zeta$$

holds for every $z \in K$. Recall that γ is the sum of some number N of horizontal or vertical oriented segments in $U \setminus K$, all having equal length r for some small $r > 0$ depending on K and U. Subdivide each of these segments into n sub-segments with consistent orientations and length r/n. Label the collection of the sub-segments thus obtained $\{\gamma_1, \ldots, \gamma_{nN}\}$, so γ is homologous to $\sum_{k=1}^{nN} \gamma_k$. Since f is uniformly continuous on the compact set $|\gamma| = \bigcup_{k=1}^{nN} |\gamma_k|$, for a given $\varepsilon > 0$ we can choose n large enough so that

$$p, q \in |\gamma_k| \qquad \text{implies} \qquad |f(p) - f(q)| < \varepsilon$$

for every $1 \leq k \leq nN$. Fix a sample point $q_k \in |\gamma_k|$ for each k. Set

$$M = \sup_{z \in |\gamma|} |f(z)| \qquad \text{and} \qquad d = \mathrm{dist}(|\gamma|, K) > 0.$$

Since $\mathrm{length}(\gamma_k) = r/n$, we see that for $z \in K$ and $\zeta \in |\gamma_k|$,

$$\left| \frac{f(\zeta)}{\zeta - z} - \frac{f(q_k)}{q_k - z} \right| \leq \left| \frac{f(\zeta)}{\zeta - z} - \frac{f(\zeta)}{q_k - z} \right| + \left| \frac{f(\zeta)}{q_k - z} - \frac{f(q_k)}{q_k - z} \right|$$

$$= \frac{|f(\zeta)| \, |\zeta - q_k|}{|\zeta - z| \, |q_k - z|} + \frac{|f(\zeta) - f(q_k)|}{|q_k - z|} \leq \frac{Mr}{nd^2} + \frac{\varepsilon}{d} < \frac{2\varepsilon}{d},$$

provided that n is large enough. It follows from the ML-inequality that for every $z \in K$,

$$(9.18) \qquad \left| \int_{\gamma_k} \frac{f(\zeta)}{\zeta - z} \, d\zeta - \frac{f(q_k)}{q_k - z} \Delta_k \right| = \left| \int_{\gamma_k} \left(\frac{f(\zeta)}{\zeta - z} - \frac{f(q_k)}{q_k - z} \right) d\zeta \right| \leq \frac{2\varepsilon r}{dn},$$

where $\Delta_k = \int_{\gamma_k} d\zeta = \gamma_k(1) - \gamma_k(0)$.

For each $1 \leq k \leq nN$, let p_k be the element of E that belongs to the same component of $\hat{\mathbb{C}} \setminus K$ as q_k does. Lemma 9.22 shows that there is a rational function R_k with a unique pole at p_k such that

$$\sup_{z \in K} \left| \frac{f(q_k)}{q_k - z} \Delta_k - R_k(z) \right| \leq \frac{\varepsilon}{n}.$$

Then, by (9.18),

$$\sup_{z \in K} \left| \int_{\gamma_k} \frac{f(\zeta)}{\zeta - z} d\zeta - R_k(z) \right| \leq \left(\frac{2r}{d} + 1 \right) \frac{\varepsilon}{n}.$$

Now the rational function $R = 1/(2\pi i) \sum_{k=1}^{nN} R_k$ has poles along E, and by (9.17),

$$\sup_{z \in K} |f(z) - R(z)| = \sup_{z \in K} \left| \frac{1}{2\pi i} \sum_{k=1}^{nN} \int_{\gamma_k} \frac{f(\zeta)}{\zeta - z} d\zeta - \frac{1}{2\pi i} \sum_{k=1}^{nN} R_k(z) \right|$$

$$\leq \frac{1}{2\pi} \sum_{k=1}^{nN} \sup_{z \in K} \left| \int_{\gamma_k} \frac{f(\zeta)}{\zeta - z} d\zeta - R_k(z) \right|$$

$$\leq \frac{nN}{2\pi} \left(\frac{2r}{d} + 1 \right) \frac{\varepsilon}{n} = \frac{N}{2\pi} \left(\frac{2r}{d} + 1 \right) \varepsilon.$$

The proof is complete since ε was arbitrary. □

Our next goal is to establish the analog of Theorem 9.18 for open sets. The proof will make use of the following topological fact already encountered in §5.1:

LEMMA 9.23 (Existence of nice exhaustions). *Every non-empty open set $U \subset \mathbb{C}$ has a "nice exhaustion" by compact subsets $\{K_n\}$, in the sense that*

(i) $\bigcup_{n=1}^{\infty} K_n = U$,
(ii) *each K_n is contained in the interior of K_{n+1}, and*
(iii) *each connected component of $\hat{\mathbb{C}} \smallsetminus K_n$ contains a connected component of $\hat{\mathbb{C}} \smallsetminus U$.*

Intuitively, property (iii) means that the holes of K_n are never redundant: They exist because of the holes of U.

PROOF. For $n \geq 1$ define

$$K_n = \{p \in \mathbb{C} : |p| \leq n \text{ and } \mathbb{D}(p, 1/n) \subset U\}.$$

As we saw in §5.1, properties (i) and (ii) hold. To verify (iii), first note that since $\hat{\mathbb{C}} \smallsetminus U \subset \hat{\mathbb{C}} \smallsetminus K_n$, every component of $\hat{\mathbb{C}} \smallsetminus U$ is contained in a unique component of $\hat{\mathbb{C}} \smallsetminus K_n$. Thus, it suffices to show that every component V of $\hat{\mathbb{C}} \smallsetminus K_n$ intersects $\hat{\mathbb{C}} \smallsetminus U$. If $\infty \in V$, then V trivially intersects $\hat{\mathbb{C}} \smallsetminus U$. Otherwise, V is a bounded domain in $\mathbb{C}$. Since

$$\mathbb{C} \smallsetminus K_n = \{p \in \mathbb{C} : |p| > n\} \cup \bigcup_{p \in \mathbb{C} \smallsetminus U} \mathbb{D}(p, 1/n),$$

V must be a union of open disks of radius $1/n$ centered at the points of $\mathbb{C} \smallsetminus U$. So, again, V intersects $\hat{\mathbb{C}} \smallsetminus U$. □

THEOREM 9.24 (Runge's theorem for open sets). *Suppose $U \subset \mathbb{C}$ is open and E is a set which contains one point from each connected component of $\hat{\mathbb{C}} \smallsetminus U$. If $f \in \mathcal{O}(U)$, there exists a sequence $\{R_n\}$ of rational functions with poles in E such that $R_n \to f$ compactly in U as $n \to \infty$. In particular, if $\hat{\mathbb{C}} \smallsetminus U$ is connected, we can take $E = \{\infty\}$ and choose each R_n to be a polynomial.*

Note that U is *not* assumed to be connected. Also, unlike Theorem 9.18, E can now be an uncountable set. This is because the complement $\hat{\mathbb{C}} \smallsetminus U$ can have uncountably many components. For example, if $\mathbb{C} \smallsetminus U$ is a ***Cantor set*** in the plane, then the prescribed set E is this Cantor set union $\{\infty\}$, which is certainly uncountable.

> A Cantor set is a compact totally disconnected topological space with no isolated point. A classical theorem in topology asserts that all Cantor sets are homeomorphic.

PROOF. Let $\{K_n\}$ be a nice exhaustion of U as in Lemma 9.23. For each n, every component of $\hat{\mathbb{C}} \smallsetminus K_n$ contains a component of $\hat{\mathbb{C}} \smallsetminus U$, hence at least one point of E. By Theorem 9.18, there exists a rational function R_n with poles in E such that $\sup_{z \in K_n} |f(z) - R_n(z)| < 1/n$.

Given a compact set $K \subset U$ and an $\varepsilon > 0$, choose an integer $N > 1/\varepsilon$ such that $K \subset K_n$ whenever $n \geq N$. Then

$$\sup_{z \in K} |f(z) - R_n(z)| \leq \sup_{z \in K_n} |f(z) - R_n(z)| < \frac{1}{n} < \varepsilon$$

whenever $n \geq N$. This proves $R_n \to f$ compactly in U. $\qquad\square$

Runge's theorem is a powerful tool for constructing holomorphic functions with peculiar properties when manufacturing them directly is very difficult. The example below serves to illustrate this point. For further examples, see problems 21–23.

EXAMPLE 9.25. We use Runge's theorem to show that there exists a sequence $\{P_n\}$ of complex polynomials such that $P_n(0) = 1$ for all n but $\lim_{n\to\infty} P_n(z) = 0$ for $z \neq 0$. It suffices to find a sequence $\{Q_n\}$ of polynomials such that $Q_n(z) \to 1$ if z is real and non-negative and $Q_n(z) \to 0$ otherwise. For then the sequence $\{S_n\}$ defined by $S_n(z) = Q_n(z)Q_n(-z)$ converges to 1 if $z = 0$ and to 0 otherwise, and the polynomials $P_n(z) = S_n(z)/S_n(0)$ will have the desired property.

The construction of $\{Q_n\}$ is quite simple. For $n \geq 2$ consider the compact sets

$$X_n = \left\{ z \in \mathbb{C} : |z| \leq n \text{ and } \mathrm{dist}(z, [0, n]) \geq \frac{1}{n} \right\} \quad \text{and} \quad Y_n = [0, n],$$

and set $K_n = X_n \cup Y_n$ (see Fig. 9.7). The function

$$f_n(z) = \begin{cases} 0 & \text{if } z \in X_n \\ 1 & \text{if } z \in Y_n \end{cases}$$

extends holomorphically to a neighborhood of K_n. Since K_n is full, Corollary 9.19 shows that there is a polynomial Q_n such that

$$\sup_{z \in K_n} |f_n(z) - Q_n(z)| \leq \frac{1}{n}.$$

It is easy to see that $Q_n(z) \to 1$ if z is real and non-negative and $Q_n(z) \to 0$ otherwise.

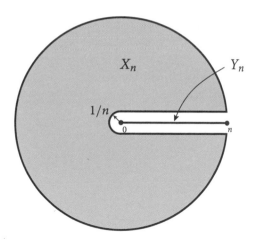

Figure 9.7. Illustration of Example 9.25. Here the compact sets $K_n = X_n \cup Y_n$ are full, with $K_n \subset K_{n+1}$ and $\bigcup K_n = \mathbb{C}$.

Let us contrast Runge's theorem with its classical counterpart in real analysis. According to the ***Stone-Weierstrass theorem***, every continuous function on an arbitrary compact set $K \subset \mathbb{C}$ can be uniformly approximated by polynomials in $z, \bar{z}$ (equivalently in x, y). In the special case where $K \subset \mathbb{R}$ is an interval, this reduces to an older theorem of Weierstrass which asserts that the approximating polynomials can be chosen to depend only on z (equivalently x). Runge's theorem is weaker than the Weierstrass theorem in this special case since it requires a holomorphic extension to some neighborhood of K, but in general neither theorem implies the other.

Mergelyan has generalized the Weierstrass theorem to all full compact sets in the plane. His result can also be viewed as a generalization of Corollary 9.19 coming from Runge's theory:

THEOREM 9.26 (Mergelyan, 1952). *Suppose $K \subset \mathbb{C}$ is a full compact set and f is continuous on K and holomorphic in the interior of K. Then there is a sequence $\{P_n\}$ of polynomials such that $P_n \to f$ uniformly on K as $n \to \infty$.*

This is a much deeper result with a substantially harder proof which we will omit (see for example [**Ru2**]). However, a special case of the theorem and some related results are discussed in problems.

9.4 Finitely connected domains

Sergei Nikitich
Mergelyan (1928–2008)

We begin with several characterizations of simply connected domains in the plane from the topological, analytic, and algebraic viewpoints. All but one of the implications in the following theorem have already been established in previous chapters or can be proved directly. The single exception is dealt with using Runge's theorem.

THEOREM 9.27 (Characterizations of simply connected domains). *The following conditions on a domain $U \subset \mathbb{C}$ are equivalent:*

(i)   *U is conformally isomorphic to $\mathbb{D}$, or $U = \mathbb{C}$.*

(ii)   *U is diffeomorphic to $\mathbb{D}$.*

(iii)   *U is homeomorphic to $\mathbb{D}$.*

(iv)   *U is simply connected.*

(v)   *U has trivial first homology: $H_1(U) = 0$.*

(vi)   *The spherical complement $\hat{\mathbb{C}} \smallsetminus U$ is connected.*

(vii)   *For every $f \in \mathscr{O}(U)$ there is a sequence $\{P_n\}$ of polynomials such that $P_n \to f$ compactly in U as $n \to \infty$.*

(viii)   *For every $f \in \mathscr{O}(U)$ and every closed curve γ in U, $\int_\gamma f(z)\,dz = 0$.*

(ix)   *Every holomorphic function in U has a primitive.*

(x)   *Every real-valued harmonic function in U is the real part of a holomorphic function in U.*

(xi)   *Every non-vanishing holomorphic function in U has a holomorphic logarithm.*

(xii)   *Every non-vanishing holomorphic function in U has holomorphic n-th roots for all $n \geq 2$.*

(xiii)   *Every non-vanishing holomorphic function in U has a holomorphic square root.*

Observe that (vi) is the only condition that is "extrinsic" in the sense that it takes into account the way U is embedded in the sphere. It justifies the intuitive idea that a simply connected domain has no "holes." Incidentally, $\hat{\mathbb{C}} \smallsetminus U$ in (vi) cannot be replaced by $\mathbb{C} \smallsetminus U$: The doubly-slit plane

$$U = \mathbb{C} \smallsetminus \{x \in \mathbb{R} : |x| \geq 1\}$$

is simply connected, but $\mathbb{C} \smallsetminus U$ has two connected components. Note also that (xii) and (xiii) are purely algebraic conditions on the ring $\mathscr{O}(U)$.

PROOF. (i) $\Longrightarrow$ (ii): If U is conformally isomorphic to $\mathbb{D}$, then it is certainly diffeomorphic to it since a holomorphic function is C^∞-smooth. The complex plane, although not conformally isomorphic to $\mathbb{D}$, is still diffeomorphic to it: the map $z \mapsto z/\sqrt{1 + |z|^2}$ provides an example of a diffeomorphism $\mathbb{C} \to \mathbb{D}$.

(ii) $\Longrightarrow$ (iii) and (iii) $\Longrightarrow$ (iv) are trivial implications, while (iv) $\Longrightarrow$ (v) is Corollary 2.40.

(v) $\Longrightarrow$ (vi): Assuming the closed set $\hat{\mathbb{C}} \smallsetminus U$ is disconnected, we can write it as the disjoint union $E \cup K$ of two non-empty closed sets, where $\infty \in E$. Then K is a compact subset of the plane and the union $U \cup K = \hat{\mathbb{C}} \smallsetminus E \subset \mathbb{C}$ is open. By Lemma 9.21, there is a cycle γ in $U \cup K$ that represents K, so $|\gamma| \cap K = \varnothing$ and $\mathrm{W}(\gamma, \cdot) = 1$ on K. It follows that γ is a cycle in U whose homology class $\langle \gamma \rangle$ is non-zero in $H_1(U)$.

(vi) $\Longrightarrow$ (vii) follows from Theorem 9.24.

(vii) $\implies$ (viii): Take $f \in \mathcal{O}(U)$ and a sequence $\{P_n\}$ of polynomials which converges compactly in U to f. Let γ be any closed curve in U. Since $\int_\gamma P_n(z)\,dz = 0$ for all n by Cauchy's theorem, and since $P_n \to f$ uniformly on $|\gamma|$, it follows that

$$\int_\gamma f(z)\,dz = \lim_{n\to\infty} \int_\gamma P_n(z)\,dz = 0.$$

(viii) $\implies$ (ix) follows from Theorem 1.29, while (ix) $\implies$ (x) follows from the proof of Theorem 7.9.

(x) $\implies$ (xi): Suppose $f \in \mathcal{O}(U)$ is non-vanishing. Then $u = \log|f|$ is harmonic in U (Example 7.7), so $u = \text{Re}(g)$ for some $g \in \mathcal{O}(U)$. Equivalently, $|f| = e^u = e^{\text{Re}(g)} = |e^g|$. The quotient f/e^g is holomorphic in U and has absolute value 1, so it must be a constant c by the open mapping theorem. After adding a suitable constant to g, we can arrange $c = 1$, so $f = e^g$.

(xi) $\implies$ (xii) is immediate since if g is a holomorphic logarithm of f, then $e^{g/n}$ is a holomorphic n-th root of f, and (xii) $\implies$ (xiii) is trivial.

Finally, (xiii) $\implies$ (i) follows from the proof of the Riemann mapping theorem (Theorem 6.1) since the only property of the domain U used in that proof was the existence of holomorphic square roots. $\qquad\square$

As an application of the above theorem, we note that *the two complementary components of a Jordan curve in $\hat{\mathbb{C}}$ are simply connected*. In fact, if γ is a Jordan curve in $\hat{\mathbb{C}}$ and U, V are the connected components of $\hat{\mathbb{C}} \smallsetminus |\gamma|$, then $\partial U = \partial V = |\gamma|$ by the Jordan curve theorem. It follows that $\hat{\mathbb{C}} \smallsetminus U = \overline{V}$ and $\hat{\mathbb{C}} \smallsetminus V = \overline{U}$ are both connected, so U, V are simply connected. We use this observation together with Carathéodory's Theorem 6.20 to give a quick proof of the following classical result in 2-dimensional topology:

THEOREM 9.28 (Schönflies, 1908). *If γ is a Jordan curve in $\hat{\mathbb{C}}$, every homeomorphism $\varphi : \mathbb{T} \to |\gamma|$ extends to a homeomorphism $\Phi : \hat{\mathbb{C}} \to \hat{\mathbb{C}}$.*

The theorem means that there are no knots in the plane.

PROOF. As we observed above, the connected components U, V of $\hat{\mathbb{C}} \smallsetminus |\gamma|$ are both simply connected. Take biholomorphisms $f : \mathbb{D} \to U$ and $g : \hat{\mathbb{C}} \smallsetminus \overline{\mathbb{D}} \to V$ given by the Riemann mapping theorem. Theorem 6.20 guarantees that both f and g extend to homeomorphisms between the closures. It follows that $f^{-1} \circ \varphi : \mathbb{T} \to \mathbb{T}$ and $g^{-1} \circ \varphi : \mathbb{T} \to \mathbb{T}$ are circle homeomorphisms and therefore extend to homeomorphisms $F : \overline{\mathbb{D}} \to \overline{\mathbb{D}}$ and $G : \hat{\mathbb{C}} \smallsetminus \mathbb{D} \to \hat{\mathbb{C}} \smallsetminus \mathbb{D}$, respectively (take, for example, their radial extensions). The map $\Phi : \hat{\mathbb{C}} \to \hat{\mathbb{C}}$ defined by

$$\Phi = \begin{cases} f \circ F & \text{in } \mathbb{D} \\ g \circ G & \text{in } \hat{\mathbb{C}} \smallsetminus \mathbb{D} \end{cases}$$

is now a homeomorphic extension of φ. $\qquad\square$

Arthur Moritz Schönflies
(1853–1928)

Some of the above results, when suitably modified, remain valid for **finitely connected** domains, namely domains which have finitely many complementary components. More quantitatively, a domain $U \subset \hat{\mathbb{C}}$ is called m-**connected** if $\hat{\mathbb{C}} \smallsetminus U$ has precisely m connected components. In this terminology the equivalence (iv) $\Longleftrightarrow$ (vi) in Theorem 9.27 states that a proper sub-domain of $\hat{\mathbb{C}}$ is 1-connected if and only if it is simply connected.

Each complementary component K of a finitely connected domain is compact and full in the sense that $\hat{\mathbb{C}} \smallsetminus K$ is connected. Moreover, either K is a single point, in which case we think of it as a puncture, or it contains more than one point (in fact, uncountably many) in which case we refer to it as **non-degenerate**.

The following statement provides convenient topological and conformal models for finitely connected domains:

THEOREM 9.29. *Every m-connected domain $U \subset \hat{\mathbb{C}}$ is homeomorphic to an m-punctured sphere. If the complementary components of U are all non-degenerate, then U is conformally isomorphic to a domain bounded by m disjoint analytic Jordan curves.*

Here an "analytic" Jordan curve is one that is locally the image of a straight line segment (or circular arc) under a conformal map, a condition that is stronger than C^∞-smooth. Analytic arcs and Jordan curves are studied in §10.3.

PROOF. For the first statement we use induction on m. For $m = 1$ the result holds since every 1-connected domain is simply connected and therefore homeomorphic to the punctured sphere $\hat{\mathbb{C}} \smallsetminus \{\infty\} = \mathbb{C}$. Assume the result holds for some $m \geq 1$. Let U be an $(m+1)$-connected domain with complementary components $K_1, \ldots, K_{m+1}$. Since $\hat{\mathbb{C}} \smallsetminus K_{m+1}$ is simply connected, there is a homeomorphism g between $\hat{\mathbb{C}} \smallsetminus K_{m+1}$ and the plane $\hat{\mathbb{C}} \smallsetminus \{\infty\}$. The domain $g(U) \cup \{\infty\} = \hat{\mathbb{C}} \smallsetminus (g(K_1) \cup \cdots \cup g(K_m))$ is m-connected, so by the induction hypothesis there is a homeomorphism h mapping this domain to an m-punctured sphere $\hat{\mathbb{C}} \smallsetminus \{p_1, \ldots, p_m\}$. Evidently $h \circ g$ maps U homeomorphically to $\hat{\mathbb{C}} \smallsetminus \{p_1, \ldots, p_m, h(\infty)\}$. This completes the induction step.

The proof of the second statement is also inductive and similar to above. For $m = 1$ the result follows from the Riemann mapping theorem. Assume the result holds for some $m \geq 1$. Let U be an $(m+1)$-connected domain with non-degenerate complementary components $K_1, \ldots, K_{m+1}$. By the Riemann mapping theorem there is a conformal map $g : \hat{\mathbb{C}} \smallsetminus K_{m+1} \to \mathbb{D}$. The domain $g(U) \cup (\hat{\mathbb{C}} \smallsetminus \mathbb{D}) = \hat{\mathbb{C}} \smallsetminus (g(K_1) \cup \cdots \cup g(K_m))$ is m-connected with non-degenerate complementary components, so by the induction hypothesis there is a conformal map h between this domain and one bounded by m disjoint analytic Jordan curves $\gamma_1, \ldots, \gamma_m$. The composition $h \circ g$ then maps U conformally onto the domain in $\hat{\mathbb{C}}$ bounded by the analytic Jordan curves $\gamma_1, \ldots, \gamma_m, h(\mathbb{T})$, which completes the induction step (Fig. 9.8 illustrates the inductive process for a 3-connected domain). $\qquad\square$

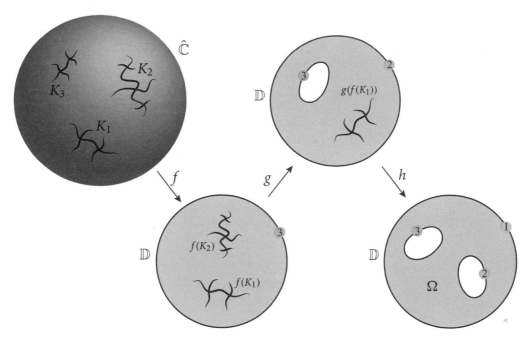

Figure 9.8. Construction of a conformal map $h \circ g \circ f$ from a 3-connected domain $\hat{\mathbb{C}} \smallsetminus (K_1 \cup K_2 \cup K_3)$ to a domain Ω bounded by three analytic Jordan curves (one being the unit circle). Here $f : \hat{\mathbb{C}} \smallsetminus K_3 \to \mathbb{D}$, $g : \hat{\mathbb{C}} \smallsetminus f(K_2) \to \mathbb{D}$, and $h : \hat{\mathbb{C}} \smallsetminus g(f(K_1)) \to \mathbb{D}$ are conformal maps given by the Riemann mapping theorem.

REMARK 9.30. A stronger version of the above result, known as ***Koebe's circle domain theorem (1920)***, asserts that an m-connected domain with non-degenerate complementary components is conformally isomorphic to a domain bounded by m disjoint Euclidean circles. Moreover, this domain is unique up to a Möbius transformation [**Ko**]. For a proof of this result in the case $m = 2$, see Theorem 13.33.

The first homology group of a finitely connected domain is easy to identify, either using the punctured sphere model or directly as follows. Suppose $K_1, \ldots, K_m$ are the complementary components of an m-connected domain U. It suffices to consider the case $m \geq 2$ since the simply connected case has trivial homology. We may also assume $\infty \in K_m$ so $U \subset \mathbb{C}$ and $K_1, \ldots, K_{m-1}$ are disjoint compact subsets of the plane. For each $1 \leq j \leq m - 1$, Lemma 9.21 applied to the open set $U \cup K_j$ gives a cycle γ_j in U with the property that $\mathrm{W}(\gamma_j, \cdot)$ is 1 on K_j and is 0 on $\mathbb{C} \smallsetminus (U \cup K_j)$. If γ is an arbitrary cycle in U, each K_j is contained in a connected component of $\mathbb{C} \smallsetminus |\gamma|$, so $\mathrm{W}(\gamma, \cdot)$ takes a constant value n_j on K_j. It follows that the cycle $\gamma - \sum_{j=1}^{m-1} n_j \gamma_j$ is null-homologous in U, so $\gamma \sim \sum_{j=1}^{m-1} n_j \gamma_j$. This proves that the homology classes $\langle \gamma_1 \rangle, \ldots, \langle \gamma_{m-1} \rangle$ generate $H_1(U)$. Moreover, if $\sum_{j=1}^{m-1} n_j \gamma_j \sim 0$ for some integers $n_1, \ldots, n_{m-1}$, then by calculating the winding number of each side on K_j, we see that $n_j = 0$ for all j. Thus, we arrive at the following

THEOREM 9.31 (Homology of finitely connected domains). *The first homology group of an m-connected domain $U \subset \mathbb{C}$ for $m \geq 2$ is a free abelian group of rank $m - 1$:*

$$H_1(U) \cong \mathbb{Z}^{m-1} = \underbrace{\mathbb{Z} \oplus \cdots \oplus \mathbb{Z}}_{m-1 \text{ copies}}.$$

If $K_1, \ldots, K_{m-1}$ are the bounded connected components of $\mathbb{C} \smallsetminus U$, we can choose a "standard basis" $\langle \gamma_1 \rangle, \ldots, \langle \gamma_{m-1} \rangle$ for $H_1(U)$ in the sense that $\mathrm{W}(\gamma_j, \cdot)$ is 1 on K_j and is 0 on K_n for $n \neq j$.

The following result gives the modified form of the conditions (ix), (x), and (xi) of Theorem 9.27 for finitely connected domains:

THEOREM 9.32. *Suppose $U \subset \mathbb{C}$ is an m-connected domain for some $m \geq 2$ with bounded complementary components $K_1, \ldots, K_{m-1}$ and a standard basis $\langle \gamma_1 \rangle, \ldots, \langle \gamma_{m-1} \rangle$ for $H_1(U)$ as in Theorem 9.31.*

(i) *A function $f \in \mathscr{O}(U)$ has a primitive in U if and only if $\int_{\gamma_j} f(z)\, dz = 0$ for all $1 \leq j \leq m - 1$.*

(ii) *A non-vanishing function $f \in \mathscr{O}(U)$ has a holomorphic logarithm if and only if $\int_{\gamma_j} f'(z)/f(z)\, dz = 0$ for all $1 \leq j \leq m - 1$.*

(iii) *For every harmonic function $u : U \to \mathbb{R}$ there are real constants $\alpha_1, \ldots, \alpha_{m-1}$ such that if $p_j \in K_j$ for $1 \leq j \leq m - 1$, then*

$$u(z) - \sum_{j=1}^{m-1} \alpha_j \log |z - p_j|$$

is the real part of a holomorphic function in U.

As in Theorem 7.13 the "periods" α_j in (iii) are uniquely determined by u (see problem 31).

PROOF. (i) Each cycle γ_j is homologous to a finite sum of closed curves in U. Conversely, every closed curve in U, viewed as a cycle, is homologous to a linear combination of the γ_j. Thus, by Cauchy's Theorem 2.41, $f \in \mathscr{O}(U)$ satisfies $\int_{\gamma_j} f(z)\, dz = 0$ for all j if and only if $\int_\gamma f(z)\, dz = 0$ for every closed curve γ in U. The result now follows from Theorem 1.29.

(ii) By part (i), $\int_{\gamma_j} f'(z)/f(z)\, dz = 0$ for all j if and only if f'/f has a primitive $g \in \mathscr{O}(U)$. The latter holds precisely when $f = \exp(g + c)$ for some constant c.

(iii) Define

$$\alpha_j = \frac{1}{\pi i} \int_{\gamma_j} u_z\, dz \qquad \text{for } 1 \leq j \leq m - 1.$$

The same argument as in the proof of Theorem 7.13 shows that every α_j is real. The function $h(z) = 2u_z - \sum_{j=1}^{m-1} \alpha_j/(z-p_j)$ is holomorphic in U and satisfies

$$\int_{\gamma_k} h(z)\, dz = 2\pi i \alpha_k - 2\pi i \sum_{j=1}^{m-1} \alpha_j \, W(\gamma_k, p_j) = 0$$

for every k. It follows from part (i) that h has a primitive $g \in \mathcal{O}(U)$. The computation

$$\frac{\partial}{\partial z}\Big(g - 2u + 2\sum_{j=1}^{m-1} \alpha_j \log |z - p_j|\Big) = g' - 2u_z + \sum_{j=1}^{m-1} \frac{\alpha_j}{z - p_j} = g' - h = 0$$

shows that the function $g - 2u + 2\sum_{j=1}^{m-1} \alpha_j \log|z - p_j|$ is anti-holomorphic, so its complex conjugate $\bar{g} - 2u + 2\sum_{j=1}^{m-1} \alpha_j \log|z - p_j|$ is holomorphic. Thus, the sum

$$g + \bar{g} - 2u + 2\sum_{j=1}^{m-1} \alpha_j \log |z - p_j|$$

is also holomorphic in U. Since this function is real-valued, it must be a constant $2c$ by the open mapping theorem. The function $f = g - c \in \mathcal{O}(U)$ now satisfies $\operatorname{Re}(f) = u - \sum_{j=1}^{m-1} \alpha_j \log|z - p_j|$. $\square$

As a byproduct of the above proof, it follows that a harmonic function $u : U \to \mathbb{R}$ is the real part of a holomorphic function in U if and only if $\int_{\gamma_j} u_z\, dz = 0$ for all $1 \le j \le m - 1$.

Problems

(1) According to Mittag-Leffler's Theorem 9.1, there is a meromorphic function in $\mathbb{C}$ with the principal part $1/(z - n)^2$ at every $n \in \mathbb{Z}$, and with no other poles.

    (i) Show that the series

$$f(z) = \frac{1}{z^2} + \sum_{n \in \mathbb{Z} \setminus \{0\}} \left(\frac{1}{(z-n)^2} - \frac{1}{n^2} \right)$$

        defines such a function.

    (ii) Verify that the familiar function $g(z) = \pi^2/\sin^2(\pi z)$ also has the above property.

    (iii) Show that f and g agree up to an additive constant and deduce the partial fraction expansion

$$\frac{\pi^2}{\sin^2(\pi z)} = \sum_{n=-\infty}^{\infty} \frac{1}{(z-n)^2}.$$

    (Hint: For (iii), show that the difference $h = f - g$ is entire, satisfies $h(z+1) = h(z)$ for all z, and is bounded on the strip $\{z : 0 \le \operatorname{Re}(z) \le 1\}$. Conclude that h is constant.)

(2) In Example 9.2 we constructed a meromorphic function in $\mathbb{C}$ with the principal part $1/(z - n)$ at every $n \in \mathbb{Z}$, and with no other poles. Verify that the function $\pi \cot(\pi z)$ also has this property. Then show that these two examples are identical:

$$\pi \cot(\pi z) = \frac{1}{z} + \sum_{n \in \mathbb{Z} \smallsetminus \{0\}} \left(\frac{1}{z-n} + \frac{1}{n} \right).$$

(Hint: Integrate the partial fraction expansion in the previous problem.)

(3)   Suppose $\{z_n\}$ is a sequence of non-zero complex numbers for which $\sum_{n=1}^{\infty} |z_n|^{-(d+1)}$ converges for some integer $d \geq 1$. Show that

$$f(z) = \sum_{n=1}^{\infty} \left(\frac{1}{(z_n - z)^d} - \frac{1}{z_n^d} \right)$$

defines a meromorphic function in $\mathbb{C}$ with poles of order d along $\{z_n\}$.

(4)   Prove that the infinite series

$$f(z) = \sum_{n=-\infty}^{\infty} \frac{1}{z^3 - n^3}$$

defines a meromorphic function in the plane. Identify the poles and principal parts of f.

(5)   Give an alternative proof of the Weierstrass product theorem using Theorem 9.1 of Mittag-Leffler along the following lines: Let $\{z_n\}$ be a sequence of distinct points in $\mathbb{C}^*$ with $|z_n| \to +\infty$ and to each z_n assign a multiplicity $m_n \geq 1$. Find $g \in \mathscr{M}(\mathbb{C})$ with the principal part $m_n/(z - z_n)$ at each z_n and with no other poles. Let

$$f(z) = \exp \left(\int_0^z g(\zeta) \, d\zeta \right),$$

where the integral is taken along any curve in $\mathbb{C} \smallsetminus \{z_n\}$ from 0 to z. Show that (i) f is well defined and holomorphic in $\mathbb{C} \smallsetminus \{z_n\}$; (ii) $f(z) \to 0$ as $z \to z_n$, so f extends to an entire function; (iii) f has a zero of order m_n at z_n and does not vanish anywhere else.

In the following problems $\Lambda = \langle 1, \tau \rangle$ is a lattice in $\mathbb{C}$, with $\mathrm{Im}(\tau) > 0$, and $\wp = \wp_\Lambda$ is the associated elliptic function.

(6)   The **Weierstrass σ-function** is defined by

$$\sigma(z) = z \prod_{\omega \in \Lambda^*} E_2 \left(\frac{z}{\omega} \right),$$

where $E_2(z) = (1 - z) \exp(z + z^2/2)$ is a Weierstrass elementary factor.
(i)   Show that the infinite product converges compactly in $\mathbb{C}$, so $\sigma \in \mathscr{O}(\mathbb{C})$.
(ii)   Use logarithmic differentiation to show that $-(\sigma'/\sigma)' = \wp$.

(7)   Verify the differential equation

(9.19) $2\wp'' + g_2 = 12\wp^2.$

Use this to show that the coefficients G_{2n+2} in (9.13) can be expressed as (universal) polynomials in g_2, g_3.

(8)   Verify the following identities between the invariants of Λ and the finite critical values of $\wp$:
(i) $g_2 = 2(e_1^2 + e_2^2 + e_3^2)$
(ii) $g_2^3 - 27g_3^2 = 16 (e_1 - e_2)^2 (e_2 - e_3)^2 (e_3 - e_1)^2.$

(9)   Suppose $f \in \mathcal{M}(\mathbb{C}, \Lambda)$ has poles of order 2 only along Λ. Show that $f = a\wp + b$ for some constants a, b with $a \neq 0$. (Hint: Use $\operatorname{res}(f, 0) = 0$ to verify that the principal part of f at every $\omega \in \Lambda$ is $a/(z - \omega)^2$ for some $a \neq 0$ independent of ω.)

(10) Show that every elliptic function is a rational function of $\wp, \wp'$. In other words, for every $f \in \mathcal{M}(\mathbb{C}, \Lambda)$ there is a rational function $R(X, Y) \in \mathbb{C}(X, Y)$ such that $f = R(\wp, \wp')$. In fact, $\mathcal{M}(\mathbb{C}, \Lambda)$ is isomorphic to the quotient field $\mathbb{C}(X, Y)/(Y^2 - 4X^3 + g_2 X + g_3)$. (Hint: First suppose f is even. Replacing f with $f + 1$ or $1/f + 1$ if necessary, we may assume that 0 is not a zero or pole of f. If f has zeros $\{\pm z_k\}$ and poles $\{\pm p_k\}$ modulo Λ in some fundamental parallelogram, consider the rational function $R(X) = \prod_k (X - \wp(z_k))/\prod_k (X - \wp(p_k))$. When f is odd, reduce to the even case by considering the function $f/\wp'$. For the general case, decompose f into its even and odd parts.)

(11) Show that the finite critical values e_1, e_2, e_3 of $\wp$ are real if and only if τ is purely imaginary. Verify that in this case
    (i) $e_1 > e_2 > e_3$.
    (ii) $\wp(z) \in \hat{\mathbb{R}} = \mathbb{R} \cup \{\infty\}$ if and only if $z \equiv \pm\bar{z} \pmod{\Lambda}$.

(12) As in Remark 9.16, the map $z \mapsto (X, Y) = (\wp(z), \wp'(z))$ parametrizes the complex cubic curve $C: Y^2 = 4X^3 - g_2 X - g_3$. This allows us to equip C with the structure of an abelian group whose addition $\oplus$ is defined as follows: If $p_1 = (\wp(z_1), \wp'(z_1))$ and $p_2 = (\wp(z_2), \wp'(z_2))$, then
$$p_1 \oplus p_2 = (\wp(z_1 + z_2), \wp'(z_1 + z_2)).$$
Here the identity element of $\oplus$ is the "point at infinity" $0_C = (\wp(0), \wp'(0))$ on C (more accurately on the closure of C in $\mathbb{CP}^2$).
    (i)   Show that if p_1, p_2, p_3 are the intersection points of C with a line $L: aX + bY + c = 0$, then $p_1 \oplus p_2 \oplus p_3 = 0_C$.
    (ii) Verify that the involution $z \mapsto -z$ on $\mathbb{C}$ corresponds to the involution $I: (X, Y) \mapsto (X, -Y)$ on C.
    (iii) Deduce the following geometric interpretation of the addition law on C: If $p_1, p_2 \in C$ and if the line L passing through p_1, p_2 intersects C at the third point p_3, then $p_1 \oplus p_2 = I(p_3)$ (see Fig. 9.9).
    (Hint: For (i), let $p_k = (\wp(z_k), \wp'(z_k))$ and use the fact that the poles of $a\wp + b\wp' + c \in \mathcal{M}(\mathbb{C}, \Lambda)$ are along Λ, so $z_1 + z_2 + z_3 \equiv 0 \pmod{\Lambda}$ by Theorem 9.8(iii).

(13) Prove the addition formula

$$\wp(z_1 + z_2) = -\wp(z_1) - \wp(z_2) + \frac{1}{4}\left(\frac{\wp'(z_2) - \wp'(z_1)}{\wp(z_2) - \wp(z_1)}\right)^2$$

for $z_1 \not\equiv \pm z_2 \pmod{\Lambda}$ and deduce the doubling formula

$$\wp(2z) = -2\wp(z) + \frac{1}{4}\left(\frac{\wp''(z)}{\wp'(z)}\right)^2$$

for $2z \not\equiv 0 \pmod{\Lambda}$. (Hint: Let $p_k = (X_k, Y_k)$ for $k = 1, 2, 3$ be the intersection points of $C: Y^2 = 4X^3 - g_2 X - g_3$ with the line $L: Y = mX + b$. Verify that $X_1 + X_2 + X_3 = m^2/4$ and use it to express X_3 in terms of X_1, X_2, Y_1, Y_2.)

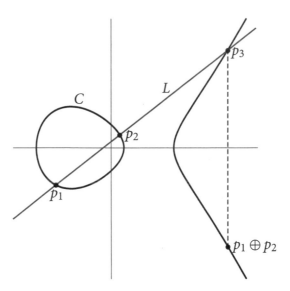

Figure 9.9. Geometry of the addition law on a cubic curve C.

(14)   Show that there is a rational function R of degree 4 which makes the following diagram commute:

$$
\begin{array}{ccc}
\mathbb{C} & \xrightarrow{z \mapsto 2z} & \mathbb{C} \\
{\scriptstyle \wp}\downarrow & & \downarrow{\scriptstyle \wp} \\
\hat{\mathbb{C}} & \xrightarrow{\ R\ } & \hat{\mathbb{C}}
\end{array}
$$

Conclude that R has a dense set of periodic points. Such R is often called a ***Lattès map*** (Hint: Use the doubling formula of the previous problem together with the equations (9.14) and (9.19).)

(15)   Consider the lattices $\Lambda = \langle \omega_1, \omega_2 \rangle = \{m\omega_1 + n\omega_2 : m, n \in \mathbb{Z}\}$ and $\Lambda' = \langle \omega'_1, \omega'_2 \rangle$, with $\mathrm{Im}(\omega_2/\omega_1) > 0$ and $\mathrm{Im}(\omega'_2/\omega'_1) > 0$.

   (i)   Show that $\Lambda = \Lambda'$ if and only if

$$
\begin{bmatrix} \omega'_2 \\ \omega'_1 \end{bmatrix} = \begin{bmatrix} a & b \\ c & d \end{bmatrix} \begin{bmatrix} \omega_2 \\ \omega_1 \end{bmatrix}
$$

   for some $a, b, c, d \in \mathbb{Z}$ with $ad - bc = 1$.

   (ii)   Show that there is an automorphism $z \mapsto \alpha z$ carrying $\Lambda' = \langle 1, \tau' \rangle$ onto $\Lambda = \langle 1, \tau \rangle$ if and only if

$$
\tau' = \frac{a\tau + b}{c\tau + d} \qquad \text{for some } a, b, c, d \in \mathbb{Z} \text{ with } ad - bc = 1.
$$

   Prove that in this case

$$
|\alpha|^2 = \frac{\mathrm{Im}\,\tau}{\mathrm{Im}\,\tau'}
$$

and

$$\wp_{\Lambda'}(z) = \alpha^2 \wp_\Lambda(\alpha z).$$

(16)   Think of the lattice invariants g_2, g_3 as functions of $\tau \in \mathbb{H}$. Show that

$$g_2\left(\frac{a\tau + b}{c\tau + d}\right) = (c\tau + d)^4 g_2(\tau)$$

$$g_3\left(\frac{a\tau + b}{c\tau + d}\right) = (c\tau + d)^6 g_3(\tau)$$

whenever $a, b, c, d \in \mathbb{Z}$ and $ad - bc = 1$.

(17)   Suppose $\Lambda = \langle 1, \tau \rangle$ and $\alpha\Lambda = \Lambda$ for some non-integer $\alpha \in \mathbb{C}$. Show that α satisfies a quadratic equation of the form

$$\alpha^2 + n\alpha + 1 = 0$$

where n is an integer with $|n| \leq 1$. Conclude that Λ is either the lattice of Gaussian integers corresponding to $\tau = i$ or the triangular lattice corresponding to $\tau = e^{\pi i/3}$ or $e^{2\pi i/3}$.

(18)   Verify that $g_3(i)$ and $g_2(e^{i\pi/3})$ are both zero.

(19)   Show that for Runge's Theorem 9.18 to hold, it is necessary to allow a pole in every component of $\hat{\mathbb{C}} \smallsetminus K$. (Hint: Look at the argument in Remark 9.20.)

(20)   Deduce Mittag-Leffler's Theorem 9.4 from Runge's Theorem 9.18 by completing the following outline: Take a nice exhaustion $\emptyset = K_0 \subset K_1 \subset K_2 \subset \cdots$ of U as in Lemma 9.23. For $n \geq 1$, let Q_n be the finite sum of the principal parts $P_k((z - z_k)^{-1})$ over all k such that $z_k \in K_n \smallsetminus K_{n-1}$. For each $n \geq 2$, find a rational function R_n with poles outside U such that $|R_n - Q_n| \leq 2^{-n}$ on K_{n-1}. Verify that $f = Q_1 + \sum_{n=2}^\infty (Q_n - R_n)$ is the desired meromorphic function.

(21)   Construct a sequence $\{P_n\}$ of polynomials such that

$$\lim_{n\to\infty} P_n(z) = \begin{cases} 1 & \text{if } \mathrm{Im}(z) > 0 \\ 0 & \text{if } \mathrm{Im}(z) = 0 \\ -1 & \text{if } \mathrm{Im}(z) < 0. \end{cases}$$

Can one arrange the additional property

$$|P_n(z)| \leq 1 \qquad \text{for all } z \in \mathbb{D} \text{ and } n \geq 1$$

in this construction?

(22)   Is there a sequence of polynomials which tends to 0 compactly in the upper half-plane but does not have a limit at any point of the lower half-plane?

(23)   Runge's theorem can be used to construct dramatic examples of holomorphic functions in $\mathbb{D}$ with no radial limit in any direction (compare Example 7.42). Construct an $f \in \mathcal{O}(\mathbb{D})$ such that the curve $\{f(re^{i\theta}) : 0 < r < 1\}$ is dense in $\mathbb{C}$ for every θ by completing the following outline [**Gm**]: Take a countable dense set $\{q_n\}$ in $\mathbb{C}$ and consider the compact sets

$$X_n = \overline{\mathbb{D}}(0, 1 - 2^{-n}) \qquad \text{and} \qquad Y_n = \{(1 - 2^{-n-1})e^{it} : 2^{-n} \leq t \leq 2\pi\}$$

for $n \geq 1$. Define the polynomials $\{P_n\}$ inductively as follows: Set $P_1 = q_1$. Having defined P_{n-1}, use Runge's theorem to find a polynomial P_n such that

$$|P_n - P_{n-1}| < 2^{-n} \quad \text{on } X_n \quad \text{and} \quad |P_n - q_n| < 2^{-n} \quad \text{on } Y_n.$$

Then the function $f = \lim_{n \to \infty} P_n$ has the desired property.

In problems 24–27 below, we will use a few function spaces associated with a compact set $K \subset \mathbb{C}$. Let $\mathscr{C}(K)$ be the Banach space of all continuous functions $f : K \to \mathbb{C}$, equipped with the sup norm $\|f\|_K = \sup_{z \in K} |f(z)|$. Consider the following closed subspaces of $\mathscr{C}(K)$:

- $\mathscr{R}(K)$, the uniform limits on K of rational functions with poles in a prescribed set which has one point in every component of $\hat{\mathbb{C}} \smallsetminus K$.
- $\mathscr{P}(K)$, the uniform limits on K of polynomials.
- $\mathscr{O}(K)$, the uniform limits on K of functions that are holomorphic in some neighborhood of K.
- $\mathscr{A}(K)$, continuous functions on K that are holomorphic in the interior of K.

We have the obvious inclusions

$$\mathscr{P}(K) \subset \mathscr{R}(K) \subset \mathscr{O}(K) \subset \mathscr{A}(K).$$

In terms of these notations, Runge's Theorem 9.18 asserts that

$$\mathscr{R}(K) = \mathscr{O}(K)$$

and Corollary 9.19 asserts that

$$\mathscr{P}(K) = \mathscr{R}(K) = \mathscr{O}(K) \quad \text{if } K \text{ is full.}$$

Mergelyan's Theorem 9.26 shows that in fact all four subspaces coincide:

$$\mathscr{P}(K) = \mathscr{R}(K) = \mathscr{O}(K) = \mathscr{A}(K) \quad \text{if } K \text{ is full.}$$

(24)   This exercise outlines a more modern proof of Runge's Theorem 9.18 which requires knowledge of two basic theorems in functional analysis. We fix $f \in \mathscr{O}(U)$ for some open set U containing K and show that $f \in \mathscr{R}(K)$. By the Hahn-Banach theorem, this holds if every bounded linear functional on $\mathscr{C}(K)$ which vanishes on $\mathscr{R}(K)$ vanishes at f. By the Riesz representation theorem, every bounded linear functional on $\mathscr{C}(K)$ is of the form $g \mapsto \int_K g \, d\mu$ for some complex Borel measure μ on K. Thus, it suffices to show that if μ is a complex Borel measure on K which satisfies $\int_K g \, d\mu = 0$ for all $g \in \mathscr{R}(K)$, then $\int_K f \, d\mu = 0$.

(i)   Consider the Cauchy transform of μ defined by

$$\hat{\mu}(z) = \int_K \frac{d\mu(\zeta)}{\zeta - z}$$

which is holomorphic in $\mathbb{C} \smallsetminus K$ (see problem 28 in chapter 1). Use Lemma 9.22 to prove that $\hat{\mu} = 0$ in $\mathbb{C} \smallsetminus K$.

(ii)   Let γ be a null-homologous cycle in U with the property $\mathrm{W}(\gamma, \cdot) = 1$ on K. Integrate Cauchy's formula

$$f(z) = \frac{1}{2\pi i} \int_\gamma \frac{f(\zeta)}{\zeta - z} \, d\zeta$$

over K against $d\mu$ and use Fubini's theorem and (i) to obtain $\int_K f \, d\mu = 0$.

(25) This exercise outlines a proof of a special case of Mergelyan's Theorem 9.26 due to Hartogs and Rosenthal (1931): "If $K \subset \mathbb{C}$ is a full compact set with area zero, then $\mathscr{P}(K) = \mathscr{C}(K)$." As in the previous problem, it suffices to show that if μ is a complex Borel measure on K which satisfies $\int_K g \, d\mu = 0$ for all $g \in \mathscr{R}(K)$, then $\mu = 0$.

   (i) Verify that the Cauchy transform $\hat{\mu}$ vanishes in $\mathbb{C} \smallsetminus K$ (hence a.e. in $\mathbb{C}$).

   (ii) Let $X \subset \mathbb{C}$ be an open rectangle whose oriented boundary ∂X meets K along a set of length zero. Use

$$\frac{1}{2\pi i} \int_{\partial X} \frac{d\zeta}{\zeta - z} = 1 \qquad \text{for } z \in X$$

and Fubini's theorem to show that $\mu(X \cap K) = 0$. Conclude that $\mu = 0$.

(26) Let $\kappa(z) = \bar{z}$. Define the **analytic content** of a compact set $K \subset \mathbb{C}$ by

$$\lambda_K = \mathrm{dist}(\kappa, \mathscr{R}(K)) = \inf_{f \in \mathscr{R}(K)} \|\kappa - f\|_K.$$

   (i) Show that $\lambda_K = 0$ if and only if $\mathscr{R}(K) = \mathscr{C}(K)$. Conclude that $\lambda_K > 0$ if K has non-empty interior.

   (ii) If K is the closure of a bounded domain with piecewise C^1 boundary, show that

$$\lambda_K \geq \frac{2 \, \mathrm{area}(K)}{\mathrm{length}(\partial K)}.$$

   (iii) What is the analytic content of a closed disk of radius r?

(Hint: For the "only if" part in (i), use the fact that by the Stone-Weierstrass theorem, polynomials in $x = \mathrm{Re}(z)$ and $y = \mathrm{Im}(z)$ are dense in $\mathscr{C}(K)$. For (ii), apply the complex form of Green's theorem (problem 14 in chapter 2) to the smooth function $\kappa - f$, where $f \in \mathscr{R}(K)$.)

(27) Let $\{D_n\}$ be an infinite sequence of open subdisks of $\mathbb{D}$ with disjoint closures whose radii $\{r_n\}$ satisfy $\sum_{n=1}^{\infty} r_n < 1$. If $K = \overline{\mathbb{D}} \smallsetminus \bigcup_{n=1}^{\infty} D_n$, prove that

$$\lambda_K \geq 1 - \sum_{n=1}^{\infty} r_n > 0.$$

This, by the previous problem, shows that $\mathscr{R}(K) \neq \mathscr{C}(K)$. The disks D_n can be arranged so that K has no interior, so $\mathscr{C}(K) = \mathscr{A}(K)$. Such K, called a **Swiss cheese set** for obvious reasons, provides an example for which $\mathscr{R}(K) \neq \mathscr{A}(K)$. Thus, the generalization of Mergelyan's theorem for rational approximation on non-full compact sets is false. (Hint: For any rational function f with poles outside K, apply Green's theorem to the function $\kappa - f$ on the compact set $\overline{\mathbb{D}} \smallsetminus \bigcup_{n=1}^{N} D_n$ for large N.)

The idea of a Swiss cheese set was pioneered by Swiss mathematician Alice Roth in 1938.

(28) Let $K \subsetneq \hat{\mathbb{C}}$ be compact and connected. Show that every connected component of $\hat{\mathbb{C}} \smallsetminus K$ is simply connected.

(29) Use Theorem 9.27 to prove that if U, V are simply connected domains in the plane with $U \cap V \neq \emptyset$, then every component of $U \cap V$ is simply connected.

(30) Show that a domain $U \subset \mathbb{C}$ for which $H_1(U) \cong \mathbb{Z}^{m-1}$ is m-connected.

(31)   In the situation of Theorem 9.32(iii), show that the periods α_j are uniquely determined by u. (Hint: By an argument similar to Example 7.8, verify that a relation of the form

$$\sum_{j=1}^{m-1} \alpha_j \log |z - p_j| - \sum_{j=1}^{m-1} \beta_j \log |z - q_j| = \mathrm{Re}(f(z))$$

for $\alpha_j, \beta_j \in \mathbb{R}$, $p_j, q_j \in K_j$, and $f \in \mathcal{O}(U)$ implies $\alpha_j = \beta_j$ for all j.)

(32)   Let $U \subset \mathbb{C}$ be an m-connected domain. If $\mathcal{O}_0(U) \subset \mathcal{O}(U)$ is the vector space of all functions which have a primitive in U, show that the quotient space $\mathcal{O}(U)/\mathcal{O}_0(U)$ is isomorphic to $\mathbb{C}^{m-1}$. Similarly, if $\mathcal{H}(U)$ is the vector space of all real-valued harmonic functions in U and $\mathcal{H}_0(U)$ is the subspace of functions which are the real parts of holomorphic functions in U, show that the quotient space $\mathcal{H}(U)/\mathcal{H}_0(U)$ is isomorphic to $\mathbb{R}^{m-1}$.

10

The holomorphic extension problem

A central question in complex analysis can be roughly stated as follows: "Given a holomorphic function f defined in an open set U, what are the possible holomorphic extensions of f to open sets containing U?" In this chapter we explore several formulations and refinements of this question that commonly occur in practice.

10.1 Regular and singular points

DEFINITION 10.1. Let $U \subset \mathbb{C}$ be open and $f \in \mathcal{O}(U)$. A boundary point $q \in \partial U$ is a **regular** point of f if there is a neighborhood V of q and a function $g \in \mathcal{O}(V)$ such that $f = g$ in $U \cap V$. Otherwise, q is a **singular** point of f.

EXAMPLE 10.2. Consider $f(z) = \sum_{n=0}^{\infty} z^n = 1/(1-z)$ in $\mathbb{D}$. Then $q = 1$ is a singular point of f while every $q \in \mathbb{T} \setminus \{1\}$ is regular. Note, in particular, that q being regular does not imply the convergence of the power series $\sum_{n=0}^{\infty} z^n$ at $z = q$. In fact, $\sum_{n=0}^{\infty} z^n$ diverges everywhere on $\mathbb{T}$.

THEOREM 10.3. *Suppose the power series $f(z) = \sum_{n=0}^{\infty} a_n (z-p)^n$ has radius of convergence $0 < R < +\infty$. Then f has at least one singular point on the circle $\mathbb{T}(p, R)$.*

PROOF. Assume by way of contradiction that every point of $\mathbb{T}(p, R)$ is regular. For each $q \in \mathbb{T}(p, R)$, choose an open disk D_q centered at q and a function $g_q \in \mathcal{O}(D_q)$ such that $f = g_q$ in $D_q \cap \mathbb{D}(p, R)$. Note that if $D_q \cap D_{q'} \neq \emptyset$, then $D_q \cap D_{q'} \cap \mathbb{D}(p, R) \neq \emptyset$ and $g_q = f = g_{q'}$ in $D_q \cap D_{q'} \cap \mathbb{D}(p, R)$, so $g_q = g_{q'}$ in $D_q \cap D_{q'}$ by the identity theorem. It follows that the function

$$F(z) = \begin{cases} f(z) & z \in \mathbb{D}(p, R) \\ g_q(z) & z \in D_q \text{ for some } q \in \mathbb{T}(p, R) \end{cases}$$

is a well-defined holomorphic extension of f to the union $U = \mathbb{D}(p, R) \cup \bigcup D_q$. Since U is an open neighborhood of the closed disk $\overline{\mathbb{D}}(p, R)$, we can choose an $\varepsilon > 0$

small enough so that $\mathbb{D}(p, R + \varepsilon) \subset U$. It follows that F is represented by a power series in $\mathbb{D}(p, R + \varepsilon)$. By Theorem 1.20(iii), this power series must coincide with $\sum_{n=0}^{\infty} a_n(z-p)^n$. This contradicts our assumption that the radius of convergence of this power series is R. $\qquad\square$

It is clear from Definition 10.1 that the set of singular points of every $f \in \mathcal{O}(U)$ is a closed subset of ∂U. It may happen to be the whole ∂U.

> **DEFINITION 10.4.** We say that ∂U is the ***natural boundary*** of $f \in \mathcal{O}(U)$ if every point of ∂U is a singular point of f.

THEOREM 10.5. *For every domain $U \subsetneq \mathbb{C}$ there exists an $f \in \mathcal{O}(U)$ with ∂U as its natural boundary.*

PROOF. Take a countable dense subset $\{q_n\}$ of ∂U and for each n choose a point $p_n \in U$ such that $|p_n - q_n| < 1/n$. The sequence $\{p_n\}$ has no accumulation point in U but accumulates at every point of ∂U. By Theorem 8.25, there exists an $f \in \mathcal{O}(U)$ which has a simple zero at every p_n and does not vanish anywhere else. It is easy to see that ∂U is the natural boundary of f. In fact, if f extended holomorphically to an open disk D centered at some $q \in \partial U$, then q would be an accumulation point of zeros of f in D. Hence, f would have to vanish identically in D and therefore in U by the identity theorem. $\qquad\square$

In particular, boundaries of disks can occur as natural boundaries. Examples of this case can also be constructed by carefully controlling the coefficients of power series.

EXAMPLE 10.6. Consider the function $f \in \mathcal{O}(\mathbb{D})$ defined by $f(z) = \sum_{n=1}^{\infty} z^{n!}$. Let $z = re^{it}$ where $t = 2\pi a/b$ and a, b are integers with $b > 1$. Then

$$f(z) = \left(\sum_{n=1}^{b-1} + \sum_{n=b}^{\infty} \right) r^{n!} e^{2\pi i n! a/b} = \text{(polynomial in } r) + \sum_{n=b}^{\infty} r^{n!},$$

from which it easily follows that $f(re^{it}) \to \infty$ as $r \to 1$. This shows that e^{it} is a singular point of f. Since the set of all such e^{it} is dense on $\mathbb{T}$ and since the set of singular points of f is closed, we conclude that $\mathbb{T}$ is the natural boundary of f.

The preceding example is a special case of a general phenomenon for power series with large gaps in their coefficients, which is stated as Theorem 10.9 below. We shall deduce this theorem from a stronger result that is the content of Theorem 10.8.

DEFINITION 10.7. A pair $(\{m_j\}, \{k_j\})$ of sequences of positive integers is an *Ostrowski pair* if

- $m_1 < k_1 \le m_2 < k_2 \le \cdots \le m_j < k_j \le \cdots$
- there is a constant $\lambda > 1$ such that $k_j > \lambda\, m_j$ for all j.

THEOREM 10.8 (Ostrowski, 1921). *Suppose $f(z) = \sum_{n=0}^{\infty} a_n z^n$ has radius of convergence 1 and there is an Ostrowski pair $(\{m_j\}, \{k_j\})$ such that $a_n = 0$ whenever $m_j < n < k_j$ for some j. If $q \in \mathbb{T}$ is a regular point of f, then the sequence $\{\sum_{n=0}^{m_j} a_n z^n\}$ converges in some open neighborhood of q as $j \to \infty$.*

Alexander Markowich
Ostrowski (1893–1986)

If $s_m(z) = \sum_{n=0}^{m} a_n z^n$ is the m-th partial sum of the power series, the theorem asserts that the subsequence $\{s_{m_j}(z)\}$ converges at some points z with $|z| > 1$, even though the full sequence $\{s_m(z)\}$ diverges whenever $|z| > 1$. This phenomenon is often referred to as *over-convergence*.

PROOF. Replacing $f(z)$ by $f(qz)$, we may assume $q = 1$. Let $\mathbb{D}(1, r)$ be a disk in which f extends holomorphically, and set $U = \mathbb{D} \cup \mathbb{D}(1, r)$. Fix a positive integer N large enough so that $\lambda > 1 + 1/N$, where λ is the constant associated with the pair $(\{m_j\}, \{k_j\})$. Consider the polynomial

$$P(w) = \frac{1}{2} w^N (w + 1).$$

It is easy to see that $P(1) = 1$ and $|P(w)| < 1$ if $|w| \le 1$ but $w \ne 1$. Continuity of P then shows that $P(\mathbb{D}(0, 1 + \varepsilon)) \subset U$ for sufficiently small $\varepsilon > 0$. The holomorphic function $f \circ P \colon \mathbb{D}(0, 1 + \varepsilon) \to \mathbb{C}$ is represented by a power series

$$(10.1) \qquad f(P(w)) = \sum_{n=0}^{\infty} b_n w^n \qquad \text{for } |w| < 1 + \varepsilon.$$

The power series of f gives the alternative representation

$$(10.2) \qquad f(P(w)) = \sum_{n=0}^{\infty} a_n (P(w))^n \qquad \text{for } |w| < 1.$$

By uniqueness, the coefficients of similar powers of w in (10.1) and (10.2) must agree. The condition $k_j > \lambda m_j > (1 + 1/N)m_j$ implies $m_j(N + 1) < k_j N$, which shows

highest power of w in $(P(w))^{m_j} <$ lowest power of w in $(P(w))^{k_j}$.

This observation, combined with the assumption $a_n = 0$ for $m_j < n < k_j$, implies that

$$\sum_{n=0}^{m_j(N+1)} b_n\, w^n = \sum_{n=0}^{m_j} a_n\, (P(w))^n \qquad \text{for all } j.$$

The left side converges for every $w \in \mathbb{D}(0, 1+\varepsilon)$ as $j \to \infty$, so the same must be true of the right side, that is, the sequence $\{\sum_{n=0}^{m_j} a_n\, z^n\}$ converges for every $z \in P(\mathbb{D}(0, 1+\varepsilon))$ as $j \to \infty$. This completes the proof since $P(\mathbb{D}(0, 1+\varepsilon))$ is an open set containing 1. $\qquad\square$

The following classical result is an immediate corollary:

The word "gap" refers to the missing coefficients a_n over the integer segments $m_j < n < m_{j+1}$. Power series with such gaps are sometimes called "lacunary."

THEOREM 10.9 (Hadamard's gap theorem, 1892). *Let $\lambda > 1$ and $\{m_j\}$ be a sequence of positive integers such that*

$$m_{j+1} > \lambda\, m_j \qquad \text{for all } j.$$

If the power series $f(z) = \sum_{n=0}^{\infty} a_n\, z^{m_n}$ has radius of convergence 1, then $\mathbb{T}$ is the natural boundary of f.

PROOF. As before, let s_m denote the m-th partial sum of the power series of f. The assumption shows that $(\{m_j\}, \{m_{j+1}\})$ is an Ostrowski pair and the subsequence $\{s_{m_j}\}$ coincides with the full sequence $\{s_m\}$ except for repetition of terms. If f had a regular point on $\mathbb{T}$, it would follow from Theorem 10.8 that $\{s_m\}$ converges in some neighborhood of that point, which is clearly impossible. $\qquad\square$

EXAMPLE 10.10. Having $\mathbb{T}$ as the natural boundary is not necessarily an indication of discontinuity or non-differentiability on $\mathbb{T}$; there could be a deeper analytic issue involved. Consider, for example, the function

$$f(z) = \sum_{n=0}^{\infty} \exp(-2^{n/2})\, z^{2^n}.$$

Since

$$\limsup_{n \to \infty} [\exp(-2^{n/2})]^{2^{-n}} = \limsup_{n \to \infty} \exp(-2^{-n/2}) = 1,$$

the series has radius of convergence 1. By Hadamard's gap theorem, with $\{m_j = 2^j\}$, the unit circle is the natural boundary of f. Term-by-term differentiation gives

$$f^{(k)}(z) = \sum_{2^n \geq k} 2^n (2^n - 1) \cdots (2^n - k + 1)\, \exp(-2^{n/2})\, z^{2^n - k}.$$

The series converges uniformly on the closed disk $\overline{\mathbb{D}}$ since

$$|2^n (2^n - 1) \cdots (2^n - k + 1)\, \exp(-2^{n/2})\, z^{2^n - k}| \leq 2^{kn} \exp(-2^{n/2}) \qquad \text{if } |z| \leq 1$$

and $\sum_{2^n \geq k} 2^{kn} \exp(-2^{n/2})$ converges for each $k \geq 1$. It follows that f and all of its derivatives extend continuously to $\overline{\mathbb{D}}$. In particular, $f|_{\mathbb{T}}$ is C^{∞}-smooth. See Fig. 10.13 and compare problem 8 for a similar example where $f(\mathbb{T})$ is a Jordan curve.

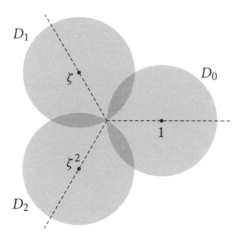

Figure 10.1. Illustration of Example 10.12. The holomorphic branches of $\sqrt{z}$ match in $D_0 \cap D_1$ and in $D_1 \cap D_2$, but not in $D_0 \cap D_2$.

10.2 Analytic continuation

DEFINITION 10.11. By a ***function element*** is meant a pair (f, D), where D is an open disk in $\mathbb{C}$ and $f \in \mathscr{O}(D)$. We say that the function element (f_1, D_1) is a ***(direct) analytic continuation*** of a function element (f_0, D_0) if $D_0 \cap D_1 \neq \emptyset$ and $f_0 = f_1$ in $D_0 \cap D_1$. In this case, we write $(f_0, D_0) \sim (f_1, D_1)$.

The essential property of round disks that we use here is that $D_0 \cap D_1$ is either empty or connected. The relation $\sim$ is clearly reflexive and symmetric but it is *not* transitive, as the following example shows.

EXAMPLE 10.12. Consider the cubic root of unity $\zeta = e^{2\pi i/3}$ and the disks $D_k = \mathbb{D}(\zeta^k, 1)$ for $k = 0, 1, 2$ (see Fig. 10.1). Define the branches $f_k \in \mathscr{O}(D_k)$ of $\sqrt{z}$ in polar coordinates by

$$f_0(re^{it}) = \sqrt{r}e^{it/2} \qquad -\pi/2 < t < \pi/2$$

$$f_1(re^{it}) = \sqrt{r}e^{it/2} \qquad \pi/6 < t < 7\pi/6$$

$$f_2(re^{it}) = \sqrt{r}e^{it/2} \qquad 5\pi/6 < t < 11\pi/6.$$

The $(f_0, D_0) \sim (f_1, D_1)$ and $(f_1, D_1) \sim (f_2, D_2)$ but $(f_0, D_0) \nsim (f_2, D_2)$ since in $D_0 \cap D_2$ we have $f_0 = -f_2$.

However, the following form of transitivity holds:

LEMMA 10.13. *If the triple intersection $D_0 \cap D_1 \cap D_2$ is non-empty, then the relations $(f_0, D_0) \sim (f_1, D_1)$ and $(f_1, D_1) \sim (f_2, D_2)$ imply $(f_0, D_0) \sim (f_2, D_2)$.*

PROOF. By the identity theorem, the assumption $f_0 = f_1 = f_2$ in $D_0 \cap D_1 \cap D_2$ implies $f_0 = f_2$ in $D_0 \cap D_2$ since $D_0 \cap D_2$ is connected. $\square$

This notion of chain is not to be confused with the one in chapter 2 in the context of homology.

> **DEFINITION 10.14.** A collection $\{(f_k, D_k)\}_{0 \leq k \leq n}$ of function elements is called a **chain** if $(f_k, D_k) \sim (f_{k+1}, D_{k+1})$ for every $0 \leq k \leq n-1$.

Observe that the chain $\{(f_k, D_k)\}_{0 \leq k \leq n}$ is uniquely determined by f_0 and the disks $\{D_k\}$. In fact, f_0 determines f_1 in $D_0 \cap D_1$, hence in D_1 by the identity theorem. Similarly, f_1 determines f_2 in $D_1 \cap D_2$, hence in D_2, and so on.

If the chain $\{(f_k, D_k)\}_{0 \leq k \leq n}$ is "efficient" in the sense that $D_k \cap D_j \neq \emptyset$ implies $|k - j| = 1$ (each disk only intersects its predecessor and successor), then f_0 extends uniquely to a well-defined holomorphic function $F : \bigcup_{k=0}^n D_k \to \mathbb{C}$; simply set $F = f_k$ in D_k for each k. But this is hardly a recipe for extending a given $f_0 \in \mathcal{O}(D_0)$ to a larger domain. For one thing, there may be no non-trivial chain starting at (f_0, D_0) (for instance ∂D_0 could be the natural boundary of f_0). But even assuming chains can be constructed freely from (f_0, D_0), there is a more serious issue to deal with. If $\{(f_k, D_k)\}_{0 \leq k \leq n}$ and $\{(g_j, B_j)\}_{0 \leq j \leq m}$ are efficient chains with $(f_0, D_0) = (g_0, B_0)$, and if $D_n \cap B_m \neq \emptyset$, there is no reason to expect the relation $(f_n, D_n) \sim (g_m, B_m)$ to hold, so the unique extensions $F : \bigcup_{k=0}^n D_k \to \mathbb{C}$ and $G : \bigcup_{j=0}^m B_j \to \mathbb{C}$ may not match in $D_n \cap B_m$ (compare Example 10.12, where the extensions of f_1 to $D_1 \cup D_0$ and to $D_1 \cup D_2$ do not match in $D_0 \cap D_2$).

To deal with this ambiguity and to somewhat simplify the combinatorics involved, we first formulate a notion of extension in which curves, and not domains, play the prominent role.

> **DEFINITION 10.15.** We say that a function element (f_0, D_0) **continues analytically** along a curve $\gamma : [0, 1] \to \mathbb{C}$ if there is a chain $\{(f_k, D_k)\}_{0 \leq k \leq n}$ and a partition $0 = t_0 < t_1 < \cdots < t_n < t_{n+1} = 1$ such that $\gamma([t_k, t_{k+1}]) \subset D_k$ for every $0 \leq k \leq n$. In this case, the chain $\{(f_k, D_k)\}_{0 \leq k \leq n}$ is said to **realize** this continuation.

Compare Fig. 10.2. It is clear from the above definition that if (f_0, D_0) continues analytically along $\gamma : [0, 1] \to \mathbb{C}$, then it continues analytically along the restriction $\gamma : [0, b] \to \mathbb{C}$ for every $0 < b < 1$.

The following lemma shows that the result of analytic continuation of a function element along a given curve is essentially unique.

LEMMA 10.16. *Suppose the chains* $\{(f_k, D_k)\}_{0 \leq k \leq n}$ *and* $\{(g_j, B_j)\}_{0 \leq j \leq m}$ *both realize the analytic continuation of* (f_0, D_0) *along* $\gamma : [0, 1] \to \mathbb{C}$, *and take partitions* $\{t_k\}$ *and* $\{s_j\}$ *of* $[0, 1]$ *such that* $\gamma([t_k, t_{k+1}]) \subset D_k$ *for all* $0 \leq k \leq n$ *and* $\gamma([s_j, s_{j+1}]) \subset B_j$ *for all* $0 \leq$

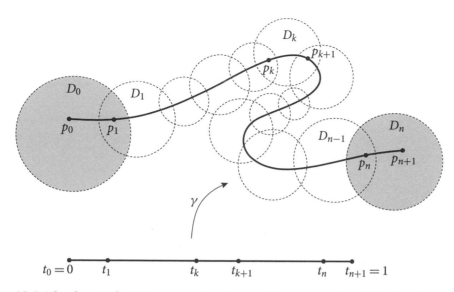

Figure 10.2. The chain $\{(f_k, D_k)\}_{0 \le k \le n}$ realizes the analytic continuation of (f_0, D_0) along γ. Here $p_k = \gamma(t_k)$.

$j \le m$. If $[t_k, t_{k+1}] \cap [s_j, s_{j+1}] \ne \emptyset$ for some k and j, then $(f_k, D_k) \sim (g_j, B_j)$. In particular, $(f_n, D_n) \sim (g_m, B_m)$.

PROOF. Suppose the lemma is false, and choose a violating pair k, j for which $k + j$ is minimal. We have $k + j > 0$ since $(f_0, D_0) = (g_0, B_0)$. Without loss of generality, assume $t_k \le s_j \le t_{k+1}$. Then $\gamma(s_j) \in B_{j-1} \cap B_j \cap D_k$. Moreover, by minimality, $(f_k, D_k) \sim (g_{j-1}, B_{j-1})$. Since $(g_{j-1}, B_{j-1}) \sim (g_j, B_j)$ and since $B_{j-1} \cap B_j \cap D_k \ne \emptyset$, Lemma 10.13 gives $(f_k, D_k) \sim (g_j, B_j)$, which contradicts our hypothesis on the pair k, j. □

Suppose (f_0, D_0) continues analytically along $\gamma : [0, 1] \to \mathbb{C}$, and take a realizing chain $\{(f_k, D_k)\}_{0 \le k \le n}$ and a partition $\{t_k\}_{0 \le k \le n+1}$ as in Definition 10.15. The curve $\tilde{\gamma} : [0, 1] \to \mathbb{C}$ given by

$$(10.3) \qquad \tilde{\gamma}(t) = f_k(\gamma(t)) \qquad \text{if } t \in [t_k, t_{k+1}] \text{ for some } 0 \le k \le n$$

is well defined since $\{(f_k, D_k)\}$ is a chain, and it does not depend on the choice of the realizing chain by Lemma 10.16. We call $\tilde{\gamma}$ the **lift** of γ determined by (f_0, D_0).

THEOREM 10.17 (Homotopy lifting property of analytic continuation). *Let $\{\gamma_s\}_{0 \le s \le 1}$ be a homotopy between two curves $\gamma_0, \gamma_1 : [0, 1] \to \mathbb{C}$. Suppose the function element (f, D) continues analytically along every γ_s. Then the family $\{\tilde{\gamma}_s\}_{0 \le s \le 1}$ of the lifts of $\{\gamma_s\}_{0 \le s \le q}$ determined by (f, D) defines a homotopy between $\tilde{\gamma}_0$ and $\tilde{\gamma}_1$.*

In particular, the lifts $\tilde{\gamma}_0, \tilde{\gamma}_1$ (which have the same initial point) must have the same end point.

PROOF. Let $a \in [0, 1]$ and take a chain $\{(f_k, D_k)\}_{0 \leq k \leq n}$ which realizes the analytic continuation of (f, D) along γ_a. Find a partition $\{t_k\}$ of $[0, 1]$ such that $\gamma_a([t_k, t_{k+1}]) \subset D_k$ for all $0 \leq k \leq n$. Let $\varepsilon_k > 0$ be the minimal distance between the disjoint compact sets $\gamma_a([t_k, t_{k+1}])$ and ∂D_k, and set $\varepsilon = \min_{0 \leq k \leq n} \varepsilon_k$. By uniform continuity of the homotopy $(t, s) \mapsto \gamma_s(t)$ on $[0, 1] \times [0, 1]$, there is a relatively open neighborhood $I_a \subset [0, 1]$ of a with the property

$$|\gamma_s(t) - \gamma_a(t)| < \varepsilon/2 \qquad \text{if } (t, s) \in [0, 1] \times I_a.$$

By the choice of ε, this implies $\gamma_s([t_k, t_{k+1}]) \subset D_k$ for all $0 \leq k \leq n$ if $s \in I_a$. In other words, the *same* chain realizes the analytic continuation of (f, D) along every γ_s as long as $s \in I_a$.

Now the definition (10.3) of the lift shows that

$$(10.4) \qquad \tilde{\gamma}_s(t) = f_k(\gamma_s(t)) \qquad \text{if } (t, s) \in [t_k, t_{k+1}] \times I_a.$$

This proves continuity of $(t, s) \mapsto \tilde{\gamma}_s(t)$ on $[0, 1] \times I_a$. Moreover, (10.4) shows that the end point $\tilde{\gamma}_s(1) = f_n(\gamma_s(1))$ does not depend on $s \in I_a$. Since the interval $[0, 1]$ can be covered by neighborhoods of the form I_a, the map $(t, s) \mapsto \tilde{\gamma}_s(t)$ must be continuous on $[0, 1] \times [0, 1]$, and $s \mapsto \tilde{\gamma}_s(1)$ must be constant on $[0, 1]$. □

"Monodromy" comes from the Greek $\mu\acute{o}\nu o$ (single) and $\delta\rho o\mu\acute{o}\varsigma$ (a running). It is meant to convey the idea that taking analytic continuation around a loop will result in a single-valued function.

THEOREM 10.18 (The monodromy theorem). *Suppose $U \subset \mathbb{C}$ is a simply connected domain and (f, D) is a function element with $D \subset U$ which continues analytically along every curve in U whose initial point is the center of D. Then f extends to a holomorphic function in U.*

PROOF. We construct the extension $F \in \mathcal{O}(U)$ as follows. Let p be the center of D. For each $z \in U$, take a curve $\gamma : [0, 1] \to U$ from p to z, and let $\tilde{\gamma} : [0, 1] \to \mathbb{C}$ be the lift of γ determined by (f, D). Define $F(z) = \tilde{\gamma}(1)$. Since U is simply connected, any curve $\eta : [0, 1] \to U$ from p to z must be homotopic to γ. By Theorem 10.17, $\tilde{\eta}$ is homotopic to $\tilde{\gamma}$, so $\tilde{\eta}(1) = \tilde{\gamma}(1)$. Thus, F is well defined.

With z and γ as above, let $\{(f_k, D_k)\}_{0 \leq k \leq n}$ be a chain that realizes the analytic continuation of (f, D) along γ. For each $w \in D_n$, let γ_w denote the product $\gamma \cdot [z, w]$, that is, the curve γ followed by the oriented segment $[z, w] \subset D_n$. Clearly, (f, D) continues analytically along γ_w and the same chain $\{(f_k, D_k)\}_{0 \leq k \leq n}$ realizes this continuation. It follows that $F(w) = \tilde{\gamma}_w(1) = f_n(w)$ for every $w \in D_n$, so $F \in \mathcal{O}(D_n)$. Since z was arbitrary, we conclude that $F \in \mathcal{O}(U)$. That $F = f$ in D follows by taking $z = p$ and considering the trivial chain $\{(f, D)\}$ in the above construction. □

EXAMPLE 10.19. Corollary 2.22 guaranteed the existence of holomorphic branches of $\log z$ and $\sqrt[n]{z}$ in every simply connected domain in $\mathbb{C}^*$; see also Example 2.23. The monodromy theorem enables us to fashion that argument in a rather different (albeit related) language. Take a point

$p = r_0 e^{it_0} \in \mathbb{C}^*$, where $r_0 \in (0, +\infty)$ and $t_0 \in [0, 2\pi)$. Every point in the disk $B_p = \mathbb{D}(p, r_0)$ can be written as re^{it} for unique values of $r \in (0, +\infty)$ and $t \in (t_0 - \pi/2, t_0 + \pi/2)$. The functions $L_{p,k} : B_p \to \mathbb{C}$ defined by

$$L_{p,k} : re^{it} \to \log r + i(t + 2\pi k) \qquad \text{for } k \in \mathbb{Z}$$

are holomorphic and satisfy $\exp \circ L_{p,k} = \mathrm{id}$ in B_p. We call them the **branches of the logarithm** in B_p. The image domains $\{L_{p,k}(B_p)\}_{k \in \mathbb{Z}}$ are easily seen to be disjoint. In particular, given any $w \in B_p$ and any z with $e^z = w$, there is a unique integer k such that $z = L_{p,k}(w)$. Furthermore, these branches are consistent in the following sense: If $p, q \in \mathbb{C}^*$ with $B_p \cap B_q \neq \emptyset$, and if $L_{p,k}(w) = L_{q,j}(w)$ for some $w \in B_p \cap B_q$, then $L_{p,k} = L_{q,j}$ everywhere in $B_p \cap B_q$. This follows from the open mapping theorem applied to the holomorphic function $1/(2\pi i)(L_{p,k} - L_{q,j})$ in $B_p \cap B_q$ which takes integer values and vanishes at w.

Now let (f, D) be any function element in which $D \subset \mathbb{C}^*$ is a disk centered at p_0 and f is a branch of the logarithm in B_{p_0}. Then (f, D) continues analytically along any curve $\gamma : [0, 1] \to \mathbb{C}^*$ with $\gamma(0) = p_0$. To see this, let $\varepsilon = \mathrm{dist}(|\gamma|, 0)$ and use uniform continuity of γ on $[0, 1]$ to find a partition $t_0 = 0 < t_1 < \cdots < t_n < t_{n+1} = 1$ such that $\mathrm{diam}\, \gamma([t_k, t_{k+1}]) < \varepsilon$ for every $0 \leq k \leq n$. Set $p_k = \gamma(t_k)$ and $D_k = \mathbb{D}(p_k, \varepsilon) \subset B_{p_k}$. By the choice of ε, the inclusion $\gamma([t_k, t_{k+1}]) \subset D_k$ holds for every $0 \leq k \leq n$. Starting with $f_0 = f$ in D_0, we choose f_1 to be the unique branch of the logarithm in D_1 which agrees with f_0 at some point of $D_0 \cap D_1$, hence everywhere in this intersection by the above consistency property. Similarly, we choose f_2 to be the unique branch of the logarithm in D_2 which agrees with f_1 in $D_1 \cap D_2$, and so on. Continuing this way, we can construct a chain which realizes the desired continuation.

It follows from the monodromy theorem that in every simply connected domain $U \subset \mathbb{C}^*$ there is a holomorphic branch of $\log z$, that is, a function $f \in \mathcal{O}(U)$ which satisfies $\exp \circ f = \mathrm{id}$ in U. This immediately proves the existence of holomorphic branches of $\sqrt[n]{z}$ in U for any integer $n \geq 2$: Simply consider the function $g = \exp(f/n)$ which satisfies $g^n = \mathrm{id}$. Figure 10.3 illustrates the case $n = 3$.

The above example is a special case of a more general result on the existence of global inverse branches, which we will discuss in chapter 12 (see Corollary 12.26).

10.3 Analytic arcs and reflections

There is a notion of analyticity for functions defined in open subsets of the real line which, apart from its intrinsic importance, is intimately connected with the problem of extending holomorphic functions. A key role in the following discussion is played by the **complex conjugation** $\kappa : z \mapsto \bar{z}$ with its two essential properties:

- $\kappa(z) = z$ if and only if $z \in \mathbb{R}$.
- κ is an involution in the sense that $\kappa \circ \kappa = \mathrm{id}$ but $\kappa \neq \mathrm{id}$.

LEMMA 10.20.

(i) $f \in \mathcal{O}(U)$ if and only if $\kappa \circ f \circ \kappa \in \mathcal{O}(\kappa(U))$.

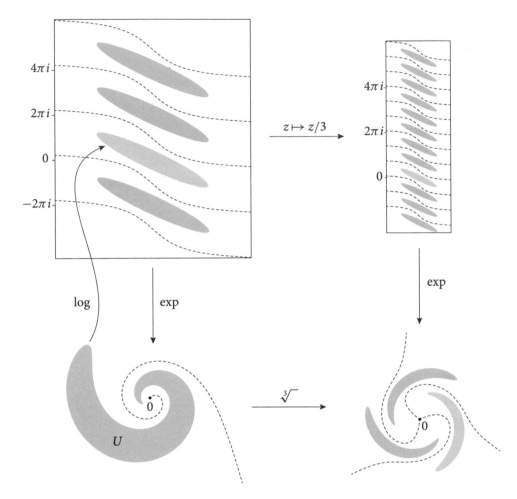

Figure 10.3. In every simply connected domain $U \subset \mathbb{C}^*$ there is a holomorphic branch of $\log z$ which is unique up to addition of an integer multiple of $2\pi i$. The composition $\exp(\log z/3)$ defines a holomorphic branch of $\sqrt[3]{z}$ in U, which is unique up to multiplication by a cube root of unity.

(ii) *Suppose $p \in \mathbb{R}$ and $f \in \mathcal{O}(\mathbb{D}(p,r))$ has the power series representation $f(z) = \sum_{n=0}^{\infty} a_n (z-p)^n$. Then f takes on real values in the interval $(p-r, p+r)$ if and only if $\kappa \circ f = f \circ \kappa$ if and only if $a_n \in \mathbb{R}$ for all n.*

(Compare Example 1.55.)

PROOF. (i) For any disk $\mathbb{D}(p,r) \subset U, f(z) = \sum a_n (z-p)^n$ in $\mathbb{D}(p,r)$ if and only if $(\kappa \circ f \circ \kappa)(z) = \sum \bar{a}_n (z - \bar{p})^n$ in $\mathbb{D}(\bar{p}, r)$. It follows that f is complex analytic in U if and only if $\kappa \circ f \circ \kappa$ is complex analytic in $\kappa(U)$.

(ii) Suppose $f((p-r, p+r)) \subset \mathbb{R}$ and let $g = \kappa \circ f \circ \kappa$. Then $g = f$ in $(p-r, p+r)$. Since $g \in \mathcal{O}(\mathbb{D}(p,r))$ by (i), the identity theorem implies $g = f$ or $\kappa \circ f = f \circ \kappa$ everywhere in $\mathbb{D}(p,r)$. Applying the latter condition on the power series of f gives $a_n = \bar{a}_n$, or $a_n \in \mathbb{R}$ for all n. Conversely, it is clear that the sequence $\{a_n\}$ being real implies $f(x) = \sum a_n (x-p)^n \in \mathbb{R}$ whenever $x \in (p-r, p+r)$.   □

DEFINITION 10.21. Let $I \subset \mathbb{R}$ be a non-empty open interval. A function $f : I \to \mathbb{R}$ is **real analytic** if for every $p \in I$ there exists a power series $\sum_{n=0}^{\infty} a_n (x-p)^n$ with real coefficients which converges to $f(x)$ for all x in some neighborhood of p.

Notice the subtle difference between the definitions of complex and real analytic functions. If $F : U \to \mathbb{C}$ is complex analytic, it has a power series representation in *every* disk that fits in U. But a real analytic $f : I \to \mathbb{R}$ is only required to have power series representations in sufficiently short intervals centered at each point of I. Example 10.26 below shows that this requirement does not guarantee convergent power series in every subinterval of I.

Holomorphic functions provide a large supply of real analytic functions, as the following example shows.

EXAMPLE 10.22. Suppose $U \subset \mathbb{C}$ is a domain which meets the real line along a non-empty open interval I. If $f \in \mathscr{O}(U)$, then $\mathrm{Re}(f)$ and $\mathrm{Im}(f)$ are real analytic in I. To see this, take any $p \in I$ and choose $r > 0$ small enough to ensure $\mathbb{D}(p, r) \subset U$. Then f has a power series representation $\sum_{n=0}^{\infty} a_n (z-p)^n$ in $\mathbb{D}(p, r)$, hence

$$\mathrm{Re}(f(x)) = \sum_{n=0}^{\infty} \mathrm{Re}(a_n) (x-p)^n \quad \text{and} \quad \mathrm{Im}(f(x)) = \sum_{n=0}^{\infty} \mathrm{Im}(a_n) (x-p)^n$$

whenever $x \in (p-r, p+r)$.

For the following discussion, it will be convenient to call a planar domain U **real symmetric** if $\kappa(U) = U$. In this case $U \cap \mathbb{R}$ is non-empty since U is connected.

THEOREM 10.23. *Every real analytic function $f : I \to \mathbb{R}$ extends to a holomorphic function $F : U \to \mathbb{C}$, where U is a real symmetric domain with $U \cap \mathbb{R} = I$ and F commutes with the complex conjugation $\kappa : z \mapsto \bar{z}$. Conversely, if U is a real symmetric domain with $U \cap \mathbb{R} = I$ and if $F : U \to \mathbb{C}$ is holomorphic and commutes with κ, then the restriction $f = F|_I : I \to \mathbb{R}$ is real analytic.*

PROOF. First suppose $f : I \to \mathbb{R}$ is real analytic. For each $p \in I$ there is an interval $(p-r, p+r) \subset I$ in which f can be expanded into a power series $\sum a_n (x-p)^n$. By Theorem 1.17, the power series $\sum a_n (z-p)^n$ with the same coefficients converges in the disk $D_p = \mathbb{D}(p, r)$, hence defines a holomorphic function F_p there. As $a_n \in \mathbb{R}$ for all n, Lemma 10.20 shows that F_p commutes with κ. If $(q-s, q+s) \subset I$ is another such interval which intersects $(p-r, p+r)$, then $F_q = f = F_p$ on $(q-s, q+s) \cap (p-r, p+r)$, hence in $D_q \cap D_p$ by the identity theorem. Thus, the F_p for various

$p \in I$ patch together unambiguously to define a holomorphic function F in the union $U = \bigcup_{p \in I} D_p$, which evidently has the desired properties.

Conversely, suppose U is real symmetric with $U \cap \mathbb{R} = I$ and $F : U \to \mathbb{C}$ is holomorphic and commutes with κ. For any disk $\mathbb{D}(p, r) \subset U$ with $p \in \mathbb{R}$, expand F into a power series $\sum a_n (z - p)^n$. Then, $f = F|_I$ has the power series expansion $\sum a_n (x - p)^n$ in $(p - r, p + r)$, where $a_n \in \mathbb{R}$ for all n since F commutes with κ. This shows that $f : I \to \mathbb{R}$ is real analytic. $\qquad\square$

REMARK 10.24. If $f : (p - r, p + r) \to \mathbb{R}$ is represented by a convergent power series in $x - p$, it must be real analytic in $(p - r, p + r)$. This can be verified directly, but Theorem 10.23 makes the proof rather trivial: The holomorphic extension $F : \mathbb{D}(p, r) \to \mathbb{C}$ of f is complex analytic in every disk $\mathbb{D}(q, s) \subset \mathbb{D}(p, r)$ with $q \in \mathbb{R}$, so $f : (q - s, q + s) \to \mathbb{R}$ is represented by a convergent power series in $x - q$.

Combining Theorems 1.20 and 10.23, we obtain the following

COROLLARY 10.25. *Every real analytic function $f : I \to \mathbb{R}$ is infinitely differentiable and all its derivatives are real analytic. If $f(x) = \sum_{n=0}^{\infty} a_n (x - p)^n$ for $x \in (p - r, p + r) \subset I$, then $f'(x) = \sum_{n=1}^{\infty} n a_n (x - p)^{n-1}$ for $x \in (p - r, p + r)$. In particular, there are holomorphic extensions F of f and G of f' defined in the same real symmetric domain U with $U \cap \mathbb{R} = I$, and $F'(z) = G(z)$ for every $z \in U$.*

Informally, the last part of the corollary says that *for a real analytic function, the complex derivative of the extension is the extension of the real derivative.* In practice, the derivative of the holomorphic extension of a real analytic f is obtained by applying rules of calculus to find $f'(x)$, and then replacing x with z in the resulting formula.

EXAMPLE 10.26. The function $f(x) = 1/(1 + x^2)$ is real analytic on $\mathbb{R}$ since it is the restriction to $\mathbb{R}$ of the holomorphic function $F(z) = 1/(1 + z^2)$ in $\mathbb{C} \smallsetminus \{\pm i\}$ which commutes with κ. Here $F'(z) = -2z/(1 + z^2)^2$ is the holomorphic extension of $f'(x) = -2x/(1 + x^2)^2$.

Because of the poles of F at $\pm i$, $(-1, 1)$ is the largest interval centered at 0 in which f is represented by a power series, even though f is perfectly smooth near ± 1. More generally, for each $p \in \mathbb{R}$, $(p - \sqrt{p^2 + 1}, p + \sqrt{p^2 + 1})$ is the largest interval centered at p in which f can be expanded into a convergent power series.

We now discuss a useful method of extending holomorphic functions by reflection through lines. To motivate the discussion, we begin with a simple example reminiscent of the situation in the proof of Lemma 6.28: Let U be a real symmetric domain, $U^{\pm}$ be the upper and lower components of $U \smallsetminus \mathbb{R}$, and $I = U \cap \mathbb{R}$. Suppose f is holomorphic in U^+ and continuous on $U^+ \cup I$, with $f(I) \subset \mathbb{R}$. Then, the function $F : U \to \mathbb{C}$ defined by

$$F(z) = \begin{cases} f(z) & z \in U^+ \cup I \\ \overline{f(\bar{z})} & z \in U^- \end{cases}$$

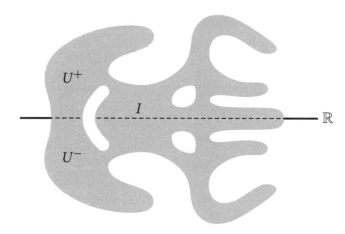

Figure 10.4. The domains $U^{\pm}$ in the Schwarz reflection principle.

is continuous in U and holomorphic in each of the components $U^{\pm}$. Since lines are removable (Example 1.40), it follows that $F \in \mathcal{O}(U)$.

The only downside of this construction is the requirement that f should extend continuously to I, something that can be difficult to verify in practice. Fortunately, the result stands if we only assume $\operatorname{Im}(f)$ extends continuously to 0 on I; continuity of $\operatorname{Re}(f)$ on I will then be automatic. This is achieved by applying Theorem 7.33 on local extensions of harmonic functions.

THEOREM 10.27 (Schwarz reflection principle, 1869). *Let $U \subset \mathbb{C}$ be a real symmetric domain, $U^+ = \{z \in U : \operatorname{Im}(z) > 0\}$, and $I = U \cap \mathbb{R}$. Suppose $f \in \mathcal{O}(U^+)$ and $\operatorname{Im}(f)(z) \to 0$ as $z \in U^+$ tends to a point of I. Then f extends uniquely to a holomorphic function $F : U \to \mathbb{C}$ which commutes with $\kappa : z \mapsto \bar{z}$.*

PROOF. Set $U^- = \kappa(U^+) = \{z \in U : \operatorname{Im}(z) < 0\}$, so $U = U^+ \cup I \cup U^-$ (see Fig. 10.4). For each $p \in I$ choose an open disk $D_p \subset U$ centered at p, and set $D_p^+ = D_p \cap U^+$. By Theorem 7.33, the harmonic function $v = \operatorname{Im}(f)$ in D_p^+ extends by reflection to a harmonic function v_p in D_p. Since disks are simply connected, Theorem 7.9 guarantees the existence of an $F_p \in \mathcal{O}(D_p)$ such that $\operatorname{Im}(F_p) = v_p$. By Lemma 10.20(ii), F_p commutes with κ since $v_p = 0$ on $D_p \cap \mathbb{R}$. Furthermore, since $\operatorname{Im}(f - F_p) = 0$ in D_p^+, the open mapping theorem shows that $f - F_p$ is a real constant c_p there. Thus, by replacing F_p by $F_p + c_p$, we may assume without loss of generality that $F_p = f$ in D_p^+.

If $p, q \in I$ and $D_p \cap D_q \neq \emptyset$, then $F_p = f = F_q$ in $D_p^+ \cap D_q^+$, hence $F_p = F_q$ in $D_p \cap D_q$ by the identity theorem. It follows that the function

$$F = \begin{cases} f & \text{in } U^+ \\ F_p & \text{in } D_p \text{ for some } p \in I \\ \kappa \circ f \circ \kappa & \text{in } U^- \end{cases}$$

is a well-defined holomorphic extension of f to U which commutes with κ. Uniqueness of F follows from the identity theorem. $\qquad\square$

REMARK 10.28. In the situation of Theorem 10.27, suppose f maps U^+ conformally onto the upper half V^+ of a real symmetric domain V. Then the extension F maps U conformally onto V. In fact, since $F = \kappa \circ f \circ \kappa : U^- \to V^-$ is conformal and $F(I) = V \cap \mathbb{R}$, we need only check that $F|_I$ is injective. Suppose $F(p_1) = F(p_2) = q$ for distinct points $p_1, p_2 \in I$. Take disjoint open disks $D_1, D_2 \subset U$ centered at p_1, p_2. By the open mapping theorem, $\Omega = F(D_1) \cap F(D_2)$ is an open neighborhood of $q \in V \cap \mathbb{R}$, so $\Omega \cap V^+ \neq \emptyset$. Under F, any $w \in \Omega \cap V^+$ has a preimage in each of $D_1 \cap U^+$ and $D_2 \cap U^+$, contradicting the hypothesis that $F|_{U^+} = f$ is injective.

Of course conformality of F implies $F' \neq 0$ everywhere in U. The assumptions $F(I) = V \cap \mathbb{R}$ and $F(U^+) = V^+$ easily imply that $F'(z) > 0$ for all $z \in I$.

It is clear that the extension scheme of Theorem 10.27 works equally well if we replace $\mathbb{R}$ and κ with any straight line and the Euclidean reflection across that line.

EXAMPLE 10.29 (Modulus of a rectangle revisited). Recall from Example 6.31 that the modulus $\mathrm{mod}(R) = b/a$ of a rectangle $R = (0, a) \times (0, b)$ is a conformal invariant. In other words, if f is a conformal map from R to $R' = (0, a') \times (0, b')$ which carries each corner of R to the corresponding corner of R', then $b/a = b'/a'$. Here we give a new proof of this invariance by using the Schwarz reflection principle. After pre- and postcomposing f with suitable complex affine maps, which do not change the modulus, we may assume $R = (0, 1) \times (0, m)$ and $R' = (0, 1) \times (0, m')$. By Carathéodory's Theorem 6.20, f extends to a homeomorphism between the closures, carrying each side of R onto the corresponding side of R'. By reflecting across the bottom sides, we find an extension of f to a conformal map $F : (0, 1) \times (-m, m) \to (0, 1) \times (-m', m')$ (conformality follows from Remark 10.28). Repeating this process, each time reflecting across the top or bottom side of the resulting rectangle, we can extend F further to a conformal map $F : (0, 1) \times \mathbb{R} \to (0, 1) \times \mathbb{R}$. This map extends to a homeomorphism between the closures, carrying each vertical boundary line to itself. Finally, we reflect repeatedly across these vertical boundary lines in a similar fashion to extend F to an automorphism $F : \mathbb{C} \to \mathbb{C}$, which by Theorem 4.11 must be of the form $F(z) = \alpha z + \beta$ for some $\alpha, \beta \in \mathbb{C}$ (see Fig. 10.5). Since $F(0) = 0$ and $F(1) = 1$, it follows that $F(z) = z$, so $m' = F(m) = m$.

As another application of the Schwarz reflection principle, we consider Riemann maps of finite-sided polygons. Throughout the following discussion the notation z^α for $z \in \mathbb{H}$ refers to the holomorphic branch of the power function in the upper half-plane defined by

$$z^\alpha = \exp(\alpha \log z),$$

where $\log z = \log(re^{it}) = \log r + it$ for $r > 0$ and $0 < t < \pi$. With this convention, the derivative formula $(z^\alpha)' = \alpha z^{\alpha-1}$ holds for all $z \in \mathbb{H}$ (compare the discussion at the end of §2.2).

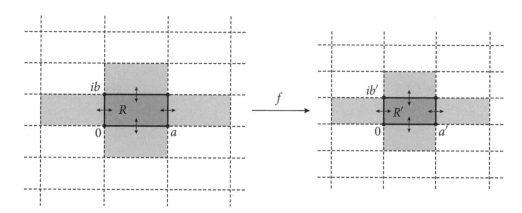

Figure 10.5. Conformal invariance of the modulus of a rectangle (Example 10.29). The initial conformal map $f : R \to R'$ can be extended to an automorphism $F : \mathbb{C} \to \mathbb{C}$ by repeatedly reflecting across the sides of rectangles. Since F is a complex affine map which preserves the vertices, we obtain $b/a = b'/a'$.

It will be convenient to first look at the problem locally. Consider the "wedge"

$$W = \{v + re^{it} : r > 0 \text{ and } t_0 < t < t_0 + \pi\theta\}$$

with the vertex v and angle $\pi\theta \in (0, 2\pi)$. The map $\varphi : \mathbb{H} \to W$ defined by

$$\varphi(z) = v + e^{it_0} z^{\theta}$$

is easily seen to be conformal with a homeomorphic extension to the boundary which sends 0 to v. Take $p \in \mathbb{R}, \varepsilon > 0$ and let $D = \mathbb{D}(p, \varepsilon)$ and $D^+ = D \cap \mathbb{H}$. Suppose f maps D^+ conformally to the intersection U of an open neighborhood of v with W. We assume that f extends homeomorphically to ∂D^+, sending p to v, $\partial D^+ \cap \mathbb{R}$ to $\partial U \cap \partial W$, and $\partial D^+ \cap \mathbb{H}$ to $\partial U \cap W$ (see Fig. 10.6). The conformal map $\varphi^{-1} \circ f : D^+ \to \varphi^{-1}(U)$ extends homeomorphically to the boundary, sending p to 0 and $\partial D^+ \cap \mathbb{R}$ into $\mathbb{R}$. By the Schwarz reflection principle, $\varphi^{-1} \circ f$ extends to a conformal map defined in D with non-zero (in fact positive) derivative at p. Thus, by shrinking D if necessary, we can find a non-vanishing $h \in \mathcal{O}(D)$ such that $(\varphi^{-1} \circ f)(z) = (z - p)h(z)$ for all $z \in D$. Applying φ on both sides then gives

$$f(z) = v + (z - p)^{\theta} g(z) \qquad \text{for } z \in D^+,$$

where the non-vanishing function $g \in \mathcal{O}(D)$ is a suitable branch of $e^{it_0} h^{\theta}$. Differentiating twice, we obtain the relations

$$f'(z) = (z - p)^{\theta-1} g_1(z) \qquad \text{and} \qquad f''(z) = (z - p)^{\theta-2} g_2(z)$$

in D^+, where $g_1, g_2 \in \mathcal{O}(D)$ with

$$g_1(p) = \theta g(p) \neq 0 \qquad \text{and} \qquad g_2(p) = \theta(\theta - 1) g(p).$$

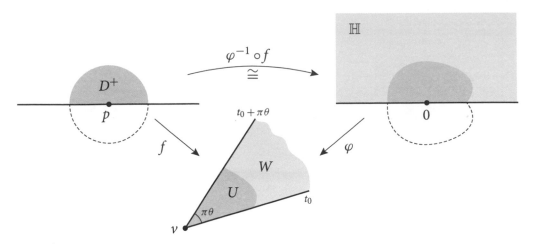

Figure 10.6. Every conformal map $f : D^+ \to U$ near the vertex of the wedge W factors as a local biholomorphism followed by a power map φ.

Thus,

$$\frac{f''(z)}{f'(z)} = \frac{1}{z-p} \cdot \frac{g_2(z)}{g_1(z)} \qquad \text{for } z \in D^+.$$

If $\theta \neq 1$, it follows that f''/f' extends to a meromorphic function in D with a simple pole of residue $\theta - 1$ at p. If $\theta = 1$ (so the wedge W is a half-plane), then f''/f' extends to a holomorphic function in D since in this case f itself extends to a conformal map of D.

A similar local description is possible even if $p = \infty$. We use the change of coordinate $w = -1/z$ which preserves $\mathbb{H}$ and sends ∞ to 0. The preceding discussion applied to the function $\psi(w) = f(-1/w)$ shows that ψ''/ψ' is meromorphic near $w = 0$ with at worst a simple pole there. It follows that

$$\frac{f''(z)}{f'(z)} = 2w + w^2 \frac{\psi''(w)}{\psi'(w)}$$

is holomorphic in a neighborhood of ∞ and tends to 0 as $z \to \infty$.

Now suppose Ω is a polygon with vertices $v_1, \ldots, v_n$, labeled counter-clockwise, and internal angles $\pi\theta_1, \ldots, \pi\theta_n$ (see Fig. 10.7 for the case $n = 5$). By Carathéodory's Theorem 6.20, any conformal map $f : \mathbb{H} \to \Omega$ extends to a homeomorphism between the closures. We may assume $f(\infty) = v_n$, so there are points $p_1 < \cdots < p_{n-1}$ in $\mathbb{R}$ such that $f(p_k) = v_k$ for $1 \leq k \leq n-1$. The intervals $I_1 = (-\infty, p_1), I_2 = (p_1, p_2), \ldots, I_n = (p_{n-1}, +\infty)$ map homeomorphically to the sides $[v_n, v_1], [v_1, v_2], \ldots, [v_{n-1}, v_n]$ of Ω. By reflecting across I_k, we obtain a holomorphic extension F_k of f to the union $\mathbb{H} \cup -\mathbb{H} \cup I_k$ whose restriction to the lower half-plane $-\mathbb{H}$ takes values in the reflection Ω_k of Ω across the side $[v_{k-1}, v_k]$. The extension F_k is still conformal and in particular $F_k' \neq 0$ in I_k (see Remark 10.28). Moreover, if I_k, I_j are adjacent, then Ω_k, Ω_j are rotated copies of one another (see Fig. 10.7). In this case $F_k = \alpha F_j + \beta$ for some

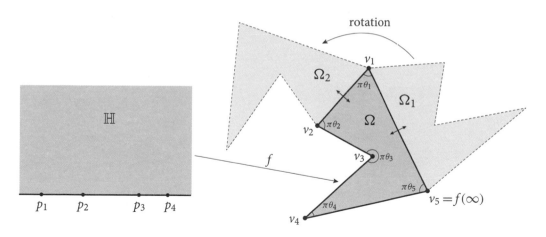

Figure 10.7. The Riemann map of a pentagon Ω and the derivation of the Schwarz-Christoffel formula (10.5).

$\alpha, \beta \in \mathbb{C}$ with $|\alpha| = 1$, so $F_k''/F_k' = F_j''/F_j'$ in $-\mathbb{H}$. This shows that f''/f' extends to a well-defined holomorphic function in $\mathbb{C} \setminus \{p_1, \ldots, p_{n-1}\}$. By the preceding local discussion, the singularity p_k of f''/f' is a simple pole of residue $\theta_k - 1$ if $\theta_k \neq 1$ and is removable if $\theta_k = 1$. Moreover, $f''(z)/f'(z)$ tends to 0 as $z \to \infty$. It follows from Theorem 3.20 that f''/f' is a rational function, and in fact

$$\frac{f''(z)}{f'(z)} = \sum_{k=1}^{n-1} \frac{\theta_k - 1}{z - p_k}.$$

Integrating this, we obtain

$$f'(z) = C \prod_{k=1}^{n-1} (z - p_k)^{\theta_k - 1}$$

Elwin Bruno Christoffel
(1829–1900)

for some constant $C \neq 0$. Another integration then proves the following result known as the ***Schwarz-Christoffel formula***:

THEOREM 10.30. *Suppose $\Omega \subset \mathbb{C}$ is a polygon with vertices $v_1, \ldots, v_n$ and internal angles $\pi\theta_1, \ldots, \pi\theta_n \in (0, 2\pi)$. Let $f : \mathbb{H} \to \Omega$ be the conformal map which sends the boundary points $p_1, \ldots, p_{n-1}, \infty$ to the vertices $v_1, \ldots, v_n$. Then,*

The formula was obtained by Christoffel in 1867 and independently by Schwarz in 1869.

$$(10.5) \qquad f(z) = f(z_0) + C \int_{z_0}^{z} \left(\prod_{k=1}^{n-1} (\zeta - p_k)^{\theta_k - 1} \right) d\zeta$$

for $z_0 \in \mathbb{H} \cup \mathbb{R}$ and a suitable constant $C \neq 0$.

Problems 13–15 discuss examples and applications of this formula.

The Schwarz reflection principle extends to other geometric situations. It turns out that what makes this principle work is not the special geometry of the real line or the formula for the complex conjugation. Rather, it is the fact that $\mathbb{R}$ is an "analytic

Figure 10.8. Left: An open arc which is not topologically embedded in the plane. Right: The analytic arc of Example 10.32.

arc" and κ is an "anti-holomorphic involution" that fixes $\mathbb{R}$ pointwise. The rest of this section will elaborate this idea.

DEFINITION 10.31. By an ***analytic arc*** $J \subset \mathbb{C}$ we mean the image of a curve $\gamma : (a, b) \to \mathbb{C}$ with the following properties:

  (i) $\mathrm{Re}(\gamma), \mathrm{Im}(\gamma) : (a, b) \to \mathbb{R}$ are real analytic;
  (ii) $\gamma'(t) \neq 0$ for all $t \in (a, b)$;
  (iii) $\gamma : (a, b) \to J$ is a homeomorphism when J is given the induced topology from $\mathbb{C}$.

We call such γ a ***real analytic parametrization*** of J.

The condition (iii) says that γ is a topological embedding which, roughly speaking, ensures that the open ends of J do not accumulate on any point of J (see Fig. 10.8 left).

EXAMPLE 10.32. Any open line segment or circular arc is trivially an analytic arc. A more interesting example is the ever-spiraling arc parametrized by

$$\gamma(t) = t \exp(i \tan t) \qquad \text{for } 0 < t < \frac{\pi}{2}$$

(see Fig. 10.8 right). The arc

$$J = \left\{ x : x \leq 0 \right\} \cup \left\{ x + i e^{-1/x^2} : x \geq 0 \right\}$$

has a C^∞ parametrization, but is not analytic. In fact, if $\gamma : (a, b) \to \mathbb{C}$ is a real analytic parametrization of J with $\gamma(t_0) = 0$, then $\mathrm{Im}(\gamma) = 0$ in (a, t_0) or (t_0, b). Since $\mathrm{Im}(\gamma)$ is real analytic, it follows that $\mathrm{Im}(\gamma) = 0$ in (a, b), which is a contradiction.

The image of the real segment $(-1, 1) = \mathbb{D} \cap \mathbb{R}$ under a biholomorphism $\varphi : \mathbb{D} \to \varphi(\mathbb{D})$ is easily seen to be an analytic arc. The following lemma shows that all analytic arcs are essentially obtained this way:

LEMMA 10.33 (Existence of global coordinate functions). *Let J be an analytic arc with a real analytic parametrization $\gamma : (a, b) \to J$. Then there is a real symmetric domain U with $U \cap \mathbb{R} = (a, b)$ and a biholomorphism $\varphi : U \to \varphi(U)$ such that $\varphi|_{(a,b)} = \gamma$.*

We call φ a **coordinate function** and the image $\Omega = \varphi(U)$ a **coordinate neighborhood** of J. Since $U \smallsetminus (a, b)$ has two connected components, it follows that $\Omega \smallsetminus J$ has two connected components which can be thought of as lying on different "sides" of J. Although the lemma is easy to prove locally, the construction of a global coordinate function takes a bit of effort.

PROOF. Every $p \in (a, b)$ is the center of an open interval $I_p \subset (a, b)$ in which $\operatorname{Re}(\gamma), \operatorname{Im}(\gamma)$ have power series expansions. It follows that $\operatorname{Re}(\gamma), \operatorname{Im}(\gamma)$ extend to holomorphic functions f_p, g_p defined in the open disk D_p centered at p with $D_p \cap \mathbb{R} = I_p$. The function $\varphi_p = f_p + ig_p : D_p \to \mathbb{C}$ is then a holomorphic extension of $\gamma|_{I_p}$. If $D_p \cap D_q \neq \emptyset$, then $\varphi_p = \gamma = \varphi_q$ on $I_p \cap I_q$, so $\varphi_p = \varphi_q$ in $D_p \cap D_q$ by the identity theorem. It follows that the φ_p patch together consistently to define a holomorphic function φ in the domain $V = \bigcup_{p \in (a,b)} D_p$ such that $\varphi|_{(a,b)} = \gamma$. Our task is to show that V contains a real symmetric domain in which φ is injective.

We claim that for any compact interval $[\alpha, \beta] \subset (a, b)$ there is an $\varepsilon > 0$ such that the closed rectangle $R = [\alpha, \beta] \times [-\varepsilon, \varepsilon] \subset V$ is "nice" in the following sense:

(i) φ is injective on R.
(ii) $\varphi(R) \cap J = \gamma([\alpha, \beta])$.

If (i) does not hold for any $\varepsilon > 0$, we can find for each $n \geq 1$ a distinct pair z_n, w_n in $[\alpha, \beta] \times [-1/n, 1/n]$ such that $\varphi(z_n) = \varphi(w_n)$. After passing to a subsequence, we may assume that $\{z_n\}, \{w_n\}$ converge to $t, s \in [\alpha, \beta]$, where $\varphi(t) = \varphi(s)$ by continuity. Since $\varphi|_{(a,b)} = \gamma$ is injective, we must have $t = s$. It follows that φ is not injective in any neighborhood of t, a contradiction since $\varphi'(t) = \gamma'(t) \neq 0$. Thus, (i) is achieved for small enough $\varepsilon > 0$.

If (ii) does not hold for any $\varepsilon > 0$, we can find for each $n \geq 1$ a point z_n in $[\alpha, \beta] \times [-1/n, 1/n]$ and some t_n in $(a, b) \smallsetminus [\alpha, \beta]$ such that $\varphi(z_n) = \gamma(t_n)$. After passing to a subsequence, we may assume that $z_n \to t \in [\alpha, \beta]$. Evidently $\{t_n\}$ cannot accumulate on t since in that case φ would not be locally injective near t, contradicting $\varphi'(t) = \gamma'(t) \neq 0$. Thus, there is a open interval $I \subset (a, b)$ centered at t such that $t_n \notin I$ for all n. Since $\gamma(I)$ is relatively open in J, we have $\gamma(I) = J \cap W$ for some open set $W \subset \mathbb{C}$. Then, for large n, W contains $\varphi(z_n) = \gamma(t_n) \in J$, so $\gamma(t_n) \in \gamma(I)$. By injectivity of γ, this implies $t_n \in I$, which is a contradiction. This proves our claim on the existence of nice rectangles.

To finish the proof of the lemma, write (a, b) as a union of compact intervals K_n, where each K_n is contained in the interior of K_{n+1}. Let $R_1 = K_1 \times [-\varepsilon_1, \varepsilon_1]$ be a nice rectangle in the above sense. We claim that there is a nice rectangle $R_2 = K_2 \times [-\varepsilon_2, \varepsilon_2]$ such that φ is injective on $R_1 \cup R_2$ (see Fig. 10.9). Otherwise for each $n \geq 1$ we can find distinct points $z_n \in R_1$ and $w_n \in (K_2 \smallsetminus K_1) \times [-1/n, 1/n]$ such that $\varphi(z_n) = \varphi(w_n)$. After passing to a subsequence, we may assume $z_n \to z \in R_1$ and

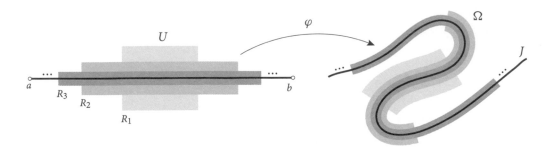

Figure 10.9. Construction of a global coordinate function for an analytic arc, as in the proof of Lemma 10.33.

$w_n \to t \in K_2$. By continuity, the point $\varphi(z) \in \varphi(R_1)$ coincides with $\varphi(t) = \gamma(t) \in J$, so $t \in K_1$ since R_1 is nice. Injectivity of φ on R_1 then implies $t = z$. This shows φ is not locally injective near t, which contradicts $\varphi'(t) = \gamma'(t) \neq 0$ and proves our claim. Continuing inductively, given nice rectangles $R_j = K_j \times [-\varepsilon_j, \varepsilon_j]$ for $1 \leq j \leq n$ such that φ is injective on $R_1 \cup \cdots \cup R_n$, we can repeat the same argument to find a nice rectangle $R_{n+1} = K_{n+1} \times [-\varepsilon_{n+1}, \varepsilon_{n+1}]$ such that φ is injective on $R_1 \cup \cdots \cup R_{n+1}$. Evidently, the union U of the interiors of the R_n will have the desired properties. $\square$

A function $f : U \to \mathbb{C}$ is **anti-holomorphic** if in every disk $\mathbb{D}(p, r) \subset U$ it has a power series representation of the form $f(z) = \sum_{n=0}^{\infty} a_n (\bar{z} - \bar{p})^n$. It is easy to see that this condition is equivalent to either of the conditions $\kappa \circ f \in \mathscr{O}(U)$ and $f \circ \kappa \in \mathscr{O}(\kappa(U))$ (compare problem 5 in chapter 1). We note the following properties whose proofs are easily supplied:

- The composition of two anti-holomorphic functions is holomorphic, while the composition of a holomorphic and an anti-holomorphic function (in either order) is anti-holomorphic.
- The identity theorem holds for anti-holomiorphic functions: If f, g are anti-holomorphic in a domain U, and if $f = g$ along a set which has an accumulation point in U, then $f = g$ everywhere in U.

LEMMA 10.34. *For any coordinate neighborhood Ω of an analytic arc J there is an anti-holomorphic involution $\zeta : \Omega \to \Omega$ which permutes the two components of $\Omega \setminus J$ and acts as the identity map on J. Moreover, any anti-holomorphic involution defined near J which keeps J fixed pointwise must coincide with ζ in some neighborhood of J.*

We call ζ the **canonical reflection** across J. The lemma asserts that modulo the choice of its domain, ζ is uniquely determined by J.

PROOF. Let $\varphi : U \to \Omega$ be a coordinate function as in Lemma 10.33. By real symmetry of U, the two components of $U \setminus \mathbb{R}$ are permuted by the complex conjugation κ. Since φ is biholomorphic, it follows that the two components of $\Omega \setminus J$ are permuted by the anti-holomorphic involution $\zeta = \varphi \circ \kappa \circ \varphi^{-1} : \Omega \to \Omega$. The fixed point set of ζ is J since the fixed point set of κ in U is $U \cap \mathbb{R}$.

To see uniqueness, suppose $\hat{\zeta}$ is any anti-holomorphic involution defined in a neighborhood V of J such that $\hat{\zeta}(z) = z$ for all $z \in J$. As ζ and $\hat{\zeta}$ agree on J, the identity theorem implies that they agree in the connected component of $\Omega \cap V$ containing J. $\qquad\square$

EXAMPLE 10.35. The canonical reflection across an interval in $\mathbb{R}$ is the complex conjugation $z \mapsto \bar{z}$. For a circular arc in $\mathbb{T}$, the canonical reflection is $z \mapsto 1/\bar{z}$.

REMARK 10.36. Any anti-holomorphic function ζ which keeps an analytic arc J fixed pointwise is necessarily an involution. This follows from the identity theorem since $\zeta \circ \zeta$ is holomorphic and agrees with the identity map on J.

We now use canonical reflections to prove the following general version of the Schwarz reflection principle:

THEOREM 10.37 (Schwarz reflection for analytic arcs). *Let $J, \hat{J}$ be analytic arcs with coordinate neighborhoods $\Omega, \hat{\Omega}$. Let $\Omega^{\pm}$ be the connected components of $\Omega \smallsetminus J$. Suppose $f : \Omega^+ \to \hat{\Omega}$ is holomorphic and extends continuously to $\Omega^+ \cup J$, with $f(J) \subset \hat{J}$. Then f extends uniquely to a holomorphic function $F : \Omega \to \hat{\Omega}$. Moreover, $F \circ \zeta = \hat{\zeta} \circ F$, where $\zeta : \Omega \to \Omega, \hat{\zeta} : \hat{\Omega} \to \hat{\Omega}$ are the canonical reflections across the arcs $J, \hat{J}$.*

In practice, the requirement that f should map Ω^+ into $\hat{\Omega}$ is not a burden, since it can always be achieved by shrinking the coordinate neighborhood Ω if necessary.

PROOF. Take coordinate functions $\varphi : U \to \Omega$ and $\hat{\varphi} : \hat{U} \to \hat{\Omega}$ for J and $\hat{J}$ as in Lemma 10.33, so $\zeta = \varphi \circ \kappa \circ \varphi^{-1}$ and $\hat{\zeta} = \hat{\varphi} \circ \kappa \circ \hat{\varphi}^{-1}$. Set $U^{\pm} = \varphi^{-1}(\Omega^{\pm})$ and consider the induced map $g = \hat{\varphi}^{-1} \circ f \circ \varphi : U^+ \to \hat{U}$. As $w \in U^+$ tends to $w_0 \in U \cap \mathbb{R}$, the image $z = \varphi(w) \in \Omega^+$ tends to $z_0 = \varphi(w_0) \in J$, hence $f(z) \to f(z_0) \in \hat{J}$ by the assumption. It follows that $g(w) = \hat{\varphi}^{-1}(f(z)) \to \hat{\varphi}^{-1}(f(z_0)) \in \hat{U} \cap \mathbb{R}$. By Theorem 10.27, g has a holomorphic extension $G : U \to \hat{U}$ which commutes with κ. It follows that $F = \hat{\varphi} \circ G \circ \varphi^{-1} : \Omega \to \hat{\Omega}$ is a holomorphic extension of f which satisfies $F \circ \zeta = \hat{\zeta} \circ F$. Uniqueness of F follows from the identity theorem. $\qquad\square$

So far we have restricted the discussion to analytic arcs, but it is easy to extend the definition of analyticity and Lemma 10.34 and Theorem 10.37 to Jordan curves (see problems 19 and 20). Analytic arcs and Jordan curves together constitute the so-called "one-dimensional analytic submanifolds" of the plane.

10.4 Two removability results

The extension problem for holomorphic functions has another formulation in terms of **holomorphic removability** of compact subsets of the plane. Suppose K is compact and U is an open set containing K. We ask: "Does every $f \in \mathscr{O}(U \smallsetminus K)$ extend to a function in $\mathcal{O}(U)$?" The example $f(z) = 1/(z - p)$ with $p \in K$ shows that without some extra assumptions the answer to this question is always negative. Two assumptions that guarantee an affirmative answer have already been encountered: If K is a point and f is bounded in $U \smallsetminus K$ (Theorem 3.5), and if K is a line segment and f is continuous in U (Example 1.40). In this section we prove two theorems that generalize these basic results.

To motivate the first theorem, let us revisit Theorem 3.5, Riemann's removable singularity theorem, in a way that can lead us to a more general result. Suppose f is bounded and holomorphic in a punctured neighborhood of p. Fix a small radius $r > 0$, let $z \in \mathbb{D}^*(p, r)$ and take any ε such that $0 < \varepsilon < |z - p|/2$. By Cauchy's integral formula,

$$f(z) = \frac{1}{2\pi i} \int_{\mathbb{T}(p,r)} \frac{f(\zeta)}{\zeta - z} \, d\zeta - \frac{1}{2\pi i} \int_{\mathbb{T}(p,\varepsilon)} \frac{f(\zeta)}{\zeta - z} \, d\zeta.$$

The absolute value of the term $1/(\zeta - z)$ in the second integral is uniformly bounded above by $2/|z - p|$ by the choice of ε. Since f is assumed bounded and the length of $\mathbb{T}(p, \varepsilon)$ tends to zero as $\varepsilon \to 0$, the ML-inequality shows that the second integral tends to zero as $\varepsilon \to 0$. It follows that

$$f(z) = \frac{1}{2\pi i} \int_{\mathbb{T}(p,r)} \frac{f(\zeta)}{\zeta - z} \, d\zeta \qquad \text{if } z \in \mathbb{D}^*(p, r).$$

Since the right-hand side is a holomorphic function of z off the circle $\mathbb{T}(p, r)$ by Corollary 1.48, it provides a holomorphic extension of f to the disk $\mathbb{D}(p, r)$.

In effect, the only property of the exceptional set $\{p\}$ that was used in the above argument was that it can be surrounded by arbitrarily short loops. Our first removability result (Theorem 10.47) asserts that $\{p\}$ can be replaced by any compact set that is "small" in a similar sense. The proper meaning of smallness in this context is captured by the notion of Hausdorff measure which we now introduce in some detail since it will also play a role in our second removability result (Theorem 10.48).

Let us say that a countable collection $\{X_k\}$ of compact sets in $\mathbb{C}$ is an ε-**cover** of a set $E \subset \mathbb{C}$ if $E \subset \bigcup_k X_k$ and $\operatorname{diam}(X_k) < \varepsilon$ for all k. For a fixed $\alpha > 0$, define

$$\Lambda_\alpha^\varepsilon(E) = \inf \left\{ \sum_k (\operatorname{diam}(X_k))^\alpha : \{X_k\} \text{ is an } \varepsilon\text{-cover of } E \right\}.$$

Clearly, $\varepsilon \mapsto \Lambda_\alpha^\varepsilon(E)$ is decreasing, so its limit as $\varepsilon \to 0$ exists (but could be $+\infty$).

We could use closed sets, open sets, Borel sets, or even arbitrary sets to form ε-covers. The resulting measures Λ_α will be the same.

> **DEFINITION 10.38.** The quantity
>
> $$\Lambda_\alpha(E) = \lim_{\varepsilon \to 0} \Lambda_\alpha^\varepsilon(E) \in [0, +\infty]$$
>
> is called the α-**dimensional Hausdorff measure** of E.

It can be shown that Λ_α is indeed a measure defined on a σ-algebra of subsets of $\mathbb{C}$ that contains all Borel sets ([**Fa**] or [**P**]). The measure Λ_1 is what we may recognize as "length" and Λ_2 is essentially "area."

EXAMPLE 10.39. For any segment $[p, q] \subset \mathbb{C}$, $\Lambda_1([p, q]) = |p - q|$. Here is why: Let $\varepsilon > 0$ and choose an integer $n > |p - q|/\varepsilon$. There is an ε-cover of $[p, q]$ by n closed sub-segments each of diameter $|p - q|/n$, which shows $\Lambda_1^\varepsilon([p, q]) \le n \cdot |p - q|/n = |p - q|$. Letting $\varepsilon \to 0$, we obtain $\Lambda_1([p, q]) \le |p - q|$. On the other hand, for any ε-cover $\{X_k\}$ of $[p, q]$, the intersections $I_k = X_k \cap [p, q]$ are linear sets covering $[p, q]$, so by countable sub-additivity of length,

$$|p - q| = \text{length}([p, q]) \le \sum_k \text{length}(I_k) \le \sum_k \text{diam}(I_k) \le \sum_k \text{diam}(X_k).$$

Taking the infimum over all ε-covers and letting $\varepsilon \to 0$, we obtain the reverse inequality $|p - q| \le \Lambda_1([p, q])$.

It is not hard to check that *for linear sets in the plane, the 1-dimensional Hausdorff measure is the same as the 1-dimensional Lebesgue measure.* For example, the standard middle-thirds Cantor set $K \subset [0, 1]$ has $\Lambda_1(K) = 0$.

EXAMPLE 10.40. The 2-dimensional Hausdorff measure is not the same as area but it is comparable to it. Let us check for example that for any square $S \subset \mathbb{C}$,

$$\text{area}(S) \le \Lambda_2(S) \le 2 \, \text{area}(S).$$

Let a denote the side length of S. Let $\varepsilon > 0$ and choose an integer $n > a\sqrt{2}/\varepsilon$. There is an ε-cover of S by n^2 closed sub-squares each of side length a/n, which shows $\Lambda_2^\varepsilon(S) \le n^2 \cdot 2a^2/n^2 = 2a^2$. Letting $\varepsilon \to 0$, we obtain $\Lambda_2(S) \le 2a^2$. On the other hand, for any ε-cover $\{X_k\}$ of S, the intersections $I_k = X_k \cap S$ cover S, so by countable sub-additivity of area,

$$a^2 = \text{area}(S) \le \sum_k \text{area}(I_k) \le \sum_k \text{area}(X_k) \le \sum_k (\text{diam}(X_k))^2,$$

where the last inequality holds since a set of diameter d is contained in a square of side length d. Taking the infimum over all ε-covers of S and letting $\varepsilon \to 0$, we obtain $a^2 \le \Lambda_2(S)$.

LEMMA 10.41. *If $\Lambda_\alpha(E) < +\infty$, then $\Lambda_\beta(E) = 0$ for every $\beta > \alpha$.*

PROOF. Since $t^\beta/t^\alpha \to 0$ as $t \to 0$, for every $\delta > 0$ we can find a $t_0 > 0$ such that $t^\beta < \delta t^\alpha$ whenever $0 < t < t_0$. If $0 < \varepsilon < t_0$ and $\{X_k\}$ is an ε-cover of E, then

$$\Lambda_\beta^\varepsilon(E) \le \sum_k (\text{diam}(X_k))^\beta \le \delta \sum_k (\text{diam}(X_k))^\alpha.$$

Taking the infimum over all ε-covers gives $\Lambda_\beta^\varepsilon(E) \le \delta \Lambda_\alpha^\varepsilon(E)$. Letting $\varepsilon \to 0$, we obtain $\Lambda_\beta(E) \le \delta \Lambda_\alpha(E)$. Letting $\delta \to 0$ then proves $\Lambda_\beta(E) = 0$. $\square$

For every square $S \subset \mathbb{C}$, $\Lambda_2(S) \le 2 \, \text{area}(S) < +\infty$ (Example 10.40), so by the above lemma $\Lambda_\alpha(S) = 0$ whenever $\alpha > 2$. Since every Borel set $E \subset \mathbb{C}$ is contained in a countable union of squares, it follows that $\Lambda_\alpha(E) = 0$ whenever $\alpha > 2$.

Felix Hausdorff
(1868–1942)

> **DEFINITION 10.42.** The *Hausdorff dimension* of E is defined by
>
> $$\dim_H(E) = \inf\{\alpha > 0 : \Lambda_\alpha(E) = 0\}.$$
>
> It satisfies $0 \le \dim_H(E) \le 2$.

By Lemma 10.41, the Hausdorff dimension is characterized by the property

$$\Lambda_\alpha(E) = \begin{cases} +\infty & \text{if } \alpha < \dim_H(E) \\ 0 & \text{if } \alpha > \dim_H(E). \end{cases}$$

The value of $\Lambda_\alpha(E)$ at $\alpha = \dim_H(E)$ could a priori be any number in $[0, +\infty]$.

EXAMPLE 10.43. Suppose $\varphi : \mathbb{C} \to \mathbb{C}$ is distance-decreasing in the sense that

$$|\varphi(p) - \varphi(q)| \le |p - q| \qquad \text{for all } p, q \in \mathbb{C}.$$

Then $\Lambda_\alpha(\varphi(E)) \le \Lambda_\alpha(E)$ for every Borel set E. To see this, let $\{X_k\}$ be any ε-cover of E. Each image $Y_k = \varphi(X_k)$ is compact since X_k is compact and φ is continuous. Moreover, $\operatorname{diam}(Y_k) \le \operatorname{diam}(X_k)$ since φ is distance-decreasing. It follows that $\{Y_k\}$ is an ε-cover of $\varphi(E)$. Thus,

$$\Lambda_\alpha^\varepsilon(\varphi(E)) \le \sum_k (\operatorname{diam}(Y_k))^\alpha \le \sum_k (\operatorname{diam}(X_k))^\alpha.$$

Taking the infimum over all ε-covers of E and then letting $\varepsilon \to 0$ proves the result.

Here is an application: Suppose $E \subset \mathbb{C}$ is connected and contains more than one point. Then the orthogonal projection of E onto one of the coordinate axes must be an interval of positive length. Since projections are distance-decreasing, this shows $\Lambda_1(E) > 0$ and therefore $\dim_H(E) \ge 1$.

It follows that *if $E \subset \mathbb{C}$ satisfies $\Lambda_1(E) = 0$ (in particular if $\dim_H(E) < 1$), then E must be totally disconnected.*

For linear sets the measure Λ_1 is the usual length or 1-dimensional Lebesgue measure (Example 10.39). This is indeed true for more general sets such as arcs. Recall that for an arc $J \subset \mathbb{C}$ parametrized by a continuous injective map $\gamma : [0, 1] \to J$, the length of J is defined by

$$\operatorname{length}(J) = \sup \sum_{k=1}^{n} |\gamma(t_k) - \gamma(t_{k-1})|$$

where the supremum is taken over all partitions $t_0 = 0 < t_1 < \cdots < t_{n-1} < t_n = 1$ of the interval $[0, 1]$. This quantity is easily seen to be independent of the parametrization γ. We call J *rectifiable* if $\operatorname{length}(J) < +\infty$. It is well known that if J has a piecewise C^1 parametrization γ, it is rectifiable and its length is given by the formula

$$\operatorname{length}(J) = \int_\gamma |dz| = \int_0^1 |\gamma'(t)|\, dt,$$

in agreement with the definition in §1.3.

THEOREM 10.44. *For every arc $J \subset \mathbb{C}$,*

$$\text{length}(J) = \Lambda_1(J).$$

Thus, J is rectifiable if and only if $\Lambda_1(J) < +\infty$, in which case $\dim_H(J) = 1$.

PROOF. First suppose $\Lambda_1(J) < +\infty$. Take a parametrization $\gamma : [0,1] \to J$ and a partition $t_0 = 0 < t_1 < \cdots < t_n = 1$, and set $p_k = \gamma(t_k)$. For each $1 \le k \le n$ let $J_k = \gamma([t_{k-1}, t_k])$ and let $E_k \supset [p_{k-1}, p_k]$ be the orthogonal projection of J_k onto the line in $\mathbb{C}$ passing through p_{k-1} and p_k. We have, by Examples 10.39 and 10.43,

$$\sum_{k=1}^{n} |p_{k-1} - p_k| = \sum_{k=1}^{n} \Lambda_1([p_{k-1}, p_k]) \le \sum_{k=1}^{n} \Lambda_1(E_k) \le \sum_{k=1}^{n} \Lambda_1(J_k) = \Lambda_1(J),$$

where the last equality holds since Λ_1 is a measure and the J_k are disjoint except at the endpoints. Taking the supremum over all partitions of $[0, 1]$ now gives the inequality $\text{length}(J) \le \Lambda_1(J)$.

Now suppose $\text{length}(J) < +\infty$. For every $\varepsilon > 0$, we can divide J into finitely many closed subarcs $J_1, \ldots, J_n$ such that $\text{diam}(J_k) \le \text{length}(J_k) < \varepsilon$. Then,

$$\Lambda_1^{\varepsilon}(J) \le \sum_{k=1}^{n} \text{diam}(J_k) \le \sum_{k=1}^{n} \text{length}(J_k) = \text{length}(J).$$

Letting $\varepsilon \to 0$, we obtain $\Lambda_1(J) \le \text{length}(J)$. □

For our subsequent arguments, it will be convenient to make use of a standard collection of squares in the plane with good covering properties. By a **dyadic square of level** $\nu \ge 0$ we mean a closed square of side length $2^{-\nu}$ whose corners belong to the lattice $2^{-\nu}(\mathbb{Z} \oplus \mathbb{Z}) = \{2^{-\nu}(m + in) : m, n \in \mathbb{Z}\}$. We say that two dyadic squares are **almost disjoint** if they have disjoint interiors (but can meet along the boundary). The most useful property here is that *two dyadic squares are either nested or almost disjoint.*

LEMMA 10.45. *For any finite collection $S_1, \ldots, S_n$ of almost disjoint dyadic squares in $\mathbb{C}$ there is a cycle η such that*

    (i) $|\eta| = \partial\left(\bigcup_{k=1}^{n} S_k\right)$,
    (ii) $\int_{\eta} f(z)\, dz = \sum_{k=1}^{n} \int_{\partial S_k} f(z)\, dz$ *whenever* $f : \bigcup_{k=1}^{n} \partial S_k \to \mathbb{C}$ *is continuous,*
    (iii) $W(\eta, p) = \begin{cases} 1 & \text{if } p \text{ is in the interior of } \bigcup_{k=1}^{n} S_k \\ 0 & \text{if } p \in \mathbb{C} \smallsetminus \bigcup_{k=1}^{n} S_k. \end{cases}$

Fig. 10.10 illustrates this lemma.

PROOF. It suffices to find an η which satisfies (i) and (ii) since property (iii) will be a consequence. In fact, applying (ii) to the function $f(z) = 1/(z - p)$ with

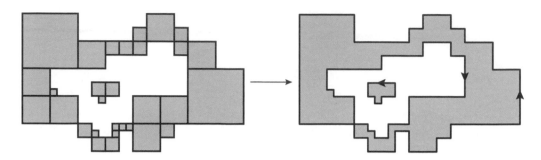

Figure 10.10. Illustration of Lemma 10.45: The boundary of a finite union of almost disjoint dyadic squares defines a cycle which has winding number 1 with respect to any point in the interior of this union, and 0 with respect to any point outside this union.

$p \notin \bigcup_{k=1}^{n} \partial S_k$ gives

$$W(\eta, p) = \sum_{k=1}^{n} W(\partial S_k, p) = \begin{cases} 1 & \text{if } p \text{ is in the interior of some } S_k \\ 0 & \text{if } p \in \mathbb{C} \setminus \bigcup_{k=1}^{n} S_k. \end{cases}$$

Every p in the interior of $\bigcup_{k=1}^{n} S_k$ is either in the interior of some S_k or accumulated by such points. Since $W(\eta, \cdot)$ is constant in a neighborhood of p, we obtain (iii).

The construction of η satisfying (i) and (ii) mimics the proof of Lemma 9.21. First suppose every S_k is of the same level ν. Write ∂S_k as the sum $\beta_k^1 + \beta_k^2 + \beta_k^3 + \beta_k^4$ of four oriented segments of length $2^{-\nu}$, let $\mathcal{C}$ be the collection of all β_k^j for $1 \leq k \leq n$ and $1 \leq j \leq 4$, and let $\mathcal{D}$ be the sub-collection obtained from $\mathcal{C}$ by removing every oriented segment ξ whose reverse ξ^- is also in $\mathcal{C}$. The chains

$$\sum_{k=1}^{n} \partial S_k = \sum_{\xi \in \mathcal{C}} \xi \qquad \text{and} \qquad \eta = \sum_{\xi \in \mathcal{D}} \xi$$

are both cycles and (ii) clearly holds. The topological boundary $\partial \left(\bigcup_{k=1}^{n} S_k \right)$ consists of points in $\bigcup_{k=1}^{n} \partial S_k$ which do not lie on the common edge of two adjacent squares in $\{S_1, \ldots, S_n\}$. This is precisely $\bigcup_{\xi \in \mathcal{D}} |\xi| = |\eta|$, so (i) holds.

Now consider the general case where the dyadic squares $\{S_1, \ldots, S_n\}$ may be of different levels, with ν being the highest (corresponding to the smallest square(s)). By subdividing each S_k of level $\nu' < \nu$ into $4^{\nu - \nu'}$ dyadic squares of level ν, we obtain a finite collection $\{Q_1, \ldots, Q_m\}$ of almost disjoint dyadic squares of the same level ν for which the above argument produces a cycle η satisfying (i) and (ii) for $\{Q_1, \ldots, Q_m\}$. It is easy to check that this η also satisfies (i) and (ii) for the original collection $\{S_1, \ldots, S_n\}$. $\qquad \square$

LEMMA 10.46. *Suppose $K \subset \mathbb{C}$ is compact and $\Lambda_1(K) < +\infty$. For every $\varepsilon > 0$ there are almost disjoint dyadic squares $S_1, \ldots, S_n$ with side lengths $a_1, \ldots, a_n$ less than ε such that*

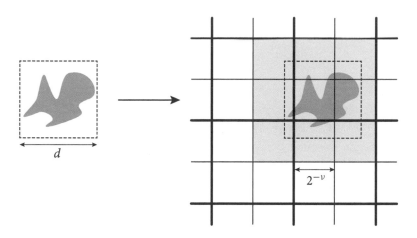

Figure 10.11. A set of diameter $0 < d < 1$ is contained in a closed square of side length d, which in turn is contained in the union of at most nine dyadic squares of level ν. Here $\nu \geq 1$ is the smallest integer for which $2^{-\nu} \leq d$.

(i) K is contained in the interior of $\bigcup_{k=1}^{n} S_k$;
(ii) $\sum_{k=1}^{n} a_k < 9\Lambda_1(K) + \varepsilon$.

PROOF. Fix $0 < \varepsilon < 1$. Choose an ε-cover $\{X_j\}$ of K such that $\sum \operatorname{diam}(X_j) < \Lambda_1^{\varepsilon}(K) + \varepsilon/10 \leq \Lambda_1(K) + \varepsilon/10$. Each X_j is contained in an open set U_j whose diameter d_j is as close as we wish to $\operatorname{diam}(X_j)$. Since K is compact, it follows that there is a finite ε-cover $\{U_1, \ldots, U_m\}$ of K such that $\sum_{j=1}^{m} d_j < \Lambda_1(K) + \varepsilon/9$. For each $1 \leq j \leq m$, let $\nu_j \geq 1$ be the smallest integer for which $2^{-\nu_j} \leq d_j$. The open set U_j is contained in a closed square of side length d_j, which in turn is contained in the union of at most nine dyadic squares of level ν_j (see Fig. 10.11). Let $S_1, \ldots, S_n$ be the collection of such dyadic squares that contain the U_j for $1 \leq j \leq m$. Discarding those squares that are contained in others, we may assume that $S_1, \ldots, S_n$ are almost disjoint. By the construction, each U_j lies in the interior of $\bigcup_{k=1}^{n} S_k$, so (i) holds. If $a_1, \ldots, a_n$ denote the side lengths of $S_1, \ldots, S_n$, then $a_k < \varepsilon$ and $\sum_{k=1}^{n} a_k \leq 9 \sum_{j=1}^{m} d_j < 9\Lambda_1(K) + \varepsilon$, so (ii) holds also.   □

We are now ready to prove our removability results. The first one is a generalization of Riemann's removable singularity theorem and concerns compact sets whose 1-dimensional Hausdorff measure is zero.

THEOREM 10.47 (Painlevé, 1888). *Suppose $K \subset \mathbb{C}$ is compact and $\Lambda_1(K) = 0$. If U is an open set containing K, and if f is bounded and holomorphic in $U \smallsetminus K$, then f extends to a holomorphic function in U.*

PROOF. By Lemma 9.21 there is a null-homologous cycle γ in U such that $|\gamma| \cap K = \emptyset$ and $\mathrm{W}(\gamma, p) = 1$ for every $p \in K$. Thus, the open set $V = \{p \in U \smallsetminus |\gamma| :$

Paul Painlevé
(1863–1933)

$W(\gamma, p) = 1\}$ contains K. Fix $z \in V \setminus K$ and take any ε such that

(10.6) $$0 < \varepsilon < \frac{1}{2} \min \{ \operatorname{dist}(z, K), \operatorname{dist}(|\gamma|, K) \}.$$

By Lemma 10.46, there are almost disjoint dyadic squares $S_1, \ldots, S_n$ with side lengths $a_1, \ldots, a_n$ such that K lies in the interior of $\bigcup_{k=1}^n S_k$ and $\sum_{k=1}^n a_k < \varepsilon$. We may assume that $K \cap S_k \neq \emptyset$ for all k. By the choice of ε in (10.6), the union $\bigcup_{k=1}^n S_k$ is contained in $V \setminus \{z\}$. Let η be the cycle corresponding to $\{S_1, \ldots, S_n\}$ given by Lemma 10.45. It is readily seen that $\gamma - \eta$ is a null-homologous cycle in $U \setminus K$ and $W(\gamma - \eta, z) = 1$. Hence, by Cauchy's integral formula,

$$f(z) = \frac{1}{2\pi i} \int_\gamma \frac{f(\zeta)}{\zeta - z} \, d\zeta - \frac{1}{2\pi i} \int_\eta \frac{f(\zeta)}{\zeta - z} \, d\zeta.$$

If $\zeta \in |\eta|$, then $\zeta \in \partial S_k$ for some k. If p is any point in $K \cap S_k$, it follows that

(10.7)
$$|\zeta - z| \geq |z - p| - |p - \zeta| \geq \operatorname{dist}(z, K) - \operatorname{diam}(S_k)$$
$$> \operatorname{dist}(z, K) - \sqrt{2}\varepsilon > \left(1 - \frac{1}{\sqrt{2}}\right) \operatorname{dist}(z, K) > \frac{1}{4} \operatorname{dist}(z, K)$$

by the choice of ε in (10.6). Thus, the absolute value of the term $1/(\zeta - z)$ in the second integral is uniformly bounded above by $4/\operatorname{dist}(z, K)$. Since f is bounded and $\operatorname{length}(\eta) \leq \sum_{k=1}^n \operatorname{length}(\partial S_k) = 4 \sum_{k=1}^n a_k < 4\varepsilon$, the ML-inequality shows that the second integral tends to zero as $\varepsilon \to 0$. This proves the formula

$$f(z) = \frac{1}{2\pi i} \int_\gamma \frac{f(\zeta)}{\zeta - z} \, d\zeta \qquad \text{for every } z \in V \setminus K.$$

An outstanding mathematician, Painlevé still had time for extracurricular activities like serving as the prime minister of France. Twice.

By Corollary 1.48, the right-hand side defines a holomorphic function of z off $|\gamma|$. It follows that f extends holomorphically to V, hence to U. $\qquad \square$

THEOREM 10.48. *Suppose $K \subset \mathbb{C}$ is compact and $\Lambda_1(K) < +\infty$. If U is an open set containing K, and if f is continuous in U and holomorphic in $U \setminus K$, then f is holomorphic in U.*

As a typical example, it follows from Theorem 10.44 that a continuous function which is holomorphic off a rectifiable arc is holomorphic everywhere.

PROOF. The bulk of the argument is similar to Theorem 10.47, but there is a new twist at the end. Use Lemma 9.21 to find a null-homologous cycle γ in U such that $|\gamma| \cap K = \emptyset$ and $W(\gamma, p) = 1$ for every $p \in K$. As before, consider the open set $V = \{p \in U \setminus |\gamma| : W(\gamma, p) = 1\}$ containing K, take $z \in V \setminus K$, and let

$$0 < \varepsilon < \frac{1}{2} \min \{ \operatorname{dist}(z, K), \operatorname{dist}(|\gamma|, K) \}.$$

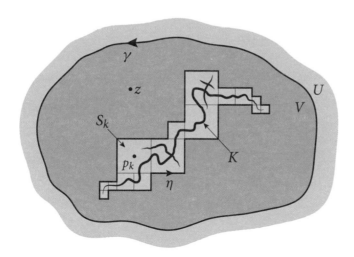

Figure 10.12. Proof of Theorem 10.48. The cycle η is the boundary of the union of the dyadic squares $\{S_k\}$ covering K. The S_k are individually small, but the sum of their side lengths is comparable to $\Lambda_1(K)$.

The closed ε-neighborhood $K_\varepsilon = \{\zeta \in \mathbb{C} : \mathrm{dist}(\zeta, K) \leq \varepsilon\}$ is then contained in $V \smallsetminus \{z\}$. By uniform continuity of f on K_ε, there is a $0 < \delta < \min\{1, \varepsilon/2\}$ such that $|f(p) - f(q)| < \varepsilon$ whenever $p, q \in K_\varepsilon$ and $|p - q| < \delta$. By Lemma 10.46, there are almost disjoint dyadic squares $S_1, \ldots, S_n$ with side lengths $a_1, \ldots, a_n < \delta$ such that K lies in the interior of $\bigcup_{k=1}^n S_k$ and $\sum_{k=1}^n a_k < 9\Lambda_1(K) + 1$. We may also assume $K \cap S_k \neq \emptyset$ for all k. Each S_k will then be contained in K_ε by the choice of δ.

Let η be the cycle corresponding to $\{S_1, \ldots, S_n\}$ given by Lemma 10.45 (see Fig. 10.12). Then $\gamma - \eta$ is a null-homologous cycle in $U \smallsetminus K$ with $\mathrm{W}(\gamma - \eta, z) = 1$, so by Cauchy's integral formula,

$$(10.8) \qquad f(z) = \frac{1}{2\pi i} \int_\gamma \frac{f(\zeta)}{\zeta - z} \, d\zeta - \frac{1}{2\pi i} \int_\eta \frac{f(\zeta)}{\zeta - z} \, d\zeta.$$

To show that the second integral tends to zero as $\varepsilon \to 0$ requires a new trick since unlike the situation in Theorem 10.47 η is no longer short. Let p_k be the center of the square S_k. For each k, we have $\int_{\partial S_k} d\zeta/(\zeta - z) = \mathrm{W}(\partial S_k, z) = 0$. Moreover, if $\zeta \in \partial S_k$, then $|f(\zeta) - f(p_k)| < \varepsilon$ since $\zeta, p_k \in S_k \subset K_\varepsilon$ and $|\zeta - p_k| < a_k < \delta$. Finally, if $\zeta \in \partial S_k$, an estimate similar to (10.7) shows that $1/|\zeta - z| \leq 4/\mathrm{dist}(z, K)$. Putting these facts together, we obtain

$$\left| \int_{\partial S_k} \frac{f(\zeta)}{\zeta - z} \, d\zeta \right| = \left| \int_{\partial S_k} \frac{f(\zeta) - f(p_k)}{\zeta - z} \, d\zeta \right|$$

$$\leq \frac{4\varepsilon}{\mathrm{dist}(z, K)} \, \mathrm{length}(\partial S_k) = \frac{16\varepsilon}{\mathrm{dist}(z, K)} \, a_k.$$

Thus, by Lemma 10.45,

$$\left| \int_\eta \frac{f(\zeta)}{\zeta - z} \, d\zeta \right| = \left| \sum_{k=1}^n \int_{\partial S_k} \frac{f(\zeta)}{\zeta - z} \, d\zeta \right|$$

$$\leq \frac{16\varepsilon}{\mathrm{dist}(z, K)} \sum_{k=1}^n a_k < \frac{16(9\Lambda_1(K) + 1)}{\mathrm{dist}(z, K)} \, \varepsilon.$$

This shows that the second integral in (10.8) tends to zero as $\varepsilon \to 0$, and as before we conclude that the first integral in (10.8), as a function of z, provides a holomorphic extension of f to V. $\qquad\square$

REMARK 10.49. The assumption $\Lambda_1(K) < +\infty$ in Theorem 10.48 is essential. For example, Denjoy has observed that if K is a Jordan curve of positive area, the Cauchy transform $z \mapsto \iint_K dx \, dy / (\zeta - z)$, where $\zeta = x + iy$, extends continuously to $\mathbb{C}$, is holomorphic off K, and tends to 0 as $z \to \infty$ without being identically zero in $\mathbb{C} \setminus K$.

Problems

(1) Is every closed subset of $\mathbb{T}$ the singular set of some holomorphic function in $\mathbb{D}$?

(2) Let $f: \mathbb{C} \to \mathbb{C}$ be an entire function which sends real numbers to real numbers and imaginary numbers to imaginary numbers. Prove that f is an odd function:

$$f(-z) = -f(z) \qquad \text{for all } z \in \mathbb{C}.$$

(3) Let f be a holomorphic function defined in a neighborhood of 0 which satisfies

$$f(2z) = (f(z))^2$$

for all z sufficiently close to 0. Use this functional equation to show that f can be extended to an entire function. Can you determine all such entire functions explicitly? (Hint: Study the cases $f(0) = 0$ and $f(0) = 1$ separately.)

(4) Suppose $f \in \mathscr{O}(\mathbb{D})$ and the sequence $\{f^{(n)}(0)\}$ grows at most exponentially fast, i.e., there is a constant $\lambda > 1$ such that $|f^{(n)}(0)| \leq \lambda^n$ for all $n \geq 0$. Show that f extends to an entire function.

(5) Suppose $f(z) = \sum_{n=0}^\infty a_n z^n$ has radius of convergence 1, and define

$$g(w) = \frac{1}{1-w} f\left(\frac{w}{1-w} \right).$$

   (i) Show that g is holomorphic in the half-plane $\{w : \mathrm{Re}(w) < 1/2\}$.
   (ii) If $g(w) = \sum_{k=0}^\infty b_k w^k$, show that

$$b_k = \sum_{n=0}^k \binom{k}{n} a_n,$$

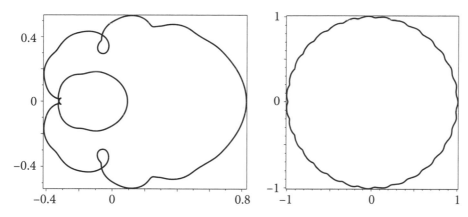

Figure 10.13. The image of the unit circle under the extension of the function $f \in \mathscr{O}(\mathbb{D})$ of Example 10.10 (left) and problem 8 (right). Both curves are C^∞-smooth but nowhere analytic.

where, as usual, $\binom{k}{n} = \dfrac{k!}{n!(k-n)!}$.

    (iii)   Show that $z = 1$ is a singular point of f if and only if $w = 1/2$ is a singular point of g if and only if $\limsup_{k \to \infty} \sqrt[k]{|b_k|} = 2$.

(6)   Let U be an open set containing the closed unit disk $\overline{\mathbb{D}}$ and $|p| = 1$. Suppose $f \in \mathscr{O}(U \smallsetminus \{p\})$ has a simple pole at p. If $f(z) = \sum_{n=0}^\infty a_n z^n$ in $\mathbb{D}$, show that

$$\lim_{n \to \infty} (a_n - p a_{n+1}) = 0.$$

(7)   (Pringsheim) Suppose $f(z) = \sum_{n=0}^\infty a_n z^n$ has radius of convergence 1, and $a_n \geq 0$ for all n. Show that $z = 1$ is a singular point of f. (Hint: Otherwise f could be holomorphically extended to a small disk around $z = 1$. Then the power series of f about $1/2$ would converge in a disk $\mathbb{D}(1/2, 1/2 + \varepsilon)$ for small $\varepsilon > 0$. Since $a_n \geq 0$, for every real t the power series of f about $e^{it}/2$ would converge in $\mathbb{D}(e^{it}/2, 1/2 + \varepsilon)$, which would lead to a contradiction.)

(8)   Use Hadamard's gap theorem to verify that the power series

$$f(z) = z + \sum_{n=5}^\infty \frac{1}{n!} z^{2^n}$$

has $\mathbb{T}$ as its natural boundary [**Z1**]. Show however that f extends homeomorphically to $\overline{\mathbb{D}}$ and $f|_{\mathbb{T}}$ is C^∞-smooth (see Fig. 10.13). Thus, by Theorem 10.37, no subarc of the Jordan curve $f(\mathbb{T})$ can be analytic. (Hint: Use problem 21 in chapter 6).

(9)   Suppose the chains $\{(f_k, D_k)\}_{0 \leq k \leq n}$ and $\{(g_j, B_j)\}_{0 \leq j \leq m}$ realize the analytic continuations of (f_0, D_0) and (g_0, B_0) along the same curve. Suppose $P(f_0, g_0) = 0$ everywhere in $D_0 \cap B_0$, where $P = P(z, w)$ is a polynomial in two complex variables. Show that $P(f_n, g_m) = 0$ everywhere in $D_n \cap B_m$. *Thus, the functional relation $P(\cdot, \cdot) = 0$ persists under analytic continuation.*

(10)   Use the monodromy theorem to give another proof for Theorem 7.9: If $U \subset \mathbb{C}$ is a simply connected domain and $u : U \to \mathbb{R}$ is harmonic, then there exists an $f \in \mathscr{O}(U)$ such that $\mathrm{Re}(f) = u$.

(11)   Are sums, products, ratios, or compositions of real analytic functions real analytic?

(12)   What can be said about a bounded function in $\mathcal{O}(\mathbb{H})$ that takes real values on the horizontal line $\{z : \operatorname{Im}(z) = 1\}$?

(13)   Let Ω be the open triangle with vertices $0, 1, \tau \in \mathbb{H}$ and internal angles $\pi\theta_1, \pi\theta_2, \pi\theta_3$ respectively, so $\theta_1 + \theta_2 + \theta_3 = 1$. Let $f : \mathbb{H} \to \Omega$ be the unique conformal map normalized so that $f(0) = 0, f(1) = 1, f(\infty) = \tau$. Show that

$$f(z) = C \int_0^z \zeta^{\theta_1 - 1}(\zeta - 1)^{\theta_2 - 1}\, d\zeta,$$

where C is uniquely determined by the condition $f(1) = 1$. Why is the parameter τ not showing up in this formula?

(14)   Consider the half-strip $\Omega = \{z : \operatorname{Re}(z) > 0 \text{ and } 0 < \operatorname{Im}(z) < \pi\}$.
   (i)   By means of elementary maps (exponential, Möbius, squaring), find the formula for the unique conformal map $g : \Omega \to \mathbb{H}$ normalized by the conditions $g(0) = 1$, $g(i\pi) = -1, g(\infty) = \infty$.
   (ii) Think of Ω as a "triangle" with vertices $0, i\pi, \infty$ and internal angles $\pi/2, \pi/2, 0$. Use a limiting case of the Schwarz-Christoffel formula (10.5) to show that

$$g^{-1}(z) = \int_1^z \frac{d\zeta}{\sqrt{\zeta^2 - 1}} = \log(\sqrt{z^2 - 1} + z)$$

   for suitable holomorphic branches of the square root and logarithm. Compare your answer to the one found in part (i).

(15)   Elliptic functions with respect to rectangular lattices (see §9.2) can be constructed by combining the Riemann mapping theorem and the Schwarz reflection principle. Fix $h > 0$ and let g be the unique conformal map from the rectangle $\Omega = \{z : 0 < \operatorname{Re}(z) < 1/2 \text{ and } 0 < \operatorname{Im}(z) < h/2\}$ onto $\mathbb{H}$ which satisfies $g(0) = \infty, g(1/2) = -1, g(ih/2) = 1$.
   (i)   By repeated reflections across the sides of Ω in the domain and across the real line in the range (similar to Example 10.29), show that g extends to an elliptic function with respect to the lattice $\Lambda = \langle 1, ih \rangle$ with poles of order 2 along Λ. Conclude that $g = a\wp_\Lambda + b$ for some real constants a, b. What is the relation between b and $v = g((1 + ih)/2) \in (-1, 1)$?
   (ii) Use the Schwarz-Christoffel formula (10.5) to show that g^{-1} is given by the elliptic integral

$$g^{-1}(z) = \frac{1}{2} + C \int_{-1}^z \frac{d\zeta}{\sqrt{(\zeta^2 - 1)(\zeta - v)}}$$

   for a suitable constant C.
   (iii) How do these computations simplify in the special case $h = 1$ where Ω is a square? (Hint: For (i), problems 9 and 11 in chapter 9 are relevant, and Corollary 9.15 can be used to show $v = 3b$. For (iii), check that $b = v = 0$ and use (9.16) to express a and C in terms of the invariant g_2 of the lattice of Gaussian integers.)

(16)   Show that $\zeta : \mathbb{C} \to \mathbb{C}$ is an anti-holomorphic involution if and only if

$$\zeta(z) = -\alpha^2 \bar{z} + r\alpha$$

with $|\alpha| = 1$ and $r \in \mathbb{R}$. Similarly, show that $\zeta : \mathbb{D} \to \mathbb{D}$ is an anti-holomorphic involution if and only if

$$\zeta(z) = \frac{r - \bar{\alpha}\bar{z}}{\alpha - r\bar{z}}$$

with $|\alpha| = 1$ and $0 \leq r < 1$. Determine the fixed point set $\{z : \zeta(z) = z\}$ in each case.

(17) Let $U \subset \mathbb{C}$ be a simply connected domain and $\zeta : U \to U$ be an anti-holomorphic involution. Show that the fixed point set $J = \{z \in U : \zeta(z) = z\}$ is an analytic arc.

(18) Suppose $J, \hat{J}$ are analytic arcs. Formulate a suitable definition for real analyticity of a map $f : J \to \hat{J}$. Prove that such f extends to a holomorphic function F between open neighborhoods of J and $\hat{J}$, and that this extension satisfies $F \circ \zeta = \hat{\zeta} \circ F$, where $\zeta, \hat{\zeta}$ are the canonical reflections across $J, \hat{J}$.

(19) Let us say that a Jordan curve in the plane is analytic if every open subarc of it is an analytic arc in the sense of Definition 10.31. Formulate and prove an analog of Lemma 10.33 for analytic Jordan curves, show the existence of anti-holomorphic involutions which fix them, and use it to prove a version of Theorem 10.37 for them.

(20) Verify that circles and ellipses in the plane are analytic Jordan curves. Then find explicit formulas for their canonical reflections. (Hint: For the case of an ellipse, use the Zhukovskii map in Example 6.6.)

(21) (Modulus of round annuli revisited) According to problem 24 of chapter 6, the modulus $\mathrm{mod}(A) = 1/(2\pi) \log(b/a)$ of the annulus $A = \{z : a < |z| < b\}$ is a conformal invariant. In other words, if $f : A \to A' = \{z : a' < |z| < b'\}$ is conformal, then $b/a = b'/a'$. Give another proof of this fact by imitating the argument in Example 10.29 for the case of a rectangle, this time using repeated reflections across the boundary circles.

(22) If $E \subset \mathbb{C}$ is a Borel set with $\mathrm{area}(E) > 0$, show that $\dim_{\mathrm{H}}(E) = 2$. Given an example where $\dim_{\mathrm{H}}(E) = 2$ but $\mathrm{area}(E) = 0$.

(23) Show that for any countable collection $\{E_n\}_{n \geq 1}$ of Borel sets in $\mathbb{C}$,

$$\dim_{\mathrm{H}} \left(\bigcup_{n \geq 1} E_n \right) = \sup_{n \geq 1} \dim_{\mathrm{H}}(E_n).$$

(24) Suppose $f : E \to \mathbb{C}$ is a **Lipschitz function** in the sense that there is an $M > 0$ such that

$$|f(p) - f(q)| \leq M|p - q| \qquad \text{for every } p, q \in E.$$

Prove that for every $\alpha > 0$,

$$\Lambda_\alpha(f(E)) \leq M^\alpha \Lambda_\alpha(E).$$

(25) Show that for a compact set $K \subset \mathbb{C}$, the following conditions are equivalent:
  (i)   For some open set $U_0 \supset K$, every bounded holomorphic function in $U_0 \smallsetminus K$ extends holomorphically to U_0.
  (ii) For every open set $U \supset K$, every bounded holomorphic function in $U \smallsetminus K$ extends holomorphically to U.
  (iii) Every bounded holomorphic function in $\mathbb{C} \smallsetminus K$ is constant.
If these conditions are met, we say that K is **holomorphically removable for bounded functions**. (Hint: Clearly (ii) $\Longrightarrow$ (iii) $\Longrightarrow$ (i). To prove (i) $\Longrightarrow$ (ii), revisit the proof of Theorem 10.47, consider the cycles γ and η, the open set V containing K, and the point $z \in V \smallsetminus K$ as in that proof, and write $f(z)$ as the sum $f_1(z) + f_2(z)$ of the two integrals in (10.8). Verify that f_1 is holomorphic in V and (by moving η arbitrarily close to K) f_2 extends to a holomorphic function in $\mathbb{C} \smallsetminus K$. Conclude that f_2 is bounded in $\mathbb{C} \smallsetminus K$, hence in $U_0 \smallsetminus K$, and therefore it extends holomorphically to U_0 by the assumption.)

(26)   Show that if K is holomorphically removable for bounded functions in the sense of the previous problem, then K is totally disconnected. (Hint: Suppose K has a connected component E which is not a point. Then each component of $\hat{\mathbb{C}} \smallsetminus E$ can be mapped conformally onto $\mathbb{D}$ by the Riemann mapping theorem, violating condition (iii) of the previous problem.)

(27)   Suppose $K \subset \mathbb{R}$ is compact. Show that for K to be holomorphically removable for bounded functions it is necessary and sufficient that $\text{length}(K) = 0$. (Hint: Sufficiency follows from Theorem 10.47. For necessity, suppose $\text{length}(K) > 0$ and verify that the Cauchy transform $z \mapsto \int_K dx/(x - z)$ is a non-constant holomorphic function in $\mathbb{C} \smallsetminus K$ that takes values in the horizontal strip $\{w \in \mathbb{C} : |\,\text{Im}(w)| < \pi\}$.)

CHAPTER

11

Ranges of holomorphic functions

This chapter presents a host of interconnected results on the ranges of holomorphic functions, most notably the celebrated theorems of Bloch, Schottky, and Picard, as well as a general version of Montel's theorem. We provide three different approaches to these central results of complex function theory to illustrate the power and diversity of the basic tools of the subject. A fourth approach, based on the idea of uniformization, will be discussed in chapter 13.

11.1 Bloch's theorem

> **DEFINITION 11.1.** Let f be a non-constant holomorphic function in a domain $U \subset \mathbb{C}$. An open disk $D \subset f(U)$ is called an ***unramified disk*** for f if there is an open set $V \subset U$ such that f maps V conformally to D.

In other words, an unramified disk is one in which a holomorphic branch of the inverse function is defined. As an example, Koebe's 1/4-theorem (Theorem 6.14) asserts that $\mathbb{D}(0, 1/4)$ is an unramified disk for every f in the class $\mathscr{S}$ of schlicht functions.

Surprisingly, normalized holomorphic functions in $\mathbb{D}$, injective or not, have unramified disks of universal size. For $f \in \mathscr{O}(\mathbb{D})$ with $f'(0) = 1$, define

$$B_f = \sup \{r > 0 : f \text{ has an unramified disk of radius } r\}.$$

We have $B_f > 0$ since f certainly has an unramified disk centered at $f(0)$. The number

$$\mathfrak{B} = \inf_{f} B_f$$

is known as ***Bloch's constant***.

THEOREM 11.2 (Bloch, 1924). $\mathfrak{B} > 0$.

Thus, every $f \in \mathscr{O}(\mathbb{D})$ with $f'(0) = 1$ has an unramified disk of radius R for every $0 < R < \mathfrak{B}$.

Unlike Koebe's $1/4$-theorem, the unramified disks guaranteed by the above theorem are generally *not* centered at $f(0)$. For example, for every $\varepsilon > 0$ the function $f(z) = \varepsilon e^{z/\varepsilon}$ is holomorphic in $\mathbb{D}$ and satisfies $f'(0) = 1$. Since $0 \notin f(\mathbb{D})$, the largest disk in $f(\mathbb{D})$ centered at $f(0) = \varepsilon$ has radius $\leq \varepsilon$, which can be arbitrarily small.

The question of finding the exact value of $\mathfrak{B}$ has a long history and is still unsolved. Our elementary approach in this section will show that $1/4 \leq \mathfrak{B} \leq \pi/4$ (see the proof of Theorem 11.2 as well as Remark 11.5). With a more substantial effort, the sharper bounds

$$\frac{\sqrt{3}}{4} = 0.433\ldots < \mathfrak{B} < 0.472$$

can be proved. The **Ahlfors-Grunsky conjecture**, open since 1936, predicts the value

$$\mathfrak{B} = \sqrt{\frac{\sqrt{3}-1}{2}} \, \frac{\Gamma(1/3)\Gamma(11/12)}{\Gamma(1/4)} \approx 0.4719\ldots,$$

where Γ is the classical gamma function defined for $x > 0$ by $\Gamma(x) = \int_0^\infty t^{x-1}e^{-t}\,dt$. See [**A4**].

PROOF OF THEOREM 11.2. We prove $\mathfrak{B} \geq 1/4$ by showing that $B_f \geq 1/4$ for every $f \in \mathscr{O}(\mathbb{D})$ with $f'(0) = 1$. First assume f is holomorphic in a neighborhood of the closed unit disk $\overline{\mathbb{D}}$. The function $z \mapsto (1 - |z|^2)|f'(z)|$ is continuous on $\overline{\mathbb{D}}$, vanishes identically on $\mathbb{T}$, and takes the value 1 at the origin. Hence,

In the language of §4.4, the quantity β_f is twice the derivative norm of the map $f : (\mathbb{D}, g_{\mathbb{D}}) \to (\mathbb{C}, g_0)$. It was first introduced by Landau in 1929.

$$(11.1) \qquad \beta_f = \sup_{z \in \mathbb{D}} (1 - |z|^2)|f'(z)|$$

is finite and satisfies $\beta_f \geq 1$.

The quantity β_f is invariant under precomposition by disk automorphisms: If $\varphi \in \mathrm{Aut}(\mathbb{D})$ and $g = f \circ \varphi$, then by Pick's Theorem 4.40,

$$(1 - |z|^2)|g'(z)| = (1 - |z|^2)|\varphi'(z)|\,|f'(\varphi(z))|$$

$$(11.2) \qquad\qquad\qquad\quad = (1 - |\varphi(z)|^2)\,|f'(\varphi(z))|$$

for every $z \in \mathbb{D}$, which shows $\beta_g \leq \beta_f$. Applying the same argument to $f = g \circ \varphi^{-1}$, we obtain the reverse inequality $\beta_g \geq \beta_f$.

Now let $p \in \mathbb{D}$ be a point where the supremum in (11.1) is achieved. Consider the disk automorphism $\varphi(z) = (z + p)/(1 + \bar{p}z)$ which sends 0 to p, and set $g = f \circ \varphi$. By (11.2),

$$|g'(0)| = (1 - |p|^2)|f'(p)| = \beta_f = \beta_g = \sup_{z \in \mathbb{D}} (1 - |z|^2)|g'(z)|,$$

which yields the estimate

$$|g'(z)| \leq \frac{|g'(0)|}{1 - |z|^2} \qquad \text{for all } z \in \mathbb{D}.$$

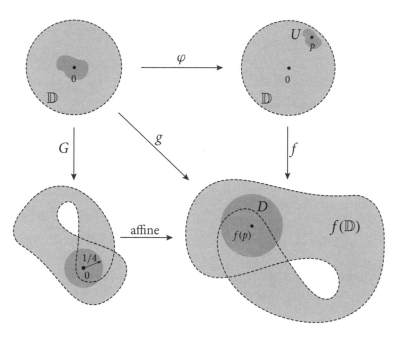

Figure 11.1. Illustration of the proof of Theorem 11.2. Here f maps the neighborhood U of p conformally to the unramified disk D of radius $\geq 1/4$.

Define $G \in \mathcal{O}(\mathbb{D})$ by

$$G(z) = \frac{g(z) - g(0)}{g'(0)},$$

which is normalized so that $G(0) = 0$ and $G'(0) = 1$, and satisfies

(11.3) $$|G'(z)| \leq \frac{1}{1 - |z|^2} \qquad \text{for all } z \in \mathbb{D}.$$

We will show that $\mathbb{D}(0, 1/4)$ is an unramified disk for G (compare Fig. 11.1). It will follow that $\mathbb{D}(g(0), |g'(0)|/4)$ is an unramified disk for g. Since f and g differ by a disk automorphism and since $|g'(0)| = \beta_f \geq 1$, this will prove that $\mathbb{D}(f(p), 1/4)$ is an unramified disk for f.

We begin with the observation that $G''(0) = 0$ [**Re**]. In fact, if $G(z) = z + \sum_{n=2}^{\infty} a_n z^n$, then (11.3) becomes

$$|1 + 2a_2 z + 3a_3 z^2 + \cdots| \leq 1 + |z|^2 + |z|^4 + \cdots.$$

Writing $z = r\lambda$, where $0 < r < 1$ and $|\lambda| = 1$, this implies

$$\mathrm{Re}\left(1 + 2a_2 r\lambda + 3a_3 r^2 \lambda^2 + \cdots\right) \leq 1 + r^2 + r^4 + \cdots,$$

or

$$2\,\mathrm{Re}(a_2 \lambda) + 3r\,\mathrm{Re}(a_3 \lambda^2) + \cdots \leq r + r^3 + \cdots.$$

Letting $r \to 0$ then gives $\mathrm{Rc}(a_2 \lambda) \leq 0$. Since this inequality holds for every λ on the unit circle, we must have $a_2 = 0$.

Now (11.3) and Cauchy's estimates (Theorem 1.42) applied to $G'(z) = 1 + \sum_{n=3}^{\infty} n a_n z^{n-1}$ on the disk $\mathbb{D}(0, 7/10)$ show that

$$|n a_n| = \frac{|G^{(n)}(0)|}{(n-1)!} \le \frac{\sup_{|z|=7/10} |G'(z)|}{(7/10)^{n-1}} \le \frac{1}{1-(7/10)^2} \left(\frac{10}{7}\right)^{n-1} < \frac{7}{5} \left(\frac{10}{7}\right)^n$$

for all $n \ge 3$. Hence, if $|z| = 7/20$,

$$|G(z) - z| = \left| \sum_{n=3}^{\infty} a_n z^n \right| \le \sum_{n=3}^{\infty} |a_n| \, |z|^n < \frac{7}{5} \sum_{n=3}^{\infty} \frac{2^{-n}}{n} < \frac{1}{10}.$$

It follows that if $|z| = 7/20$ and $|q| < 1/4$, then

$$|(G(z) - q) - (z - q)| = |G(z) - z| < \frac{1}{10} < |z - q|.$$

By Rouché's theorem, $G - q$ has a unique zero in the disk $\mathbb{D}(0, 7/20)$. In other words, for every $q \in \mathbb{D}(0, 1/4)$ there exists a unique $z \in \mathbb{D}(0, 7/20)$ such that $G(z) = q$. If V denotes the open set $G^{-1}(\mathbb{D}(0, 1/4)) \cap \mathbb{D}(0, 7/20)$, it follows that G maps V bijectively onto $\mathbb{D}(0, 1/4)$. This proves that $\mathbb{D}(0, 1/4)$ is an unramified disk for G, as required.

For a general $f \in \mathcal{O}(\mathbb{D})$ with $f'(0) = 1$, consider the function $z \mapsto f(rz)/r$ for $0 < r < 1$ which extends holomorphically to $\mathbb{D}(0, 1/r) \supset \overline{\mathbb{D}}$ and has derivative 1 at the origin. By the special case treated above, this function has an unramified disk of radius $1/4$. It follows that f has an unramified disk of radius $r/4$. Since this holds for every $0 < r < 1$, we obtain $B_f \ge 1/4$. $\qquad\square$

COROLLARY 11.3. *Let f be holomorphic in the disk $\mathbb{D}(p, r)$ and $f'(p) \neq 0$. Then f has an unramified disk of radius $Rr|f'(p)|$ for every $0 < R < \mathfrak{B}$.*

PROOF. The function $g(z) = f(rz + p)/(rf'(p))$ is holomorphic in $\mathbb{D}$ and $g'(0) = 1$, so it has an unramified disk of radius R by Theorem 11.2. $\qquad\square$

The following is an immediate corollary:

This corollary is sometimes attributed to G. Valiron (1884–1955).

COROLLARY 11.4. *A non-constant entire function has arbitrarily large unramified disks.*

Observe that this contains Liouville's theorem as a special case.

REMARK 11.5. While finding lower bounds for $\mathfrak{B}$ requires non-trivial work, it is very easy to find upper bounds. For example, take the Möbius map $\varphi(z) = (1 + z)/(1 - z)$ which sends $\mathbb{D}$ conformally to $\{z : \text{Re}(z) > 0\}$, and let g be the unique branch of the logarithm in this half-plane which satisfies $g(1) = 0$. The composition $f = (1/2)(g \circ \varphi)$ then maps $\mathbb{D}$ conformally to the strip $\{z : |\text{Im}(z)| < \pi/4\}$, with $f(0) = 0$ and $f'(0) = (1/2)g'(1) \, \varphi'(0) = 1$. Clearly every unramified disk for f has radius $\le \pi/4$. It follows that $\mathfrak{B} \le \pi/4 \approx 0.785 \cdots$.

11.2 Picard's theorems

This section will give elementary proofs of Picard's little and great theorems, which should be viewed as far-reaching generalizations of the theorems of Liouville (Theorem 1.44) and Casorati-Weierstrass (Theorem 3.7(iii)), respectively. The proof of Picard's great theorem is based on a theorem of Schottky which is also used in a general version of Montel's theorem on normal families. Our presentation is indebted to the beautiful book by Remmert [**Re**].

It will be convenient to say that $f \in \mathcal{O}(U)$ **omits** the value q if $q \notin f(U)$.

Reinhold Remmert
(1930–2016)

THEOREM 11.6 (Picard's little theorem, 1879). *An entire function which omits two distinct values must be constant.*

Here is the main idea behind the proof: Suppose $f \in \mathcal{O}(\mathbb{C})$ omits, say, the values 0 and 1. The trick is to write f as a composition $\Phi \circ g$ of entire functions, where g now omits all values in $E = \Phi^{-1}(\{0, 1\})$. If we can choose Φ such that the points in E occur more or less regularly like a rectangular lattice, then $\mathbb{C} \smallsetminus E$ cannot contain large disks and it follows from Corollary 11.4 that g, hence f, must be constant.

The task of finding a suitable Φ is accomplished in Lemma 11.8 below; it turns out that a double-cosine function does the job.

LEMMA 11.7 (Existence of holomorphic arccosines). *Suppose $U \subset \mathbb{C}$ is a simply connected domain and $f \in \mathcal{O}(U)$ omits the values ± 1. Then there exists a $g \in \mathcal{O}(U)$ such that $f = \cos(g)$.*

PROOF. By Corollary 2.22 the non-vanishing function $1 - f^2 \in \mathcal{O}(U)$ has a holomorphic square root, so $f^2 + h^2 = 1$ for some $h \in \mathcal{O}(U)$. The function $f + ih \in \mathcal{O}(U)$ is non-vanishing since $(f + ih)(f - ih) = f^2 + h^2 = 1$. Hence, by another application of Corollary 2.22, $f + ih = \exp(ig)$ for some $g \in \mathcal{O}(U)$. It follows that $f - ih = 1/(f + ih) = \exp(-ig)$, so $f = (\exp(ig) + \exp(-ig))/2 = \cos(g)$. □

LEMMA 11.8. *Suppose $U \subset \mathbb{C}$ is a simply connected domain and $f \in \mathcal{O}(U)$ omits the values 0 and 1. Then there exists a $g \in \mathcal{O}(U)$ such that*

$$(11.4) \qquad f = \frac{1}{2}\big(1 + \cos(\pi \cos(\pi g))\big).$$

Moreover, the image $g(U)$ does not contain any disk of radius 1.

PROOF. Since $2f - 1$ omits ± 1, Lemma 11.7 shows that $f = (1 + \cos(\pi h))/2$ for some $h \in \mathcal{O}(U)$. As h omits all integer values, another application of Lemma 11.7 shows that $h = \cos(\pi g)$ for some $g \in \mathcal{O}(U)$, so (11.4) holds.

To verify the claim on the image of g, let $a_n = \log(n + \sqrt{n^2 - 1})$ for every integer $n \geq 1$, and consider the "grid"

$$E = \left\{ m \pm \frac{i}{\pi} a_n : m, n \in \mathbb{Z} \text{ and } n \geq 1 \right\}.$$

The grid E doesn't come out of thin air! It is essentially $\Phi^{-1}(\{0, 1\})$, where $\Phi(z) = \frac{1}{2}(1 + \cos(\pi \cos(\pi z)))$.

Let us check that $g(U) \cap E = \varnothing$. If $g(z) = m \pm (i/\pi)a_n$ for some $z \in U$, then

$$e^{\pi i g(z)} = (-1)^m e^{\mp a_n} = (-1)^m (n + \sqrt{n^2 - 1})^{\mp 1}$$

$$e^{-\pi i g(z)} = (-1)^m e^{\pm a_n} = (-1)^m (n + \sqrt{n^2 - 1})^{\pm 1},$$

so

$$e^{\pi i g(z)} + e^{-\pi i g(z)} = (-1)^m \left(n + \sqrt{n^2 - 1} + \frac{1}{n + \sqrt{n^2 - 1}} \right) = (-1)^m 2n.$$

This would imply

$$h(z) = \cos(\pi g(z)) = \frac{1}{2}(e^{\pi i g(z)} + e^{-\pi i g(z)}) = (-1)^m n,$$

which would be a contradiction since h omits all integer values.

The horizontal separation of the grid E is clearly 1. The vertical separation is uniformly less than 1 since

$$a_{n+1} - a_n = \log \left(\frac{(n+1) + \sqrt{(n+1)^2 - 1}}{n + \sqrt{n^2 - 1}} \right) = \log \left(\frac{1 + \frac{1}{n} + \sqrt{1 + \frac{2}{n}}}{1 + \sqrt{1 - \frac{1}{n^2}}} \right)$$

$$\leq \log \left(1 + \frac{1}{n} + \sqrt{1 + \frac{2}{n}} \right) \leq \log(2 + \sqrt{3}) < \pi.$$

This shows that for every $w \in \mathbb{C}$ there is a $q \in E$ such that $|\operatorname{Re}(w) - \operatorname{Re}(q)| \leq 1/2$ and $|\operatorname{Im}(w) - \operatorname{Im}(q)| \leq 1/2$. This implies $|w - q| < 1$ and shows that the disk $\mathbb{D}(w, 1)$ intersects E. Since $g(U) \cap E = \varnothing$, we conclude that $\mathbb{D}(w, 1)$ is not contained in $g(U)$. ☐

We are now ready to prove Picard's little theorem.

PROOF OF THEOREM 11.6. Suppose $f \in \mathcal{O}(\mathbb{C})$ omits the distinct values $q_1, q_2 \in \mathbb{C}$. After replacing f by $(f - q_1)/(q_2 - q_1)$, we may assume $q_1 = 0, q_2 = 1$. By Lemma 11.8, $f = (1 + \cos(\pi \cos(\pi g)))/2$ for some $g \in \mathcal{O}(\mathbb{C})$, and the image of g does not contain any disk of radius 1. By Corollary 11.4, g is constant. Hence f must be constant as well. ☐

EXAMPLE 11.9. It is not hard to verify directly that the complex sine function takes on every complex value (compare the argument in Lemma 11.13 below for the cosine). An alternative proof is provided by Picard's little theorem: If the sine function omits a value q, then it must also omit the value $-q$ since $\sin(-z) = -\sin z$. As $q \neq 0$, this produces two distinct omitted values, contradicting Theorem 11.6. Of course the same argument proves that *every odd entire function takes on all complex values*. The similar statement for even entire functions is false; for example, $\exp(z^2)$ omits 0.

COROLLARY 11.10 (Picard's little theorem for meromorphic functions). *A meromorphic function in $\mathbb{C}$ which omits three distinct values in $\hat{\mathbb{C}}$ is constant.*

PROOF. Suppose $f \in \mathscr{M}(\mathbb{C})$ omits the distinct values $q_1, q_2, q_3 \in \hat{\mathbb{C}}$. Take the Möbius map φ which sends (q_1, q_2, q_3) to $(0, 1, \infty)$. Then $\varphi \circ f$ is an entire function which omits $0, 1$. We conclude that $\varphi \circ f$, hence f, is constant. ☐

EXAMPLE 11.11. The *Fermat equation* $f^n + g^n = 1$ has no non-constant entire solutions f, g if $n \geq 3$. Here is why: Assuming such entire functions exist, consider the meromorphic function $h = f/g$ and note that $g \neq 0$ implies $h^n + 1 = 1/g^n \neq 0$ and $f \neq 0$ implies $h^{-n} + 1 = 1/f^n \neq 0$. Since f, g have no common zeros, it follows that $h^n \neq -1$ everywhere in $\mathbb{C}$. That means h omits the n-th roots of -1. Since there are $n \geq 3$ such roots, h must be constant by Corollary 11.10. Hence $f = cg$ for some $c \in \mathbb{C}$. But then $g^n = 1/(c^n + 1)$ is constant, hence g and f are both constant, which contradicts our assumption.

There are, however, non-constant *meromorphic* functions in the plane which satisfy the Fermat equation for $n \geq 3$. Here is an example, following [**Re**], that uses elliptic functions introduced in §9.2: Let $\tau = e^{\pi i/3}$ and consider the triangular lattice $\Lambda = \langle 1, \tau \rangle$ whose invariant g_2 vanishes. To see this, simply note that the rotation $z \mapsto \tau z$ preserves Λ, so

$$g_2 = 60 \sum_{\omega \in \Lambda^*} \frac{1}{(\tau \omega)^4} = \frac{60}{\tau^4} \sum_{\omega \in \Lambda^*} \frac{1}{\omega^4} = \frac{1}{\tau^4} g_2,$$

which implies $g_2 = 0$ since $\tau^4 \neq 1$. By Theorem 9.14, the associated Weierstrass $\wp$-function satisfies the differential equation $(\wp')^2 = 4\wp^3 - g_3$. It follows that

$$\left(\frac{a + b\wp'}{\wp} \right)^3 + \left(\frac{a - b\wp'}{\wp} \right)^3 = 1,$$

where

$$a = \frac{1}{2} g_3^{1/3} \qquad \text{and} \qquad b = \frac{1}{2\sqrt{3}} g_3^{-1/6}.$$

The following is an important result in its own right, but it only plays a supporting role in the present discussion:

THEOREM 11.12 (Schottky, 1904). *Let $M > 0$ and $0 < r < 1$. There exists a constant $C = C(r, M) > 0$ such that every holomorphic function $f : \mathbb{D} \to \mathbb{C} \setminus \{0, 1\}$ with $|f(0)| \leq M$ satisfies*

$$|f(z)| \leq C \qquad \text{for } |z| \leq r.$$

The proof is based on Corollary 11.3 and an elementary lemma on the complex cosine function:

LEMMA 11.13.

> (i) *For every* $z = x + iy \in \mathbb{C}$, $|\sinh y| \leq |\cos z| \leq \cosh y \leq \cosh |z|$.
>
> (ii) *For every pair* $z, w \in \mathbb{C}$ *with* $\cos z = w$, *there exists an integer* n *such that* $|z + 2\pi n| \leq |w| + \pi$.

PROOF. (i) follows from the inequalities

$$|\cos z| \leq \frac{|e^{iz}| + |e^{-iz}|}{2} = \frac{e^{-y} + e^{y}}{2} = \cosh y \leq \cosh |z|$$

and

$$|\cos z| \geq \frac{||e^{iz}| - |e^{-iz}||}{2} = \frac{|e^{-y} - e^{y}|}{2} = |\sinh y|.$$

For (ii), write $z = x + iy$ and find an integer n such that $|x + 2\pi n| \leq \pi$. By part (i), $|y| \leq |\sinh y| \leq |w|$. It follows that $|z + 2\pi n| \leq |x + 2\pi n| + |y| \leq |w| + \pi$. $\quad\square$

Friedrich Hermann
Schottky (1851–1935)

PROOF OF THEOREM 11.12. Fix a holomorphic function $f : \mathbb{D} \to \mathbb{C} \smallsetminus \{0, 1\}$ with $|f(0)| \leq M$. Use Lemma 11.8 to represent f as in (11.4), so $2f - 1 = \cos(\pi h)$ where $h = \cos(\pi g)$ for some $g \in \mathscr{O}(\mathbb{D})$. By Lemma 11.13, after adding a suitable even integer to h, we may arrange

$$|h(0)| \leq \frac{|2f(0) - 1|}{\pi} + 1 \leq \frac{2M + 1}{\pi} + 1 < M + 2.$$

Again by Lemma 11.13 and after adding an even integer to g, we may arrange

$$|g(0)| \leq \frac{|h(0)|}{\pi} + 1 \leq \frac{M + 2}{\pi} + 1 < M + 2.$$

Since by Lemma 11.8 the image $g(\mathbb{D})$ does not contain any disk of radius 1, for every $\zeta \in \mathbb{D}$ we can apply Corollary 11.3 to the restriction of g to the disk $\mathbb{D}(\zeta, 1 - |\zeta|)$ to arrive at the estimate

$$\mathfrak{B}(1 - |\zeta|)|g'(\zeta)| \leq 1 \qquad \text{or} \qquad |g'(\zeta)| \leq \frac{1}{\mathfrak{B}(1 - |\zeta|)},$$

where $\mathfrak{B} > 0$ is Bloch's constant. Integrating radially from 0 to $z \in \mathbb{D}$, we obtain

$$|g(z) - g(0)| = \left| \int_{[0,z]} g'(\zeta) \, d\zeta \right| \leq \frac{1}{\mathfrak{B}} \int_{0}^{|z|} \frac{dt}{1 - t} = \frac{1}{\mathfrak{B}} \log\left(\frac{1}{1 - |z|}\right).$$

This gives the estimate $|g(z)| \leq C_1$ for $|z| \leq r$, where

$$C_1 = C_1(r, M) = M + 2 + \frac{1}{\mathfrak{B}} \log\left(\frac{1}{1 - r}\right).$$

Now apply Lemma 11.13(i) twice to conclude that for $|z| \leq r$,

$$|f(z)| \leq \frac{1}{2}\left(1 + |\cos(\pi \cos(\pi g(z)))|\right) \leq \frac{1}{2}\left(1 + \cosh(\pi \cosh(\pi |g(z)|))\right) \leq C,$$

where

$$C = C(r, M) = \frac{1}{2} \left(1 + \cosh(\pi \cosh(\pi C_1)) \right). \qquad \square$$

REMARK 11.14. The above proof makes no attempt at finding sharp bounds for the constant $C(r, M)$. Ahlfors has proved the bound

$$C(r, M) = \exp\left[\left(\frac{1+r}{1-r} \right) (7 + \log^+ M) \right],$$

which is not optimal but has the right asymptotic orders in r and M. See [**A4**].

We are now ready for Picard's spectacular generalization of the Casorati-Weierstrass theorem:

THEOREM 11.15 (Picard's great theorem, 1880). *Suppose f is holomorphic in the punctured disk $\mathbb{D}^*(p, r)$ and has an essential singularity at p. Then, with at most one exception, f takes on every value in $\mathbb{C}$ infinitely often.*

The condition means that there is at most one $q \in \mathbb{C}$ for which the equation $f(z) = q$ has finitely many (or no) solutions in $\mathbb{D}^*(p, r)$.

A non-constant entire function f is either a polynomial, in which case it has no omitted value by the fundamental theorem of algebra, or it is transcendental, in which case the function $z \mapsto f(1/z)$ has an essential singularity at $z = 0$. Thus, Picard's great theorem implies the following stronger version of Picard's little theorem:

Charles Émile Picard
(1856–1941)

COROLLARY 11.16. *Suppose f is a transcendental entire function. Then, with at most one exception, f takes on every value in $\mathbb{C}$ infinitely often.*

PROOF OF THEOREM 11.15. Assume by way of contradiction that there are distinct values $q_1, q_2 \in \mathbb{C}$ for which the equations $f(z) = q_1$ and $f(z) = q_2$ have at most finitely many solutions in $\mathbb{D}^*(p, r)$. By shrinking r if necessary, we may assume that f omits both values q_1, q_2 to begin with. After pre and postcomposing f with affine maps, we can reduce to the case where $f \in \mathcal{O}(\mathbb{D}^*)$ has an essential singularity at the origin and omits the values 0 and 1.

Under this assumption, the function $F(w) = f(e^w)$ is holomorphic in the left half-plane $\{ w : \operatorname{Re}(w) < 0 \}$. Since $f(\mathbb{D}^*)$ is dense in $\mathbb{C}$ by the Casorati-Weierstrass theorem, there is a sequence $z_n \to 0$ such that $|z_{n+1}| < |z_n|$ and $|f(z_n)| \leq 1$ for all n. Let $z_n = e^{w_n}$, so $\operatorname{Re}(w_n) \to -\infty$, $\operatorname{Re}(w_{n+1}) < \operatorname{Re}(w_n)$, and $|F(w_n)| \leq 1$ for all n. The holomorphic functions $F_n : \mathbb{D} \to \mathbb{C} \setminus \{0, 1\}$ defined by $F_n(w) = F(8w + w_n)$ satisfy $|F_n(0)| \leq 1$. Schottky's Theorem 11.12 with $M = 1$ and $r = \pi/4 < 1$ then gives the upper bound

$$\sup_{t \in [0,1]} |F(w_n + 2\pi i t)| = \sup_{t \in [0,1]} \left| F_n\left(\frac{\pi i t}{4} \right) \right| \leq C$$

for a constant $C > 0$ independent of n. This means $|f| \leq C$ on every circle $|z| = |z_n|$, and therefore on every closed annulus $|z_{n+1}| \leq |z| \leq |z_n|$ by the maximum principle. It follows that $|f| \leq C$ in a punctured neighborhood of 0, which implies the singularity at 0 is removable. This contradicts the hypothesis that 0 is an essential singularity. $\square$

EXAMPLE 11.17. The exponential function has no fixed point on the real line (since $e^x > x$ for all $x \in \mathbb{R}$) but it has infinitely many fixed points in the complex plane. This can be checked most easily by invoking Picard's great theorem: Since the non-constant entire function ze^{-z} takes on the value 0 only at $z = 0$, it has to take every other value infinitely many times. In particular, the equation $ze^{-z} = 1$, or equivalently $e^z = z$, has infinitely many solutions.

As another application of Schottky's theorem, we prove the following

THEOREM 11.18 (Generalized Montel's theorem). *Suppose $U \subset \mathbb{C}$ is open and $q_1, q_2, q_3 \in \hat{\mathbb{C}}$ are distinct. Then the family $\mathscr{F}$ of all meromorphic functions $U \to \hat{\mathbb{C}} \smallsetminus \{q_1, q_2, q_3\}$ is normal. Moreover, every limit function of $\mathscr{F}$ either belongs to $\mathscr{F}$ or is a constant function taking its value in $\{q_1, q_2, q_3\}$.*

PROOF. Since normality is a local property (Theorem 5.5), it suffices to consider the case where $U = \mathbb{D}$. After postcomposing elements of $\mathscr{F}$ with the Möbius map which sends (q_1, q_2, q_3) to $(0, 1, \infty)$, we may assume that $\mathscr{F}$ is the family of all holomorphic functions $\mathbb{D} \to \mathbb{C} \smallsetminus \{0, 1\}$. Take a sequence $\{f_n\}$ in $\mathscr{F}$ and consider two cases depending on the behavior of the sequence $\{f_n(0)\}$.

First suppose $\{f_n(0)\}$ is bounded. Then, by Schottky's Theorem 11.12, $\{f_n\}$ is compactly bounded in $\mathbb{D}$. It follows from the basic version of Montel's theorem (Theorem 5.25) that $\{f_n\}$ is precompact in $\mathscr{O}(\mathbb{D})$, so it has a compactly convergent subsequence. By Corollary 5.20 the limit function either maps $\mathbb{D}$ to $\mathbb{C} \smallsetminus \{0, 1\}$ or is a constant function taking its value in $\{0, 1\}$.

Now consider the complementary case where $f_{n_k}(0) \to \infty$ for some subsequence $\{n_k\}$ of integers. Then the functions $g_{n_k} = 1/f_{n_k} \in \mathscr{F}$ satisfy $g_{n_k}(0) \to 0$. By the first case treated above, some subsequence of $\{g_{n_k}\}$ converges compactly in $\mathbb{D}$ to a limit $g \in \mathscr{O}(\mathbb{D})$. Since g vanishes at the origin, we must have $g = 0$ everywhere. It follows that the corresponding subsequence of $\{f_{n_k}\}$ converges compactly to ∞ in the spherical metric. $\square$

REMARK 11.19. In the above proof, the general version of Montel's theorem was deduced from its basic version together with Schottky's theorem. Conversely, the latter two results can be deduced from the general version of Montel's theorem. The implication Theorem 11.18 $\implies$ Theorem 5.25 is trivial. To prove Theorem 11.18 $\implies$ Theorem 11.12, suppose there are constants $M > 0$ and $0 < r < 1$, a sequence $\{z_n\}$ in $\mathbb{D}(0, r)$, and a sequence of holomorphic functions $f_n : \mathbb{D} \to \mathbb{C} \smallsetminus \{0, 1\}$ such that $|f_n(0)| \leq M$ but $|f_n(z_n)| > n$ for all n. After passing to a subsequence, we may assume

$z_n \to p \in \overline{\mathbb{D}}(0, r)$. Since $\{f_n\}$ is normal by Theorem 11.18, it has a subsequence which converges compactly in the spherical metric to a limit function f. On the one hand, $|f_n(0)| \leq M$ shows that $|f(0)| \leq M$. On the other hand, $|f_n(z_n)| > n$ implies $f(p) = \infty$ and therefore $f = \infty$ everywhere. This is a contradiction.

It is also instructive to see how Picard's great theorem can be deduced from the general version of Montel's theorem, without the direct intervention of theorems of Schottky and Casorati-Weierstrass:

SECOND PROOF OF THEOREM 11.15. It suffices to show that every $f \in \mathcal{O}(\mathbb{D}^*)$ which omits the values 0 and 1 has either a removable singularity or a pole at the origin. Consider the holomorphic functions $f_n : \mathbb{D}^* \to \mathbb{C} \setminus \{0, 1\}$ defined by $f_n(z) = f(z/n)$. By Theorem 11.18, there is a subsequence $\{f_{n_k}\}$ which converges compactly in the spherical metric to a limit function f. Moreover, either $f \in \mathcal{O}(\mathbb{D}^*)$ or $f = \infty$ everywhere.

If $f \in \mathcal{O}(\mathbb{D}^*)$, since $f_{n_k} \to f$ uniformly on the circle $|z| = 1/2$, there is a constant $C > 0$ such that $|f_{n_k}(z)| \leq C$ if $|z| = 1/2$ and $k \geq 1$. It follows that $|f(z)| \leq C$ on the circle $|z| = 1/(2n_k)$, and therefore on the closed annulus $1/(2n_{k+1}) \leq |z| \leq 1/(2n_k)$ for every $k \geq 1$ by the maximum principle. This shows that f is bounded in $\mathbb{D}^*$, so has a removable singularity at 0.

If $f = \infty$, then $1/f_{n_k} \to 0$ uniformly on the circle $|z| = 1/2$. A similar argument using the maximum principle then shows that $\lim_{z\to 0} 1/f(z) = 0$, or $\lim_{z\to 0} f(z) = \infty$, which means 0 is a pole of f. $\qquad\square$

11.3 A rescaling approach to Picard and Montel

We now introduce an alternative route to the theorems of Picard and Montel based on a characterization of normality due to Zalcman. This result has the curious feature that it produces convergence (after rescaling) precisely when normality *fails*.

THEOREM 11.20 (Zalcman, 1975). *A family $\mathscr{F}$ of meromorphic functions in a domain $U \subset \mathbb{C}$ is **not** normal if and only if the following holds: There exist sequences $\{f_n\}$ in $\mathscr{F}$, $\{z_n\}$ in U converging to some point of U, and radii $\{r_n\}$ converging to 0 such that the rescaled functions*

$$(11.5) \qquad\qquad g_n(z) = f_n(r_n z + z_n)$$

converge compactly to a non-constant $g \in \mathscr{M}(\mathbb{C})$ with $g^{\#} = |g'|/(1 + |g|^2) \leq 1$.

The crucial point of the theorem is that the limit g is non-constant. Observe that if every f_n is holomorphic in U, then g is an entire function.

Lawrence Allen Zalcman
(1943–)

PROOF. One direction is easy: Suppose sequences $\{f_n\}, \{z_n\}, \{r_n\}$ with the above properties exist but $\mathscr{F}$ is normal. By Marty's Theorem 5.37, the sequence $\{f_n^\#\}$ of spherical derivative norms is compactly bounded in U. If $\lim_{n\to\infty} z_n = p \in U$, then $\lim_{n\to\infty}(r_n z + z_n) = p$ for each $z \in \mathbb{C}$, so $\{f_n^\#(r_n z + z_n)\}$ is a bounded sequence. The rescaled functions g_n defined by (11.5) satisfy

$$g_n^\#(z) = r_n f_n^\#(r_n z + z_n),$$

hence $g^\#(z) = \lim_{n\to\infty} g_n^\#(z) = 0$. Since z was arbitrary, this implies g being constant, which is a contradiction.

Conversely, suppose $\mathscr{F}$ is not normal. By Marty's Theorem 5.37, the family $\{f^\# : f \in \mathscr{F}\}$ is not compactly bounded in U, so we can find a sequence $\{f_n\}$ in $\mathscr{F}$ and a sequence $\{w_n\}$ in a compact subset of U such that $f_n^\#(w_n) \to \infty$ as $n \to \infty$. After passing to a subsequence, we may assume that $w_n \to p \in U$. Precomposing the f_n with an affine map of the form $z \mapsto \alpha z + p$, we may assume without losing generality that $p = 0$ and $\overline{\mathbb{D}} \subset U$. For each n, the function $z \mapsto (1 - |z|) f_n^\#(z)$ is continuous on $\overline{\mathbb{D}}$ and vanishes on the unit circle. Hence it reaches its maximum value M_n at some $z_n \in \mathbb{D}$:

$$M_n = \max_{z \in \mathbb{D}} (1 - |z|) f_n^\#(z) = (1 - |z_n|) f_n^\#(z_n).$$

Since $M_n \geq (1 - |w_n|) f_n^\#(w_n) \geq (1/2) f_n^\#(w_n)$ for large n, we see that M_n and therefore $f_n^\#(z_n)$ tends to $+\infty$ as $n \to \infty$. Set

$$r_n = \frac{1}{f_n^\#(z_n)}.$$

Since $|z| \leq M_n$ is easily seen to imply $|r_n z + z_n| \leq 1$, the rescaled function g_n defined by (11.5) is well defined and meromorphic in the disk $\mathbb{D}(0, M_n)$.

To prove compact convergence of $\{g_n\}$ in $\mathbb{C}$, fix a radius $R > 0$ and find $N \geq 1$ such that $M_n > 2R$ whenever $n \geq N$. Then, for $|z| \leq R$ and $n \geq N$,

$$1 - |r_n z + z_n| \geq 1 - r_n R - |z_n| = r_n(M_n - R),$$

so

$$g_n^\#(z) = r_n f_n^\#(r_n z + z_n)$$
$$= \frac{r_n(1 - |r_n z + z_n|) f_n^\#(r_n z + z_n)}{1 - |r_n z + z_n|} \leq \frac{r_n M_n}{r_n(M_n - R)}.$$

This proves

$$(11.6) \qquad g_n^\#(z) \leq \frac{M_n}{M_n - R} \qquad \text{if } |z| \leq R \text{ and } n \geq N.$$

Since $M_n/(M_n - R) \leq 2$ for $n \geq N$, we conclude that $\{g_n^\#\}$ is uniformly bounded on the disk $\mathbb{D}(0, R)$. Since this holds for every $R > 0$, another application of Marty's theorem shows that a subsequence of $\{g_n\}$ converges compactly in $\mathbb{C}$ to some limit function g.

As $g_n^\#(0) = r_n f_n^\#(z_n) = 1$ for all n, we have $g^\#(0) = 1$, so g is non-constant. By Theorem 5.30, $g \in \mathcal{M}(U)$. Finally, $g^\#(z) \le 1$ follows from (11.6) since $M_n/(M_n - R) \to 1$ as $n \to \infty$. $\qquad\square$

REMARK 11.21. A slightly sharper version of Zalcman's Theorem 11.20 asserts that if $\mathcal{F}$ fails to be normal in any neighborhood of $p \in U$, the sequence $\{z_n\}$ can be chosen to tend to p. See problem 19.

EXAMPLE 11.22. Since the sequence of powers $\{z^n\}_{n \ge 1}$ is not normal in $\mathbb{C}$ (see Examples 5.36 and 5.38), Theorem 11.20 guarantees the existence of sequences $\{z_n\}$ (tending to $\mathbb{T}$) and $\{r_n\}$ (tending to 0) such that some subsequence of $\{(r_n z + z_n)^n\}$ converges compactly to a non-constant entire function. Clearly the choice $z_n = 1$ and $r_n = 1/n$ will do since $(z/n + 1)^n \to \exp(z)$.

To illustrate the power of Zalcman's theorem, we use it to give alternative proofs of Picard's little theorem and the general version of Montel's theorem.

SECOND PROOF OF THEOREM 11.6. Assume by way of contradiction that f is a non-constant entire function with two omitted values, which can be taken to be 0 and 1. Precomposing f with a translation, we may also assume $f'(0) \ne 0$. Since non-vanishing entire functions have holomorphic n-th roots (Corollary 2.22), for each $n \ge 1$ there is an entire function h_n which satisfies $(h_n)^{2^n} = f$. Since $h_n^\#(0) \ne 0$, the function

$$f_n(z) = h_n\left(\frac{z}{h_n^\#(0)}\right)$$

is entire with $f_n^\#(0) = 1$, and it omits the 2^k-th roots of unity for every $0 \le k \le n$.

First suppose $\{f_n\}$ is normal, so it has a subsequence which converges compactly to a limit g in the spherical metric. Since $g^\#(0) = 1$, the function g cannot be identically ∞, so it must be a non-constant entire function. By Corollary 5.20, g omits the 2^k-th roots of unity for all k. As these roots form a dense subset of the unit circle, it follows that $g(\mathbb{C}) \cap \mathbb{T} = \varnothing$. Since $g(\mathbb{C})$ is connected, we must have $g(\mathbb{C}) \subset \mathbb{D}$ or $g(\mathbb{C}) \subset \mathbb{C} \smallsetminus \overline{\mathbb{D}}$. But either case is impossible by Liouville's theorem.

If $\{f_n\}$ is not normal, we can repeat essentially the same argument after rescaling. In fact, Theorem 11.20 gives sequences $\{z_n\}$ and $\{r_n\}$ such that a subsequence of $\{f_n(r_n z + z_n)\}$ converges compactly in the spherical metric to a non-constant entire function g. Again by Corollary 5.20, g omits the 2^k-th roots of unity for all k, hence it omits the unit circle, which leads to a contradiction as before. $\qquad\square$

SECOND PROOF OF THEOREM 11.18. After the usual normalization, we may assume $\mathcal{F}$ is the family of all holomorphic functions $U \to \mathbb{C} \smallsetminus \{0, 1\}$. Suppose $\mathcal{F}$ is not normal. By Theorem 11.20, there is a sequence $\{f_n\}$ in $\mathcal{F}$ and sequences $\{z_n\}$ and

$\{r_n\}$ such that $\{f_n(r_n z + z_n)\}$ converges compactly in the spherical metric to a non-constant $g \in \mathscr{O}(\mathbb{C})$. By Corollary 5.20, g has to omit 0 and 1. This contradicts Picard's little theorem. $\square$

REMARK 11.23. Zalcman's result provides a neat explanation for a heuristic principle in complex analysis known as **Bloch's principle**. Roughly speaking, this principle states that if P is a property that implies an entire function is constant, then any family of holomorphic functions in a domain whose elements share the property P must be normal. Examples of this principle include the implications (Liouville's theorem $\implies$ basic Montel's theorem) and (Picard's little theorem $\implies$ generalized Montel's theorem). We refer to [**Z2**] for an excellent account of this point of view.

As a final application of Zalcman's theorem, we include the following striking result:

THEOREM 11.24 (Ubiquity of **exp**). *For every non-constant non-vanishing entire function f there exist sequences $\{\alpha_n\}$ and $\{\beta_n\}$ of complex numbers such that*

$$f(\alpha_n z + \beta_n) \to e^z$$

A version of this result (for curves in $\mathbb{C}P^n$) seems to have first appeared in [**BD**]. I have learned this formulation and the idea of its proof from F. Berteloot.

compactly in $\mathbb{C}$ as $n \to \infty$.

The proof will make use of the following generalization of the formula $\lim_{n \to \infty}(1 + z/n)^n = e^z$:

LEMMA 11.25. *Let U be a neighborhood of 0 and $\{f_n\}$ be a compactly convergent sequence in $\mathscr{O}(U)$ such that $f_n(0) = 1$ and $\lim_{n \to \infty} f_n'(0) = 1$. Then, $(f_n(z/n))^n \to e^z$ compactly in $\mathbb{C}$ as $n \to \infty$.*

PROOF. Denote the limit of $\{f_n\}$ by f. Consider the expansions

$$f_n(z) = 1 + f_n'(0)\, z + z^2\, R_n(z) \quad \text{and} \quad f(z) = 1 + z + z^2 R(z),$$

where $R_n, R \in \mathscr{O}(U)$. Fix $\delta > 0$ such that $\overline{\mathbb{D}}(0, \delta) \subset U$. The expression

$$R_n(z) - R(z) = \frac{f_n(z) - f(z) + (1 - f_n'(0))\, z}{z^2}$$

shows that $R_n \to R$ uniformly for $|z| = \delta$, hence for $|z| \le \delta$ by the maximum principle. Setting $w_n(z) = f_n(z/n) - 1$, it follows that

$$n w_n(z) = f_n'(0)\, z + \frac{z^2}{n}\, R_n\left(\frac{z}{n}\right) \to z$$

and therefore $w_n \to 0$ compactly in $\mathbb{C}$ as $n \to \infty$.

Recall that Log is the principal branch of the logarithm defined in Example 2.23. By the power series representation (2.3), there is a constant $C > 0$ such that

$$(11.7) \qquad |\operatorname{Log}(1+w) - w| \leq C|w|^2 \qquad \text{if } |w| \leq \frac{1}{2}.$$

Given $\varepsilon > 0$ and $r > 0$, find $N \geq 1/\varepsilon$ such that $|w_n(z)| < 1/2$ and $|nw_n(z) - z| < \varepsilon$ whenever $|z| \leq r$ and $n \geq N$. Then, by (11.7),

$$|n\operatorname{Log}(1+w_n(z)) - z| \leq |n\operatorname{Log}(1+w_n(z)) - nw_n(z)| + \varepsilon$$

$$\leq \frac{C}{n}|nw_n(z)|^2 + \varepsilon \leq \frac{C}{N}(r+\varepsilon)^2 + \varepsilon$$

$$\leq \left(C(r+\varepsilon)^2 + 1\right)\varepsilon$$

whenever $|z| \leq r$ and $n \geq N$. This shows $n\operatorname{Log}(f_n(z/n)) \to z$ uniformly in $\mathbb{D}(0,r)$ as $n \to \infty$. Taking the exponential of each side now proves the result. $\qquad \square$

PROOF OF THEOREM 11.24. We may assume $f'(0) \neq 0$. For each n, take a holomorphic n-th root of f, that is, a function $h_n \in \mathcal{O}(\mathbb{C})$ such that $(h_n)^n = f$. The function $h_n(nz/h_n^\#(0))$ has spherical derivative norm n at the origin, so by Marty's Theorem 5.37 the sequence $\{h_n(nz/h_n^\#(0))\}$ is not normal in any neighborhood of 0. By Zalcman's Theorem 11.20, after passing to a subsequence of $\{h_n\}$, we can find sequences $\{a_n\}$ and $\{b_n\}$ of complex numbers such that $\{g_n(z) = h_n(a_nz + b_n)\}$ converges compactly to a non-constant $g \in \mathcal{O}(\mathbb{C})$.

Choose a point $p \in \mathbb{C}$ such that $q = g(p)$ is on the unit circle and $g'(p) \neq 0$. Precomposing our functions with a suitable affine map, we may assume without losing generality that $p = 0$ and $g(0) = g'(0) = q$. Since $g_n \to g$ compactly in $\mathbb{C}$, by Lemma 5.16 we can find a sequence $p_n \to 0$ such that $g_n(p_n) = q$ for all large n. Define

$$f_n(z) = \frac{1}{q}g_n(z + p_n).$$

Evidently $\{f_n\}$ converges compactly to g/q, with $f_n(0) = 1$ and $f_n'(0) = g_n'(p_n)/q \to g'(0)/q = 1$. By Lemma 11.25,

$$\left(f_n\left(\frac{z}{n}\right)\right)^n = \frac{1}{q^n}f\left(\frac{a_n}{n}z + a_np_n + b_n\right)$$

converges compactly in $\mathbb{C}$ to e^z. The result follows by considering a further subsequence along which $q^n \to 1$ and setting $\alpha_n = a_n/n$, $\beta_n = a_np_n + b_n$. $\qquad \square$

11.4 Ahlfors's generalization of the Schwarz-Pick lemma

This section will provide yet another approach to the theorems of Picard and Montel through the geometric notion of hyperbolicity. Recall from §4.5 that a conformal metric $g = \rho(z)\,|dz|$ in a domain U is determined by its density function $\rho : U \to$

$(0, +\infty)$. For convenience *we assume ρ to be C^2-smooth* even though this assumption is not necessary and can be relaxed.

> **DEFINITION 11.26.** The *(Gaussian) curvature* of a conformal metric $g = \rho(z)\,|dz|$ is defined by
> $$K_g = -\frac{\Delta \log \rho}{\rho^2}.$$

Here, as before, Δ is the Laplace operator

$$\Delta = \frac{\partial^2}{\partial x^2} + \frac{\partial^2}{\partial y^2} = 4 \frac{\partial}{\partial z}\frac{\partial}{\partial \bar z} = 4 \frac{\partial}{\partial \bar z}\frac{\partial}{\partial z}.$$

Computations involving curvature are facilitated by the following formula for the Laplacian of a composition:

LEMMA 11.27. *If h is C^2-smooth and f is holomorphic, then*

$$\Delta(h \circ f) = (\Delta h \circ f)\,|f'|^2.$$

PROOF. We apply the product and chain rules in complex-variable notation, keeping in mind that $f_{\bar z} = \bar f_z = 0$, $f_z = f'$, and $\bar f_{\bar z} = \overline{f_z} = \overline{f'}$ since f is holomorphic (compare problem 3 in chapter 1). Taking $\partial/\partial z$ of the composition $h \circ f$ gives

$$(h \circ f)_z = (h_z \circ f)f_z + (h_{\bar z} \circ f)\bar f_z = (h_z \circ f) f_z,$$

and taking $\partial/\partial \bar z$ of the resulting expression yields

$$\begin{aligned}
\Delta(h \circ f) = 4\,(h \circ f)_{z\bar z} &= 4\,((h_z \circ f)f_z)_{\bar z} \\
&= 4\,\big((h_z \circ f)_{\bar z}\,f_z + (h_z \circ f)\,f_{z\bar z}\big) \\
&= 4\,(h_z \circ f)_{\bar z}\,f_z \\
&= 4\,\big((h_{zz} \circ f)\,f_{\bar z} + (h_{z\bar z} \circ f)\,\bar f_{\bar z}\big)f_z \\
&= 4\,(h_{z\bar z} \circ f)\,f_z\overline{f_{\bar z}} = (\Delta h \circ f)\,|f'|^2. \qquad \square
\end{aligned}$$

A fundamental property of curvature is its conformal invariance. Recall from §4.4 that the pull-back of a conformal metric $g = \rho(z)\,|dz|$ under a holomorphic map f is the conformal metric $f^* g = \rho(f(z))\,|f'(z)|\,|dz|$.

LEMMA 11.28 (Curvature is a conformal invariant). *Let $f : \tilde U \to U$ be a conformal map, g be a conformal metric in U, and $\tilde g = f^* g$. Then*

$$K_{\tilde g} = K_g \circ f.$$

PROOF. Let $g = \rho(z)\,|dz|$ and $\tilde{g} = \tilde{\rho}(z)\,|dz|$, where $\tilde{\rho} = (\rho \circ f)\,|f'|$, so

$$\log \tilde{\rho} = \log \rho \circ f + \log|f'|.$$

As a conformal map, f has non-vanishing derivative, so $\log|f'|$ is harmonic. Hence,

$$\Delta \log \tilde{\rho} = \Delta(\log \rho \circ f) = ((\Delta \log \rho) \circ f)\,|f'|^2$$

by Lemma 11.27. It follows that

$$K_{\tilde{g}} = -\frac{\Delta \log \tilde{\rho}}{\tilde{\rho}^2} = -\frac{((\Delta \log \rho) \circ f)\,|f'|^2}{(\rho^2 \circ f)\,|f'|^2} = K_g \circ f. \qquad \square$$

EXAMPLE 11.29. The hyperbolic metric $g_{\mathbb{D}} = 2|dz|/(1 - |z|^2)$ in $\mathbb{D}$ has constant curvature -1. More generally, for each $r > 0$ the conformal metric

$$\omega(z)\,|dz| = \frac{2r}{r^2 - |z|^2}\,|dz|$$

in the disk $\mathbb{D}(0, r)$ has constant curvature -1. This can be seen from the computation

$$\Delta \log \omega(z) = -\Delta \log(r^2 - |z|^2) = -4\,\frac{\partial}{\partial \bar{z}}\frac{\partial}{\partial z}[\,\log(r^2 - z\bar{z})\,]$$

$$= 4\,\frac{\partial}{\partial \bar{z}}\Big[\frac{\bar{z}}{r^2 - z\bar{z}}\Big] = \frac{4r^2}{(r^2 - z\bar{z})^2} = \omega^2(z).$$

It follows from Lemma 11.28, or by direct computation, that the hyperbolic metric $g_{\mathbb{H}} = |dz|/\mathrm{Im}(z)$ in the upper half-plane $\mathbb{H}$ also has constant curvature -1 (compare Example 4.37).

EXAMPLE 11.30. The spherical metric $\sigma = 2|dz|/(1 + |z|^2)$ has constant curvature $+1$ since

$$\Delta \log\Big[\frac{2}{1 + |z|^2}\Big] = -\Delta \log(1 + |z|^2) = -4\,\frac{\partial}{\partial \bar{z}}\frac{\partial}{\partial z}[\,\log(1 + z\bar{z})\,]$$

$$= -4\,\frac{\partial}{\partial \bar{z}}\Big[\frac{\bar{z}}{1 + z\bar{z}}\Big] = -\frac{4}{(1 + z\bar{z})^2} = -\Big[\frac{2}{1 + |z|^2}\Big]^2.$$

THEOREM 11.31 (Ahlfors, 1938). *Suppose $g = \rho(z)\,|dz|$ is a conformal metric in a domain U whose curvature K_g is everywhere bounded above by a negative constant $-K$. If $f : (\mathbb{D}, g_{\mathbb{D}}) \to (U, g)$ is holomorphic, then for all $z \in \mathbb{D}$,*

$$\|f'(z)\| \le \frac{1}{\sqrt{K}}, \qquad \text{or} \qquad |f'(z)| \le \frac{2}{\sqrt{K}\,\rho(f(z))\,(1 - |z|^2)}.$$

Lars Valerian Ahlfors
(1907–1996)

When $U = \mathbb{D}$ and $g = g_{\mathbb{D}}$, we can take $K = 1$ and the theorem reduces to Pick's version of the Schwarz lemma (Theorem 4.40).

PROOF. We may assume f is non-constant, since otherwise the result is trivial. Take any $0 < r < 1$ and consider the hyperbolic metric

$$\omega(z)\,|dz| = \frac{2r}{r^2 - |z|^2}\,|dz|$$

in the disk $\mathbb{D}(0, r)$ which has constant curvature -1 (Example 11.29). It suffices to prove that the ratio

$$u(z) = \frac{\rho(f(z))\,|f'(z)|}{\omega(z)}$$

is bounded above in $\mathbb{D}(0, r)$ by $1/\sqrt{K}$; the theorem will follow by letting $r \to 1$. Since u depends continuously on z and tends to zero as $|z| \to r$, it must take a positive maximum value at some point $p \in \mathbb{D}(0, r)$. Our task then is to prove that $u(p) \leq 1/\sqrt{K}$.

Let $V \subset \mathbb{D}(0, r)$ be a neighborhood of p in which $f' \neq 0$. Then $\log u$ is a C^2-smooth function in V reaching a maximum at p. In particular, $(\Delta \log u)(p) \leq 0$. On the other hand, $\log |f'|$ is harmonic in V since $f' \neq 0$ there. Hence, by Lemma 11.27,

$$\Delta \log u = \Delta(\log \rho \circ f) + \Delta \log |f'| - \Delta \log \omega$$

$$= \Delta(\log \rho \circ f) - \omega^2$$

$$= (\Delta(\log \rho) \circ f)\,|f'|^2 - \omega^2 \qquad \text{in } V.$$

But the assumption $K_g \leq -K$ translates into

$$\Delta(\log \rho) \circ f \geq K(\rho \circ f)^2,$$

which shows that throughout V,

$$\Delta \log u \geq K(\rho \circ f)^2 |f'|^2 - \omega^2 = \omega^2(Ku^2 - 1).$$

In view of the inequality $(\Delta \log u)(p) \leq 0$, this proves $u(p) \leq 1/\sqrt{K}$. $\square$

The type of domains that can occur in the above theorem are of fundamental importance:

> **DEFINITION 11.32.** A domain in $\mathbb{C}$ is called *hyperbolic* if it admits a conformal metric whose curvature has a negative upper bound.

Observe that by Lemma 11.28 hyperbolicity is preserved under conformal maps. Evidently, all subdomains of a hyperbolic domain are hyperbolic.

EXAMPLE 11.33. Every disk or half-plane is hyperbolic since it admits a metric of constant curvature -1. More generally, *every domain U whose complement contains a disk must be hyperbolic.* In fact, if $\mathbb{D}(p, r) \subset \mathbb{C} \smallsetminus U$, then the image of U under the Möbius map $z \mapsto r/(z - p)$ is contained in $\mathbb{D}$, hence is a hyperbolic domain. It follows that U is hyperbolic as well.

THEOREM 11.34 (Hyperbolic Liouville's theorem). *If $U \subset \mathbb{C}$ is a hyperbolic domain, every holomorphic function $\mathbb{C} \to U$ must be constant.*

PROOF. Let $g = \rho(z) \, |dz|$ be a conformal metric in U with $K_g \leq -K < 0$, and let $f : \mathbb{C} \to U$ be holomorphic. Take any $p \in \mathbb{C}$ and any radius $R > |p|$. Define $\varphi : \mathbb{D} \to U$ by $\varphi(z) = f(Rz)$. By Ahlfors's Theorem 11.31,

$$|\varphi'(z)| \leq \frac{2}{\sqrt{K} \, \rho(\varphi(z)) \, (1 - |z|^2)}$$

or

$$|f'(Rz)| \leq \frac{2}{\sqrt{K} \, R \, \rho(f(Rz)) \, (1 - |z|^2)}$$

whenever $|z| < 1$. Setting $z = p/R$, we obtain the estimate

$$|f'(p)| \leq \frac{2R}{\sqrt{K} \, \rho(f(p)) \, (R^2 - |p|^2)}.$$

Letting $R \to +\infty$ then gives $f'(p) = 0$. Since p was arbitrary, we conclude that f is constant. $\square$

To illustrate the power of hyperbolicity in proving the main results of §11.2, we need to show that the twice punctured plane $\mathbb{C} \smallsetminus \{0, 1\}$ is hyperbolic. The following lemma does this in an ad hoc fashion by finding an explicit metric that does the trick. In chapter 13 we will see that in fact $\mathbb{C} \smallsetminus \{0, 1\}$ admits a *complete* metric of *constant* negative curvature (see §13.3).

LEMMA 11.35. *The twice punctured plane* $\mathbb{C} \smallsetminus \{0, 1\}$ *is hyperbolic.*

PROOF. Take the metric $g = \rho(z) \, |dz|$ with the density

(11.8)
$$\rho(z) = \frac{1 + |z|^{0.2}}{|z|^{0.9}} \cdot \frac{1 + |z - 1|^{0.2}}{|z - 1|^{0.9}}$$

(see Fig. 11.2 and compare [**Kr**]). A brief computation, using the fact that $\log |z|$ is harmonic away from 0, shows that

$$\Delta \log \left(\frac{1 + |z|^{0.2}}{|z|^{0.9}} \right) = \Delta \log(1 + |z|^{0.2}) = \frac{0.04 |z|^{-1.8}}{(1 + |z|^{0.2})^2}$$

if $z \neq 0$. Similarly,

$$\Delta \log \left(\frac{1 + |z - 1|^{0.2}}{|z - 1|^{0.9}} \right) = \frac{0.04 |z - 1|^{-1.8}}{(1 + |z - 1|^{0.2})^2}$$

if $z \neq 1$. It follows that for $z \in \mathbb{C} \smallsetminus \{0, 1\}$,

$$K_g(z) = -\frac{\Delta \log \rho(z)}{\rho^2(z)}$$

$$= -0.04 \left[\frac{|z - 1|^{1.8}}{(1 + |z|^{0.2})^4 (1 + |z - 1|^{0.2})^2} + \frac{|z|^{1.8}}{(1 + |z|^{0.2})^2 (1 + |z - 1|^{0.2})^4} \right].$$

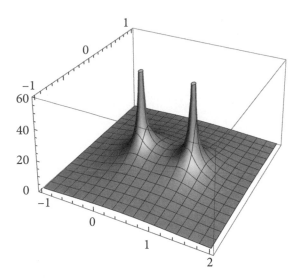

Figure 11.2. Graph of the density (11.8) of a metric of negative curvature in $\mathbb{C} \smallsetminus \{0, 1\}$.

It is clear that $K_g < 0$ everywhere,

$$\lim_{z \to 0} K_g(z) = \lim_{z \to 1} K_g(z) = -0.01, \quad \text{and} \quad \lim_{z \to \infty} K_g(z) = -\infty.$$

Continuity of K_g now shows that it is bounded above by a negative constant. This proves hyperbolicity of $\mathbb{C} \smallsetminus \{0, 1\}$. $\qquad \square$

THEOREM 11.36. *A domain $U \subset \mathbb{C}$ is hyperbolic if and only if $\mathbb{C} \smallsetminus U$ contains at least two points.*

PROOF. Since there are non-constant holomorphic functions $\mathbb{C} \to \mathbb{C}$ and $\mathbb{C} \to \mathbb{C} \smallsetminus \{\text{point}\}$, Theorem 11.34 shows that every hyperbolic domain must miss at least two points. Conversely, every twice punctured plane is conformally isomorphic to $\mathbb{C} \smallsetminus \{0, 1\}$, so it must be hyperbolic by Lemma 11.35. It follows that any domain missing at least two points is hyperbolic. $\qquad \square$

We are now ready for the alternative proofs of Picard's little theorem and generalized Montel's theorem based on the notion of hyperbolicity.

THIRD PROOF OF THEOREM 11.6. An entire function f that omits distinct values a, b can be viewed as a holomorphic function $\mathbb{C} \to \mathbb{C} \smallsetminus \{a, b\}$. By Theorem 11.36, $\mathbb{C} \smallsetminus \{a, b\}$ is a hyperbolic domain. By Theorem 11.34, f must be constant. $\qquad \square$

THIRD PROOF OF THEOREM 11.18. It suffices to show that the family $\mathscr{F}$ of all holomorphic functions $\mathbb{D} \to \mathbb{C} \smallsetminus \{0, 1\}$ is normal. By Marty's Theorem 5.37, this is accomplished once we prove that the family $\mathscr{F}^{\#} = \{f^{\#} : f \in \mathscr{F}\}$ is compactly bounded in $\mathbb{D}$.

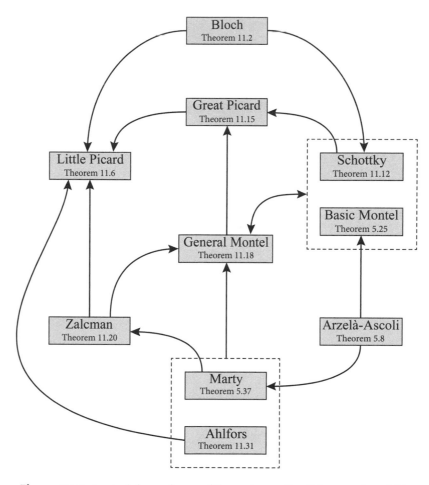

Figure 11.3. Logical dependence of the main results of chapters 5 and 11.

Let $g = \rho(z)\,|dz|$ be the metric of negative curvature in $\mathbb{C} \smallsetminus \{0, 1\}$ constructed in Lemma 11.35 and take $K > 0$ so that $K_g \leq -K$. Let $\sigma(z) = 2/(1 + |z|^2)$ denote the density of the spherical metric. By (11.8), $\rho(z) \to +\infty$ as $z \to 0$ or 1, and $|z|^{1.4}\rho(z) \to 1$ as $z \to \infty$. Hence $\sigma(z)/\rho(z)$ tends to 0 as $z \to 0$ or 1 or ∞. By continuity, this ratio is bounded above in $\mathbb{C} \smallsetminus \{0, 1\}$ by some constant $M > 0$. By Ahlfors's Theorem 11.31, for every $z \in \mathbb{D}$,

$$f^{\#}(z) = \frac{|f'(z)|}{1 + |f(z)|^2} = \frac{1}{2}\,|f'(z)|\,\sigma(f(z))$$

$$\leq \frac{M}{2}\,|f'(z)|\,\rho(f(z)) \leq \frac{M}{\sqrt{K}(1 - |z|^2)},$$

from which it easily follows that $\mathscr{F}^{\#}$ is compactly bounded in $\mathbb{D}$. $\square$

Figure 11.3 summarizes the logical dependence of the main results of this chapter and the related ones from chapter 5.

Problems

(1)   Suppose $f \in \mathcal{O}(\mathbb{D})$ and $f'(0) = 1$. Show that if $0 < R < \mathfrak{B}$, there is a $p \in \mathbb{D}$ such that

$$(1 - |p|^2)|f'(p)| \geq R.$$

(Hint: Look at the branch of f^{-1} defined in an unramified disk of radius R.)

(2)   Suppose f is holomorphic in a neighborhood of $\overline{\mathbb{D}}$ and $f'(0) = 1$. Show that

$$\sup_{z \in \mathbb{D}} (1 - |z|^2)|f'(z)| \leq \frac{B_f}{\mathfrak{B}}.$$

(Hint: Revisit the proof of Theorem 11.2 and note that G has unramified disks of radii arbitrarily close to $\mathfrak{B}$.)

(3)   Complete the following outline of a conceptual proof of Bloch's theorem without an explicit bound for $\mathfrak{B}$: It suffices to work with the family

$$\mathscr{F} = \left\{ f \in \mathcal{O}(\mathbb{D}) : f(0) = 0, \ \sup_{z \in \mathbb{D}} (1 - |z|^2)|f'(z)| = f'(0) = 1 \right\}.$$

By Montel's Theorem 5.25, $\mathscr{F}$ is compact. Since the assignment $f \mapsto B_f$ is lower semi-continuous on $\mathscr{F}$, it must attain a positive minimum.

(4)   Use Theorem 9.5 to show that there exists an entire function with a dense set of critical values. Contrast this with Corollary 11.4.

(5)   Prove that every $f \in \mathcal{O}(\mathbb{D}^*(p, r))$ with a pole or essential singularity at p has arbitrarily large unramified disks. (Hint: Consider $g(z) = f(r/z + p)$, which is holomorphic in the annulus $\{z : |z| > 1\}$, and use Corollary 11.3.)

(6)   Prove that Picard's little theorem is equivalent to the assertion that there are no non-constant entire functions f, g which satisfy the equation $e^f + e^g = 1$.

(7)   (i)   Show that for every $q \in \mathbb{C}$ the equation $w^3 + w^2 + q = 0$ has at least two distinct solutions in $w \in \mathbb{C}$.
            (ii) Show that if f is a non-constant entire function and $g = f^3 + f^2$, then $g(\mathbb{C}) = \mathbb{C}$.

(8)   Let f and g be entire functions with $f(0) = g(0)$, and let P and Q be polynomials such that

$$e^f + P = e^g + Q.$$

Show that $f = g$ and $P = Q$. (Hint: If $P \neq Q$, the entire function e^{f-g} takes the value 1 finitely many times.)

(9)   Suppose f is a periodic entire function, i.e., $f(z + \omega) = f(z)$ for some $\omega \neq 0$. Show that f has a fixed point.

(10)   Let f be an entire function such that $f \circ f$ has no fixed point. In other words, $f(f(z)) \neq z$ for all $z \in \mathbb{C}$. Prove that $f(z) = z + c$ for some $c \neq 0$. (Hint: Use Picard's little theorem to show that the entire function $(f(f(z)) - z)/(f(z) - z)$ is constant. Another application of the same theorem then shows that f' must be constant.)

(11)   Suppose f is a non-constant entire function which omits the value q. Let P be a polynomial which is not identically q. Prove that the equation $f(z) = P(z)$ has infinitely many solutions.

(12)   Show that for every $c \in \mathbb{C}$ the equations $\cos z = cz$ and $\sin z = cz$ have infinitely many solutions.

(13)   Prove that every holomorphic map $f : \mathbb{C} \smallsetminus \{0, 1\} \to \mathbb{C} \smallsetminus \{0, 1\}$ has a fixed point. (Hint: If one of $0, 1, \infty$ is an essential singularity of f, apply Picard's great theorem. Otherwise f must be a rational map with $f^{-1}(\{0, 1, \infty\}) = \{0, 1, \infty\}$.)

(14)   Give an example of a family of holomorphic functions $\mathbb{C} \to \mathbb{C}^*$ that fails to be normal.

(15)   Suppose $\psi_1, \psi_2, \psi_3 \in \mathscr{M}(U)$ and $\psi_1(z), \psi_2(z), \psi_3(z)$ are distinct points of $\hat{\mathbb{C}}$ for every $z \in U$. Let $\mathscr{F}$ be the family of all $f \in \mathscr{M}(U)$ for which $f(z) \neq \psi_j(z)$ for every $z \in U$ and $1 \leq j \leq 3$. Show that $\mathscr{F}$ is normal.

(16)   Let $\varepsilon > 0$ and $\mathscr{F}$ be the family of all $f \in \mathscr{M}(U)$ for which there are points $a = a(f), b = b(f), c = c(f)$ in $\hat{\mathbb{C}} \smallsetminus f(U)$ with mutual spherical distance $> \varepsilon$. Prove that $\mathscr{F}$ is normal.

(17)   Prove:
    (i)   The family $\{f \in \mathscr{O}(\mathbb{D}) : f \text{ is injective in } \mathbb{D}\}$ is not normal.
    (ii)   The family $\{f \in \mathscr{O}(\mathbb{D}) : f \text{ is injective and non-zero in } \mathbb{D}\}$ is normal.
(Hint: For (ii), take an arbitrary sequence $\{f_n\}$ in the family, find $p_n \in \mathbb{C} \smallsetminus f_n(\mathbb{D})$ with $|p_n| = |f_n(0)|$ and consider $g_n = f_n/p_n$.)

(18)   Let $U \subset \mathbb{C} \smallsetminus \{0, 1\}$ be a domain, $f : U \to U$ be holomorphic, and $f(p) = p$ and $|f'(p)| < 1$ for some $p \in U$. Show that the sequence of iterates $\{f^{\circ n}\}$ converges compactly in U to the constant function p.

(19)   In the situation of Zalcman's Theorem 11.20, suppose the family $\mathscr{F}$ fails to be normal in any neighborhood of a point $p \in U$. Show that the corresponding sequence $\{z_n\}$ can be chosen to tend to p. (Hint: We may assume $p = 0$ and $\overline{\mathbb{D}} \subset U$. Find a sequence $\{f_n\}$ in $\mathscr{F}$ such that $\sup_{z \in \mathbb{D}(0, 1/(2n))} f_n^\#(z) > n^2$ for all n. Apply Theorem 11.20 to the sequence $\{h_n\}$ in $\mathscr{O}(\mathbb{D})$ defined by $h_n(z) = f_n(z/n)$.)

12

Holomorphic (branched) covering maps

12.1 Covering spaces

We begin this chapter with a brief account of covering spaces and their lifting properties akin to what we saw in §2.2 for the exponential map. The presentation will treat covering maps between topological spaces, even though in our applications we are primarily concerned with spherical domains. The choice of the general setting is easy to justify. First, it requires virtually no additional work compared to the spherical domain case. Second, and more importantly, it allows us to vividly see which properties of the underlying spaces are essential for the validity of each result. The reader is advised to review the notions of homotopy of curves and the fundamental group of a topological space in §2.1, as they naturally appear in the context of covering spaces.

Unless otherwise stated, all topological spaces in this chapter are Hausdorff, path-connected, and locally path-connected. All neighborhoods are meant to be open.

The idea of a covering space (mainly for surfaces) can be traced back to Riemann, Poincaré, and Weyl among others, but the first systematic treatment of the theory seems to have appeared in Seifert-Threlfall's 1934 classical textbook [**ST**].

DEFINITION 12.1. A continuous map $f : Y \to X$ between topological spaces is a ***covering map*** if every $x \in X$ has a neighborhood W which is ***evenly covered*** by f in the sense that $f^{-1}(W)$ is the disjoint union of a collection $\{O_\alpha\}$ of open sets such that the restriction $f : O_\alpha \to W$ is a homeomorphism for every α. In this case Y (or $f : Y \to X$) is called a ***covering space*** of X.

The evenly covered neighborhood W can be taken to be path-connected, in which case $\{O_\alpha\}$ will be the path-components of $f^{-1}(W)$. It is immediate from the above definition that every covering map is a local homeomorphism and it is surjective.

Two covering spaces $f_0 : Y_0 \to X$ and $f_1 : Y_1 \to X$ are ***isomorphic*** if there is a homeomorphism $\varphi : Y_0 \to Y_1$ such that $f_0 = f_1 \circ \varphi$, so the following diagram commutes:

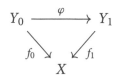

If W is evenly covered by a covering map $f: Y \to X$, the cardinality of the **fiber** $f^{-1}(x)$ is clearly independent of $x \in W$. Thus, as a locally constant function on the connected space X, the cardinality of fibers must be globally constant. This number, a positive integer or ∞, is called the **(mapping) degree** of f and is denoted by $\deg(f)$.

A covering map of degree n is also called "n-sheeted" or "n-to-1."

EXAMPLE 12.2. For every integer $n \neq 0$ the map $f: \mathbb{C}^* \to \mathbb{C}^*$ defined by $f(z) = z^n$ is a covering of degree $|n|$. The map $\exp: \mathbb{C} \to \mathbb{C}^*$ is a covering of infinite degree.

EXAMPLE 12.3 (Covering maps of degree 1). A covering map $f: Y \to X$ with $\deg(f) = 1$ is a bijection and the inverse map $f^{-1}: X \to Y$ is automatically continuous since f, being a local homeomorphism, is an open map. Thus, *every covering map of degree 1 is a homeomorphism*.

Let $f: Y \to X$ and $h: Z \to X$ be continuous. A continuous map $\tilde{h}: Z \to Y$ is called a **lift** of h under f if $f \circ \tilde{h} = h$:

$$\begin{array}{ccc} & & Y \\ & \overset{\tilde{h}}{\nearrow} & \downarrow f \\ Z & \underset{h}{\longrightarrow} & X \end{array}$$

LEMMA 12.4 (Uniqueness of lifts). *Suppose $\tilde{h}_1, \tilde{h}_2: Z \to Y$ are lifts of $h: Z \to X$ under a local homeomorphism $f: Y \to X$. If $\tilde{h}_1 = \tilde{h}_2$ somewhere in Z, then $\tilde{h}_1 = \tilde{h}_2$ everywhere in Z.*

PROOF. The set $E = \{z \in Z : \tilde{h}_1(z) = \tilde{h}_2(z)\}$ is non-empty by the assumption and is closed since $\tilde{h}_1, \tilde{h}_2$ are continuous. If $z_0 \in E$, set $y_0 = \tilde{h}_1(z_0) = \tilde{h}_2(z_0)$. Take neighborhoods O of y_0 and W of $f(y_0) = h(z_0)$ such that the restriction $f: O \to W$ is a homeomorphism. The neighborhood $\Omega = \tilde{h}_1^{-1}(O) \cap \tilde{h}_2^{-1}(O)$ of z_0 satisfies $\tilde{h}_j(\Omega) \subset O$ for $j = 1, 2$. It follows that $\tilde{h}_1 = g \circ h = \tilde{h}_2$ in Ω, where $g: W \to O$ is the inverse of $f: O \to W$, so $\Omega \subset E$. This proves E is open. Connectivity of Z now shows that $E = Z$. $\square$

DEFINITION 12.5. A continuous surjective map $f: Y \to X$ has the **curve lifting property** if for every curve $\gamma: [0, 1] \to X$ and every point $y_0 \in Y$ with $f(y_0) = \gamma(0)$ there exists a lift $\tilde{\gamma}: [0, 1] \to Y$ of γ under f with $\tilde{\gamma}(0) = y_0$.

Under a local homeomorphism with the curve lifting property, every curve has a *unique* lift once the initial point is given (Lemma 12.4). On the other hand, the squaring map $f: \mathbb{C} \to \mathbb{C}$ defined by $f(z) = z^2$ has the curve lifting property (verify!) without uniqueness. In fact, every non-constant curve starting at 0 has at least two lifts under squaring (see Fig. 12.1).

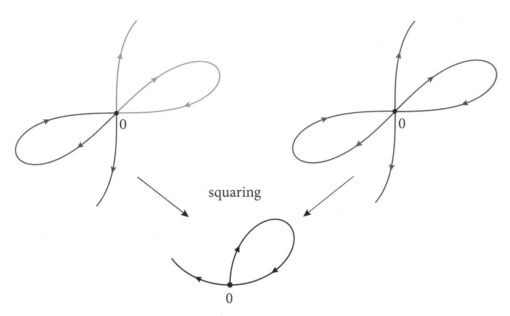

Figure 12.1. The squaring map $f : z \mapsto z^2$ has the curve lifting property without uniqueness. The curve in black with the initial point 0 has four lifts under f, distinguished by different colors.

LEMMA 12.6. *Every covering map has the curve lifting property.*

PROOF. The proof is virtually identical to that of Theorem 2.16(i) for the exponential map. Let $f : Y \to X$ be a covering map and take a curve $\gamma : [0,1] \to X$ and $y_0 \in Y$ with $\gamma(0) = f(y_0)$. For each $t \in [0,1]$ the point $\gamma(t)$ belongs to a neighborhood evenly covered by f. By continuity, the preimages under γ of these neighborhoods from an open cover of $[0,1]$. Let $\delta > 0$ be a Lebesgue number for this cover, given by Lemma 2.17, and take any partition $0 = t_0 < t_1 < \cdots < t_n = 1$ for which $t_k - t_{k-1} < \delta$ for all $1 \leq k \leq n$. The choice of δ guarantees that each image $\gamma([t_{k-1}, t_k])$ lies entirely in a neighborhood W_k evenly covered by f.

Now let O_1 be the open set in the decomposition of $f^{-1}(W_1)$ that contains y_0, and $g_1 : W_1 \to O_1$ be the corresponding local inverse of f. Define $\tilde{\gamma} = g_1 \circ \gamma$ on $[t_0, t_1]$. Let O_2 be the open set in the decomposition of $f^{-1}(W_2)$ that contains $\tilde{\gamma}(t_1)$, and $g_2 : W_2 \to O_2$ be the corresponding local inverse of f. Define $\tilde{\gamma} = g_2 \circ \gamma$ on $[t_1, t_2]$. Repeating this process n times, we construct the desired lift $\tilde{\gamma} : [0,1] \to Y$. ☐

EXAMPLE 12.7. The squaring map $f : \mathbb{C} \smallsetminus \{0,1\} \to \mathbb{C}^*$ defined by $f(z) = z^2$ is a surjective local homeomorphism but not a covering map since no neighborhood of 1 is evenly covered. Alternatively, the curve $\gamma : [0,1] \to \mathbb{C}^*$ defined by $\gamma(t) = e^{\pi i (1-t)}$ does not have a lift under f that starts at i.

The next result deals with the question of existence of lifts under covering maps. It is formulated for the a priori larger class of local homeomorphisms with the curve lifting property. Later we will prove that covering maps of reasonable spaces are precisely local homeomorphisms with the curve lifting property (see Theorem 12.15).

THEOREM 12.8. *Let $f : Y \to X$ be a local homeomorphism with the curve lifting property. Fix some $y_0 \in Y$.*

(i) If $\gamma_0, \gamma_1 : [0, 1] \to X$ are homotopic curves with $\gamma_0(0) = \gamma_1(0) = f(y_0)$, and if $\tilde{\gamma}_0, \tilde{\gamma}_1 : [0, 1] \to Y$ are their unique lifts under f with $\tilde{\gamma}_0(0) = \tilde{\gamma}_1(0) = y_0$, then $\tilde{\gamma}_0, \tilde{\gamma}_1$ are homotopic curves in Y. In particular, $\tilde{\gamma}_0(1) = \tilde{\gamma}_1(1)$.

(ii) Assume Z is simply connected. If $h : Z \to X$ is continuous and $h(z_0) = f(y_0)$ for some $z_0 \in Z$, then there is a unique lift $\tilde{h} : Z \to Y$ of h under f with $\tilde{h}(z_0) = y_0$.

PROOF. (i) Take a homotopy $H : [0, 1] \times [0, 1] \to X$ between γ_0 and γ_1, so $\gamma_0 = H(\cdot, 0)$ and $\gamma_1 = H(\cdot, 1)$. We construct a lift $\tilde{H} : [0, 1] \times [0, 1] \to Y$ of H with $\tilde{H}(0, s) = y_0$ for all s. Once this is accomplished, uniqueness of lifts shows that $\tilde{\gamma}_0 = \tilde{H}(\cdot, 0)$ and $\tilde{\gamma}_1 = \tilde{H}(\cdot, 1)$. Since $f(\tilde{H}(1, s)) = H(1, s) = H(1, 0)$ is independent of s, the continuous map $s \mapsto \tilde{H}(1, s)$ takes values in the fiber $f^{-1}(H(1, 0))$, a discrete set since f is a local homeomorphism. It follows that this map is constant, $\tilde{H}$ is a homotopy between $\tilde{\gamma}_0$ and $\tilde{\gamma}_1$, and $\tilde{\gamma}_0(1) = \tilde{H}(1, 0) = \tilde{H}(1, 1) = \tilde{\gamma}_1(1)$.

> A set is **discrete** if all of its points are isolated. Discrete sets are totally disconnected.

The definition of $\tilde{H}$ is easy: Let $\tilde{\gamma}_s$ be the unique lift of the curve $\gamma_s = H(\cdot, s)$ with $\tilde{\gamma}_s(0) = y_0$ and set $\tilde{H}(t, s) = \tilde{\gamma}_s(t)$. Evidently $\tilde{H}(0, s) = y_0$ and $f(\tilde{H}(t, s)) = f(\tilde{\gamma}_s(t)) = \gamma_s(t) = H(t, s)$.

Checking continuity of $\tilde{H}$ is a bit tricky (see [**Fo**]). First take neighborhoods O of y_0 and W of $f(y_0) = \gamma_s(0)$ such that the restriction $f : O \to W$ is a homeomorphism with the inverse $g : W \to O$. By continuity, $H^{-1}(W)$ is an open set in $[0, 1] \times [0, 1]$ which contains $\{0\} \times [0, 1]$, hence it contains $[0, \varepsilon) \times [0, 1]$ for some $\varepsilon > 0$. For each $s \in [0, 1]$ the curve $g \circ \gamma_s$ is a lift of γ_s on $[0, \varepsilon)$, with $(g \circ \gamma_s)(0) = y_0$. It follows from uniqueness of lifts that $\tilde{\gamma}_s = g \circ \gamma_s$ on $[0, \varepsilon)$, or $\tilde{H} = g \circ H$ on $[0, \varepsilon) \times [0, 1]$. Thus, $\tilde{H}$ is continuous on $[0, \varepsilon) \times [0, 1]$.

To prove continuity of $\tilde{H}$ everywhere, assume by way of contradiction that $\tilde{H}$ has a point of discontinuity, say on the line $[0, 1] \times \{b\}$. Let a be the infimum of all t such that $\tilde{H}$ is discontinuous at (t, b). By the previous paragraph, $\varepsilon \le a \le 1$. Take neighborhoods O of $\tilde{\gamma}_b(a)$ and W of $\gamma_b(a)$ such that $f : O \to W$ is a homeomorphism with the local inverse $g : W \to O$. By continuity of H, there are interval neighborhoods $I \subset [0, 1]$ of a and $J \subset [0, 1]$ of b such that $H(I \times J) \subset W$. We have

$$(12.1) \qquad \tilde{\gamma}_b = g \circ \gamma_b \qquad \text{on } I$$

since both curves are lifts of γ_b on I and they coincide at $t = a$. Choose any $c \in I$ with $0 < c < a$. By the choice of a, $\tilde{H}$ is continuous at (c, b). Since $\tilde{H}(c, b) = \tilde{\gamma}_b(c) = g(\gamma_b(c)) \in O$ by (12.1), there is an interval neighborhood $J' \subset J$ of b such that $\tilde{H}(\{c\} \times J') \subset O$ and therefore $\tilde{\gamma}_s(c)$ is the unique element of the fiber $f^{-1}(\gamma_s(c))$ in O as long as $s \in J'$. It follows that

$$\tilde{\gamma}_s = g \circ \gamma_s \qquad \text{on } I \text{ whenever } s \in J'$$

since both curves are lifts of γ_s on I and they coincide at $t = c$. In other words, $\tilde{H} = g \circ H$ on $I \times J'$. This shows continuity of $\tilde{H}$ on $I \times J'$, which contradicts the choice of (a, b).

(ii) This is very similar to the proof of Theorem 2.21 for the exponential map. Since uniqueness is guaranteed by Lemma 12.4, we only need to show the existence of a lift. Take $p \in Z$ and let $\gamma : [0, 1] \to Z$ be any curve from z_0 to p. The curve $h \circ \gamma : [0, 1] \to X$ has a unique lift $\xi : [0, 1] \to Y$ under f with $\xi(0) = y_0$. Define $\tilde{h}(p) = \xi(1)$. The definition does not depend on the choice of γ since by simple connectivity of Z any curve $\hat{\gamma} : [0, 1] \to Z$ from z_0 to p is homotopic to γ, so $h \circ \hat{\gamma}$ is homotopic to $h \circ \gamma$, so their unique lifts with the same initial point y_0 have the same end point by part (i). Since $f(\tilde{h}(p)) = f(\xi(1)) = h(\gamma(1)) = h(p)$, it remains to check that the map $\tilde{h}$ defined this way is continuous.

Fix $p \in Z$ and take neighborhoods O of $\tilde{h}(p)$ and W of $h(p)$ such that $f : O \to W$ is a homeomorphism. By continuity of h and local path-connectivity of Z, there is a path-connected neighborhood V of p such that $h(V) \subset W$. For any $z \in V$, take a curve η in V from p to z and let γ and ξ be as above. Then the product $\gamma \cdot \eta$ is a curve from z_0 to z, and the lift of $h \circ (\gamma \cdot \eta) = (h \circ \gamma) \cdot (h \circ \eta)$ with the initial point y_0 is given by $\xi \cdot (g \circ h \circ \eta)$, where $g : W \to O$ is the local inverse of f. It follows that $\tilde{h}(z)$ is the end point of $g \circ h \circ \eta$, which belongs to O. This shows $\tilde{h}(V) \subset O$ and proves continuity of $\tilde{h}$ at p. $\qquad \square$

REMARK 12.9. The homotopy lifting property in the above theorem shows that if $f : Y \to X$ is a covering map and $f(y_0) = x_0$, the induced homomorphism $f_* : \pi_1(Y, y_0) \to \pi_1(X, x_0)$ between the fundamental groups (see §2.1) is injective. In fact, if $f_*([\gamma])$ is trivial, we can take a homotopy between the image $f \circ \gamma$ and the constant curve ε_{x_0} and lift it to a homotopy between γ and the constant curve ε_{y_0}.

COROLLARY 12.10. *Suppose X is simply connected and $f : Y \to X$ is a local homeomorphism with the curve lifting property. Then f is a homeomorphism.*

PROOF. Fix $y_0 \in Y$ and let $x_0 = f(y_0)$. By Theorem 12.8(ii) the identity map $\mathrm{id}_X : X \to X$ has a unique lift $g : X \to Y$ under f with $g(x_0) = y_0$. The relation $f \circ g = \mathrm{id}_X$ makes g a right inverse of f. By path-connectivity, for every $y \in Y$ there is a curve γ in Y from y_0 to y. The curve $g \circ f \circ \gamma$ is then a lift of $f \circ \gamma$ which starts at y_0. By uniqueness, $g \circ f \circ \gamma = \gamma$, and in particular the endpoints $g(f(y))$ and y must coincide. The relation $g \circ f = \mathrm{id}_Y$ now shows that g is a left inverse of f, and we conclude that f is a homeomorphism. $\qquad \square$

COROLLARY 12.11. *If $f : Y \to X$ is a covering map and $W \subset X$ is open and simply connected, then W is evenly covered by f.*

PROOF. Let $\{O_\alpha\}$ be the connected components of $f^{-1}(W)$. It is easy to see that the restriction $f : O_\alpha \to W$ is a covering map for each α (see problem 1). By Corollary 12.10, this covering map must be a homeomorphism. □

Our next goal is to give alternative topological characterizations of covering maps that are often easier to verify. We first need two definitions.

> **DEFINITION 12.12.** An open-ended curve $\gamma : [a, b) \to Y$ is **escaping** if $\gamma(t)$ eventually leaves every compact subset of Y as $t \to b$, that is, if for every compact set $K \subset Y$ there is an $\varepsilon > 0$ such that $\gamma(t) \notin K$ whenever $b - \varepsilon < t < b$.

This is reminiscent of the definition of escaping sequences in §6.3.

Being escaping, of course, depends on the underlying space. For example, $\gamma(t) = t$ is escaping as a curve $[0, 1) \to \mathbb{D}$, but not as a curve $[0, 1) \to \mathbb{C}$.

It is not hard to see that when Y is locally compact,

$$\gamma \text{ is escaping} \iff \bigcap_{a < t < b} \overline{\gamma((t, b))} = \emptyset \iff \lim_{t \to b} \gamma(t) = \infty.$$

Here ∞ is the point at infinity in the one-point compactification of Y (see problem 9).

> **DEFINITION 12.13.** A point $q \in X$ is called an **asymptotic value** of a continuous map $f : Y \to X$ if there is an escaping curve $\gamma : [a, b) \to Y$ such that $\lim_{t \to b} f(\gamma(t)) = q$.

When Y is compact, no continuous map $Y \to X$ can have an asymptotic value because there are no escaping curves in Y.

EXAMPLE 12.14. As maps $\mathbb{C} \to \mathbb{C}$, non-constant polynomials have no asymptotic values, while the exponential function has 0 as its (unique) asymptotic value.

A topological space is **locally simply connected** if every neighborhood of every point contains a simply connected neighborhood of that point. Any open subset of such a space is again locally simply connected and therefore locally path-connected as well.

THEOREM 12.15 (Characterizations of covering maps). *Assuming X is locally simply connected, the following conditions on a local homeomorphism $f : Y \to X$ are equivalent:*

> *(i) f is a covering map.*
> *(ii) f has no asymptotic values in X.*
> *(iii) f has the curve lifting property.*

The implications

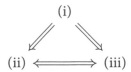

hold without extra assumptions on X. Local simple connectivity of X is only needed in the proof of (iii) $\implies$ (i) (compare problems 2 and 3).

PROOF. (i) $\implies$ (ii): Suppose f is a covering map with an asymptotic value $q \in X$. Take an escaping curve $\gamma : [0, 1) \to Y$ whose image $\eta = f \circ \gamma$ satisfies $\lim_{t \to 1} \eta(t) = q$. Take an evenly covered neighborhood W of q and find $0 < a < 1$ such that $\eta(t) \in W$ whenever $a \leq t < 1$. Let O be the open set in the decomposition of $f^{-1}(W)$ which contains $\gamma(a)$ and denote by $g : W \to O$ the corresponding local inverse of f. Both $g \circ \eta$ and γ are lifts of η on $[a, 1)$ and they coincide at $t = a$. By uniqueness, $g \circ \eta = \gamma$ on $[a, 1)$. In particular, $\lim_{t \to 1} \gamma(t) = g(q)$ and γ has a continuous extension to $[0, 1]$. This contradicts the hypothesis that γ is escaping.

(ii) $\implies$ (iii): The image $f(Y)$ is connected since Y is, and it is open since f is a local homeomorphism. First we prove that $f : Y \to f(Y)$ has the curve lifting property, then we show that $f(Y) = X$. Take a curve $\gamma : [0, 1] \to f(Y)$ with $\gamma(0) = x_0$ and choose any $y_0 \in Y$ such that $f(y_0) = x_0$. Define

$$T = \left\{ t \in [0, 1] : \gamma|_{[0,t]} \text{ has a lift under } f \text{ that starts at } y_0 \right\}.$$

Notice that T contains a neighborhood of 0 in $[0, 1]$ since f is a local homeomorphism. Moreover, $t \in T$ implies $[0, t] \subset T$. Our goal is to show that $T = [0, 1]$.

Let $s = \sup T$, so $0 < s \leq 1$. For each $t \in (0, s)$ the restriction $\gamma|_{[0,t]}$ has a unique lift that starts at y_0. By Lemma 12.4, these lifts are compatible on their common domain, so they give rise to a well-defined lift $\tilde{\gamma} : [0, s) \to Y$. Since $\lim_{t \to s} f(\tilde{\gamma}(t)) = \lim_{t \to s} \gamma(t) = \gamma(s) \in X$, and since f has no asymptotic value, it follows that $\tilde{\gamma}|_{[0,s)}$ is not escaping. Thus, there is a compact set $K \subset Y$ and a sequence $t_n \to s$ such that $\tilde{\gamma}(t_n) \in K$ for all n. We can then find some $p \in K$ such that every neighborhood of p contains $\tilde{\gamma}(t_n)$ for infinitely many n (this is trivial if $\{\tilde{\gamma}(t_n)\}$ is a finite set; otherwise, let p be an accumulation point of $\{\tilde{\gamma}(t_n)\}$). By continuity, $f(p) = \gamma(s)$. Take neighborhoods O of p and W of $\gamma(s)$ such that the restriction $f : O \to W$ is a homeomorphism with the inverse $g : W \to O$. For small enough $\varepsilon > 0$, the curve $g \circ \gamma$ is a lift of γ on $[s - \varepsilon, s)$ which coincides with $\tilde{\gamma}$ at $t = t_n$ for some large n. It follows from uniqueness that $\tilde{\gamma} = g \circ \gamma$ on $[s - \varepsilon, s)$. If $s < 1$, we could extend $\tilde{\gamma}$ further by setting $\tilde{\gamma} = g \circ \gamma$ on $[s, s + \delta]$ for a small $\delta > 0$, which would contradict $s = \sup T$. Thus $s = 1$ and we can extend $\tilde{\gamma}$ continuously to $[0, 1]$ by setting $\tilde{\gamma}(1) = (g \circ \gamma)(1) = p$.

To prove surjectivity of f, assume $f(Y) \neq X$. Since X is path-connected, there is a curve $\gamma : [0, 1] \to X$ such that $\gamma(0) \in f(Y)$ but $\gamma(1) \in X \setminus f(Y)$. The preimage $\gamma^{-1}(f(Y))$ is relatively open in $[0, 1]$, so its connected component containing 0 has the form $[0, a)$ for some $0 < a < 1$. Thus, $\gamma(t) \in f(Y)$ for all $t \in [0, a)$ but $\gamma(a) \in X \setminus f(Y)$. Fix some $y_0 \in f^{-1}(\gamma(0))$. By what we showed above, for each $0 < t < a$ there is a lift $\tilde{\gamma} : [0, t] \to Y$ of $\gamma|_{[0,t]}$ starting at y_0, and these lifts are compatible, so they give rise to a lift $\tilde{\gamma} : [0, a) \to Y$ of $\gamma|_{[0,a)}$. Since f has no asymptotic value, $\tilde{\gamma}|_{[0,a)}$ is not escaping. An argument similar to the above paragraph then gives a point $p \in Y$ and a sequence $t_n \to a$ such that every neighborhood of p contains $\tilde{\gamma}(t_n)$ for infinitely many n. By continuity, $\gamma(a) = f(p) \in f(Y)$, a contradiction.

(iii) $\Longrightarrow$ (i): Suppose f has the curve lifting property. We claim that any simply connected neighborhood W of a point in X is evenly covered. To see this, decompose the open set $f^{-1}(W)$ into its path components $\{O_\alpha\}$. Each O_α is path-connected by definition, and is open since Y is locally path-connected. Moreover, each restriction $f : O_\alpha \to W$ is surjective. In fact, take any $x \in W$, choose any $y_0 \in O_\alpha$, and let $\gamma : [0, 1] \to W$ be a curve with $\gamma(0) = f(y_0)$ and $\gamma(1) = x$. Let $\tilde{\gamma} : [0, 1] \to Y$ be the unique lift of γ with $\tilde{\gamma}(0) = y_0$. As a curve in $f^{-1}(W)$ with its initial point in O_α, the lift $\tilde{\gamma}$ must remain in O_α. It follows that $x = f(\tilde{\gamma}(1)) \in f(O_\alpha)$, which proves surjectivity. Now each restriction $f : O_\alpha \to W$ is a local homeomorphism with the curve lifting property, so by Corollary 12.10 it must be a homeomorphism. □

THEOREM 12.16. *Consider the commutative diagram*

$$\begin{array}{ccc} & & Y \\ & \overset{g}{\nearrow} & \downarrow{\scriptstyle f} \\ Z & \underset{h}{\longrightarrow} & X \end{array}$$

in which the maps are continuous and the spaces are locally simply connected.

   (i) If g and f are covering maps, so is h.
   (ii) If h and f are covering maps, so is g.
   (iii) If h is a covering map and g is open and surjective, then both f and g are covering maps.

In (i) the assumption of local simple connectivity is essential when working with covering maps of infinite degree, while (ii) holds even without this assumption (see problems 3 and 4).

PROOF. In view of Theorem 12.15, it is enough to show that the relevant map in each case is a local homeomorphism with the curve lifting property.

(i) This is easy: Since f and g are local homeomorphisms with the curve lifting property, so is their composition h.

(ii) Let us first check that g is a local homeomorphism. Take any $z_0 \in Z$ and let $y_0 = g(z_0)$ and $x_0 = f(y_0) = h(z_0)$. Find a path-connected neighborhood W of x_0 that is evenly covered by both h and f and let O_1 and O_2 be the corresponding path-connected neighborhoods of z_0 and y_0 such that the restrictions $h : O_1 \to W$ and $f : O_2 \to W$ are homeomorphisms. We have $g(O_1) \subset O_2$ since $g(O_1)$ is a path-connected subset of $f^{-1}(W)$ containing y_0. It follows that $g = f^{-1} \circ h$ in O_1, where $f^{-1} : W \to O_2$ is the local inverse of f. Thus, $g : O_1 \to O_2$ is a homeomorphism.

Now let $y \in Y$ and $\gamma : [0, 1] \to Y$ be any curve with $\gamma(0) = y_0, \gamma(1) = y$. Let η be the unique lift of $f \circ \gamma$ under h with $\eta(0) = z_0$. Then both γ and $g \circ \eta$ are lifts of $f \circ \gamma$ under f with the initial point y_0. By uniqueness, $\gamma = g \circ \eta$, so η is a lift of γ under

g. In particular, $y = g(\eta(1)) \in g(Z)$. This proves that g is surjective and has the curve lifting property.

(iii) By part (ii) it suffices to prove that f is a covering map. The relation $f \circ g = h$ shows that f is surjective. Let us check that f is a local homeomorphism. Take any $y_0 \in Y$ and use surjectivity of g to find $z_0 \in Z$ such that $g(z_0) = y_0$. Set $x_0 = f(y_0) = h(z_0)$. Find a neighborhood W of x_0 that is evenly covered by h and let O be the corresponding neighborhood of z_0 such that the restriction $h : O \to W$ is a homeomorphism. The image $V = g(O)$ is a neighborhood of y_0 since g is an open map, and the relation $f \circ g = h$ in O shows that f maps V bijectively onto W. The local inverse $f^{-1} : W \to V$ is continuous since it is the composition of the local inverse $h^{-1} : W \to O$ followed by g.

Now let $\gamma : [0, 1] \to X$ be any curve with $\gamma(0) = x_0$. Take the unique lift η of γ under h with $\eta(0) = z_0$. Then $g \circ \eta$ is a lift of γ under f with $(g \circ \eta)(0) = y_0$. This shows that f has the curve lifting property.   □

DEFINITION 12.17. A covering map $f : Y \to X$ is called **_universal_** if Y is simply connected.

The term "universal" is suggested by the fact that if $f : Y \to X$ is a covering map with Y simply connected, and if $g : Z \to X$ is any covering map, then f has a lift $\tilde{f} : Y \to Z$ under g (by Theorem 12.8(ii)) which is also a covering map (by Theorem 12.16(ii); see also problem 4). In other words, Y is a covering space of any covering space of X.

REMARK 12.18. Universal covering maps are unique up to isomorphism: If $f_0 : Y_0 \to X$ and $f_1 : Y_1 \to X$ are universal coverings and $y_0 \in Y_0$ and $y_1 \in Y_1$ satisfy $f_0(y_0) = f_1(y_1) = x_0$, then there is a unique homeomorphism $\varphi : Y_0 \to Y_1$ with $\varphi(y_0) = y_1$ such that $f_0 = f_1 \circ \varphi$. Uniqueness of such φ follows from Lemma 12.4. For the existence, use Theorem 12.8(ii) to find a lift $\varphi : Y_0 \to Y_1$ of f_0 under f_1 with $\varphi(y_0) = y_1$ and a lift $\psi : Y_1 \to Y_0$ of f_1 under f_0 with $\psi(y_1) = y_0$:

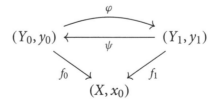

Since $f_0 \circ \psi \circ \varphi = f_1 \circ \varphi = f_0$, we see that $\psi \circ \varphi$ is a lift of f_0 under f_0 which fixes y_0. By Lemma 12.4, $\psi \circ \varphi$ is the identity map of Y_0. Similarly, $\varphi \circ \psi$ is the identity map of Y_1, which shows φ, ψ are homeomorphisms with $\varphi = \psi^{-1}$.

Presumably "deck" comes from the German *decken* which means "to cover."

DEFINITION 12.19. A **_deck transformation_** of a covering map $f : Y \to X$ is a homeomorphism $\varphi : Y \to Y$ which preserves each fiber of f, that is,

$$f \circ \varphi = f.$$

The collection of all such φ form a group under composition called the **deck group** of the covering $f : Y \to X$.

EXAMPLE 12.20. The deck group of the covering map $\exp : \mathbb{C} \to \mathbb{C}^*$ is infinite cyclic generated by the translation $z \mapsto z + 2\pi i$. For any integer $n \neq 0$, the deck group of the covering map $f : \mathbb{C}^* \to \mathbb{C}^*$ defined by $f(z) = z^n$ is cyclic of order $|n|$ generated by the rotation $z \mapsto e^{2\pi i/n} z$. Both statements are easy exercises (compare Example 12.23).

The following theorem summarizes basic properties of the deck group of a universal covering. For an analogous result on general covering spaces, see problem 6.

THEOREM 12.21. *Let $f : Y \to X$ be a universal covering with the associated deck group G.*

(i) *If $f(y_0) = f(y_1)$ for some $y_0, y_1 \in Y$, there is a unique $\varphi \in G$ such that $\varphi(y_0) = y_1$.*
(ii) *$\varphi(y) \neq y$ for all $y \in Y$ and $\varphi \in G \smallsetminus \{\mathrm{id}_Y\}$.*
(iii) *More generally, every point in Y has a neighborhood U with the property that $\varphi(U) \cap U = \varnothing$ for all $\varphi \in G \smallsetminus \{\mathrm{id}_Y\}$.*
(iv) *The quotient space Y/G, consisting of all orbits $G \cdot y = \{\varphi(y) : \varphi \in G\}$ for $y \in Y$, is homeomorphic to X.*
(v) *For any base point $x_0 \in X$, the fundamental group $\pi_1(X, x_0)$ is isomorphic to G.*

One expresses (i) by saying that the action of G on the fibers of f is **simply transitive**, and (ii) by saying that G acts **freely** on Y. Part (iii) is a strong form of what is known as a **properly discontinuous** action.

It follows from (i) that $f^{-1}(x) = G \cdot y$ if $f(y) = x$. In other words, *the fibers of f are precisely the orbits of G.* Thus the assignment $G \to f^{-1}(x)$ given by $\varphi \mapsto \varphi(y)$ is bijective and G has the same cardinality as the fiber $f^{-1}(x)$, which is $\deg(f)$.

The topology on Y/G is the smallest topology which makes the natural projection $\Pi : Y \to Y/G$ defined by $\Pi(y) = G \cdot y$ continuous, that is, $U \subset Y/G$ is open precisely when $\Pi^{-1}(U)$ is open in Y. The proof will show that there is a homeomorphism $h : Y/G \to X$ which makes the following diagram commute:

$$
\begin{array}{ccc}
 & Y & \\
{\scriptstyle \Pi} \swarrow & & \searrow {\scriptstyle f} \\
Y/G & \xrightarrow{\quad h \quad} & X
\end{array}
$$

In particular, $\Pi : Y \to Y/G$ itself is a covering map.

PROOF. (i) follows from Remark 12.18 since both (Y, y_0) and (Y, y_1) are universal covering spaces of (X, x_0), where $x_0 = f(y_0) = f(y_1)$.

(ii) If $\varphi(y) = y$ for some $\varphi \in G$ and $y \in Y$, then $\varphi = \mathrm{id}_Y$ by the uniqueness part in (i).

(iii) Let $y_0 \in Y$ and $x_0 = f(y_0)$. Take an evenly covered path-connected neighborhood W of x_0, and write $f^{-1}(W)$ as the disjoint union of the path components $\{O_\alpha\}$ each mapping homeomorphically by f onto W. We claim that every deck transformation φ permutes the components O_α. In fact, $\varphi(O_\alpha)$ is a path-connected subset of $f^{-1}(W)$, so $\varphi(O_\alpha) \subset O_\beta$ for some β. Applying φ^{-1} gives $O_\alpha \subset \varphi^{-1}(O_\beta)$, and since $\varphi^{-1}(O_\beta)$ is a path-connected subset of $f^{-1}(W)$, we must have $O_\alpha = \varphi^{-1}(O_\beta)$, or $\varphi(O_\alpha) = O_\beta$. It follows that if U is the component O_α containing y_0 and if $\varphi(U) \cap U \neq \emptyset$, then $\varphi(U) = U$. In particular, $\varphi(y_0)$ lies in U and belongs to the fiber $f^{-1}(x_0)$. This implies $\varphi(y_0) = y_0$, so $\varphi = \mathrm{id}_Y$ by (ii).

(iv) Define $h : Y/G \to X$ by $h(G \cdot y) = f(y)$. Since the fibers of f are the orbits of G, we see that h is well defined and bijective. If $W \subset X$ is open, then $\Pi^{-1}(h^{-1}(W)) = f^{-1}(W)$ is open, so by the definition of the quotient topology, $h^{-1}(W)$ is open. This proves that h is continuous. Similarly, if $W \subset Y/G$ is open, then $h(W) = f(\Pi^{-1}(W))$ is open since Π is continuous and f, being a local homeomorphism, is an open map. This proves continuity of h^{-1}.

(v) Fix $y_0 \in f^{-1}(x_0)$. Given $\varphi \in G$, let γ be any curve in Y from y_0 to $\varphi(y_0)$. Since Y is simply connected, the homotopy class $[\gamma]$ is independent of the choice of γ (Theorem 2.7) and therefore the same must be true of the homotopy class $f_*([\gamma]) = [f \circ \gamma] \in \pi_1(X, x_0)$. We show that the map $\Phi : G \to \pi_1(X, x_0)$ defined by $\Phi(\varphi) = [f \circ \gamma]$ is a group isomorphism.

Given $\varphi, \psi \in G$, take curves γ and η in Y from y_0 to $\varphi(y_0)$ and $\psi(y_0)$, respectively. The product $\gamma \cdot (\varphi \circ \eta)$ is a curve from y_0 to $(\varphi \circ \psi)(y_0)$. Hence,

$$\Phi(\varphi \circ \psi) = [f \circ (\gamma \cdot (\varphi \circ \eta))] = [(f \circ \gamma) \cdot (f \circ \varphi \circ \eta)]$$

$$= [f \circ \gamma] \cdot [f \circ \eta] = \Phi(\varphi) \cdot \Phi(\psi).$$

Thus, Φ is a homomorphism. For any homotopy class $[\xi] \in \pi_1(X, x_0)$, let $\tilde{\xi}$ be the unique lift of ξ under f with the initial point y_0. The end point y_1 of $\tilde{\xi}$ depends only on $[\xi]$ by Theorem 12.8(i) and the unique $\varphi \in G$ which sends y_0 to y_1 satisfies $\Phi(\varphi) = [f \circ \tilde{\xi}] = [\xi]$. This proves surjectivity of Φ. In the special case where $[\xi] = [\varepsilon_{x_0}]$ is trivial, we have $y_1 = y_0$ and $\varphi = \mathrm{id}_Y$, which proves Φ is injective. $\square$

REMARK 12.22. The question of the existence of a universal covering for X has a definitive answer under our standing assumption that X is locally path connected. In this case, it is not hard to show that there is a universal covering $u : \tilde{X} \to X$ if and only if X is ***semi-locally simply connected*** in the sense that every $x \in X$ has a neighborhood W such that the natural homomorphism $\pi_1(W, x) \to \pi_1(X, x)$ is trivial. The points of $\tilde{X}$ are the homotopy classes of curves in X that start at a given base point and the map u assigns to each such homotopy class its endpoint (see [**Fu**] or [**H**] for details). As a special case, it follows that every domain $U \subset \hat{\mathbb{C}}$ has a universal covering $u : \tilde{U} \to U$, where the topological space $\tilde{U}$ is a "surface" since it is locally homeomorphic to an open disk in the plane. We will meet this fact again in §13.2.

12.2 Holomorphic coverings and inverse branches

We now specialize to our familiar setting of *holomorphic* covering maps between spherical domains. Let $f : U \to V$ be a holomorphic covering map between domains in $\hat{\mathbb{C}}$ with the deck group G. By definition, each $\varphi \in G$ is a homeomorphism of U which satisfies $f \circ \varphi = f$. For every $p \in U$ we can write $\varphi = g \circ f$ in a small neighborhood of p, where g is the local inverse of f mapping $f(p)$ to $\varphi(p)$. This shows that φ is holomorphic in U. Thus, the deck group G is a subgroup of the automorphism group $\mathrm{Aut}(U)$ consisting of all biholomorphisms $U \to U$.

The case of a holomorphic universal covering $f : U \to V$ where U is simply connected is of special interest. In this case the Riemann mapping theorem tells us that U is conformally isomorphic to $\hat{\mathbb{C}}$, $\hat{\mathbb{C}} \smallsetminus \{\text{point}\} \cong \mathbb{C}$, or $\mathbb{D} \cong \mathbb{H}$. Whichever the case, it follows that $G \subset \mathrm{Aut}(U)$ is isomorphic to a subgroup of $\mathrm{M\ddot{o}b} \cong \mathrm{PSL}_2(\mathbb{C})$ (see §4.1). The possibilities for G and the resulting domains $V \cong U/G$ are discussed in chapter 13, where we prove the uniformization theorem according to which every spherical domain has $\hat{\mathbb{C}}$, $\mathbb{C}$, or $\mathbb{D}$ as its holomorphic universal covering.

EXAMPLE 12.23 (The universal covering of a round annulus). Let V denote the round annulus $\{w : e^a < |w| < e^b\}$, where $-\infty \leq a < b \leq +\infty$. The deck group G of the universal covering

$$\exp : U = \{z : a < \mathrm{Re}(z) < b\} \to V$$

is the infinite cyclic group generated by the translation $T : z \mapsto z + 2\pi i$. In fact, G contains T since $\exp \circ T = \exp$. On the other hand, if $\varphi \in G$, the condition $\exp \circ \varphi = \exp$ implies that $z \mapsto (\varphi(z) - z)/(2\pi i)$ is an integer-valued function, hence a constant $n \in \mathbb{Z}$ by continuity, which shows $\varphi = T^{\circ n}$. Note that the universal covering domain U is conformally isomorphic to $\mathbb{D}$ unless $a = -\infty$, $b = +\infty$, in which case it is isomorphic to $\mathbb{C}$.

It is hard to overlook the similarity between the lifting arguments in covering space theory and the process of analytic continuation along curves as described in §10.2. The following theorem gives this similarity some context by providing an alternative characterization of holomorphic covering maps in terms of analytic continuation:

THEOREM 12.24. *A surjective holomorphic map $f : U \to V$ without critical points is a covering map if and only if each local branch of f^{-1} in V continues analytically along every curve in V.*

By a local branch of f^{-1} in V we mean a function element (g, D), where D is a disk in V, $g(D) \subset U$, and $f(g(z)) = z$ for all $z \in D$.

PROOF. First suppose f is a covering map. Take any local branch (g_0, D_0) of f^{-1} in V and set $B_0 = g(D_0)$. Let γ be a curve in V starting at the center of D_0. As in the proof of Lemma 12.6, there is a partition $t_0 = 0 < t_1 < \cdots < t_n = 1$ such that $\gamma([t_k, t_{k+1}])$ is contained in an evenly covered disk D_k for $0 \le k \le n - 1$. Define the function elements (g_k, D_k) for $1 \le k \le n - 1$ inductively as follows: Suppose (g_{k-1}, D_{k-1}) is defined. Let B_k be the connected component of $f^{-1}(D_k)$ that contains $g_{k-1}(D_{k-1} \cap D_k)$ and let $g_k : D_k \to B_k$ be the inverse of the biholomorphic restriction $f : B_k \to D_k$. Since $f \circ g_{k-1} = f \circ g_k$ in $D_{k-1} \cap D_k$ and since $g_{k-1}(D_{k-1} \cap D_k)$ and $g_k(D_{k-1} \cap D_k)$ are both contained in B_k in which f is injective, we must have $g_{k-1} = g_k$ in $D_{k-1} \cap D_k$. This completes the induction step and shows that (g_0, D_0) continues analytically along γ.

Conversely, suppose each local branch of f^{-1} in V continues analytically along every curve in V. Take a curve $\gamma : [0, 1] \to V$ and some $p \in f^{-1}(\gamma(0))$. Since p is not a critical point, there is a local branch (g, D) of f^{-1} where $\gamma(0)$ is the center of D and $g(\gamma(0)) = p$. By the assumption, the function element (g, D) continues analytically along γ. The lift $\tilde{\gamma}$ of γ determined by (g, D) in the sense of §10.2 is a curve in U which starts at p and satisfies $f \circ \tilde{\gamma} = \gamma$. This shows that f has the curve lifting property. Since f is a local homeomorphism by the absence of critical points, we conclude from Theorem 12.15 that f is a covering map. □

THEOREM 12.25 (Covering restrictions of holomorphic maps). *Let $f : \Omega \to f(\Omega)$ be non-constant holomorphic, $V \subset f(\Omega)$ be a domain which contains no asymptotic value of f, and $U \subset \Omega$ be a connected component of $f^{-1}(V)$ which contains no critical point of f. Then the restriction $f : U \to V$ is a covering map.*

PROOF. The absence of critical points in U guarantees that $f|_U$ is a local homeomorphism. By Theorem 12.15 it suffices to show that $f|_U$ has no asymptotic value in V. Suppose $\gamma : [0, 1) \to U$ is escaping and $f(\gamma(t))$ tends to some $q \in V$ as $t \to 1$. Since by the assumption there are no asymptotic values of f in V, it follows that γ, viewed as a curve $[0, 1) \to \Omega$, is not escaping. Thus, $\gamma(t)$ must have an accumulation point $p \in \Omega$ as $t \to 1$, where $f(p) = q$ by continuity. The open mapping theorem then shows that some neighborhood of p maps into V and therefore is contained in U. In particular $p \in U$, which contradicts $\gamma : [0, 1) \to U$ being escaping. □

Recall from Corollary 2.22 (see also Example 10.19) that there are holomorphic branches of $\log z$ and $\sqrt[n]{z}$ in every simply connected domain in $\mathbb{C}^*$. Theorem 12.25 gives the following important generalization of this fact:

In some classical literature this result is called the "monodromy theorem."

COROLLARY 12.26 (Existence of global inverse branches). *Let $f : \Omega \to f(\Omega)$ be non-constant holomorphic and $V \subset f(\Omega)$ be a simply connected domain which contains no asymptotic or critical value of f. Then, every connected component U of $f^{-1}(V)$ is simply connected and $f : U \to V$ is a biholomorphism. In particular, there are holomorphic branches of the inverse f^{-1} defined in V.*

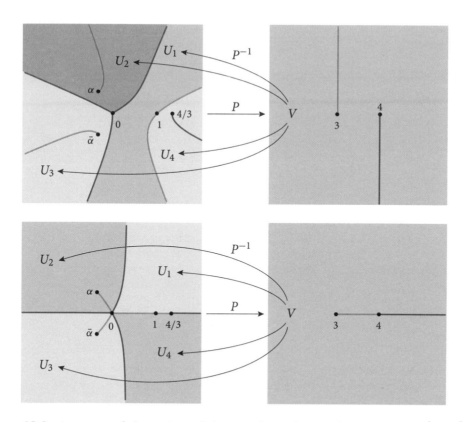

Figure 12.2. Anatomy of the action of the complex polynomial map $P(z) = 3z^4 - 4z^3 + 4$, described in Example 12.27. Here the critical points $0, 1$ have the corresponding critical values $P(0) = 4, P(1) = 3$, and $P^{-1}(4) = \{0, 4/3\}$, $P^{-1}(3) = \{1, \alpha, \bar{\alpha}\}$, where $\alpha = (-1 + \sqrt{2}i)/3$. In every simply connected domain $V \subset \mathbb{C} \smallsetminus \{3, 4\}$ there are four holomorphic branches of P^{-1} which map V onto simply connected domains U_k, $1 \le k \le 4$. Top: V is a doubly-slit plane. Bottom: V is the slit plane $\mathbb{C} \smallsetminus [3, +\infty)$.

PROOF. By Theorem 12.25, $f : U \to V$ is a covering map. Since V is simply connected, Corollary 12.10 shows that $f : U \to V$ is a biholomorphism. □

EXAMPLE 12.27. The polynomial $P(z) = 3z^4 - 4z^3 + 4$ has one critical point of local degree 3 at $z = 0$ and another of local degree 2 at $z = 1$, with critical values $P(0) = 4$ and $P(1) = 3$. P has no asymptotic value as a map $\mathbb{C} \to \mathbb{C}$ since $\lim_{z \to \infty} P(z) = \infty$. Take two disjoint embedded arcs Σ_1, Σ_2 from the critical values to ∞; we refer to Σ_1, Σ_2 as "slits" (in the top right frame of Fig. 12.2, these slits are taken to be vertical lines). Let V denote the simply connected domain $\mathbb{C} \smallsetminus (\Sigma_1 \cup \Sigma_2)$. By Corollary 12.26, every connected component of $P^{-1}(V)$ is simply connected and maps conformally onto V, resulting in a holomorphic branch of P^{-1} defined in V. It is not hard to see that there are four such components, separated from one another by the components of $P^{-1}(\Sigma_1 \cup \Sigma_2)$, which are slits from the points in $P^{-1}(\{3, 4\})$ to ∞. For comparison, the bottom right frame of Fig. 12.2 shows a different choice of V obtained by removing from $\mathbb{C}$ a horizontal slit passing through both critical values.

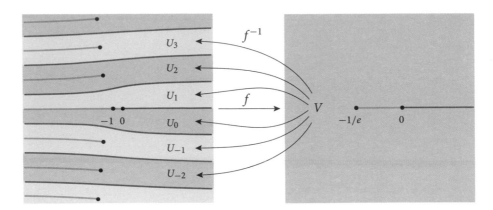

Figure 12.3. Anatomy of the action of the transcendental entire function $f(z) = z\,e^z$, described in Example 12.28. Here the unique critical point at -1 has the critical value $f(-1) = -1/e$, and 0 is the unique asymptotic value. There are countably many holomorphic branches of f^{-1} mapping the slit plane $V = \mathbb{C} \smallsetminus [-1/e, +\infty)$ onto the simply connected domains U_k for $k \in \mathbb{Z}$.

EXAMPLE 12.28. Let us now construct global inverse branches for the transcendental entire function $f(z) = z\,e^z$ which has a unique critical point of local degree 2 at $z = -1$. Unlike the polynomial case, there is now a unique asymptotic value at 0. To see uniqueness, suppose $\gamma : [0, 1) \to \mathbb{C}$ is a curve with $\lim_{t \to 1} |\gamma(t)| = +\infty$ such that $\lim_{t \to 1} f(\gamma(t)) = \lim_{t \to 1} \gamma(t) \exp(\gamma(t)) = q \neq 0$. Then $\lim_{t \to 1} |\gamma(t)| \exp(\mathrm{Re}(\gamma(t))) = |q|$, which implies $\lim_{t \to 1} \mathrm{Re}(\gamma(t)) = -\infty$. In particular, for t close to 1, we can write $\gamma(t) = \exp(\eta(t))$ where η is continuous and $\pi/2 < \mathrm{Im}(\eta(t)) < 3\pi/2$. The estimates

$$\mathrm{Re}(\eta(t)) \leq |\eta(t)| \leq \mathrm{Re}(\eta(t)) + \frac{3\pi}{2}$$

then show that

(12.2)
$$\lim_{t \to 1} \frac{|\eta(t)|}{\log |\gamma(t)|} = \lim_{t \to 1} \frac{|\eta(t)|}{\mathrm{Re}(\eta(t))} = 1.$$

On the other hand,

$$\lim_{t \to 1} \exp(\eta(t) + \gamma(t)) = \lim_{t \to 1} \gamma(t) \exp(\gamma(t)) = q \neq 0,$$

which shows that $\gamma(t) + \eta(t)$ has a well-defined limit in $\mathbb{C}$ as $t \to 1$. Thus, $\lim_{t \to 1} |\eta(t)|/|\gamma(t)| = 1$, in clear contradiction with (12.2).

To construct holomorphic branches of f^{-1}, take a slit Σ which joins the critical value $f(-1) = -1/e$ to the asymptotic value 0 and then goes out to infinity (in Fig. 12.3, this slit is taken to be the real interval $[-1/e, +\infty)$ for simplicity). Let V be the simply connected domain $\mathbb{C} \smallsetminus \Sigma$. By Corollary 12.26, every connected component of $f^{-1}(V)$ is simply connected and maps conformally onto V, resulting in a holomorphic branch of f^{-1} defined in V. There are countably many such components, separated from one another by the components of $f^{-1}(\Sigma)$, which are either slits starting from a preimage of $-1/e$ and going out to ∞, or open arcs that go out to ∞ in both directions.

The branches of f^{-1} collectively are called the *Lambert W-function*, a multi-valued special function with wide-ranging applications.

Notice how this topological picture provides an illustration of Picard's theorem (Corollary 11.16), as f takes on the value 0 only once, but every other value infinitely often.

12.3 Proper maps and branched coverings

We begin with a simple characterization of finite degree covering maps. We continue assuming that U, V are domains in $\hat{\mathbb{C}}$.

> **DEFINITION 12.29.** A continuous map $f : U \to V$ is **proper** if for every compact set $K \subset V$ the preimage $f^{-1}(K) \subset U$ is compact.

The choice of the target space in this definition is crucial. For example, $f(z) = z^2$ is proper as a map $\mathbb{D} \to \mathbb{D}$, but it is not proper as a map $\mathbb{D} \to \mathbb{C}$.

It is an easy exercise to show that $f : U \to V$ is proper if and only if for every espacing sequence $\{p_n\}$ in U (in the sense of §6.3), the image $\{f(p_n)\}$ is an escaping sequence in V. Equivalently, if $p_n \to \partial U$ implies $f(p_n) \to \partial V$, where the boundaries are taken in $\hat{\mathbb{C}}$. Thus, *proper maps do not have asymptotic values* in the sense of Definition 12.13.

EXAMPLE 12.30 (Proper holomorphic self-maps of the plane). Every non-constant polynomial map $f : \mathbb{C} \to \mathbb{C}$ is proper. In fact, if $K \subset \mathbb{C}$ is compact, then $f^{-1}(K)$ is closed since f is continuous, and is bounded since $f(z) \to \infty$ as $z \to \infty$. Conversely, if $f : \mathbb{C} \to \mathbb{C}$ is a proper entire function, for every $R > 0$ the preimage $f^{-1}(\{w \in \mathbb{C} : |w| > R\})$ is the complement of a compact set, so it contains $\{z \in \mathbb{C} : |z| > r\}$ for some $r = r(R) > 0$. This means $\lim_{z \to \infty} f(z) = \infty$, which shows that f is a non-constant polynomial.

EXAMPLE 12.31 (Proper holomorphic self-maps of the disk). A holomorphic map $\mathbb{D} \to \mathbb{D}$ is proper if and only if it is a finite Blaschke product of the form

$$(12.3) \qquad B(z) = \alpha \prod_{n=1}^{d} \left(\frac{z - p_n}{1 - \overline{p_n} z} \right) \qquad \text{with } |\alpha| = 1 \text{ and } |p_n| < 1 \text{ for } 1 \le n \le d.$$

In fact, each B in the above form is holomorphic in some neighborhood of $\overline{\mathbb{D}}$, with $|B(z)| < 1$ if $|z| < 1$ and $|B(z)| = 1$ if $|z| = 1$. This shows that $B : \mathbb{D} \to \mathbb{D}$ is proper. The converse statement was proved in Lemma 7.51, but here is an alternative argument. Suppose $f : \mathbb{D} \to \mathbb{D}$ is proper and holomorphic. Then $|f(z)| \to 1$ as $|z| \to 1$, so we can use the Schwarz reflection principle to extend f to a holomorphic map $F : \hat{\mathbb{C}} \to \hat{\mathbb{C}}$ which satisfies

$$(12.4) \qquad F\left(\frac{1}{\overline{z}} \right) = \frac{1}{\overline{F(z)}} \qquad \text{for all } z.$$

By Theorem 3.20, F is a rational function which satisfies $F(\mathbb{D}) = \mathbb{D}$ and $F(\hat{\mathbb{C}} \smallsetminus \overline{\mathbb{D}}) = \hat{\mathbb{C}} \smallsetminus \overline{\mathbb{D}}$. In particular, all zeros of F are in $\mathbb{D}$ and all poles are outside $\overline{\mathbb{D}}$. If $p_1, \ldots, p_d$ are the zeros of F

repeated according to their orders, it follows from (12.4) that $1/\overline{p_1}, \ldots, 1/\overline{p_d}$ are the poles of F repeated according to their orders. The Blaschke product $B(z) = \prod_{n=1}^{d}(z - p_n)/(1 - \overline{p_n}z)$ has the same zeros and poles of the same orders as F. Hence F/B is a rational function without zeros and poles, so it must be constant. This shows $F = \alpha B$ for some $\alpha \in \mathbb{C}$. Since $|F(z)| = |B(z)| = 1$ when $|z| = 1$, we must have $|\alpha| = 1$.

EXAMPLE 12.32. There are no proper holomorphic maps $\mathbb{C} \to \mathbb{D}$ or $\mathbb{D} \to \mathbb{C}$. The former follows from Liouville's theorem. For the latter, suppose $f : \mathbb{D} \to \mathbb{C}$ is proper and holomorphic. Then $\lim_{|z| \to 1} f(z) = \infty$ and in particular f has only finitely many zeros. If B is a Blaschke product of the form (12.3) with the same zeros as f, then B/f extends to a non-vanishing holomorphic function in $\mathbb{D}$ which satisfies $\lim_{|z| \to 1} B(z)/f(z) = 0$. The maximum principle then implies that $B/f = 0$ everywhere, which is a contradiction.

As a consequence, we see that up to biholomorphic change of coordinates, every proper holomorphic map $f : U \to V$ between *simply connected* domains is either a polynomial or a finite Blaschke product. Take conformal maps $\varphi : U \to U_0$ and $\psi : V \to V_0$, where U_0, V_0 are either $\mathbb{C}$ or $\mathbb{D}$. The induced map $g = \psi \circ f \circ \varphi^{-1} : U_0 \to V_0$ is proper and holomorphic. By the above paragraph, $U_0 = V_0$. It follows from Examples 12.30 and 12.31 that g is a polynomial or a finite Blaschke product according as $U_0 = V_0 = \mathbb{C}$ or $U_0 = V_0 = \mathbb{D}$.

THEOREM 12.33. $f : U \to V$ *is a finite degree covering map if and only if it is a proper local homeomorphism.*

PROOF. Suppose $f : U \to V$ is a covering map with $\deg(f) = d < \infty$. For every $q \in V$, take an evenly covered neighborhood W and decompose $f^{-1}(W)$ into the disjoint union of d open sets $O_1, \ldots, O_d$, where $f : O_k \to W$ is a homeomorphism for each k. If D is a neighborhood of q with compact closure $\overline{D} \subset W$, it follows that $f^{-1}(\overline{D}) \cap O_k$ is compact for each k. Thus, as a union of d compact sets, $f^{-1}(\overline{D})$ itself is compact. Now any compact set $K \subset V$ can be covered by finitely many such neighborhoods $D_1, \ldots, D_s$, so $f^{-1}(K)$ is a closed subset of the compact set $\bigcup_{k=1}^{s} f^{-1}(\overline{D_k})$. This shows $f^{-1}(K)$ is compact and proves that f is proper.

Conversely, if f is a proper local homeomorphism, it clearly has no asymptotic values in V, so by Theorem 12.15 it must be a covering map. For $q \in V$ the fiber $f^{-1}(q)$ is compact by properness and is discrete since f is a local homeomorphism. It follows that $f^{-1}(q)$ is a finite set, proving that $\deg(f) < \infty$. $\square$

COROLLARY 12.34. *A holomorphic map between spherical domains is a finite degree covering map if and only if it is proper and has no critical points.*

It would be desirable to study continuous maps between spherical domains that globally behave like covering maps but allow mild branching near isolated points, similar to the behavior of non-constant holomorphic functions near their critical points. These "branched covering" maps occur frequently in the natural setting of complex analysis, for example as proper holomorphic maps (Corollary 12.41 below). Along with their higher dimensional cousins, they have also become common tools in topology and geometry.

We begin with the local topological model of branched coverings. It will be convenient to adopt the notation $f : (A, a) \to (B, b)$ to indicate a function $f : A \to B$ with $f(a) = b$.

DEFINITION 12.35. Let O, W be simply connected domains in $\hat{\mathbb{C}}$. We call a continuous map $f : (O, p) \to (W, q)$ **powerlike** if there are homeomorphisms $\varphi : (O, p) \to (\mathbb{D}, 0)$ and $\psi : (W, q) \to (\mathbb{D}, 0)$ and an integer $m \geq 1$ such that $(\psi \circ f \circ \varphi^{-1})(w) = w^m$ for all $w \in \mathbb{D}$:

$$
\begin{array}{ccc}
(O, p) & \xrightarrow{\ f\ } & (W, q) \\
{\scriptstyle \varphi} \downarrow & & \downarrow {\scriptstyle \psi} \\
(\mathbb{D}, 0) & \xrightarrow[\ w \mapsto w^m\ ]{} & (\mathbb{D}, 0)
\end{array}
$$

It follows from the definition that every point in $W \smallsetminus \{q\}$ has m distinct preimages in $O \smallsetminus \{p\}$. The point q itself has a single preimage p, which should be thought of as having multiplicity m (compare the similar situation in Fig. 1.4 where $m = 3$). The integer m, which is independent of the choice of the coordinates φ and ψ, is called the **local degree** of f at p and is denoted by $\deg(f, p)$. Note that $\deg(f, p) = 1$ if and only if $f : (O, p) \to (W, q)$ is a homeomorphism.

It is natural to call a map $f : U \to V$ **locally powerlike** if for every $p \in U$ there are simply connected neighborhoods $O \subset U$ of p and $W \subset V$ of $q = f(p)$ such that $f : (O, p) \to (W, q)$ is powerlike. Evidently such a map f is open and has discrete fibers. Theorem 1.59 shows that *all non-constant holomorphic functions are locally powerlike* and the above notion of local degree for them agrees with the one in Definition 1.60 based on power series.

DEFINITION 12.36. A continuous map $f : U \to V$ is a **branched covering** if every $q \in V$ has a **ramified neighborhood** W in the following sense: The preimage $f^{-1}(W)$ is the disjoint union of open sets $\{O_k\}$, each O_k contains a single preimage p_k of q, and the restriction $f : (O_k, p_k) \to (W, q)$ is powerlike.

We call p_k a **branch point** of f if $\deg(f, p_k) > 1$. It follows from this definition that the set of branch points of f has no accumulation point in U and that f reduces to a covering map in the usual sense if it has no branch point. When f is holomorphic, its branch points are precisely its critical points.

Observe that every branched covering map is locally powerlike but not vice versa. This is similar to the fact that every covering map is a local homeomorphism but not vice versa.

LEMMA 12.37 (Existence of mapping degree). *For every branched covering map $f : U \to V$, the number*

$$N(q) = \sum_{p \in f^{-1}(q)} \deg(f, p)$$

of preimages of $q \in V$ counting multiplicities, which may be finite or infinite, is independent of q.

As in the case of covering maps, we call this number the ***(mapping) degree*** of f and denote it by $\deg(f)$.

PROOF. Let $q \in V$, let W be a ramified neighborhood of q, and let O_k and p_k be as in Definition 12.36. Then $N(q) = \sum_k m_k$, where $m_k = \deg(f, p_k)$. If $z \in W \smallsetminus \{q\}$ and if $D \subset W \smallsetminus \{q\}$ is an open disk centered at z, then D is evenly covered by f. In fact, looking at the action of $w \mapsto w^{m_k}$ on $\mathbb{D}$ tells us that $f^{-1}(D)$ has m_k distinct components in O_k, each mapping homeomorphically onto D. Thus, z has m_k distinct preimages in O_k and f has local degree 1 at each of them. It follows that $N(z) = \sum_k m_k = N(q)$. This shows that the function $q \mapsto N(q)$ is locally constant on V. Since V is connected, this function must be constant. $\square$

THEOREM 12.38. *Let $f : U \to V$ be a branched covering, C be the set of branch points of f, and $Q = f(C)$. Then $U \smallsetminus f^{-1}(Q)$ and $V \smallsetminus Q$ are domains and the restriction $f : U \smallsetminus f^{-1}(Q) \to V \smallsetminus Q$ is a covering map of the same degree as f.*

PROOF. We may assume $C \neq \emptyset$. The definition of a ramified neighborhood shows that Q has no accumulation point in V, so $f^{-1}(Q)$ has no accumulation point in U (here we use the fact that branched coverings have discrete fibers). An easy exercise then shows that the open sets $U \smallsetminus f^{-1}(Q)$ and $V \smallsetminus Q$ are both path-connected and therefore connected. For every $q \in V \smallsetminus Q$, take a ramified neighborhood $W \subset V \smallsetminus Q$ and let O_k and p_k be as in Definition 12.36. We have $\deg(f, p_k) = 1$ for every k because $O_k \cap C = \emptyset$, so the powerlike restriction $f : (O_k, p_k) \to (W, q)$ is a homeomorphism. This shows that W is evenly covered by f. $\square$

It is not hard to find a characterization of *finite degree* branched covering maps similar to that in Theorem 12.33. This will be done in Theorem 12.40 below, whose proof uses the following

LEMMA 12.39. *Let $f : U \to V$ be a proper map. Then*

(i) *f is a closed map, i.e., $f(E)$ is closed in V whenever E is closed in U.*

(ii) *Given $q \in V$ and a neighborhood $\Omega \subset U$ of the fiber $f^{-1}(q)$, there is a neighborhood $W \subset V$ of q such that $f^{-1}(W) \subset \Omega$.*

Both parts of the lemma suggest some kind of "continuity of inverse" for proper maps even though these maps are generally not invertible.

PROOF. (i) Let E be closed in U. Let $q \in V$ be in the closure of $f(E)$, so there is a sequence $\{p_n\}$ in E such that $f(p_n) \to q$. Take an open disk D centered at q with $\overline{D} \subset V$. The sequence $\{p_n\}$ eventually lies in the compact set $E \cap f^{-1}(\overline{D})$, so it has a subsequence which converges to some $p \in E$. We have $f(p) = q$ by continuity, so $q \in f(E)$.

(ii) The image $f(U \setminus \Omega)$ is closed in V by (i) and does not contain q. Hence $W = V \setminus f(U \setminus \Omega)$ is a neighborhood of q with the desired property $f^{-1}(W) \subset \Omega$. $\square$

THEOREM 12.40. *$f : U \to V$ is a finite degree branched covering if and only if it is proper and locally powerlike.*

PROOF. First suppose f is a branched covering with $\deg(f) < \infty$. By definition, f is locally powerlike. We show properness of f by an argument almost identical to the proof of Theorem 12.33. For every $q \in V$, take a ramified neighborhood W and decompose $f^{-1}(W)$ into the disjoint union of $n \leq \deg(f)$ open sets $O_1, \ldots, O_n$ with $O_k \cap f^{-1}(q) = \{p_k\}$ such that $f : (O_k, p_k) \to (W, q)$ is powerlike for every $1 \leq k \leq n$. Since the preimage of every compact subset of $\mathbb{D}$ under $w \mapsto w^m$ is compact, the definition of powerlike maps shows that for every neighborhood D of q with compact closure $\overline{D} \subset W$ the intersection $f^{-1}(\overline{D}) \cap O_k$ is compact. Thus, as a union of n compact sets, $f^{-1}(\overline{D})$ itself must be compact. Now any compact set $K \subset V$ can be covered by finitely many such neighborhoods $D_1, \ldots, D_s$, so $f^{-1}(K)$ is a closed subset of the compact set $\bigcup_{k=1}^s f^{-1}(\overline{D_k})$. This proves compactness of $f^{-1}(K)$.

Conversely, suppose f is proper and locally powerlike. The image $f(U)$ is open since f is an open map, and it is closed in V by Lemma 12.39(i). Since V is connected, it follows that $f(U) = V$. Take $q \in V$. The fiber $f^{-1}(q)$ is a compact discrete subset of U, so it must be a finite set $\{p_1, \ldots, p_n\}$. For each $1 \leq k \leq n$ there are neighborhoods U_k of p_k and V_k of q such that $f : (U_k, p_k) \to (V_k, q)$ is powerlike. By shrinking the U_k if necessary, we may assume $U_1, \ldots, U_n$ are disjoint. Lemma 12.39(ii) gives a simply connected neighborhood $W \subset \bigcap_{k=1}^n V_k$ of q such that $f^{-1}(W) \subset \bigcup_{k=1}^n U_k$. It follows that $f^{-1}(W)$ is the disjoint union of the open sets $O_k = f^{-1}(W) \cap U_k$ for $1 \leq k \leq n$. We show that W is a ramified neighborhood of q by verifying that each $f : (O_k, p_k) \to (W, q)$ is powerlike. Fix k and take homeomorphisms $\varphi : (U_k, p_k) \to (\mathbb{D}, 0)$ and $\psi : (V_k, q) \to (\mathbb{D}, 0)$ such that $\psi \circ f \circ \varphi^{-1} = B$, where $B(w) = w^{m_k}$ and $m_k = \deg(f, p_k)$. Evidently $\psi(W)$ is a simply connected neighborhood of 0, so the same is true of its preimage $B^{-1}(\psi(W)) = \varphi(O_k)$ (this is a fun exercise in topology; see problem 11). If $\zeta : (\varphi(O_k), 0) \to (\mathbb{D}, 0)$ and $\xi : (\psi(W), 0) \to (\mathbb{D}, 0)$ are biholomorphisms given by the Riemann mapping theorem, the map $\xi \circ B \circ \zeta^{-1} : (\mathbb{D}, 0) \to (\mathbb{D}, 0)$ is proper and holomorphic, with a single zero of order m_k at 0, and therefore $\xi \circ B \circ \zeta^{-1} = \alpha B$ for some $\alpha \in \mathbb{C}$ with $|\alpha| = 1$ (Example 12.31). Thus, the homeomorphisms $\varphi_0 = \zeta \circ \varphi : (O_k, p_k) \to (\mathbb{D}, 0)$ and $\psi_0 = \alpha^{-1} \xi \circ \psi : (W, q) \to (\mathbb{D}, 0)$ satisfy the relation $\psi_0 \circ f \circ \varphi_0^{-1} = B$. $\square$

Proper holomorphic maps were first studied in 1919 by Fatou and a little later by Radó.

COROLLARY 12.41. *A holomorphic map between spherical domains is a finite degree branched covering if and only if it is proper. In particular, every proper holomorphic map has a well-defined finite mapping degree.*

EXAMPLE 12.42. Every non-constant polynomial P of algebraic degree d is a holomorphic branched covering $\mathbb{C} \to \mathbb{C}$ or $\hat{\mathbb{C}} \to \hat{\mathbb{C}}$ with the mapping degree $\deg(P) = d$ (in the latter case, ∞ is a branch point with $P^{-1}(\infty) = \infty$, so $\deg(P, \infty) = d$). More generally, every non-constant rational function $f = P/Q$, P and Q being polynomials, is a holomorphic branched covering $\hat{\mathbb{C}} \to \hat{\mathbb{C}}$. Again, the mapping degree $\deg(f)$ coincides with the algebraic degree of f, which is $\max\{\deg(P), \deg(Q)\}$.

EXAMPLE 12.43. The map $g : \mathbb{D} \to \mathbb{D}^*$ defined by

$$g(z) = \exp\left(\frac{z^2 + 1}{z^2 - 1}\right)$$

is an infinite-degree branched covering. It will be useful to think of g as the composition

$$\mathbb{D} \xrightarrow{\; s \;} \mathbb{D} \xrightarrow{\; \varphi \;} U = \{z \in \mathbb{C} : \operatorname{Re}(z) < 0\} \xrightarrow{\; \exp \;} \mathbb{D}^*,$$

where $s(z) = z^2$ and $\varphi(z) = (z+1)/(z-1)$. Notice that g has a single critical point at 0 with the critical value $g(0) = 1/e$. We show that every $w \in \mathbb{D}^*$ has a ramified neighborhood which is evenly covered by g if $w \neq 1/e$.

First suppose $w \neq 1/e$. Then any disk $D \subset \mathbb{D}^* \smallsetminus \{1/e\}$ centered at w is evenly covered by exp. The countably many components $U_k \subset U$ of $\exp^{-1}(D)$ avoid -1 and pull back biholomorphically under the Möbius map φ to disjoint simply connected domains $V_k = \varphi^{-1}(U_k) \subset \mathbb{D}$ that avoid 0. Each V_k is evenly covered by s, and $s^{-1}(V_k)$ consists of a disjoint pair $\pm O_k$ of simply connected domains in $\mathbb{D}$ that also avoid 0. It follows that $g^{-1}(D)$ is the disjoint union $\bigcup_k \pm O_k$ and each restriction $g : \pm O_k \to D$ is a biholomorphism. In other words, D is evenly covered by g.

Now let $w = 1/e$. Then any disk $D \subset \mathbb{D}^*$ centered at w is evenly covered by exp. As before, let the $\{U_k\}_{k \in \mathbb{Z}}$ be the components of $\exp^{-1}(D)$ and $V_k = \varphi^{-1}(U_k)$. Now one of the V_k, say V_0, contains 0 while the others avoid it (see Fig. 12.4). Thus, $s^{-1}(V_0) = O_0$ is a simply connected neighborhood of the critical point 0 and $s : (O_0, 0) \to (V_0, 0)$ is a powerlike map (compare the proof of Theorem 12.40), while for $k \neq 0$, $s^{-1}(V_k)$ consists of a disjoint pair $\pm O_k$ of simply connected domains in $\mathbb{D}$ that avoid 0. It follows that $g^{-1}(D)$ is the disjoint union $O_0 \cup \bigcup_{k \neq 0} \pm O_k$, where the restriction $g : (O_0, 0) \to (D, 1/e)$ is powerlike and $g : \pm O_k \to D$ is a biholomorphism for $k \neq 0$.

12.4 The Riemann-Hurwitz formula

The number of branch points of a finite degree branched covering is related to the topology of its domain and range. This section is devoted to a proof of this classical result and some of its applications. The oft-cited proof uses triangulations of surfaces.

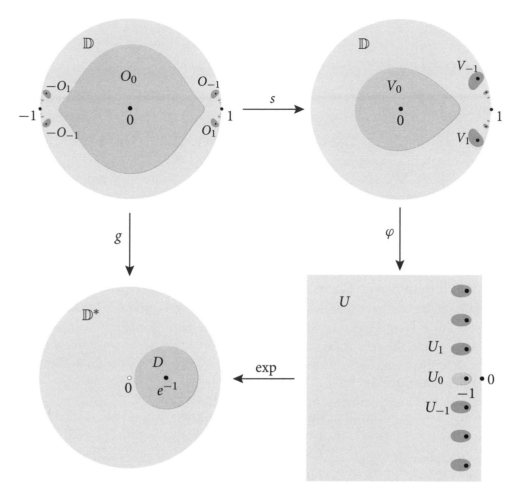

Figure 12.4. Construction of ramified neighborhoods for the map $g(z) = \exp((z^2 + 1)/(z^2 - 1))$ of Example 12.43.

The idea of the following elementary approach is adopted from [**S**], which avoids triangulations by introducing slits.

Recall from §9.4 that a domain $U \subset \hat{\mathbb{C}}$ is m-connected if the complement $\hat{\mathbb{C}} \smallsetminus U$ has precisely m connected components.

DEFINITION 12.44. The ***Euler characteristic*** of an m-connected domain $U \subset \hat{\mathbb{C}}$ is the integer

$$\chi(U) = 2 - m.$$

This definition agrees with the traditional one based on homology groups and triangulations.

Thus, $\chi(\hat{\mathbb{C}}) = 2$ while any simply connected domain $U \subsetneq \hat{\mathbb{C}}$ has connected complement by Theorem 9.27, so $\chi(U) = 1$.

THEOREM 12.45. *Let U and V be finitely connected domains in $\hat{\mathbb{C}}$. If there is a proper map $U \to V$, then $\chi(U) \leq \chi(V)$. As a result, U and V are homeomorphic if and only if they have the same Euler characteristic.*

PROOF. By Theorem 9.29, U and V are homeomorphic to spheres U' and V' with $2 - \chi(U)$ and $2 - \chi(V)$ punctures, respectively. A proper map $U \to V$ induces a proper map $U' \to V'$ which extends continuously to $\hat{\mathbb{C}}$ and maps the puncture set $\partial U'$ onto the puncture set $\partial V'$ (see problem 17). This proves $2 - \chi(U) \geq 2 - \chi(V)$ or $\chi(U) \leq \chi(V)$. It follows immediately that if U and V are homeomorphic, then $\chi(U) = \chi(V)$. Conversely, if $\chi(U) = \chi(V)$, then U' and V' have the same number of punctures. An elementary exercise then shows that U' and V', and therefore U and V, are homeomorphic. $\square$

Notice the similarity between slits and cross-cuts defined in §6.3.

By a ***slit*** Σ in a domain $\Omega \subset \hat{\mathbb{C}}$ we mean the image of a curve $\gamma : [0,1] \to \hat{\mathbb{C}}$ which maps $(0,1)$ injectively into Ω and the endpoints $0, 1$ to $\partial \Omega$. Thus Σ is either homeomorphic to a closed interval, in which case $\hat{\mathbb{C}} \smallsetminus \Sigma$ is connected and therefore simply connected (this is a precursor to the Jordan curve theorem; see [**Fu**]), or it is homeomorphic to a circle, in which case $\hat{\mathbb{C}} \smallsetminus \Sigma$ has two simply connected components by the Jordan curve theorem. Two slits in Ω are considered disjoint if they do not intersect in Ω (but may share endpoints).

Suppose Σ is a slit in an m-punctured sphere U. If Σ joins two distinct punctures, then $U \smallsetminus \Sigma$ has a single component U_1 with $m-2$ punctures, so $\chi(U_1) = 2 - (m-1) = \chi(U) + 1$. If, on the other hand, Σ connects some puncture to itself, then $U \smallsetminus \Sigma$ has two components U_1, U_2 containing n and $m - n - 1$ punctures for some $0 \leq n \leq m - 1$, so $\chi(U_1) + \chi(U_2) = (2 - (n+1)) + (2 - (m-n)) = \chi(U) + 1$.

Now every finitely connected domain Ω is homeomorphic to a finitely punctured sphere U by Theorem 9.29, and slits in Ω are easily seen to map to slits in U. The above observation then shows that for any slit Σ in Ω, the complement $\Omega \smallsetminus \Sigma$ consists of either one or two components whose Euler characteristics add up to $\chi(\Omega) + 1$. In short, *cutting a slit raises the total Euler characteristic by 1.* Applying this principle inductively k times, we obtain

LEMMA 12.46. *If k disjoint slits in a finitely connected domain Ω separate Ω into subdomains $\Omega_1, \ldots, \Omega_N$, then*

$$\sum_{j=1}^{N} \chi(\Omega_j) = \chi(\Omega) + k.$$

Compare Fig. 12.5.

THEOREM 12.47 (The Riemann-Hurwitz formula). *Let $U, V \subset \hat{\mathbb{C}}$ be finitely connected domains and $f : U \to V$ be a degree d branched covering with finitely many branch*

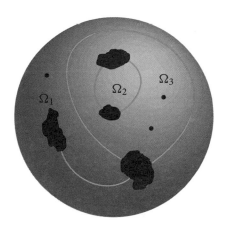

Figure 12.5. Illustration of Lemma 12.46. A 7-connected domain $\Omega \subset \hat{\mathbb{C}}$ is separated by four slits into subdomains $\Omega_1, \Omega_2, \Omega_3$ with Euler characteristics $0, 1, -2$, respectively. Here $\sum_{i=1}^{3} \chi(\Omega_i) = -1 = \chi(\Omega) + 4$.

points. Then,

$$(12.5) \qquad number\ of\ branch\ points\ of\ f = d\,\chi(V) - \chi(U).$$

Here a branch point of local degree $j \geq 2$ is counted $j-1$ times. In particular, if f is a covering map, then $\chi(U) = d\chi(V)$.

The formula was stated by Riemann and proved by Hurwitz in 1891.

We note that finiteness of the number of branch points is a superfluous assumption when f is holomorphic (see problem 25).

PROOF. By Theorem 9.29, U and V are homeomorphic to spheres with $m = 2 - \chi(U) \geq 0$ and $n = 2 - \chi(V) \geq 0$ punctures, respectively. Since the property of being a degree d branched covering and the number of branch points persist under pre- and postcomposition with homeomorphisms, we may assume at the outset that U and V themselves are spheres with m and n punctures.

First consider the special case where $f : U \to V$ is a covering map (so has no branch points). If $0 \leq n \leq 1$, then V is simply connected and f is a homeomorphism by Corollary 12.10. This gives $d = 1$ and $\chi(U) = \chi(V)$, so (12.5) trivially holds. Let us then assume $n \geq 2$. Fix a puncture of V and take $n - 1$ disjoint slits $\{\Sigma_j\}$ in V from each of the remaining punctures to this one. Each Σ_j lifts under f to d disjoint slits in U. Moreover, $W = V \setminus \bigcup \Sigma_j$ is a simply connected domain, so by Corollary 12.10 f maps each component of $f^{-1}(W)$ homeomorphically to W. It follows that there are precisely d components $\Omega_1, \ldots, \Omega_d$ of $f^{-1}(W) = U \setminus \bigcup f^{-1}(\Sigma_j)$, and each Ω_j is simply connected with $\chi(\Omega_j) = 1$. By Lemma 12.46,

$$d = \sum_{j=1}^{d} \chi(\Omega_j) = \chi(U) + d(n-1) = \chi(U) + d(1 - \chi(V)),$$

from which we obtain $d\chi(V) - \chi(U) = 0$, as required.

To prove (12.5) in general, denote by C the finite set of branch points of f. Each $v \in Q = f(C)$ has d preimages counting multiplicities, where a branch point $c \in f^{-1}(v)$ contributes with multiplicity $\deg(f, c)$. Thus,

$$\text{number of elements in } f^{-1}(v) = d - \sum_{c \in f^{-1}(v) \cap C} (\deg(f, c) - 1).$$

If $n' \geq 1$ and $m' \geq 1$ denote the number of elements in Q and $f^{-1}(Q)$ respectively, it follows that

$$(12.6) \qquad m' = dn' - \sum_{c \in C}(\deg(f, c) - 1).$$

The domains $U' = U \smallsetminus f^{-1}(Q)$ and $V' = V \smallsetminus Q$ are spheres with $m + m'$ and $n + n'$ punctures, respectively, and $f : U' \to V'$ is a degree d covering map by Theorem 12.38. The special case treated above then gives

$$2 - m - m' = d(2 - n - n') \quad \text{or} \quad dn' - m' = d(2 - n) - (2 - m),$$

and we conclude from (12.6) that

$$\sum_{c \in C}(\deg(f, c) - 1) = dn' - m' = d\chi(V) - \chi(U). \qquad \square$$

EXAMPLE 12.48. According to the Riemann-Hurwitz formula, every polynomial map $P : \mathbb{C} \to \mathbb{C}$ of degree d has $d\chi(\mathbb{C}) - \chi(\mathbb{C}) = d - 1$ critical points counting multiplicities. This is of course consistent with the algebraic count of the roots of the equation $P'(z) = 0$. If we think of P as a map $\hat{\mathbb{C}} \to \hat{\mathbb{C}}$, the total number of critical points becomes $d\chi(\hat{\mathbb{C}}) - \chi(\hat{\mathbb{C}}) = 2d - 2$, which shows ∞ is a critical point of multiplicity $d - 1$.

Similarly, every rational function $\hat{\mathbb{C}} \to \hat{\mathbb{C}}$ of degree d has $2d - 2$ critical points. The case of a degree d Blaschke product B of the form (12.3) is of special interest. For such B, the restriction $B|_{\mathbb{D}} : \mathbb{D} \to \mathbb{D}$ is proper of degree d, so B has $d\chi(\mathbb{D}) - \chi(\mathbb{D}) = d - 1$ critical points in $\mathbb{D}$. By the symmetry relation $B(1/\bar{z}) = 1/\overline{B(z)}$, the remaining $d - 1$ critical points of B lie outside $\mathbb{D}$.

EXAMPLE 12.49. Let us construct a degree 2 holomorphic branched covering from the round annulus $A = \{w \in \mathbb{C} : R^{-1} < |w| < R\}$ onto the unit disk $\mathbb{D}$. The Zhukovskii map $Z(w) = w + w^{-1}$ carries A onto the region

$$U = \{z \in \mathbb{C} : |z - 2| + |z + 2| < R + R^{-1}\},$$

sending the boundary circles in ∂A onto the ellipse ∂U (see Example 6.6). Postcomposing Z with a conformal map $g : U \to \mathbb{D}$ given by the Riemann mapping theorem, we obtain a proper holomorphic map $f = g \circ Z : A \to \mathbb{D}$, which is therefore a branched covering of degree 2. Notice that the critical points of f are those of Z, namely $w = \pm 1$. This is consistent with the count $2\chi(\mathbb{D}) - \chi(A) = 2$ coming from the Riemann-Hurwitz formula.

The above construction involves the Riemann map of the interior of an ellipse, which is a non-elementary function and difficult to compute. If we do not insist on having the round disk as the target, the map $Z : A \to U$ already provides a degree 2 branched covering. Similarly, if we do

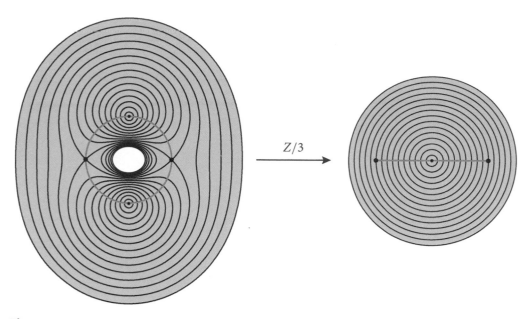

Figure 12.6. The scaled Zhukovskii map $f = Z/3$ restricts to a degree 2 holomorphic branched covering map from the annulus $f^{-1}(\mathbb{D})$ to $\mathbb{D}$. Here the critical points are at ± 1 and the unit circle collapses onto the segment $[-2/3, 2/3]$ (both shown in blue).

not insist on having round annuli as the domain, we can consider the scaled map $Z_\lambda = Z/\lambda$ for $\lambda > 2$ for which $A_\lambda = Z_\lambda^{-1}(\mathbb{D})$ is an annulus and the restriction $Z_\lambda : A_\lambda \to \mathbb{D}$ is a degree 2 branched covering (compare Fig. 12.6).

More generally, for every m-connected domain $U \subset \hat{\mathbb{C}}$ with non-degenerate complementary components and every $p \in U$ there exists a holomorphic branched covering $(U, p) \to (\mathbb{D}, 0)$ of degree m. For $m = 1$, this is just the Riemann mapping theorem. The case $m > 1$ was settled by Ahlfors in 1947, who constructed such branched coverings as solutions of the extremal problem $\max |f'(p)|$, where f ranges over all holomorphic functions $(U, p) \to (\mathbb{D}, 0)$ [**A1**]. The extremal solution, which turns out to be unique up to a rotation, is called the ***Ahlfors function*** of the pair (U, p).

Problems

In keeping with the convention of this chapter, all topological spaces are assumed Hausdorff, path-connected, and locally path-connected, unless otherwise stated.

(1)   Let $f : Y \to X$ be a covering map. If $X' \subset X$ is connected and if $Y' \subset Y$ is a connected component of $f^{-1}(X')$, verify that the restriction $f : Y' \to X'$ is a covering map.

(2)   Show that the implication (iii) $\Longrightarrow$ (ii) in Theorem 12.15 holds without the assumption of local simple connectivity.

(3)   Let X be the union of the circles $|z - 2^{-n}i| = 2^{-n}$ in the plane for $n \geq 1$ (topologists call X a ***Hawaiian earring***). Construct an infinite degree covering map $f : Y \to X$ and a degree

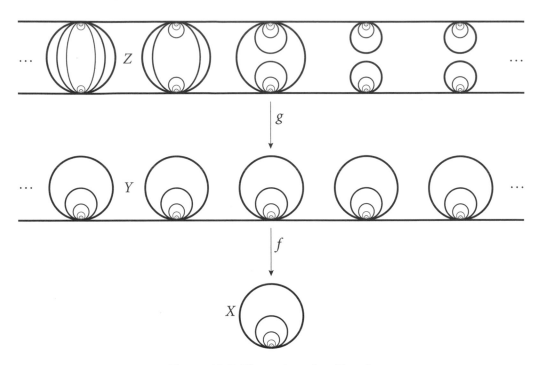

Figure 12.7. Illustration of problem 3.

2 covering map $g : Z \to Y$, as suggested by Fig. 12.7, such that $f \circ g : Z \to X$ is not a covering map (note however that $f \circ g$ must be a local homeomorphism with the curve lifting property). Contrast this with Theorem 12.15 and Theorem 12.16.

(4) The goal of this exercise is to show that parts of Theorem 12.16 remain valid without the assumption of local simple connectivity of the underlying spaces. Let $g : Z \to Y$ and $f : Y \to X$ be given maps and set $h = f \circ g : Z \to X$. Verify, by constructing evenly covered neighborhoods, the following assertions:
   (i)   If h and f are covering maps, so is g.
   (ii) If h is a covering map and g is open and surjective, then g is a covering map.

(5) (Lifting criterion) This is a more general version of Theorem 12.8(ii). Let $f : (Y, y_0) \to (X, x_0)$ be a covering map and $h : (Z, z_0) \to (X, x_0)$ be continuous. Show that there is a lift $\tilde{h} : (Z, z_0) \to (Y, y_0)$ of h under f if and only if $h_*(\pi_1(Z, z_0)) \subset f_*(\pi_1(Y, y_0))$.

(6) A covering map $f : (Y, y_0) \to (X_0, x_0)$ is **Galois** if the action of its deck group G on the fibers of f is simply transitive (for example, every universal covering is Galois). Show that f is Galois if and only if the image $f_*(\pi_1(Y, y_0))$ is a normal subgroup of $\pi_1(X, x_0)$. In this case, verify that G acts freely and properly discontinuously on Y (in the sense of Theorem 12.21(ii) and (iii)), that Y/G is homeomorphic to X, and that G is isomorphic to $\pi_1(X, x_0)/f_*(\pi_1(Y, y_0))$.

(7) Let $u : \tilde{X} \to X$ be a universal covering with the deck group G. There is a correspondence between Galois coverings of X and normal subgroups of G as described below.
   (i)   Let $f : Y \to X$ be a Galois covering. The map u has a lift $\tilde{u} : \tilde{X} \to Y$ under f which is also a universal covering (problem 4). Let H and K be the deck groups of $\tilde{u}$:

$\tilde{X} \to Y$ and $f : Y \to X$, respectively. Verify that H is a normal subgroup of G and $K \cong G/H$.

(ii)   Let H be any normal subgroup of G and $\Pi : \tilde{X} \to \tilde{X}/H$ be the natural projection $\Pi(z) = H \cdot z$. The induced map $f : \tilde{X}/H \to X$ given by $f(H \cdot z) = u(z)$ is well defined and satisfies $u = f \circ \Pi$. Verify that f is a Galois covering with the deck group isomorphic to G/H.

(8)   Suppose X, Y are locally simply connected. Let $u : \tilde{X} \to X$ and $v : \tilde{Y} \to Y$ be universal coverings and $f : X \to Y$ be continuous.

(i)   Show that there is a continuous map $F : \tilde{X} \to \tilde{Y}$ which makes the following diagram commute:

$$\begin{array}{ccc} \tilde{X} & \xrightarrow{\ F\ } & \tilde{Y} \\ u\downarrow & & \downarrow v \\ X & \xrightarrow{\ f\ } & Y \end{array}$$

(ii)   Show that f is a covering map if and only if F is a homeomorphism. Verify that in this case the conjugation $\varphi \mapsto F \circ \varphi \circ F^{-1}$ defines an isomorphism between the deck groups of $u : \tilde{X} \to X$ and $v : \tilde{Y} \to Y$.

(9)   The **one-point compactification** $\hat{X}$ of a locally compact Hausdorff space X is the disjoint union of X together with an additional point $\infty = \infty_X$. The topology of $\hat{X}$ is generated by all open sets in X as well as the neighborhoods of ∞ of the form $\hat{X} \smallsetminus K$, where $K \subset X$ is compact.

(i)   Show that a curve $\gamma : [a, b) \to X$ is escaping (in the sense of Definition 12.12) if and only if $\bigcap_{a < t < b} \gamma((t, b)) = \varnothing$ if and only if $\lim_{t \to b} \gamma(t) = \infty$.

(ii)   We call a continuous map $Y \to X$ proper if $f^{-1}(K)$ is compact whenever $K \subset X$ is compact. Verify that $f : Y \to X$ is proper if and only if it extends to a continuous map $\hat{f} : \hat{Y} \to \hat{X}$ with $\hat{f}(\infty_Y) = \infty_X$.

(iii)   Prove Lemma 12.39 for proper maps between locally compact Hausdorff spaces.

(10)   Give an example of a proper map $\mathbb{R} \to \mathbb{R}$ which is not surjective. Give an example of a closed map $\mathbb{R} \to \mathbb{R}$ which is not proper. Show however that a non-constant holomorphic map between spherical domains is proper if and only if it is a closed map.

(11)   Justify the following topological claim used in the proof of Theorem 12.40: Let $B(z) = z^m$ for an integer $m \geq 1$. If $\Omega \subset \mathbb{D}$ is a simply connected neighborhood of 0, so is $B^{-1}(\Omega)$. (Hint: Verify that $B^{-1}(\Omega)$ is path-connected and that any bounded component of $\mathbb{C} \smallsetminus B^{-1}(\Omega)$ would give rise to a bounded component of $\mathbb{C} \smallsetminus \Omega$. Use the equivalence (iv) $\iff$ (vi) of Theorem 9.27.)

(12)   Verify that the entire function

$$f(z) = \int_{[0,z]} \exp(-\zeta^2) \, d\zeta$$

maps $\mathbb{C}$ *onto* $\mathbb{C}$ and satisfies $f'(z) \neq 0$ for all z, but it is not a biholomorphism. Contrast this with Theorem 12.24 and Corollary 12.26. (Observation: f is not a covering map since it has asymptotic values $\pm\sqrt{\pi}/2$.)

(13)   Verify that the complex sine and cosine functions have no asymptotic values in $\mathbb{C}$. Then use Corollary 12.26 to construct holomorphic branches of $\arcsin(z)$ and $\arccos(z)$ in every simply connected domain $U \subset \mathbb{C} \smallsetminus \{-1, 1\}$. In other words, construct functions

$f, g \in \mathcal{O}(U)$ such that $\sin(f(z)) = \cos(g(z)) = z$ for all $z \in U$. To what extent are f and g unique?

(14)   Is an omitted value of a transcendental entire function necessarily an asymptotic value?

(15)   Let P, Q be complex polynomials with $\deg(Q) > 0$. Show that the entire function $P \exp(Q)$ has a unique asymptotic value at 0. (Hint: Adapt the reasoning in Example 12.28.)

(16)   Show that every branched covering has the curve lifting property but uniqueness of lifts cannot be guaranteed. (Hint: Examine the proof of Lemma 12.6 and see Fig. 12.1.)

(17)   Prove that every proper map $f : U \to V$ between finitely punctured spheres extends to a continuous surjective map $\hat{f} : \hat{\mathbb{C}} \to \hat{\mathbb{C}}$ which sends punctures to punctures. (Hint: For small $\varepsilon > 0$ the set $\Omega = \{z \in \hat{\mathbb{C}} : \mathrm{dist}_\sigma(z, \hat{\mathbb{C}} \smallsetminus V) < \varepsilon\}$ is a disjoint union of disk neighborhoods of the punctures of V. A sufficiently small connected neighborhood of each puncture of U maps under f to Ω.)

(18)   Suppose $P : \mathbb{C} \to \mathbb{C}$ is a monic polynomial of degree $d \geq 1$ such that $P^{-1}(\mathbb{D}) = \mathbb{D}$. Show that $P(z) = z^d$ for all $z \in \mathbb{C}$.

(19)   Suppose $f : \mathbb{D} \to \mathbb{D}$ is a proper holomorphic map of degree 2. Prove that there exist $\varphi, \psi \in \mathrm{Aut}(\mathbb{D})$ such that $(\psi \circ f \circ \varphi)(z) = z^2$ for all $z \in \mathbb{D}$.

(20)   Suppose $f : \mathbb{D} \to \mathbb{D}$ is a proper holomorphic map such that $f(0) = 1/4$ and

$$\frac{1}{2\pi i} \int_{\mathbb{T}} \frac{f'(z)}{f(z)} \, dz = 2 \qquad \frac{1}{2\pi i} \int_{\mathbb{T}} \frac{z f'(z)}{f(z)} \, dz = 0 \qquad \frac{1}{2\pi i} \int_{\mathbb{T}} \frac{z^2 f'(z)}{f(z)} \, dz = \frac{1}{2}.$$

Find a formula for $f(z)$. (Hint: The generalized argument principle in problem 27 of chapter 3 may be useful.)

(21)   Let $f : \mathbb{D} \to \mathbb{D}$ be a proper holomorphic map. Show that the map $g : \mathbb{D} \to \mathbb{D}^*$ defined by

$$g(z) = \exp\left(\frac{f(z) + 1}{f(z) - 1}\right)$$

is an infinite-degree branched covering with finitely many critical points. (Hint: Revisit Example 12.43.)

(22)   Show that $\cos : \mathbb{C} \to \mathbb{C}$ is an infinite degree branched covering with infinitely many critical points and only two critical values. (Hint: cos is the composition of the exponential map $z \mapsto e^{iz}$ followed by the scaled Zhukovskii map $z \mapsto (z + z^{-1})/2$.)

(23)   Construct an example of a holomorphic infinite degree branched covering $\mathbb{D} \to \mathbb{D}$. (Hint: Think of a restriction of exp mapping a strip to an annulus, followed by the Zhukovskii map.)

(24)   Suppose $U \subset \hat{\mathbb{C}}$ is a finitely connected domain and $f : U \to U$ is a finite degree branched covering with $1 \leq N < \infty$ branch points. Show that U is simply connected.

(25)   Every proper holomorphic map $f : U \to V$ between finitely connected domains has finitely many critical points. Complete the following outline of a proof of this statement: By Theorem 9.29 we can map U, V conformally onto domains U', V' whose boundary components are punctures or analytic Jordan curves. The induced map $g : U' \to V'$ is proper and holomorphic, has removable singularities at punctures in $\partial U'$, and extends analytically across the Jordan curve components of $\partial U'$ by the Schwarz reflection principle. Hence g has finitely many critical points in U'.

(26)   Suppose $U \subset \hat{\mathbb{C}}$ is a finitely connected domain with $\chi(U) < 0$. Show that every proper holomorphic map $U \to U$ is a biholomorphism. Does the result hold when $\chi(U) = 0$?

(27)   Prove that every proper holomorphic map between 2-connected domains is a finite degree covering map.

(28)   Does there exist a proper holomorphic map between spherical domains that has infinitely many critical points?

13

Uniformization of spherical domains

13.1 The modular group and thrice punctured spheres

Let $\kappa : \hat{\mathbb{C}} \to \hat{\mathbb{C}}$ denote the complex conjugation $z \mapsto \bar{z}$ extended so that $\kappa(\infty) = \infty$. Evidently κ is an involution, in the sense that $\kappa = \kappa^{-1}$, and $\kappa(z) = z$ if and only if z belongs to the extended real line $\hat{\mathbb{R}} = \mathbb{R} \cup \{\infty\}$.

For any Möbius map $\varphi(z) = (az + b)/(cz + d)$, the conjugate map

$$\varphi^*(z) = (\kappa \circ \varphi \circ \kappa)(z) = \frac{\bar{a}z + \bar{b}}{\bar{c}z + \bar{d}}$$

is also Möbius. It is easy to see that

$$\varphi^* = \varphi \Longleftrightarrow \kappa \circ \varphi = \varphi \circ \kappa \Longleftrightarrow a, b, c, d \text{ are real} \Longleftrightarrow \varphi(\hat{\mathbb{R}}) = \hat{\mathbb{R}}.$$

Let α be a circle in $\hat{\mathbb{C}}$ (that is, a Euclidean circle in $\mathbb{C}$ or a straight line in $\mathbb{C}$ with ∞ added). Take any $\varphi \in \text{Möb}$ that sends α to $\hat{\mathbb{R}}$. The **reflection in α** is the involution $R_\alpha : \hat{\mathbb{C}} \to \hat{\mathbb{C}}$ defined by

$$R_\alpha = \varphi^{-1} \circ \kappa \circ \varphi.$$

This definition is independent of the choice of φ since if $\psi \in \text{Möb}$ also sends α to $\hat{\mathbb{R}}$, then $\varphi \circ \psi^{-1}$ is a Möbius map that preserves $\hat{\mathbb{R}}$, so by what we have seen it must commute with κ. Note that $R_\alpha = \varphi^{-1} \circ \varphi^* \circ \kappa$, so every reflection in a circle is the composition of a Möbius map with the complex conjugation. In particular, such reflections preserve the family of circles in $\hat{\mathbb{C}}$.

In the language of §10.3, R_α is the canonical reflection across the analytic Jordan curve α, that is, *R_α is the unique anti-holomorphic involution of $\hat{\mathbb{C}}$ which restricts to the identity map on α.* This characterization of R_α is often more useful than its explicit formula (see problem 1 as well as problem 2 for a geometric interpretation and problem 3 for a variant based on cross ratios).

> **DEFINITION 13.1.** The **reflection group** Ref is the group of homeomorphisms of $\hat{\mathbb{C}}$ generated by reflections in arbitrary circles. Thus, $\varphi \in \text{Ref}$ if and only if $\varphi = R_{\alpha_1} \circ \cdots \circ R_{\alpha_n}$ for some circles $\alpha_1, \ldots, \alpha_n$.

LEMMA 13.2. Möb *is an index* 2 *subgroup of* Ref.

PROOF. By Theorem 4.2, Möb is generated by translations $T : z \mapsto z + b$, linear maps $L : z \mapsto az$, and the inversion $\iota : z \mapsto 1/z$. To show Möb $\subset$ Ref it suffices to check that these three models belong to Ref. This is easy to verify: If $b \neq 0$, then $T = R \circ S$, where $R, S \in$ Ref are defined by

$$R(z) = b - \frac{b}{\bar{b}} \bar{z}, \quad S(z) = -\frac{b}{\bar{b}} \bar{z}.$$

Similarly, $L = R \circ S \circ U \circ \kappa$ and $\iota = S \circ \kappa$, where $R, S, U \in$ Ref are defined by

$$R(z) = \frac{|a|}{\bar{z}}, \quad S(z) = \frac{1}{\bar{z}}, \quad U(z) = \frac{a}{|a|} \bar{z}.$$

To prove the index 2 statement, take any two reflections R_α, R_β and write $R_\alpha = \varphi \circ \kappa, R_\beta = \psi \circ \kappa$ with $\varphi, \psi \in$ Möb. Then

$$R_\alpha \circ R_\beta = \varphi \circ \kappa \circ \psi \circ \kappa = \varphi \circ \kappa \circ \kappa \circ \psi^* = \varphi \circ \psi^* \in \text{Möb}.$$

It follows that the composition of an even number of reflections belongs to Möb, while the composition of an odd number of reflections belongs to the coset κ Möb. Thus, Ref is the disjoint union of the cosets Möb and κ Möb. □

LEMMA 13.3. *Suppose* $\varphi \in$ Ref *and* α *is a circle in* $\hat{\mathbb{C}}$. *Then* φ *conjugates the reflections* R_α *and* $R_{\varphi(\alpha)}$:

$$R_{\varphi(\alpha)} = \varphi \circ R_\alpha \circ \varphi^{-1}.$$

PROOF. Each side of the above equation is an anti-holomorphic involution that restricts to the identity map on the circle $\varphi(\alpha)$. □

Now consider the special case where the circle α meets the unit circle $\partial \mathbb{D}$ orthogonally, so $\alpha \cap \mathbb{D}$ is a hyperbolic geodesic in the sense of §4.5. Take a $\varphi \in \text{Aut}(\mathbb{D})$ which sends the points $\alpha \cap \partial \mathbb{D}$ to ± 1, and therefore $\alpha \cap \mathbb{D}$ to $\mathbb{R} \cap \mathbb{D} = (-1, 1)$. Since the complex conjugation κ maps $\mathbb{D}$ to $\mathbb{D}$, the reflection $R_\alpha = \varphi^{-1} \circ \kappa \circ \varphi$ too maps $\mathbb{D}$ to $\mathbb{D}$, swapping the two "half-planes" in $\mathbb{D} \smallsetminus \alpha$. Moreover, since κ and φ preserve the family of hyperbolic geodesics in $\mathbb{D}$, we see that $R_\alpha(\gamma)$ is a hyperbolic geodesic in $\mathbb{D}$ whenever γ is.

DEFINITION 13.4. Ref$(\mathbb{D})$ is the subgroup of Ref consisting of maps which leave the unit disk invariant:

$$\text{Ref}(\mathbb{D}) = \{\varphi \in \text{Ref} : \varphi(\mathbb{D}) = \mathbb{D}\}.$$

It follows from Lemma 13.2 that Aut$(\mathbb{D})$ is an index 2 subgroup of Ref$(\mathbb{D})$. By the previous paragraph, Ref$(\mathbb{D})$ contains all reflections in circles orthogonal to $\partial \mathbb{D}$. It is not hard to check that Ref$(\mathbb{D})$ is in fact generated by all such reflections (problem 4).

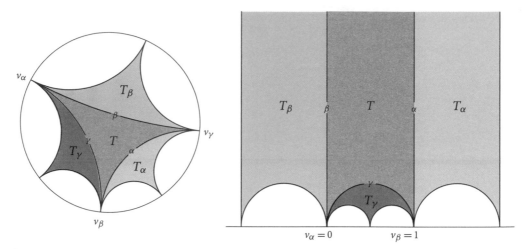

Figure 13.1. Left: An ideal triangle $T = T(\alpha, \beta, \gamma)$ in $\mathbb{D}$ and its reflections $T_\alpha = R_\alpha(T)$, $T_\beta = R_\beta(T)$, $T_\gamma = R_\gamma(T)$. Right: Same picture in the upper half-plane, normalized so that T has vertices at $0, 1, \infty$. Here the reflections in the sides of T take the simple form $R_\alpha(z) = 2 - \bar{z}$, $R_\beta(z) = -\bar{z}$, and $R_\gamma(z) = \bar{z}/(2\bar{z} - 1)$.

We define $\mathrm{Ref}(\mathbb{H})$ similarly.

Let $v_\alpha, v_\beta, v_\gamma$ be distinct points on $\partial\mathbb{D}$ labeled counterclockwise. Take the geodesics α, β, γ in $\mathbb{D}$ that connect v_β to v_γ, v_γ to v_α, and v_α to v_β, respectively. The closed region $T = T(\alpha, \beta, \gamma) \subset \mathbb{D}$ bounded by these geodesics is called the ***ideal triangle*** with vertices $v_\alpha, v_\beta, v_\gamma$ and sides α, β, γ (compare Fig. 13.1 left). The vertices themselves are not part of T and should be thought of as sitting infinitely far in the hyperbolic metric. As α, β, γ are orthogonal to $\partial\mathbb{D}$ at the vertices, the internal angles of T are zero. The sides α, β, γ determine three disjoint closed "half-planes" in $\mathbb{D}$ which do not meet the interior of T.

If $\varphi \in \mathrm{Aut}(\mathbb{D})$ sends the ordered triple (α, β, γ) to $(\alpha', \beta', \gamma')$, then

$$(13.1) \qquad \varphi(T(\alpha, \beta, \gamma)) = T(\alpha', \beta', \gamma').$$

Conversely, for any ideal triangle $T(\alpha', \beta', \gamma')$, there is a unique $\varphi \in \mathrm{Aut}(\mathbb{D})$ which sends $(v_\alpha, v_\beta, v_\gamma)$ to $(v_{\alpha'}, v_{\beta'}, v_{\gamma'})$ (see problem 10 in chapter 4) and therefore (α, β, γ) to $(\alpha', \beta', \gamma')$, so (13.1) holds. Thus, up to a disk automorphism there is only one ideal triangle in $\mathbb{D}$. The situation when φ is a reflection in a geodesic in $\mathbb{D}$ is similar except for the issue of orientation. In this case, (13.1) should be replaced by $\varphi(T(\alpha, \beta, \gamma)) = T(\alpha', \gamma', \beta')$.

It will be convenient to call an ideal triangle a ***tile*** from now on. Fix a tile $T = T(\alpha, \beta, \gamma) \subset \mathbb{D}$.

DEFINITION 13.5. The subgroup of $\mathrm{Ref}(\mathbb{D})$ generated by the reflections $R_\alpha, R_\beta, R_\gamma$ in the sides of T is called the ***reflection group*** of T and is denoted by G_T. The orbit of T under the action of G_T will be denoted by $\mathscr{T}_T$.

Thus, $\mathscr{T}_T$ is the collection of all tiles obtained from T by successively reflecting in the sides of T. Our goal is to show that $\mathscr{T}_T$ is a **tessellation** of $\mathbb{D}$ in the sense that its tiles cover $\mathbb{D}$ and have pairwise disjoint interiors.

LEMMA 13.6. *Take a tile* $T \subset \mathbb{D}$ *and let* $T' \in \mathscr{T}_T$. *Then,*

(i) $G_{T'} = G_T$ *and* $\mathscr{T}_{T'} = \mathscr{T}_T$.
(ii) *If* $\varphi(T) \in \mathscr{T}_T$ *for some* $\varphi \in \mathrm{Ref}(\mathbb{D})$, *then* $\varphi(T') \in \mathscr{T}_T$.

Part (i) means there is no preferred tile in $\mathscr{T}_T$: Picking T in $\mathscr{T}_T$ is simply a matter of choosing a base point. Part (ii) implies that if some element of $\mathrm{Ref}(\mathbb{D})$ carries one tile in $\mathscr{T}_T$ to another, then it preserves the whole collection $\mathscr{T}_T$.

PROOF. Let $T = T(\alpha, \beta, \gamma)$. It suffices to prove the assertions when $T' = R_j(T)$ for some $j \in \{\alpha, \beta, \gamma\}$. The general case follows by induction on the minimum number of reflections it takes to get from T to T'.

(i) Let $\alpha' = R_j(\alpha), \beta' = R_j(\beta), \gamma' = R_j(\gamma)$, so $T' = T(\alpha', \gamma', \beta')$. By Lemma 13.3, $R_{\alpha'} = R_j \circ R_\alpha \circ R_j^{-1}, R_{\beta'} = R_j \circ R_\beta \circ R_j^{-1}$, and $R_{\gamma'} = R_j \circ R_\gamma \circ R_j^{-1}$, so they all belong to G_T. This gives $G_{T'} \subset G_T$. Since $T = R_j(T') \in \mathscr{T}_{T'}$, the same argument shows $G_T \subset G_{T'}$. This proves $G_T = G_{T'}$ and the equality $\mathscr{T}_{T'} = \mathscr{T}_T$ follows immediately.

(ii) Let $R = \varphi \circ R_j \circ \varphi^{-1}$. By Lemma 13.3, R is the reflection in the side $\varphi(j)$ of $\varphi(T)$. Thus, $\varphi(T') = (\varphi \circ R_j)(T) = (R \circ \varphi)(T) \in \mathscr{T}_{\varphi(T)}$. The result follows since by (i), $\mathscr{T}_{\varphi(T)} = \mathscr{T}_T$. □

In what follows we will simplify the notations G_T and $\mathscr{T}_T$ to G and $\mathscr{T}$. The next definition gives a convenient labeling for the tiles in $\mathscr{T}$.

DEFINITION 13.7. An ***admissible word*** of length $n \geq 1$ in the alphabet $\{\alpha, \beta, \gamma\}$ is a string $j_1 \cdots j_n$ of letters α, β, γ in which no two consecutive letters are alike. For each admissible word $j_1 \cdots j_n$ we define

$$T_{j_1 \cdots j_n} = (R_{j_1} \circ \cdots \circ R_{j_n})(T).$$

The tile $T_{j_1 \cdots j_n} \in \mathscr{T}$ inherits its own α, β, γ-sides which are the images of the α, β, γ-sides of T under $R_{j_1} \circ \cdots \circ R_{j_n} \in G$.

There are $3 \cdot 2^{n-1}$ admissible words of length n. By convention, the only word of length 0 is the empty string and the corresponding tile is $T_\emptyset = T$.

LEMMA 13.8. $T_{j_1 \cdots j_n}$ *is obtained by reflecting* $T_{j_1 \cdots j_{n-1}}$ *in its* j_n*-side.*

PROOF. For $n = 1$ the assertion trivially holds since $T_\alpha = R_\alpha(T)$ is obtained by reflecting $T_\emptyset = T$ in its α-side and similarly for T_β and T_γ. For $n > 1$, we note that $T_{j_1 \cdots j_{n-1}}$ is the image of T under the composition $\varphi = R_{j_1} \circ \cdots \circ R_{j_{n-1}}$, so by

Lemma 13.3 the reflection in the j_n-side of $T_{j_1\cdots j_{n-1}}$ is $\varphi \circ R_{j_n} \circ \varphi^{-1}$. The lemma follows since

$$(\varphi \circ R_{j_n} \circ \varphi^{-1})(T_{j_1\cdots j_{n-1}}) = (\varphi \circ R_{j_n})(T) = T_{j_1\cdots j_n}. \qquad \square$$

Starting with $T_\emptyset = T$ and reflecting in its sides, we obtain the tiles

$$T_\alpha = R_\alpha(T), \quad T_\beta = R_\beta(T), \quad T_\gamma = R_\gamma(T)$$

which we think of as the "children" of T (Fig. 13.1). These tiles are disjoint because they are contained in disjoint closed half-planes in $\mathbb{D}$ determined by the sides α, β, γ. Moreover, these children only share a side with their "parent" T. Each of these tiles can in turn be reflected across its own sides. Evidently reflecting T_α in its α-side gives back the parent T, while by Lemma 13.8 reflecting T_α in its β-side and γ-side gives its two children $T_{\alpha\beta}$ and $T_{\alpha\gamma}$. Similarly, reflecting T_β in its sides gives $T, T_{\beta\alpha}, T_{\beta\gamma}$ and reflecting T_γ in its sides gives $T, T_{\gamma\alpha}, T_{\gamma\beta}$. Again, the six new tiles obtained this way are disjoint from one another and from T, and each of them only meets its parent along a common side. In general, for each admissible word $j_1\cdots j_n$ of length $n \geq 1$, reflecting $T_{j_1\cdots j_n}$ in its j_n-side gives the parent $T_{j_1\cdots j_{n-1}}$ while reflecting in its other sides gives its two children $T_{j_1\cdots j_n j}$ for $j \in \{\alpha, \beta, \gamma\} \setminus \{j_n\}$.

The important feature of this way of labeling tiles is that for each admissible word $j_1\cdots j_n$ and each $k \leq n$, the tile $T_{j_1\cdots j_k}$ is contained in a closed half-plane H_k determined by the j_k-side of its parent $T_{j_1\cdots j_{k-1}}$, and the condition of admissibility implies the nested property $H_1 \supset H_2 \supset \cdots \supset H_n$. It follows that

(13.2) $\qquad T_\emptyset = T, \ T_{j_1}, \ T_{j_1 j_2}, \ldots, \ T_{j_1\cdots j_n}$ have pairwise disjoint interiors.

LEMMA 13.9. *The assignment $j_1\cdots j_n \mapsto T_{j_1\cdots j_n}$ is a bijection between the set of all admissible words and $\mathscr{T}$. Moreover, distinct tiles in $\mathscr{T}$ have disjoint interiors.*

PROOF. Evidently $j_1\cdots j_n \mapsto T_{j_1\cdots j_n}$ is surjective. We show that tiles associated with different words have disjoint interiors. This will prove injectivity of the labeling as well as the second claim of the lemma. By (13.2), for any admissible word $j_1\cdots j_n$ of length $n \geq 1$ the tiles $T = T_\emptyset$ and $T_{j_1\cdots j_n}$ have disjoint interiors. More generally, take distinct admissible words $j_1\cdots j_n$ and $i_1\cdots i_m$ of lengths $n \geq m \geq 1$. If $j_k = i_k$ for all $1 \leq k \leq m$, then necessarily $n > m$, and $T_{i_1\cdots i_m} = T_{j_1\cdots j_m}$ and $T_{j_1\cdots j_n}$ have disjoint interiors by (13.2). Otherwise, let $1 \leq k \leq m$ be the smallest integer for which $j_k \neq i_k$. We know that $T_{j_1\cdots j_n}$ is contained in the closed half-plane determined by the j_k-side of $T_{j_1\cdots j_{k-1}}$ while $T_{i_1\cdots i_m}$ is contained in the closed half-plane determined by the i_k-side of $T_{i_1\cdots i_{k-1}} = T_{j_1\cdots j_{k-1}}$. As the two half-planes are disjoint, so are the tiles $T_{j_1\cdots j_n}$ and $T_{i_1\cdots i_m}$. $\qquad \square$

By the above lemma, there is a natural tree structure on $\mathscr{T}$ defined by the family lineage, where two tiles are connected by an edge if and only if they are parent and child (see Fig. 13.2). This allows us to define the ***generation*** of a tile in $\mathscr{T}$ unambiguously as the length of the unique word associated with it. Thus, T is of generation 0, $T_\alpha, T_\beta, T_\gamma$ are of generation 1, and so on.

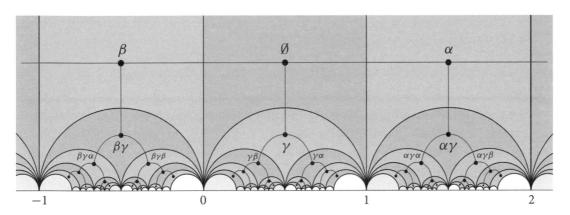

Figure 13.2. The tiles generated by the reflection group of the ideal triangle $T \subset \mathbb{H}$ with vertices $0, 1, \infty$ and the associated tree structure. To simplify the picture, only the words associated with selected tiles are shown.

At this point it is convenient to switch to the upper half-plane model and normalize the base tile T so it has vertices $v_\alpha = 0, v_\beta = 1, v_\gamma = \infty$. The sides α, β, γ of T are then contained in the lines $\mathrm{Re}(z) = 1$ and $\mathrm{Re}(z) = 0$ and the semicircle $|2z - 1| = 1$, respectively (compare Fig. 13.1 right). Consider $A, B \in \mathrm{Aut}(\mathbb{H})$ defined by

$$(13.3) \qquad A(z) = z + 1 \quad \text{and} \quad B(z) = -\frac{1}{z}.$$

It is easy to check that $A(T) = T_\alpha$ and $B(T) = T_\beta$. By Lemma 13.6(ii), the subgroup of $\mathrm{Aut}(\mathbb{H})$ generated by A, B preserves $\mathscr{T}$. The next lemma identifies this subgroup:

LEMMA 13.10. *The subgroup* $\langle A, B \rangle \subset \mathrm{Aut}(\mathbb{H})$ *generated by* A, B *in* (13.3) *is the* **modular group**

$$\Gamma = \left\{ z \mapsto \frac{az + b}{cz + d} : a, b, c, d \in \mathbb{Z} \text{ and } ad - bc = 1 \right\}.$$

The modular group and its "congruence subgroups" are important objects in number theory.

Under the isomorphism $\mathrm{Aut}(\mathbb{H}) \cong \mathrm{PSL}_2(\mathbb{R})$ (see Theorem 4.18), the modular group Γ can be identified with the matrix group

$$\mathrm{PSL}_2(\mathbb{Z}) = \left\{ \begin{bmatrix} a & b \\ c & d \end{bmatrix} : a, b, c, d \in \mathbb{Z} \text{ and } ad - bc = 1 \right\} / \{\pm I\}.$$

PROOF. Evidently $\langle A, B \rangle \subset \Gamma$ since $A, B \in \Gamma$. To see the reverse inclusion, use the matrix representations in which

$$A = \begin{bmatrix} 1 & 1 \\ 0 & 1 \end{bmatrix} \quad \text{and} \quad B = \begin{bmatrix} 0 & -1 \\ 1 & 0 \end{bmatrix}.$$

Let $M_0 = \begin{bmatrix} a_0 & b_0 \\ c_0 & d_0 \end{bmatrix}$ represent an element of Γ. If $c_0 = 0$, the condition $\det M_0 = 1$ gives $a_0 d_0 = 1$ and (after replacing M_0 by $-M_0$ if necessary) we may assume $a_0 =$

$d_0 = 1$. In this case, $M_0 = A^{b_0} \in \langle A, B \rangle$. If $c_0 \neq 0$, Euclid's division algorithm gives an integer n_1 such that $c_1 = d_0 + n_1 c_0$ satisfies $|c_1| < |c_0|$. Then

$$M_1 = M_0 A^{n_1} B = \begin{bmatrix} a_1 & b_1 \\ c_1 & d_1 \end{bmatrix}$$

for some integers a_1, b_1, d_1. If $c_1 = 0$, the above observation shows that M_1 and therefore M_0 belongs to $\langle A, B \rangle$. Otherwise, $c_1 \neq 0$ and the same argument can be repeated for M_1 in place of M_0. Continuing this process, we obtain a strictly decreasing sequence of integers $|c_0| > |c_1| > |c_2| > \cdots$ which must eventually terminate at $c_k = 0$ for some k. It follows that there are integers $n_1, \ldots, n_k$ and a_k, b_k, d_k such that

$$M_k = M_0 A^{n_1} B A^{n_2} B \cdots A^{n_k} B = \begin{bmatrix} a_k & b_k \\ 0 & d_k \end{bmatrix}.$$

Since M_k belongs to $\langle A, B \rangle$, the same must be true of M_0. $\qquad \square$

THEOREM 13.11. $\mathcal{T}$ is a tessellation of $\mathbb{H}$.

PROOF. As before, we take the tile T to be the ideal triangle with vertices $0, 1, \infty$. Since we have already seen that distinct tiles in $\mathcal{T}$ have pairwise disjoint interiors (Lemma 13.9), it remains to show that the union of all tiles in $\mathcal{T}$ is $\mathbb{H}$. As noted before, the modular group $\Gamma = \langle A, B \rangle$ preserves $\mathcal{T}$. The map $z \mapsto (az + b)/(cz + d)$ in Γ sends the vertex 0 of T to the vertex b/d of the image tile. Since for every reduced fraction b/d there are integers a, c such that $ad - bc = 1$, it follows that *every rational number appears as a vertex of some tile in $\mathcal{T}$*.

Because the translation A preserves $\mathcal{T}$, it suffices to prove that the strip $S = \{z \in \mathbb{H} : 0 \leq \mathrm{Re}(z) \leq 1\}$ is covered by tiles. The union P_n of all tiles of generation $\leq n$ in S is an ideal polygon with two vertical sides $\mathrm{Re}(z) = 0$ and $\mathrm{Re}(z) = 1$ and 2^n semicircular sides with centers along $[0, 1]$ (see Fig. 13.2 for P_4). If r_n denotes the largest of the radii of these semicircular sides, it is clear that P_n contains all $z \in S$ with $\mathrm{Im}(z) > r_n$. Thus, the theorem follows once we verify that the decreasing sequence $\{r_n\}$ tends to 0 as $n \to \infty$. Given an integer $N \geq 1$ consider the rational points $x_j = j/N$ for $j = 1, \ldots, N - 1$. By the previous paragraph, there is a tile $T^j \subset S$ with a vertex at x_j. Let n be the maximum generation among $T^1, \ldots, T^{N-1}$. Then every T^j belongs to P_n, so every x_j appears as a vertex of P_n. This shows $r_n \leq 1/(2N)$. Since N was arbitrary, we conclude that $\lim_{n \to \infty} r_n = 0$. $\qquad \square$

REMARK 13.12. Two reduced fractions $0 \leq b/d < a/c \leq 1$ are said to be ***Farey neighbors*** if $ad - bc = 1$. In this case their ***Farey sum*** $(a + b)/(c + d)$ satisfies $b/d < (a + b)/(c + d) < a/c$ and the pairs $b/d, (a + b)/(c + d)$ and $(a + b)/(c + d), a/c$ are again Farey neighbors. Starting with the initial pair $0/1, 1/1$ and building up Farey sums of neighbors successively, one produces the list in which every rational number will eventually show up. The fractions along the n-th row are precisely the vertices of the ideal polygon P_n in the above proof (see problem 9 and compare Fig. 13.3).

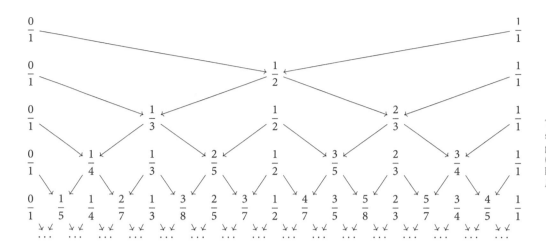

This combinatorial structure is named after geologist John Farey (1766–1826) who published his observation in 1816 in *Philosophical Magazine*.

The existence of a tessellation of $\mathbb{D}$ or $\mathbb{H}$ by ideal triangles leads to a quick proof of the main result of this section:

THEOREM 13.13 (Uniformization of the thrice-punctured sphere). *If $q_1, q_2, q_3 \in \hat{\mathbb{C}}$ are distinct, there exists a holomorphic covering map $\lambda : \mathbb{D} \to \hat{\mathbb{C}} \smallsetminus \{q_1, q_2, q_3\}$.*

PROOF. It suffices to construct a holomorphic covering map $\lambda : \mathbb{H} \to \mathbb{C} \smallsetminus \{0, 1\}$. Consider the ideal triangle $T \subset \mathbb{H}$ with vertices $v_\alpha = 0, v_\beta = 1, v_\gamma = \infty$ and the tessellation $\mathscr{T} = \mathscr{T}_T$. By the Riemann mapping theorem, there is a conformal map λ from the interior of T onto $\mathbb{H}$. Carathéodory's Theorem 6.20 guarantees that λ extends to a homeomorphism $T \to \overline{\mathbb{H}}$. After postcomposing with an element of $\mathrm{Aut}(\mathbb{H})$, we may assume $\lambda(0) = 0, \lambda(1) = 1, \lambda(\infty) = \infty$. Under this normalization, λ sends the α-side of T (contained in the line $\mathrm{Re}(z) = 1$) to the interval $(1, +\infty)$, the β-side of T (contained in the line $\mathrm{Re}(z) = 0$) to the interval $(-\infty, 0)$, and the γ-side of T (contained in the circle $|2z - 1| = 1$) to the interval $(0, 1)$ (see Fig. 13.3). By the Schwarz reflection principle, λ extends to the first generation tiles $T_\alpha, T_\beta, T_\gamma \in \mathscr{T}$, mapping their interiors conformally to the lower half-plane $-\mathbb{H}$. Applying the reflection principle inductively then allows us to extend λ to every tile in $\mathscr{T}$, mapping its interior conformally to $\mathbb{H}$ or $-\mathbb{H}$ depending on whether the tile is of even or odd generation. By the construction, for every pair of adjacent (that is, parent and child) tiles $T', T'' \in \mathscr{T}$, the interior of $T' \cup T''$ is mapped conformally onto

$$W_\alpha = \mathbb{C} \smallsetminus (-\infty, 1], \quad W_\beta = \mathbb{C} \smallsetminus [0, +\infty), \quad \text{or} \quad W_\gamma = \mathbb{C} \smallsetminus ((-\infty, 0] \cup [1, +\infty))$$

according as the common side of T' and T'' is α, β, or γ. Since the tiles in $\mathscr{T}$ cover $\mathbb{H}$ by Theorem 13.11, we obtain a function $\lambda \in \mathscr{O}(\mathbb{H})$ which takes values in $\mathbb{C} \smallsetminus \{0, 1\}$ since $0, 1, \infty$ can only be the images of the tile vertices.

We claim that λ is a covering map. With W_α as above, the preimage $\lambda^{-1}(W_\alpha)$ is $\mathbb{H}$ with the β- and γ-side of every tile in $\mathscr{T}$ removed, so the components of $\lambda^{-1}(W_\alpha)$ are precisely the interior of the unions $T' \cup T''$ where T', T'' are adjacent tiles sharing

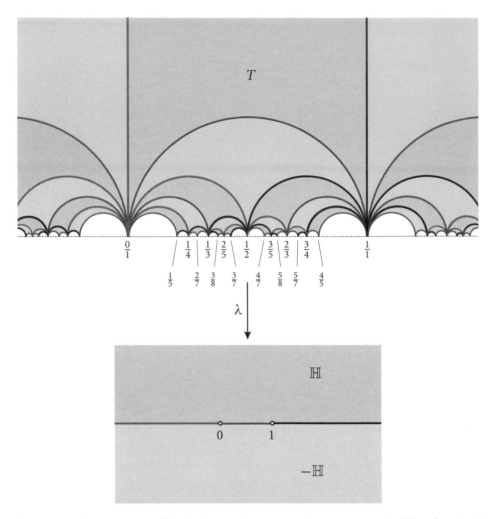

Figure 13.3. Construction of the holomorphic universal covering map $\lambda : \mathbb{H} \to \mathbb{C} \smallsetminus \{0, 1\}$.

their α-side, and we have seen that such open sets map conformally onto W_α. This shows that W_α is evenly covered. The same observation shows that W_β and W_γ are evenly covered. The claim follows since $W_\alpha \cup W_\beta \cup W_\gamma = \mathbb{C} \smallsetminus \{0, 1\}$. $\qquad\square$

REMARK 13.14. The particular covering map λ constructed above is called an ***elliptic modular function***. This terminology is due to the following classical construction. Think of each point $\tau \in \mathbb{H}$ as determining the lattice $\Lambda = \langle 1, \tau \rangle$. As in §9.2, the finite critical values

$$e_1(\tau) = \wp\left(\frac{1}{2}\right), \quad e_2(\tau) = \wp\left(\frac{1+\tau}{2}\right), \quad e_3(\tau) = \wp\left(\frac{\tau}{2}\right)$$

of the associated Weierstrass $\wp$-function are distinct. The cross ratio

$$\lambda(\tau) = [e_1(\tau), \infty, e_2(\tau), e_3(\tau)] = \frac{e_2(\tau) - e_1(\tau)}{e_2(\tau) - e_3(\tau)}$$

then agrees with the function constructed geometrically in the above proof. See for example [**A2**] or [**MM**] for details, and compare problem 8 for some elementary investigations.

Theorem 13.13 leads to quick proofs of some of the main theorems of chapter 11. Below we illustrate this idea for Picard's little theorem and the general version of Montel's theorem.

FOURTH PROOF OF THEOREM 11.6. We may assume $f \in \mathcal{O}(\mathbb{C})$ omits the values $0, 1$. Then f can be thought of as a holomorphic map $f : \mathbb{C} \to \mathbb{C} \smallsetminus \{0, 1\}$. Let $\lambda : \mathbb{D} \to \mathbb{C} \smallsetminus \{0, 1\}$ be a holomorphic covering map given by Theorem 13.13 and $\tilde{f} : \mathbb{C} \to \mathbb{D}$ be any lift of f under λ. Liouville's theorem now shows that $\tilde{f}$ and therefore f must be constant. $\square$

FOURTH PROOF OF THEOREM 11.18. It suffices to show that the family $\mathscr{F}$ of all holomorphic functions $\mathbb{D} \to \mathbb{C} \smallsetminus \{0, 1\}$ is normal. Take a sequence $\{f_n\}$ in $\mathscr{F}$. After passing to a subsequence, we may assume $f_n(0) \to q \in \hat{\mathbb{C}}$. First consider the case where $q \notin \{0, 1, \infty\}$. Let $\lambda : \mathbb{D} \to \mathbb{C} \smallsetminus \{0, 1\}$ be a holomorphic covering map and fix some $p \in \mathbb{D}$ with $\lambda(p) = q$. Let $\tilde{f}_n : \mathbb{D} \to \mathbb{D}$ be a lift of f_n under λ such that $\tilde{f}_n(0) \to p$. Since $\{\tilde{f}_n\}$ is uniformly bounded, Montel's Theorem 5.25 gives a subsequence $\{\tilde{f}_{n_k}\}$ which converges compactly to some $\tilde{f} \in \mathcal{O}(\mathbb{D})$ with $\tilde{f}(0) = p$. By Corollary 5.20, $\tilde{f}(\mathbb{D}) \subset \mathbb{D}$. It follows that $f_{n_k} = \lambda \circ \tilde{f}_{n_k} \to \lambda \circ \tilde{f}$ compactly in $\mathbb{D}$.

Next consider the case where $q \in \{0, 1, \infty\}$. After postcomposing the f_n with a Möbius map we may assume $q = 1$. As a non-vanishing holomorphic function in $\mathbb{D}$, each f_n has a holomorphic square root $g_n : \mathbb{D} \to \mathbb{C} \smallsetminus \{0, -1, 1\}$. Of the two choices for g_n, let us pick the one such that $g_n(0) \to -1$. Then, by the first case treated above, there is a subsequence $\{g_{n_k}\}$ which converges compactly to some $g \in \mathcal{O}(\mathbb{D})$. Since $g_n(\mathbb{D}) \subset \mathbb{C} \smallsetminus \{0, -1, 1\}$ but $g(0) = -1$, it follows from Corollary 5.20 that $g = -1$ everywhere. We conclude that $f_{n_k} \to 1$ compactly in $\mathbb{D}$. $\square$

13.2 The uniformization theorem

The central result of this chapter is the following generalization of the Riemann mapping theorem:

THEOREM 13.15 (The uniformization theorem). *For every domain $U \subset \hat{\mathbb{C}}$ there is a holomorphic universal covering map $f : X \to U$ where X is one of the three simply connected domains $\hat{\mathbb{C}}, \mathbb{C},$ or $\mathbb{D}$. Moreover, U must be of one of the following types:*

(i) **Spherical,** *where $U = X = \hat{\mathbb{C}}$;*
(ii) **Euclidean,** *where $\hat{\mathbb{C}} \smallsetminus U$ contains one or two points and $X = \mathbb{C}$;*
(iii) **Hyperbolic,** *where $\hat{\mathbb{C}} \smallsetminus U$ contains three or more points and $X = \mathbb{D}$.*

The main contributors to the eventual proof of the uniformization theorem (for Riemann surfaces) were Poincaré and Koebe. For an excellent account, see [**DeS**].

Since the holomorphic universal covering space is unique up to biholomorphism (see Remark 12.18), it is clear that the above three types are mutually exclusive. Every punctured sphere is conformally isomorphic to $\mathbb{C}$ and every twice punctured sphere is conformally isomorphic to $\mathbb{C}^*$ which has a holomorphic universal covering map $\exp : \mathbb{C} \to \mathbb{C}^*$. Thus, the main thrust of Theorem 13.15 is the assertion that if $\hat{\mathbb{C}} \smallsetminus U$ contains three or more points, there is a holomorphic covering map $\mathbb{D} \to U$. The maximal case where U is a thrice punctured sphere was dealt with in the previous section. We prove the general case using this special case and a normal family argument similar to the proof of the Riemann mapping theorem. Our presentation is partially inspired by [**FHW**] and [**McM**].

The proof will use the existence of a topological universal covering $u : \tilde{U} \to U$ which is unique up to covering space isomorphism (see Remark 12.22). The space $\tilde{U}$ is of course locally homeomorphic to an open disk in the plane. More precisely, we have the following

LEMMA 13.16. *The universal covering space $\tilde{U}$ is a **countable** union of open sets $\{\Omega_k\}$ such that $U_k = u(\Omega_k)$ is an open disk and $u : \Omega_k \to U_k$ is a homeomorphism for each k.*

PROOF. The domain U is a countable union of open disks U_k, each evenly covered by u. The preimage $u^{-1}(U_k)$ is the disjoint union of a collection of open sets which map homeomorphically onto U_k, and the cardinality of this collection is the same as that of any fiber of u. Thus, to prove the lemma, it suffices to show that every fiber of u is countable. By Theorem 12.21 there is a bijection between each fiber of u and the deck group G of the universal covering $u : \tilde{U} \to U$, and G in turn is isomorphic to the fundamental group $\pi_1(U)$. This reduces the lemma to verifying that $\pi_1(U)$ is countable. The latter is a consequence of the fact that U has a countable dense subset $\{p_k\}$. In fact, the collection P of all piecewise affine curves in U that start and end at p_1 and have their corners in $\{p_k\}$ is countable. Every element of $\pi_1(U, p_1)$ has a piecewise affine representative γ (see the end of §2.5) and by moving the finitely many corners of γ slightly, it follows that γ is homotopic to an element of P. $\square$

PROOF OF THEOREM 13.15. We assume that $U \subset \mathbb{C} \smallsetminus \{0, 1\}$ and we show the existence of a holomorphic covering map $\mathbb{D} \to U$. Fix a base point $q \in U$ and let $\mathscr{F}$ be the family of all holomorphic functions $f : \mathbb{D} \to \mathbb{C} \smallsetminus \{0, 1\}$ with $f(0) = q$ for which there is a subdomain $V_f \subset \mathbb{D}$ containing 0 such that the restriction $f|_{V_f} : (V_f, 0) \to (U, q)$ is a covering map. (An easy exercise shows that such V_f must be unique.)

• STEP 1. $\mathscr{F} \neq \varnothing$. In fact, the universal covering $\lambda : \mathbb{D} \to \mathbb{C} \smallsetminus \{0, 1\}$ of Theorem 13.13, normalized so that $\lambda(0) = q$, belongs to $\mathscr{F}$ since if $V_\lambda \subset \mathbb{D}$ is the connected component of $\lambda^{-1}(U)$ containing 0, then $\lambda|_{V_\lambda} : (V_\lambda, 0) \to (U, q)$ is a covering map.

• STEP 2. Let $r = \mathrm{dist}(q, \partial U) > 0$ so $\mathbb{D}(q, r) \subset U$. For every $f \in \mathscr{F}$, since $f : V_f \to U$ is a covering map and $\mathbb{D}(q, r)$ is simply connected, there is a holomorphic branch of f^{-1} defined in $\mathbb{D}(q, r)$ which takes values in V_f, with $f^{-1}(q) = 0$. By Theorem 1.43,

$|(f^{-1})'(q)| \leq 1/r$. Thus,

(13.4) $|f'(0)| \geq r$ for every $f \in \mathscr{F}$.

• STEP 3. $\mathscr{F}$ is a compact family. Take a sequence $\{f_n\}$ in $\mathscr{F}$. By generalized Montel's Theorem 11.18, after passing to a subsequence, we may assume that $\{f_n\}$ converges compactly to some $f \in \mathscr{O}(\mathbb{D})$ which is not constant since by (13.4), $|f'(0)| = \lim_{n\to\infty} |f_n'(0)| \geq r > 0$. It follows from Corollary 5.20 that f maps $\mathbb{D}$ to $\mathbb{C} \smallsetminus \{0, 1\}$.

To prove $f \in \mathscr{F}$, we use the existence of a topological universal covering map $u : (\tilde{U}, p) \to (U, q)$. Let $\tilde{u}_n : (\tilde{U}, p) \to (V_{f_n}, 0)$ be the unique lift of u under f_n. Consider the countable collection of open sets $\{\Omega_k\}$ in $\tilde{U}$ and their images $\{U_k\}$ in U as in Lemma 13.16, and let $\zeta_k : U_k \to \Omega_k$ be the local inverse of $u : \Omega_k \to U_k$. The composition $\tilde{u}_n \circ \zeta_k : U_k \to V_{f_n}$ is holomorphic since it is a local branch of f_n^{-1}. Hence, for fixed k, $\{\tilde{u}_n \circ \zeta_k\}$ is a sequence in $\mathscr{O}(U_k)$ which is uniformly bounded by 1. By Montel's theorem, some subsequence of it converges compactly in U_k, so the corresponding subsequence of $\{\tilde{u}_n\}$ converges compactly in Ω_k. Since every compact subset of $\tilde{U}$ is covered by finitely many of the Ω_k, it follows from a standard diagonal argument that $\{\tilde{u}_n\}$ has a subsequence, denoted again by $\{\tilde{u}_n\}$, which converges compactly in $\tilde{U}$ to a continuous function $\tilde{u} : \tilde{U} \to \overline{\mathbb{D}}$.

Let us verify that $\tilde{u}(\tilde{U}) \subset \mathbb{D}$. Set $\Omega = \{z \in \tilde{U} : \tilde{u}(z) \in \partial\mathbb{D}\}$. If $z_0 \in \Omega$, choose a neighborhood Ω_k containing z_0 and note that $\tilde{u}_n \circ \zeta_k : U_k \to \mathbb{D}$ converges compactly to $\tilde{u} \circ \zeta_k$, which takes the value $w_0 = \tilde{u}(z_0) \in \partial\mathbb{D}$ at $u(z_0) \in U_k$. It follows from Corollary 5.20 that $\tilde{u} \circ \zeta_k = w_0$ in U_k, or $\tilde{u} = w_0$ in Ω_k. This shows that Ω is open. Since Ω is trivially closed, connectedness of $\tilde{U}$ shows that either $\Omega = U$ or $\Omega = \emptyset$. The former cannot occur since $\tilde{u}(p) = 0$, so $\Omega = \emptyset$.

Now for each $z \in \tilde{U}$ and $n \geq 1$ the triangle inequality gives

$$|f(\tilde{u}(z)) - u(z)| \leq |f(\tilde{u}(z)) - f(\tilde{u}_n(z))| + |f(\tilde{u}_n(z)) - f_n(\tilde{u}_n(z))|.$$

As $n \to \infty$, the first term on the right tends to 0 by continuity of f, while the second term tends to 0 since $f_n \to f$ compactly in $\mathbb{D}$. We conclude that $f \circ \tilde{u} = u$. This shows in particular that the restriction of $\tilde{u}$ to each Ω_k is injective. Since $\tilde{u} \circ \zeta_k = \lim_{n\to\infty} \tilde{u}_n \circ \zeta_k$ is holomorphic in U_k, it follows that each restriction $\tilde{u}|_{\Omega_k} : \Omega_k \to \mathbb{D}$ and therefore $\tilde{u} : \tilde{U} \to \mathbb{D}$ is an open map. Hence $V = \tilde{u}(\tilde{U})$ is a subdomain of $\mathbb{D}$ containing 0 and we have the following commutative diagram:

$$
\begin{array}{ccc}
 & & V \\
 & \overset{\tilde{u}}{\nearrow} & \downarrow{\scriptstyle f|_V} \\
\tilde{U} & \underset{u}{\longrightarrow} & U
\end{array}
$$

Theorem 12.16(iii) now shows that both $\tilde{u}$ and $f|_V$ are covering maps. Thus, $f \in \mathscr{F}$ with $V_f = V$.

● STEP 4. If $f \in \mathscr{F}$ and $V_f \neq \mathbb{D}$, there is a $g \in \mathscr{F}$ such that $|g'(0)| < |f'(0)|$. This follows from a square root trick similar to the one in the proof of the Riemann mapping theorem. Let $a \in \mathbb{D} \smallsetminus V_f$ and $b \in \mathbb{D}^*$ be one of the two square roots of $-a$. Define $g : \mathbb{D} \to \mathbb{C} \smallsetminus \{0, 1\}$ by

$$g = f \circ (\varphi_{-a} \circ s \circ \varphi_{-b}),$$

where, as usual, $\varphi_p : z \mapsto (z - p)/(1 - \bar{p}z) \in \mathrm{Aut}(\mathbb{D})$ and $s : z \mapsto z^2$ is the squaring map. The image $\varphi_a(V_f)$ is a subdomain of $\mathbb{D}^*$ that contains $-a$. Let $W \subset \mathbb{D}^*$ be the connected component of $s^{-1}(\varphi_a(V_f))$ containing b. Since $s : \mathbb{D}^* \to \mathbb{D}^*$ is a covering map, the restriction $s : W \to \varphi_a(V_f)$ is also a covering map (of degree 1 or 2). It follows that the restriction $\varphi_{-a} \circ s \circ \varphi_{-b} : \varphi_b(W) \to V_f$ is a covering map. Thus $g \in \mathscr{F}$ with $V_g = \varphi_b(W)$. The map $\varphi_{-a} \circ s \circ \varphi_{-b} : \mathbb{D} \to \mathbb{D}$ fixes the origin and is not a rotation since it is a degree 2 branched covering. By the Schwarz lemma, $|(\varphi_{-a} \circ s \circ \varphi_{-b})'(0)| < 1$, hence

$$|g'(0)| = |f'(0)| \cdot |(\varphi_{-a} \circ s \circ \varphi_{-b})'(0)| < |f'(0)|.$$

● STEP 5. Let

$$\delta = \inf \left\{ |f'(0)| : f \in \mathscr{F} \right\}.$$

By (13.4), $\delta \geq r > 0$. Take a sequence $\{f_n\}$ in $\mathscr{F}$ with $|f_n'(0)| \to \delta$. By compactness of $\mathscr{F}$, there is a subsequence of $\{f_n\}$ which converges compactly to some $f \in \mathscr{F}$. Since $|f'(0)| = \delta$ is the smallest possible in the family $\mathscr{F}$, STEP 4 guarantees that $V_f = \mathbb{D}$. Thus $f : \mathbb{D} \to U$ is a holomorphic covering map, as required. □

13.3 Hyperbolic domains

The uniformization theorem shows that hyperbolicity in the sense of Definition 11.32 is equivalent to being holomorphically covered by the unit disk. We record this fact in the following sharper version of Theorem 11.36:

THEOREM 13.17 (Characterization of hyperbolic domains). *The following conditions on a domain $U \subset \hat{\mathbb{C}}$ are equivalent:*

(i) $\hat{\mathbb{C}} \smallsetminus U$ has at least three points.
(ii) U admits a conformal metric whose curvature has a negative upper bound.
(iii) There is a holomorphic covering map $\mathbb{D} \to U$.

As a matter of fact, more is true: Every hyperbolic domain $U \subset \hat{\mathbb{C}}$ carries a natural conformal metric which is *complete* and has *constant* negative curvature. This is a consequence of the fact that the unit disk $\mathbb{D}$ carries such a metric, which is invariant under the action of the full automorphism group $\mathrm{Aut}(\mathbb{D})$ (see §4.5) and therefore under the action of the deck group of the universal covering $\mathbb{D} \to U$. Let us explain this in some detail.

Take a holomorphic covering map $f : \mathbb{D} \to U$ and let $G \subset \text{Aut}(\mathbb{D})$ be the associated deck group. The hyperbolic metric $g_{\mathbb{D}} = 2|d\zeta|/(1 - |\zeta|^2)$ in $\mathbb{D}$ induces a well-defined conformal metric $g_U = \rho_U(z)\,|dz|$ in U characterized by the property $f^*g_U = g_{\mathbb{D}}$. In fact, this property is equivalent to the relation $\rho_U(f(\zeta))\,|f'(\zeta)| = 2/(1 - |\zeta|^2)$, or

$$(13.5) \qquad \rho_U(f(\zeta)) = \frac{2}{(1 - |\zeta|^2)\,|f'(\zeta)|} \qquad \text{for all } \zeta \in \mathbb{D}.$$

If $f(\zeta) = f(\hat{\zeta})$, there is a unique $\varphi \in G$ such that $\varphi(\zeta) = \hat{\zeta}$, and we have $|\varphi'(\zeta)| = (1 - |\hat{\zeta}|^2)/(1 - |\zeta|^2)$ by Corollary 4.36 since φ is a hyperbolic isometry. It follows that

$$\frac{2}{(1 - |\hat{\zeta}|^2)\,|f'(\hat{\zeta})|} = \frac{2}{(1 - |\zeta|^2)\,|f'(\hat{\zeta})|\,|\varphi'(\zeta)|} = \frac{2}{(1 - |\zeta|^2)\,|(f \circ \varphi)'(\zeta)|}$$

$$= \frac{2}{(1 - |\zeta|^2)\,|f'(\zeta)|} \qquad \text{(since } \varphi \in G\text{)}.$$

Thus, (13.5) defines the density ρ_U unambiguously at every point of U. Note that the map $f : (\mathbb{D}, g_{\mathbb{D}}) \to (U, g_U)$ is automatically a local isometry. The metric g_U does not depend on the choice of the covering map $f : \mathbb{D} \to U$ and therefore is intrinsic to U. This follows from the observation that any two universal covering maps $\mathbb{D} \to U$ differ by an automorphism of $\mathbb{D}$ which necessarily preserves $g_{\mathbb{D}}$.

Here is an interpretation for the density of the metric g_U: If $z \in U$ and if we choose the covering map $f : \mathbb{D} \to U$ such that $f(0) = z$, then (13.5) reduces to

$$(13.6) \qquad \rho_U(z) = \frac{2}{|f'(0)|}.$$

Thus, $\rho_U(z)$ is proportional to the size of the derivative at z of the local branch of f^{-1} which sends z to 0.

Recalling that $g_{\mathbb{D}}$ has constant curvature -1 (Example 11.29) and that the curvature is a conformal invariant (Lemma 11.28), it is immediate that g_U also has constant curvature -1.

We summarize the above remarks in the following

THEOREM 13.18. *Every hyperbolic domain $U \subset \hat{\mathbb{C}}$ is equipped with a unique conformal metric g_U characterized by the property $f^*g_U = g_{\mathbb{D}}$ for every holomorphic covering map $f : \mathbb{D} \to U$. This metric is C^∞-smooth and has constant curvature -1.*

We call g_U the **hyperbolic metric** of the domain U.

Observe that if $\psi : U \to V$ is a biholomorphism between hyperbolic domains, then $\psi^*g_V = g_U$. In fact, given any covering map $f : \mathbb{D} \to U$ the composition $\psi \circ f : \mathbb{D} \to V$ is also a covering, so

$$g_{\mathbb{D}} = (\psi \circ f)^*g_V = f^*(\psi^*g_V).$$

It follows from the characterizing property of hyperbolic metrics that $\psi^* g_V = g_U$. The special case of $U = V$ shows that *all automorphisms of a hyperbolic domain preserve its hyperbolic metric and therefore are hyperbolic isometries.*

EXAMPLE 13.19 (Hyperbolic metric of a strip). When U is simply connected, its hyperbolic metric can be found by taking the pull-back of $g_{\mathbb{D}}$ under any conformal map $U \to \mathbb{D}$. In Example 4.37 this idea was used to find $g_{\mathbb{H}}$:

$$g_{\mathbb{H}} = \frac{1}{\operatorname{Im}(z)} |dz| \qquad (\operatorname{Im}(z) > 0).$$

As another example, consider the horizontal strip $S = \{z : 0 < \operatorname{Im}(z) < a\}$. Using the conformal map $f : S \to \mathbb{H}$ given by $f(z) = \exp(\pi z / a)$, we see that

$$\rho_S(z) = \rho_{\mathbb{H}}(f(z)) \, |f'(z)| = \frac{|f'(z)|}{\operatorname{Im}(f(z))} = \frac{\frac{\pi}{a} \left| \exp\left(\frac{\pi}{a} z\right) \right|}{\operatorname{Im}\left(\exp\left(\frac{\pi}{a} z\right)\right)},$$

which yields

(13.7)
$$g_S = \frac{\frac{\pi}{a}}{\sin\left(\frac{\pi}{a} \operatorname{Im}(z)\right)} |dz| \qquad (0 < \operatorname{Im}(z) < a).$$

As expected, the density of this metric depends only on $\operatorname{Im}(z)$ since all translations $z \mapsto z + b$ for $b \in \mathbb{R}$ belong to $\operatorname{Aut}(S)$ and therefore are isometries of S.

EXAMPLE 13.20 (Hyperbolic metric of a round annulus). Let us now determine the explicit form of the hyperbolic metric of the round annulus $A = \{z : e^{-2\pi m} < |z| < 1\}$ for $m > 0$. The strip $S = \{\zeta : 0 < \operatorname{Im}(\zeta) < 2\pi m\}$ is a universal covering space of A via the map $f : S \to A$ given by $z = f(\zeta) = \exp(i\zeta)$. The characterizing property $f^* g_A = g_S$ and (13.7) for $a = 2\pi m$ show that

$$\rho_A(f(\zeta)) \, |f'(\zeta)| = \frac{\frac{1}{2m}}{\sin\left(\frac{1}{2m} \operatorname{Im}(\zeta)\right)} \qquad \text{or} \qquad \rho_A(z) = \frac{\frac{1}{2m}}{|f'(\zeta)| \sin\left(\frac{1}{2m} \operatorname{Im}(\zeta)\right)}.$$

Since

(13.8)
$$\operatorname{Im}(\zeta) = -\log|z| \qquad \text{and} \qquad |f'(\zeta)| = |i \exp(i\zeta)| = |z|,$$

we conclude that

(13.9)
$$g_A = \frac{-\frac{1}{2m}}{|z| \sin\left(\frac{1}{2m} \log|z|\right)} |dz| \qquad (e^{-2\pi m} < |z| < 1).$$

As expected, the density of this metric depends only on $|z|$ since all rotations around the origin are isometries of A.

EXAMPLE 13.21 (Hyperbolic metric of the punctured disk). As a final example, let us determine the hyperbolic metric of the punctured disk $\mathbb{D}^* = \{z : 0 < |z| < 1\}$. The function $z = f(\zeta) = \exp(i\zeta)$ provides a universal covering map $\mathbb{H} \to \mathbb{D}^*$ and the characterizing property $f^* g_{\mathbb{D}^*} = g_{\mathbb{H}}$

translates to

$$\rho_{\mathbb{D}^*}(f(\zeta))\,|f'(\zeta)| = \frac{1}{\mathrm{Im}(\zeta)} \qquad \text{or} \qquad \rho_{\mathbb{D}^*}(z) = \frac{1}{|f'(\zeta)|\,\mathrm{Im}(\zeta)}.$$

In view of (13.8), we obtain

(13.10)
$$g_{\mathbb{D}}^* = \frac{-1}{|z|\log|z|}\,|dz| \qquad (0 < |z| < 1).$$

Observe that this is the limit of the hyperbolic metric (13.9) as $m \to +\infty$.

The hyperbolic metric $g_U = \rho_U(z)\,|dz|$ turns U into a metric space with the distance defined by

$$\mathrm{dist}_U(p,q) = \inf_\gamma \, \mathrm{length}_U(\gamma) = \inf_\gamma \int_\gamma \rho_U(z)\,|dz|,$$

where the infimum is taken over all piecewise C^1 curves γ in U from p to q (compare the discussion in §4.4). A curve γ which realizes this infimum is a ***minimal geodesic*** for the metric g_U.

THEOREM 13.22. *Each pair of distinct points in a hyperbolic domain can be joined by at least one minimal geodesic.*

PROOF. Let $U \subset \hat{\mathbb{C}}$ be hyperbolic and take distinct points $p, q \in U$. Take a holomorphic covering map $f : \mathbb{D} \to U$, fix some $z_0 \in f^{-1}(p)$, and let w_0 be a closest point of the fiber $f^{-1}(q)$ to z_0 in the hyperbolic metric of $\mathbb{D}$. Such w_0 exists because $f^{-1}(q)$ is discrete and $\mathrm{dist}_{\mathbb{D}}(z_0, w) \to +\infty$ as $w \to \partial\mathbb{D}$. Consider the unique minimal geodesic $\tilde{\gamma}$ in $\mathbb{D}$ which joins z_0 and w_0. We claim that $\gamma = f \circ \tilde{\gamma}$ is a minimal geodesic which joins p and q. To see this, take any piecewise C^1 curve η in U from p to q, and let $\tilde{\eta}$ be the unique lift of η that starts at z_0. The end point w_1 of $\tilde{\eta}$ belongs to the fiber $f^{-1}(q)$. By Theorem 4.31, $\mathrm{length}_U(\gamma) = \mathrm{length}_{\mathbb{D}}(\tilde{\gamma})$ and $\mathrm{length}_U(\eta) = \mathrm{length}_{\mathbb{D}}(\tilde{\eta})$ since f is a local isometry. Thus

$$\mathrm{length}_U(\gamma) = \mathrm{length}_{\mathbb{D}}(\tilde{\gamma}) = \mathrm{dist}_{\mathbb{D}}(z_0, w_0) \le \mathrm{dist}_{\mathbb{D}}(z_0, w_1)$$

$$\le \mathrm{length}_{\mathbb{D}}(\tilde{\eta}) = \mathrm{length}_U(\eta). \qquad \square$$

As a corollary of the above proof, we see that

(13.11)
$$\mathrm{dist}_U(p,q) = \min\big\{\, \mathrm{dist}_{\mathbb{D}}(z,w) : z \in f^{-1}(p),\ w \in f^{-1}(q)\,\big\}.$$

EXAMPLE 13.23. Unlike the case of the unit disk (and therefore every simply connected hyperbolic domain), where there is only one minimal geodesic joining a given pair, uniqueness of minimal geodesics can no longer be guaranteed if the domain has a non-trivial fundamental group. For example, in a round annulus every pair of antipodal points can be joined by precisely two minimal geodesics which map to one another under the 180° rotation of the annulus (see Fig. 13.4 and problem 13).

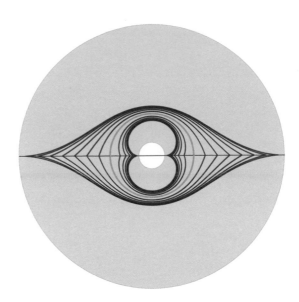

Figure 13.4. Each pair of antipodal points in a round annulus is joined by precisely two minimal hyperbolic geodesics.

It is not hard to show that every hyperbolic domain U equipped with the distance dist_U is a complete metric space. The proof is based on the following

LEMMA 13.24. *Closed balls in* $(U, \operatorname{dist}_U)$ *are compact.*

PROOF. Denote by $N_U(p, r)$ the closed ball $\{z \in U : \operatorname{dist}_U(z, p) \le r\}$ and similarly define $N_{\mathbb{D}}(p, r)$. We claim that for any holomorphic universal covering map $f : \mathbb{D} \to U$,

$$(13.12) \qquad\qquad f(N_{\mathbb{D}}(p, r)) = N_U(f(p), r).$$

In fact, if $\operatorname{dist}_{\mathbb{D}}(z, p) \le r$, then $\operatorname{dist}_U(f(z), f(p)) \le r$ by (13.11), which shows $f(N_{\mathbb{D}}(p, r)) \subset N_U(f(p), r)$. Conversely, if $\operatorname{dist}_U(w, f(p)) \le r$, then by (13.11) there is a $z \in f^{-1}(w)$ such that $\operatorname{dist}_{\mathbb{D}}(z, p) = \operatorname{dist}_U(w, f(p)) \le r$. This shows the reverse inclusion $f(N_{\mathbb{D}}(p, r)) \supset N_U(f(p), r)$ and proves (13.12).

Now, closed balls in $(\mathbb{D}, \operatorname{dist}_{\mathbb{D}})$ are closed Euclidean disks in $\mathbb{D}$ and therefore are compact (see problem 25 in chapter 4). As continuous images of compact sets, it follows from (13.12) that closed balls in $(U, \operatorname{dist}_U)$ are also compact. $\square$

COROLLARY 13.25 (Boundary points are infinitely far)**.** *For each fixed* $p \in U$, $\operatorname{dist}_U(p, z) \to +\infty$ *as* $z \to \partial U$.

COROLLARY 13.26. *The metric space* $(U, \operatorname{dist}_U)$ *is complete.*

PROOF. Let $\{p_n\}$ be a Cauchy sequence in U. Find an integer j such that $\operatorname{dist}_U(p_n, p_j) \le 1$ whenever $n \ge j$. Since the sequence $\{p_n\}_{n \ge j}$ is contained in the closed unit ball $N_U(p_j, 1)$, which is compact by Lemma 13.24, some subsequence of $\{p_n\}_{n \geq j}$ must be convergent. Since $\{p_n\}$ is a Cauchy sequence, it follows that the full sequence $\{p_n\}$ converges as well.   □

For every holomorphic map $f : U \to V$ between hyperbolic domains, we shall measure the derivative norm $\|f'(z)\|$ with respect to the hyperbolic metrics g_U and g_V:

$$\|f'(z)\| = \frac{\rho_V(f(z))}{\rho_U(z)} |f'(z)|.$$

THEOREM 13.27 (The Schwarz lemma for hyperbolic domains). *For every holomorphic map $f : U \to V$ between hyperbolic domains the inequality $\|f'(z)\| \leq 1$ holds for all $z \in U$. Moreover, the following conditions are equivalent:*

(i) $\|f'(z)\| = 1$ *for some* $z \in U$.
(ii) f *is a local isometry.*
(iii) f *is a covering map.*

PROOF. Take a pair of holomorphic universal covering maps $u : \mathbb{D} \to U$ and $v : \mathbb{D} \to V$. Since $\mathbb{D}$ is simply connected, the map $f \circ u : \mathbb{D} \to V$ lifts under v to a holomorphic map $F : \mathbb{D} \to \mathbb{D}$ (Theorem 12.8), so the following diagram commutes:

$$
\begin{array}{ccc}
\mathbb{D} & \xrightarrow{\ F\ } & \mathbb{D} \\
{\scriptstyle u}\downarrow & & \downarrow{\scriptstyle v} \\
U & \xrightarrow{\ f\ } & V
\end{array}
$$

Notice that $\|u'\| = \|v'\| = 1$ everywhere since u and v are local isometries. It follows from the chain rule

$$\| f'(u(\zeta)) \| \, \| u'(\zeta) \| = \| v'(F(\zeta)) \| \, \| F'(\zeta) \|$$

that

(13.13) $$\| f'(u(\zeta)) \| = \| F'(\zeta) \| \qquad \text{for all } \zeta \in \mathbb{D}.$$

Since $\|F'(\zeta)\| \leq 1$ by Pick's Theorem 4.40, we conclude that $\|f'(z)\| \leq 1$ for all $z \in U$.

Let us now prove the equivalence of (i)–(iii):

(i) $\Longleftrightarrow$ (ii): By Pick's theorem, $\|F'\| = 1$ somewhere in $\mathbb{D}$ if and only if $\|F'\| = 1$ everywhere in $\mathbb{D}$. It follows from (13.13) that $\|f'\| = 1$ somewhere in U if and only if $\|f'\| = 1$ everywhere in U.

(ii) $\Longrightarrow$ (iii): If $\|f'\| = 1$ in U, (13.13) shows that $\|F'\| = 1$ in $\mathbb{D}$, so by Pick's theorem $F \in \mathrm{Aut}(\mathbb{D})$. Since u and $v \circ F$ are covering maps, so is f, by Theorem 12.16.

(iii) $\Longrightarrow$ (ii): If f is a covering, so is $f \circ u$ by Theorem 12.16. The map $v : \mathbb{D} \to V$ has a unique lift $G : \mathbb{D} \to \mathbb{D}$ under $f \circ u : \mathbb{D} \to V$ with $G(F(0)) = 0$. Since

$$f \circ u \circ G \circ F = v \circ F = f \circ u,$$

we see that $G \circ F$ is a lift of $f \circ u$ under $f \circ u$ that fixes 0. By uniqueness of lifts, $G \circ F = \mathrm{id}_{\mathbb{D}}$. This shows that F is invertible and proves $F \in \mathrm{Aut}(\mathbb{D})$. It follows that $\|F'\| = 1$ in $\mathbb{D}$ and therefore $\|f'\| = 1$ in U. □

EXAMPLE 13.28. The power map $f : \mathbb{D} \to \mathbb{D}$ defined by $f(z) = z^n$ is not a covering map when $n > 1$. By Theorem 13.27 the strict inequality $\|f'(z)\| < 1$ must hold everywhere in $\mathbb{D}$, as pointed out in Example 4.41. However, the same f as a function $\mathbb{D}^* \to \mathbb{D}^*$ is a covering map, so the equality $\|f'(z)\| = 1$ must hold for all $z \in \mathbb{D}^*$. We can verify this directly using the formula (13.10) for the hyperbolic metric $g_{\mathbb{D}^*}$:

$$\|f'(z)\| = \frac{\rho_{\mathbb{D}^*}(f(z)) \, |f'(z)|}{\rho_{\mathbb{D}^*}(z)} = \frac{|z| \log |z| \cdot n|z|^{n-1}}{|z|^n \log |z^n|} = 1.$$

COROLLARY 13.29 (Holomorphic maps are distance-decreasing). *Let $f : U \to V$ be a holomorphic map between hyperbolic domains. Then*

$$\mathrm{dist}_V(f(p), f(q)) \leq \mathrm{dist}_U(p, q) \qquad \text{for all } p, q \in U.$$

If equality holds for a pair $p \neq q$, then f is a local isometry, or equivalently a covering map.

PROOF. Take a minimal geodesic γ that joins a distinct pair $p, q \in U$. Since $\|f'\| \leq 1$ everywhere by Theorem 13.27, it follows from Lemma 4.30 that

$$\mathrm{dist}_V(f(p), f(q)) \leq \mathrm{length}_V(f \circ \gamma) \leq \sup_{z \in |\gamma|} \|f'(z)\| \; \mathrm{length}_U(\gamma)$$

$$\leq \mathrm{length}_U(\gamma) = \mathrm{dist}_U(p, q).$$

If equality holds, then $\|f'(z)\| = 1$ for some $z \in |\gamma|$, which by Theorem 13.27 is equivalent to f being a local isometry or a covering map. □

COROLLARY 13.30. *Suppose U and V are hyperbolic domains in $\hat{\mathbb{C}}$ with $U \subset V$. Then*

$$\rho_V(z) \leq \rho_U(z) \qquad \text{for all } z \in U,$$

with equality at some z if and only if $U = V$.

PROOF. The inclusion map $\iota : U \to V$ is holomorphic, so by Theorem 13.27

$$\|\iota'(z)\| = \frac{\rho_V(z)}{\rho_U(z)} \leq 1$$

for all $z \in U$. If equality holds at some z, then ι is a covering map and therefore surjective, which proves $U = V$. □

We end this section with a discussion of the behavior of the hyperbolic metric near the boundary of a domain. Our goal is to show that $\rho_U(z) \to +\infty$ as z

approaches a point on ∂U and to obtain some quantitative estimates. It will be convenient to first consider the case of the twice punctured plane $\mathbb{C} \smallsetminus \{0, 1\}$ near $z = 0$. To simplify the notation, we denote $\mathbb{C} \smallsetminus \{0, 1\}$ by $\mathbb{C}_{0,1}$ and the hyperbolic metric of $\mathbb{C}_{0,1}$ by $g_{0,1} = \rho_{0,1}(z)\, |dz|$. Let $g = \rho(z)\, |dz|$ be the conformal metric in $\mathbb{C}_{0,1}$ given by (11.8). By the proof of Lemma 11.35, the curvature K_g is bounded above by a negative constant $-K$. Let $\lambda : (\mathbb{D}, g_{\mathbb{D}}) \to (\mathbb{C}_{0,1}, g_{0,1})$ be a holomorphic universal covering map, and $\mathrm{id} : (\mathbb{C}_{0,1}, g_{0,1}) \to (\mathbb{C}_{0,1}, g)$ be the identity map. The composition $f = \mathrm{id} \circ \lambda : (\mathbb{D}, g_{\mathbb{D}}) \to (\mathbb{C}_{0,1}, g)$ is holomorphic, so by Ahlfors's Theorem 11.31, $\|f'(\zeta)\| \leq 1/\sqrt{K}$ for all $\zeta \in \mathbb{D}$. We have, for $z = \lambda(\zeta)$,

$$\|f'(\zeta)\| = \|\,\mathrm{id}'(z)\|\, \|\lambda'(\zeta)\| = \|\,\mathrm{id}'(z)\| = \frac{\rho(z)}{\rho_{0,1}(z)}.$$

This leads to the lower bound

$$\rho_{0,1}(z) \geq \sqrt{K}\rho(z) \geq \frac{\text{const.}}{|z|^{0.9}}$$

if $0 < |z| < 1/2$, which shows that $\rho_{0,1}(z)$ blows up as $z \to 0$. Since the automorphism $z \mapsto 1 - z$ of $\mathbb{C}_{0,1}$ is a hyperbolic isometry, we have the symmetry relation $\rho_{0,1}(1 - z) = \rho_{0,1}(z)$. Thus,

$$(13.14) \qquad \lim_{z \to 0} \rho_{0,1}(z) = \lim_{z \to 1} \rho_{0,1}(z) = +\infty.$$

REMARK 13.31. It can be shown that $\rho_{0,1}(z)$ blows up at a rate similar to the density $\rho_{\mathbb{D}^*}(z) = -1/(|z| \log|z|)$ as $z \to 0$. More precisely, $C < \rho_{0,1}(z)/\rho_{\mathbb{D}^*}(z) < 1$ for a positive constant C if $0 < |z| < 1/2$ [**A4**].

We now turn to a general hyperbolic domain $U \subset \mathbb{C}$. For $z \in U$, consider the Euclidean distance

$$\delta_U(z) = \mathrm{dist}(z, \partial U) = \inf\, \{|z - w| : w \in \partial U\}.$$

In other words, $\delta_U(z)$ is the radius of the largest open disk centered at z and contained in U. Recall that ρ_U denotes the density of the hyperbolic metric of U.

THEOREM 13.32 (Boundary behavior of the hyperbolic metric).

(i) *For every $p \in \partial U$, $\rho_U(z) \to +\infty$ as $z \to p$.*
(ii) *$\rho_U(z) \leq 2/\delta_U(z)$ for all $z \in U$.*
(iii) *If U is simply connected, then $\rho_U(z) \geq 1/(2\delta_U(z))$ for all $z \in U$.*
(iv) *If U is bounded, there is a constant $C > 0$ such that*

$$\rho_U(z) \geq \frac{-C}{\delta_U(z) \log \delta_U(z)}$$

for all $z \in U$ with $\delta_U(z) < 1/2$.

Combining (ii) and (iii), we conclude that *in a simply connected hyperbolic domain, the density ρ_U is comparable to the inverse distance-to-the-boundary $1/\delta_U$.*

PROOF. We shall simplify the notations ρ_U, δ_U to ρ, δ.

(i) After an affine change of coordinates, we may assume that $p = 0$ and U is contained in $\mathbb{C}_{0,1} = \mathbb{C} \smallsetminus \{0, 1\}$. By Corollary 13.30,

$$\rho(z) \geq \rho_{0,1}(z) \qquad \text{for all } z \in U.$$

It follows from (13.14) that $\lim_{z \to 0} \rho(z) = +\infty$.

(ii) Take $z \in U$ and a holomorphic covering map $f : \mathbb{D} \to U$ with $f(0) = z$, so $\rho(z) = 2/|f'(0)|$ by (13.6). The disk $\mathbb{D}(z, \delta(z)) \subset U$, being simply connected, is evenly covered by f, so there is a holomorphic branch of f^{-1} defined in $\mathbb{D}(z, \delta(z))$ and taking values in $\mathbb{D}$ which sends z back to 0. By Theorem 1.43, $1/|f'(0)| = |(f^{-1})'(z)| \leq 1/\delta(z)$, and the result follows.

(iii) Take $z \in U$ and a Riemann map $f : \mathbb{D} \to U$ with $f(0) = z$, so again $\rho(z) = 2/|f'(0)|$. Apply Koebe's 1/4-theorem to the schlicht function $\zeta \mapsto (f(\zeta) - z)/f'(0)$ to conclude that U contains the open disk centered at z of radius $|f'(0)|/4$. This gives the estimate $|f'(0)|/4 \leq \delta(z)$, or $2/|f'(0)| \geq 1/(2\delta(z))$, as required.

(iv) Pick some $r > 1$ larger than the Euclidean diameter of U. For $z \in U$ let p be a closest point of ∂U to z, so $|z - p| = \delta(z)$. Since $U \subset \mathbb{D}^*(p, r)$, we have $\rho(z) \geq \rho_{\mathbb{D}^*(p,r)}(z)$. Using the formula (13.10) and the affine change of coordinate $\mathbb{D}^*(p, r) \to \mathbb{D}^*$ defined by $z \mapsto (z - p)/r$, it is easy to see that

$$g_{\mathbb{D}^*(p,r)} = \frac{-1}{|z - p| \log(|z - p|/r)} |dz|.$$

It follows that if $\delta(z) < 1/2$,

$$\rho(z) \geq \frac{-1}{\delta(z) \log(\delta(z)/r)} \geq \frac{-C}{\delta(z) \log \delta(z)},$$

where C is any constant that satisfies $0 < C < (\log r / \log 2 + 1)^{-1}$. □

13.4 Conformal geometry of topological annuli

Any 2-connected domain in $\hat{\mathbb{C}}$ is homeomorphic to a twice punctured sphere (Theorem 9.29). Because of this, such a domain is often called a ***topological annulus***. The fundamental group of a topological annulus is isomorphic to that of $\mathbb{C}^*$ and therefore it is the infinite cyclic group generated by the homotopy class of a closed curve that winds once around one of the two complementary components.

There are three possible conformal types for a topological annulus A (see Fig. 13.5):

(i) ***Euclidean***, where $\hat{\mathbb{C}} \smallsetminus A$ consists of two points. In this case A is biholomorphic to the punctured plane $\mathbb{C}^*$.

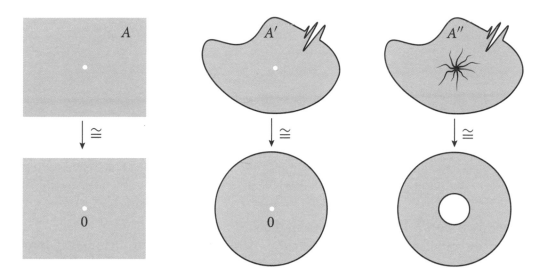

Figure 13.5. Euclidean (A), degenerate hyperbolic (A'), and non-degenerate hyperbolic (A'') topological annuli. They are conformally isomorphic, respectively, to the punctured plane $\mathbb{C}^*$, the punctured disk $\mathbb{D}^*$, and the round annulus $\{z : e^{-2\pi m} < |z| < 1\}$ for some $m > 0$.

(ii) **Degenerate hyperbolic**, where one component of $\hat{\mathbb{C}} \smallsetminus A$ is a single point p and the other is non-degenerate. In this case, A is biholomorphic to the punctured disk $\mathbb{D}^*$. In fact, the union $A \cup \{p\}$ is a 1-connected domain and therefore is simply connected (Theorem 9.27). By the Riemann mapping theorem, there is a conformal map $A \cup \{p\} \to \mathbb{D}$ sending p to 0. The restriction of this map to A would be a biholomorphism $A \to \mathbb{D}^*$.

(iii) **Non-degenerate hyperbolic**, where both components of $\hat{\mathbb{C}} \smallsetminus A$ are non-degenerate. In this case, Theorem 13.33 below will show that A is biholomorphic to a round annulus. Note that this case is conformally different from the degenerate case. In fact, any biholomorphism $f : \mathbb{D}^* \to A$ extends to a holomorphic map in $\mathbb{D}$ since by Picard's great theorem the singularity at 0 cannot be essential. The image $f(0)$ necessarily belongs to ∂A, and as $A \cup \{f(0)\} = f(\mathbb{D})$ is open, we see that $f(0)$ must be an isolated boundary point of A. This is impossible if A is non-degenerate.

THEOREM 13.33. *Every non-degenerate hyperbolic topological annulus A is conformally isomorphic to the round annulus*

$$A_m = \left\{z : e^{-2\pi m} < |z| < 1\right\}$$

for a unique $m > 0$.

The quantity m is called the **modulus** of A and is denoted by $\text{mod}(A)$. The uniqueness claim shows that the modulus is a conformal invariant, with

$$\text{mod}\left(\left\{z \in \mathbb{C} : R_1 < |z| < R_2\right\}\right) = \frac{1}{2\pi} \log\left(\frac{R_2}{R_1}\right)$$

Figure 13.6. Modulus as a measure of conformal "thickness." The annulus on the left has a "big" modulus, while the other two have "small" moduli.

whenever $0 < R_1 < R_2 < +\infty$ (compare problem 24 in chapter 6 and problem 21 in chapter 10 for alternative approaches to the conformal invariance of modulus). Intuitively, the modulus is a measure of how conformally "thick" or "fat" the annulus looks (see Fig. 13.6).

PROOF. Let $f : \mathbb{H} \to A$ be a holomorphic universal covering map with the deck group G. Recall that G is a subgroup of $\mathrm{Aut}(\mathbb{H})$ isomorphic to $\pi_1(A) \cong \mathbb{Z}$, so it is generated by a single element $\varphi \in \mathrm{Aut}(\mathbb{H})$ without fixed points in $\mathbb{H}$ (Theorem 12.21). In other words, φ is a parabolic or hyperbolic Möbius map in the sense of §4.3.

If φ is parabolic, after conjugating by a Möbius map (which does not alter the conformal type of A) we can assume $\varphi(z) = z + 1$. Since the deck group of the universal covering $\mathbb{H} \to \mathbb{D}^*$ given by $z \mapsto e^{2\pi i z}$ is also generated by φ, the map $e^{2\pi i z} \mapsto f(z)$ is a well-defined biholomorphism $\mathbb{D}^* \to A$. This is a contradiction since A is non-degenerate.

Thus, φ is necessarily hyperbolic and after conjugating by a Möbius map we can assume $\varphi(z) = e^{\pi/m} z$ for some $m > 0$. Consider the universal covering map $h : \mathbb{H} \to A_m$ given by $h(z) = \exp(2mi \log z)$, where log is the branch of the logarithm in $\mathbb{H}$ defined by

$$\log(re^{it}) = \log r + it \qquad \text{with } r > 0,\ 0 < t < \pi.$$

It is easily seen that the deck group of $h : \mathbb{H} \to A_m$ is also generated by φ, so the map $h(z) \mapsto f(z)$ is a well-defined biholomorphism $A_m \to A$.

For uniqueness, it suffices to check that if the round annuli A_m and $A_{\hat{m}}$ are biholomorphic, then $m = \hat{m}$. Let $h : \mathbb{H} \to A_m$ and $\hat{h} : \mathbb{H} \to A_{\hat{m}}$ be the respective universal covering maps with the deck groups G and $\hat{G}$. As in the proof of Theorem 13.27, for any biholomorphism $\psi : A_m \to A_{\hat{m}}$ there is an element $\Psi \in \mathrm{Aut}(\mathbb{H})$ which makes the following diagram commute:

$$
\begin{array}{ccc}
\mathbb{H} & \xrightarrow{\ \Psi\ } & \mathbb{H} \\
{\scriptstyle h}\downarrow & & \downarrow{\scriptstyle \hat{h}} \\
A_m & \xrightarrow{\ \psi\ } & A_{\hat{m}}
\end{array}
$$

The map $\nu \mapsto \Psi \circ \nu \circ \Psi^{-1}$ is easily seen to be a group isomorphism $G \to \hat{G}$. In particular, Ψ must conjugate the generator $\varphi(z) = e^{\pi/m}z$ of G to one of the two generators $\hat{\varphi}(z) = e^{\pm \pi/\hat{m}}z$ of $\hat{G}$. By Corollary 4.27, $\tau(\varphi) = \tau(\hat{\varphi})$, or

$$e^{\pi/m} + e^{-\pi/m} + 2 = e^{\pi/\hat{m}} + e^{-\pi/\hat{m}} + 2,$$

from which it follows that $m = \hat{m}$. $\qquad\square$

As another illustration of the ideas developed here, we classify all holomorphic covering maps of topological annuli, beginning with the non-degenerate hyperbolic case. It will be convenient to use the following notations:

$$E(z) = e^{2\pi i z}, \quad \Pi_d(z) = z^d,$$

$$S_m = \{z : 0 < \text{Im}(z) < m\},$$

$$A_m = \{z : e^{-2\pi m} < |z| < 1\} \quad \text{(as before)}.$$

THEOREM 13.34 (Holomorphic covering spaces of non-degenerate annuli). *Let $U \subset \hat{\mathbb{C}}$ be a domain and $f : U \to A_m$ be a holomorphic covering map of degree d.*

(i) If d is infinite, there is a biholomorphism $\varphi : U \to S_m$ such that $f = E \circ \varphi$:

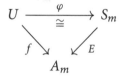

(ii) If d is finite, there is a biholomorphism $\varphi : U \to A_{m/d}$ such that $f = \Pi_d \circ \varphi$:

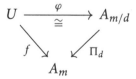

PROOF. Since $E : S_m \to A_m$ is a universal covering, there is a holomorphic covering map $g : S_m \to U$ such that $f \circ g = E$. The deck group $G \cong \pi_1(U)$ of $g : S_m \to U$ is a subgroup of the deck group $\langle z \mapsto z + 1 \rangle$ of $E : S_m \to A_m$ since if $\varphi \in G$, then

$$E \circ \varphi = (f \circ g) \circ \varphi = f \circ (g \circ \varphi) = f \circ g = E.$$

Thus, as a subgroup of the infinite cyclic group $\langle z \mapsto z + 1 \rangle$, G itself is cyclic. We distinguish two cases:

If G is trivial, then g is a biholomorphism, so we are in case (i) with $\varphi = g^{-1}$.

If G is non-trivial, it is generated by the translation $z \mapsto z + d$ for some integer $d \geq 1$. The universal covering $\tilde{E} : S_m \to A_{m/d}$ defined by $\tilde{E}(z) = E(z/d) = e^{2\pi i z/d}$ has the same deck group G, so the map $\varphi : U \to A_{m/d}$ which sends $g(z)$ to $\tilde{E}(z)$ is a well-defined biholomorphism. Evidently,

$$\Pi_d \circ \varphi \circ g = \Pi_d \circ \tilde{E} = E = f \circ g.$$

Since g is surjective, this implies $\Pi_d \circ \varphi = f$, so we are in case (ii). $\qquad\square$

COROLLARY 13.35. *Suppose $f : A \to \hat{A}$ is a proper holomorphic map between non-degenerate hyperbolic annuli. Then f is a finite-degree covering map and*

$$(13.15) \qquad\qquad \operatorname{mod}(\hat{A}) = \deg(f) \operatorname{mod}(A).$$

PROOF. f is a finite-degree branched covering by Corollary 12.41. The number of critical points of f is $\deg(f) \chi(\hat{A}) - \chi(A) = 0$ by the Riemann-Hurwitz formula. This proves that f is a covering map. The relation (13.15) follows from Theorem 13.34(ii) and the conformal invariance of modulus. $\qquad\square$

The analog of Theorem 13.34 for degenerate hyperbolic annuli can be viewed as the limiting case $m \to +\infty$:

THEOREM 13.36 (Holomorphic covering spaces of punctured disks). *Let $U \subset \hat{\mathbb{C}}$ be a domain and $f : U \to \mathbb{D}^*$ be a holomorphic covering map of degree d.*

(i) If d is infinite, there is a biholomorphism $\varphi : U \to \mathbb{H}$ such that $f = E \circ \varphi$:

$$
\begin{array}{ccc}
U & \xrightarrow{\ \varphi\ } & \mathbb{H} \\
 & {\scriptstyle f}\searrow \quad \swarrow {\scriptstyle E} & \\
 & \mathbb{D}^* &
\end{array}
$$

(ii) If d is finite, there is a biholomorphism $\varphi : U \to \mathbb{D}^*$ such that $f = \Pi_d \circ \varphi$:

$$
\begin{array}{ccc}
U & \xrightarrow{\ \varphi\ } & \mathbb{D}^* \\
 & {\scriptstyle f}\searrow \quad \swarrow {\scriptstyle \Pi_d} & \\
 & \mathbb{D}^* &
\end{array}
$$

We skip the proof, as it is virtually identical to the non-degenerate hyperbolic case above. For the Euclidean case, which is also similar to Theorem 13.36, see problem 17.

Problems

(1)   If α is the circle of radius r centered at $p \in \mathbb{C}$, show that the reflection R_α is given by

$$R_\alpha(z) = \frac{r^2}{\bar{z} - \bar{p}} + p.$$

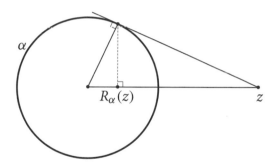

Figure 13.7. Geometric interpretation of the reflection in a circle α.

Similarly, if α is the straight line with the equation $\bar{p}z + p\bar{z} + c = 0$, where $p \in \mathbb{C}$ is non-zero and $c \in \mathbb{R}$, then

$$R_\alpha(z) = -\frac{p\bar{z} + c}{\bar{p}}.$$

(2)   Justify the geometric interpretation of the reflection R_α depicted in Fig. 13.7.

(3)   Let $[p_1, p_2, p_3, p_4] = (p_3 - p_1)(p_4 - p_2)/((p_2 - p_1)(p_4 - p_3))$ be the cross ratio of an ordered quadruple as in Definition 4.6. If α is a circle in $\hat{\mathbb{C}}$ and $p_1, p_2, p_3 \in \alpha$, show that

$$[p_1, p_2, R_\alpha(z), p_3] = \overline{[p_1, p_2, z, p_3]}$$

for all $z \in \hat{\mathbb{C}} \smallsetminus \{p_1, p_2, p_3\}$. Verify that this relation characterizes R_α uniquely and therefore can be taken as the definition of the reflection in α.

(4)   Let α be a circle in $\hat{\mathbb{C}}$ that meets $\partial\mathbb{D}$ orthogonally. Use the result of problem 1 to show that if α is a Euclidean circle centered at p, then

$$R_\alpha(z) = -\frac{\bar{q}}{q}\left(\frac{\bar{z} - q}{1 - \bar{q}\bar{z}}\right),$$

where $q = 1/p \in \mathbb{D}^*$. Similarly, if α is the straight line with the equation $\bar{p}z + p\bar{z} = 0$, then

$$R_\alpha(z) = -\frac{p}{\bar{p}}\bar{z}.$$

Either way, R_α has the form $\varphi \circ \kappa$ for some $\varphi \in \mathrm{Aut}(\mathbb{D})$. Conclude that $\mathrm{Ref}(\mathbb{D})$ is generated by the reflections in circles that are orthogonal to $\partial\mathbb{D}$.

(5)   Show that $\mathbb{T}$ is the natural boundary of any holomorphic covering map $\mathbb{D} \to \mathbb{C} \smallsetminus \{0, 1\}$. (Hint: It suffices to look at the special case of the map $\lambda : \mathbb{H} \to \mathbb{C} \smallsetminus \{0, 1\}$ constructed in Theorem 13.13.)

(6)   Show that the deck group of the covering map $\lambda : \mathbb{H} \to \mathbb{C} \smallsetminus \{0, 1\}$ constructed in Theorem 13.13 is generated by

$$z \mapsto z + 2 \qquad \text{and} \qquad z \mapsto \frac{z}{2z + 1}.$$

Verify that this group coincides with the **level 2 congruence subgroup** Γ_2 of the modular group Γ consisting of all Möbius maps $z \mapsto (az + b)/(cz + d)$ for which a, d are odd and b, c are even. (Hint: For the first statement, look at the lifts of the closed curves γ_1, γ_2 of Fig. 13.8 under λ. You will need the fact that $\pi_1(\mathbb{C} \smallsetminus \{0, 1\})$ is generated by the homotopy

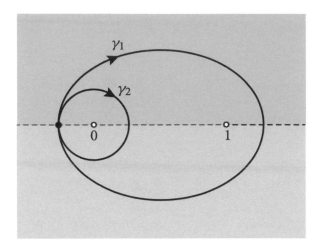

Figure 13.8. Illustration of problem 6.

classes of γ_1, γ_2. For the second statement, imitate the proof of Lemma 13.10 to show that if

$$C = \begin{bmatrix} 1 & 2 \\ 0 & 1 \end{bmatrix} \quad \text{and} \quad D = \begin{bmatrix} 1 & 0 \\ 2 & 1 \end{bmatrix},$$

and if $M = \begin{bmatrix} a_0 & b_0 \\ c_0 & d_0 \end{bmatrix}$ represents an arbitrary element of Γ_2, then there are integers $n_1, \ldots, n_k$ and $m_1, \ldots, m_{k-1}$ such that

$$MC^{n_1} D^{m_1} \cdots C^{n_{k-1}} D^{m_{k-1}} C^{n_k} \quad \text{has the form} \quad \begin{bmatrix} a & 0 \\ c & d \end{bmatrix}$$

and therefore is a power of D.)

(7)   With the notation of the previous problem, verify the following assertions:
   (i) Γ_2 is a normal subgroup of Γ.
   (ii)   The quotient group Γ / Γ_2 has order 6 and is isomorphic to the permutation group S_3. To this end, take the generators $A(z) = z + 1$, $B(z) = -1/z$ of Γ and verify that the cosets of Γ / Γ_2 are represented by the following six matrices:

$$I = \begin{bmatrix} 1 & 0 \\ 0 & 1 \end{bmatrix} \qquad A = \begin{bmatrix} 1 & 1 \\ 0 & 1 \end{bmatrix} \qquad B = \begin{bmatrix} 0 & -1 \\ 1 & 0 \end{bmatrix}$$

$$AB = \begin{bmatrix} 1 & -1 \\ 1 & 0 \end{bmatrix} \qquad BA = \begin{bmatrix} 0 & -1 \\ 1 & 1 \end{bmatrix} \qquad BAB = \begin{bmatrix} 1 & 0 \\ -1 & 1 \end{bmatrix}.$$

(8)   For each $\tau \in \mathbb{H}$, consider the lattice $\Lambda = \langle 1, \tau \rangle$ and the associated Weierstrass $\wp$-function. By Theorem 9.11, the finite critical values $e_1 = \wp(1/2)$, $e_2 = \wp((1 + \tau)/2)$, $e_3 = \wp(\tau/2)$ of $\wp$ are distinct, so the cross ratio

$$\lambda = [e_1, \infty, e_2, e_3] = \frac{e_2 - e_1}{e_2 - e_3}$$

is well defined and takes values in $\mathbb{C} \smallsetminus \{0, 1\}$. Prove the following assertions:
   (i) e_1, e_2, e_3 and therefore λ depend holomorphically on τ.

(ii)   Suppose $\tau' = \varphi(\tau)$ for some $\varphi \in \Gamma_2$ and consider the corresponding critical values e_1', e_2', e_3' and cross ratio λ'. Then there exists an $\alpha \neq 0$ such that $e_j' = \alpha^2 e_j$ for $j = 1, 2, 3$, hence $\lambda' = \lambda$.

(iii) Suppose $\tau' = \varphi(\tau)$ for some $\varphi \in \Gamma$. Then there exists an $\alpha \neq 0$ such that $\{e_1', e_2', e_3'\} = \{\alpha^2 e_1, \alpha^2 e_2, \alpha^2 e_3\}$. In other words, there is a permutation $\sigma \in S_3$ such that $e_{\sigma(j)}' = \alpha^2 e_j$ for $j = 1, 2, 3$. The action of Γ/Γ_2 on the cross ratio can be summarized in the following table:

representative $\varphi \in \Gamma/\Gamma_2$	I	A	B	AB	BA	BAB
permutation $\sigma \in S_3$	(1)	(23)	(13)	(123)	(132)	(12)
cross ratio λ'	λ	$1 - \lambda$	$\frac{1}{\lambda}$	$\frac{\lambda-1}{\lambda}$	$\frac{1}{1-\lambda}$	$\frac{\lambda}{\lambda-1}$

(Hint: For (ii) and (iii), use problem 15 in chapter 9.)

(9)   Recall from Remark 13.12 that two reduced fractions $0 \leq b/d < a/c \leq 1$ are Farey neighbors if $ad - bc = 1$, in which case their Farey sum is defined by

$$\frac{a}{c} \oplus \frac{b}{d} = \frac{a+b}{c+d}.$$

(i)   Show that if $x < y$ are Farey neighbors, so are $x < x \oplus y$ and $x \oplus y < y$. Conversely, if $x < z < y$ is a triple of Farey neighbors, then $z = x \oplus y$.

(ii)   Given a triple $x < z < y$ of Farey neighbors, let α and β be the circles with the intervals $[x, z]$ and $[z, y]$ as their diameters, respectively. Show that

$$R_\alpha(y) = x \oplus z \qquad \text{and} \qquad R_\beta(x) = z \oplus y.$$

(iii) Let $T \subset \mathbb{H}$ be the ideal triangle with vertices $0, 1, \infty$. Verify that every triple of Farey neighbors appears as the vertex set of some tile in $\mathscr{T}_T$.

(iv)   Use (ii) to show that the union of the vertices of all tiles of generation $\leq n$ in $\mathscr{T}_T$ that lie in the strip $0 \leq \mathrm{Re}(z) \leq 1$ is precisely the n-th row of the list in Remark 13.12.

(10) For every reduced fraction $0 \leq p/q \leq 1$, consider the **Ford circle** $C_{p/q}$ in the upper half-plane which is tangent to $\mathbb{R}$ at p/q and has radius $1/(2q^2)$. Show that if $0 \leq b/d < a/c \leq 1$ are Farey neighbors, then the Ford circles $C_{b/d}$ and $C_{a/c}$ are tangent to each other. Deduce the geometric construction of the Farey sum $b/d \oplus a/c$ suggested in Fig. 13.9.

(11) Suppose $f : U \to U$ is a holomorphic self-map of a hyperbolic domain. Prove the following assertions:

(i)   If f has a fixed point at which its derivative is an n-th root of unity, then $f^{\circ n} = \mathrm{id}$.

(ii)   If f has two distinct fixed points, then $f^{\circ n} = \mathrm{id}$ for some $n \geq 1$.

(Hint: In both cases, take a universal covering $u : \mathbb{D} \to U$ which sends 0 to a fixed point of f, and consider the unique holomorphic map $F : \mathbb{D} \to \mathbb{D}$ with $F(0) = 0$ such that $u \circ F = f \circ u$.)

(12) Suppose $U \subset \hat{\mathbb{C}}$ is a hyperbolic domain and $u : \mathbb{D} \to U$ is a holomorphic universal covering map with the deck group G. Show that

$$\mathrm{Aut}(U) \cong N(G)/G,$$

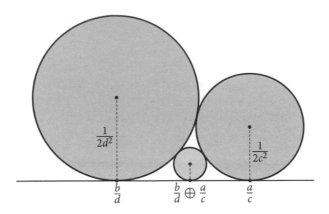

Figure 13.9. Geometric construction of Farey sums using Ford circles (see problem 10).

where $N(G)$ is the ***normalizer subgroup*** of G defined by

$$N(G) = \{\varphi \in \mathrm{Aut}(\mathbb{D}) : \varphi G \varphi^{-1} = G\}.$$

(13)   Consider the round annulus $A_m = \{z : e^{-2\pi m} < |z| < 1\}$ for some $m > 0$ and the universal covering map $h : \mathbb{H} \to A_m$ defined by $h(\zeta) = \exp(2mi \log \zeta)$.
   (i)   Show that the circle $|z| = e^{-\pi m}$ is the image under h of a hyperbolic geodesic in $\mathbb{H}$, and is the only smooth closed curve in A_m with this property. In other words, this circle is the unique "closed hyperbolic geodesic" in A_m.
   (ii)   Show that every pair $\pm p$ of antipodal points in A_m can be joined by precisely two minimal geodesics which are mapped to each other under the rotation $z \mapsto -z$. The union of these geodesics is a smooth closed curve if and only if $|p| = e^{-\pi m}$ (compare Fig. 13.4, where the unique closed geodesic is distinguished in blue).

(14)   Let $A \subset \hat{\mathbb{C}}$ be a non-degenerate hyperbolic annulus. Use the previous problem and the formula (13.9) to show that A contains a unique closed hyperbolic geodesic γ with

$$\mathrm{length}_A(\gamma) = \frac{\pi}{\mathrm{mod}(A)}.$$

Here length_A is the hyperbolic length defined by $\mathrm{length}_A(\gamma) = \int_\gamma \rho_A(z)\,|dz|$.

(15)   Suppose $U \subset \hat{\mathbb{C}}$ is a hyperbolic domain whose fundamental group is abelian. Prove that U is biholomorphic to $\mathbb{D}$, $\mathbb{D}^*$, or the annulus A_m for some $m > 0$. (Hint: The deck group G of the universal covering $\mathbb{H} \to U$ is abelian. Assuming G is non-trivial, this implies that all elements of G share the same fixed points on $\mathbb{R} \cup \{\infty\}$ (compare problem 24 in chapter 4). After conjugation by a suitable Möbius map, it follows that G is a subgroup of $\{z \mapsto tz\}_{t>0}$ or $\{z \mapsto z+t\}_{t \in \mathbb{R}}$.)

(16)   Show that the density $\rho = \rho_U$ of the hyperbolic metric of a simply connected domain $U \subsetneq \mathbb{C}$ satisfies
$$|\log \rho(z) - \log \rho(w)| \le 2 \,\mathrm{dist}_U(z, w) \qquad \text{for } z, w \in U.$$
In other words, $\log \rho$ is a Lipschitz function in (U, dist_U). (Hint: Consider a Riemann map $f : \mathbb{D} \to U$ with $f(0) = z$ and apply Koebe's distortion bound (6.9) to the schlicht function $\zeta \mapsto (f(\zeta) - z)/f'(0)$.)

(17)   Prove the following analog of Theorem 13.36: Let $f : U \to \mathbb{C}^*$ be a holomorphic covering map of degree $1 \le d \le \infty$, where U is a domain in $\hat{\mathbb{C}}$.

(i)   If $d = \infty$, there is a biholomorphism $\varphi : U \to \mathbb{C}$ such that $f = E \circ \varphi$.

(ii)   If $d < \infty$, there is a biholomorphism $\varphi : U \to \mathbb{C}^*$ such that $f = \Pi_d \circ \varphi$.

(18)   Let $A \subset \hat{\mathbb{C}}$ be a non-degenerate hyperbolic annulus and $f : A \to A$ be a proper holomorphic map. Prove that $f \in \mathrm{Aut}(A)$.

(19)   Show that there is no proper holomorphic map between any two of the three models $\mathbb{C}^*$, $\mathbb{D}^*$, and A_m for topological annuli.

(20)   Find an exact expression for the modulus of the topological annulus

$$\{z : 2 < |z - 1| + |z + 1| < 2 + r\}$$

in terms of $r > 0$. What happens as $r \to 0$ or $r \to +\infty$?

Bibliography

[A1]   L. Ahlfors, *Bounded analytic functions*, Duke Math. J. **14** (1947) 1–11.

[A2]   L. Ahlfors, *Complex Analysis*, 3rd ed., McGraw-Hill, 1979.

[A3]   L. Ahlfors, *Lectures on Quasiconformal Mappings*, 2nd ed., American Mathematical Society, 2006.

[A4]   L. Ahlfors, *Conformal Invariants: Topics in Geometric Function Theory*, AMS Chelsea Publishing, 2010.

[Bo]   R. Boas, *Invitation to Complex Analysis*, 2nd ed. (revised by H. P. Boas), Mathematical Association of America, 2010.

[BD]   F. Berteloot and J. Duval, *Sur l'hyperbolicité de certains complémentaires*, L'Enseign. Math., **47** (2001) 253–267.

[DeB]   L. de Branges, *A proof of the Bieberbach conjecture*, Acta Mathematica, **154** (1985) 137–152.

[DeS]   H. P. de Saint-Gervais, *Uniformisation des surfaces de Riemann: Retour sur un théorème centenaire*, ENS Éditions, Lyon, 2010.

[Di]   J. Dieudonné, *A History of Algebraic and Differential Topology, 1900–1960*, Birkhäuser, 2009.

[Dx]   J. Dixon, *A brief proof of Cauchy's integal theorem*, Proc. Amer. Math. Soc., **29** (1971) 625–626.

[doC]   M. do Carmo, *Differential Geometry of Curves and Surfaces*, revised 2nd ed., Dover, 2016.

[Fa]   K. Falconer, *The Geometry of Fractal Sets*, Cambridge University Press, 1990.

[FHW]   Y. Fisher, J. Hubbard and B. Wittner, *A proof of the uniformization theorem for arbitrary plane domains*, Amer. Math. Monthly, **104** (1988) 413–418.

[FK]   O. Fomenko and G. Kuz'mina, *The last 100 days of the Bieberbach conjecture*, The Mathematical Intelligencer, **8** (1986) 40–47.

[Fo]   O. Forster, *Lectures on Riemann Surfaces*, Springer, 1981.

[Fu]   W. Fulton, *Algebraic Topology, A First Course*, Springer, 1995.

[Gm]   T. Gamelin, *Complex Analysis*, Springer, 2001.

[Gd]   L. Gårding, *Some Points of Analysis and Their History*, American Mathematical Society, 1997.

[Gn]   J. Garnett, *Analytic Capacity and Measure*, Lecture Notes in Mathematics **297**, Springer, 1972.

[Gl]   I. Glicksberg, *A remark on Rouché's theorem*, Amer. Math. Monthly, **83** (1976) 186–187.

[GM]   J. Gray and S. Morris, *When is a function that satisfies Cauchy-Riemann equations analytic?*, Amer. Math. Monthly, **85** (1978) 246–256.

[H]   A. Hatcher, *Algebraic Topology*, Cambridge University Press, 2002.

[Ko]   P. Koebe, *Abhandlungen zur theorie der konformen abbildung: VI. Abbildung mehrfach zusammenhangender bereiche auf kreisbereiche etc.*, Math. Z. **7** (1920) 235–301.

[Kr]   S. Krantz, Complex Analysis: The Geometric Viewpoint, 2nd ed., The Mathematical Association of America, 2004.

[Ma]   P. Mattila, *Geometry of Sets and Measures in Euclidean Spaces*, Cambridge University Press, 1995.

[MM]   H. McKean and V. Moll, *Elliptic Curves: Function Theory, Geometry, Arithmetic*, Cambridge University Press, 1999.

[McM]   C. McMullen, *Advanced Complex Analysis*, Harvard University course notes, 2017.

[Mi]   J. Milnor, *Dynamics in One Complex Variable*, Annals of Math. Studies **160**, Princeton University Press, 2006.

[NP]   R. Nevanlinna and V. Paatero, *Introduction to Complex Analysis*, Addison-Wesley, 1969.

[P]   C. Pommerenke, *Boundary Behaviour of Conformal Maps*, Springer, 1992.

[PS]   G. Pólya and G. Szegö, *Problems and Theorems in Analysis, vol. I*, Springer-Verlag, 1972.

[Re]   R. Remmert, *Classical Topics in Complex Function Theory*, Springer, 1998.

[Ru1]   W. Rudin, *Principles of Mathematical Analysis*, 3rd ed., McGraw-Hill, 1976.

[Ru2]   W. Rudin, *Real and Complex Analysis*, 3rd ed., McGraw-Hill, 1987.

[S]   N. Steinmetz, *The formula of Riemann-Hurwitz and iteration of rational functions*, Complex Variables **22** (1993) 203–206.

[ST]   H. Seifert and W. Threlfall, *Lehrbuch der Topologie*, Teubner, 1934.

[T]   E. Titchmarsh, *The Theory of Functions*, Oxford University Press, 1939.

[V]   W. Veech, *A Second Course in Complex Analysis*, Dover Publications, 2008.

[W]   C. T. C. Wall, *A Geometric Introduction to Topology*, Dover, 1993.

[Z1]   L. Zalcman, *Real proofs of complex theorems (and vice versa)*, Amer. Math. Monthly **81** (1974) 115–137.

[Z2]   L. Zalcman, *A heuristic principle in complex function theory*, Amer. Math. Monthly **82** (1975) 813–817.

Image credits

Chapter 1

Portrait of Goursat, from MacTutor History of Mathematics Archive, http://mathshistory.st-andrews.ac.uk/PictDisplay/Goursat.html

Portrait of Morera, from Wikipedia, https://commons.wikimedia.org/wiki/File:Giacinto_Morera.JPG

Portrait of Liouville, from MacTutor History of Mathematics Archive, http://mathshistory.st-andrews.ac.uk/PictDisplay/Liouville.html

Chapter 2

Portrait of Poincaré, from the private collection of Félix Potin et Cie, https:/commons.wikimedia.org/wiki/File:CFP_Poincar%c3%a9,_Henri.jpg

Portrait of Jordan, from MacTutor History of Mathematics Archive, http://mathshistory.st-andrews.ac.uk/PictDisplay/Jordan.html

Portrait of Artin by Konrad Jacobs, from Archives of the Mathematisches Forschungsinstitut Oberwolfach, https://opc.mfo.de/detail?photo_id=116, used with permission

Portrait of Cauchy by C. H. Reutlinger, from the Scientific Identity collection of Smithsonian Libraries, https://library.si.edu/image-gallery/73510

Chapter 3

Portrait of Casorati, from MacTutor History of Mathematics Archive, http://mathshistory.st-andrews.ac.uk/PictDisplay/Casorati.html

Portrait of Laurent, from MacTutor History of Mathematics Archive, http://mathshistory.st-andrews.ac.uk/PictDisplay/Laurent_Pierre.html

Portrait of Rouché, from MacTutor History of Mathematics Archive, http://mathshistory.st-andrews.ac.uk/PictDisplay/Rouche.html

Chapter 4

Portrait of Möbius by Adolf Neumann, https://commons.wikimedia.org/wiki/File:August_Ferdinand_M%c3%b6bius.jpg

Portrait of Schwarz by Ludwig Zipfel, from ETH-Bibliothek collections, https://commons.wikimedia.org/wiki/File:ETH-BIB-Schwarz,_Hermann_Amand_(1843-1921)-Portrait-Portr_11921.tif_(cropped).jpg

Sphere Spirals by M. C. Escher, ©2020 The M. C. Escher Company–The Netherlands. All rights reserved. www.mcescher.com

Portrait of Pick, from Wikipedia, https://commons.wikimedia.org/wiki/File:GeorgPick.png

Chapter 5

Portrait of Weierstrass, from the Scientific Identity collection of Smithsonian Libraries, https://library .si.edu/image-gallery/73709

Portrait of Hurwitz, from Wikipedia, https://commons.wikimedia.org/wiki/File:Adolf_Hurwitz.jpg

Portrait of Montel, from MacTutor History of Mathematics Archive, http://mathshistory.st-andrews .ac.uk/PictDisplay/Montel.html

Portrait of Marty, courtesy of M. Michel Roussel from his family tree album at https://Geneanet.org, used with permission

Chapter 6

Portrait of Riemann, from the Scientific Identity collection of Smithsonian Libraries, https://library.si .edu/image-gallery/73213

Portrait of Zhukovskii, from MacTutor History of Mathematics Archive, http://mathshistory.st -andrews.ac.uk/PictDisplay/Zhukovsky.html

Portrait of Grönwall, from MacTutor History of Mathematics Archive, http://mathshistory.st-andrews .ac.uk/PictDisplay/Gronwall.html

Portrait of Bieberbach. Courtesy of Gerhard Hund, licensed under the Creative Commons Attribution 3.0 (https://commons.wikimedia.org/wiki/File:Bieberbach,Ludwig_1930_Jena.jpg)

Portrait of de Branges, courtesy of Purdue University, used with permission (https://www.math .purdue.edu/people/˜branges/site).

Portrait of Koebe, from MacTutor History of Mathematics Archive, http://mathshistory.st-andrews.ac .uk/PictDisplay/Koebe.html

Portrait of Carathéodory, from MacTutor History of Mathematics Archive, http://mathshistory.st -andrews.ac.uk/PictDisplay/Caratheodory.html

Chapter 7

Portrait of Laplace by James Posselwhite, from the Scientific Identity collection of Smithsonian Libraries, https://library.si.edu/image-gallery/72854

Portrait of Poisson by François-Seraphin Delpech, from the Scientific Identity collection of Smithsonian Libraries, https://library.si.edu/image-gallery/74036

Portrait of Harnack, from MacTutor History of Mathematics Archive, http://mathshistory.st-andrews .ac.uk/PictDisplay/Harnack.html

Portrait of Fatou, from a family collection, https://commons.wikimedia.org/wiki/File:Pierre_Fatou.jpg

Portrait of Blaschke by Konrad Jacobs, from Archives of the Mathematisches Forschungsinstitut Oberwolfach, https://opc.mfo.de/detail?photo_id=362, used with permission

Chapter 8

Portrait of Jensen by Vilhelm Rieger, Royal Library of Copenhagen, https://commons.wikimedia.org /wiki/File:Johan_Ludvig_William_Valdemar_Jensen_by_Vilhelm_Rieger.jpg

Portrait of Hadamard, courtesy of École Polytechnique Bibliothèque Centrale, used with permission

Chapter 9

Portrait of Mittag-Leffler, courtesy of Institut Mittag-Leffler, used with permission

Portrait of Abel, from MacTutor History of Mathematics Archive, http://mathshistory.st-andrews.ac.uk/PictDisplay/Abel.html

Portrait of Runge by Peter Matzen, from Voit Collection, https://commons.wikimedia.org/wiki/File:Voit_202_Karl_Runge.jpg

Portrait of Mergelyan, from Soviet Armenian Encyclopedia, https://commons.wikimedia.org/wiki/File:SN_Mergelyan_SAE.jpg

Portrait of Schönflies, from MacTutor History of Mathematics Archive, http://mathshistory.st-andrews.ac.uk/PictDisplay/Schonflies.html

Chapter 10

Portrait of Ostrowski, from Archives of P. Roquette, Heidelberg. Source: Archives of the Mathematisches Forschungsinstitut Oberwolfach, https://opc.mfo.de/detail?photo_id=9254. Used with permission

Portrait of Christoffel, from Wikipedia, https://commons.wikimedia.org/wiki/File:Elwin_Bruno_Christoffel.JPG

Portrait of Hausdorff, University of Bonn Library, https://de.wikipedia.org/wiki/Datei:Hausdorff_1913-1921.jpg

Portrait of Painlevé, Agence de presse Meurisse, Bibliothèque Nationale de France, https://commons.wikimedia.org/wiki/File:Paul_Painlev%c3%a9_1923.jpg

Chapter 11

Portrait of Remmert by Andreas Daniel Matt, from Archives of the Mathematisches Forschungsinstitut Oberwolfach, https://opc.mfo.de/detail?photo_id=9959, used with permission

Portrait of Schottky, from MacTutor History of Mathematics Archive, http://mathshistory.st-andrews.ac.uk/PictDisplay/Schottky.html

Portrait of Picard, from MacTutor History of Mathematics Archive, http://mathshistory.st-andrews.ac.uk/PictDisplay/Picard_Emile.html

Portrait of Zalcman, courtesy of Erez Sheiner and Bar-Ilan University, used with permission

Portrait of Ahlfors by Konrad Jacobs, from Archives of the Mathematisches Forschungsinstitut Oberwolfach, https://opc.mfo.de/detail?photo_id=20, used with permission

Index